Loose Boundary Hydraulics

THIRD EDITION

Pergamon Titles of Related Interest

JAPAN SOCIETY OF MECHANICAL ENGINEERS
Visualized Flow

MAKSIMOVIC & RADOJKOVIC
Urban Drainage Catchments
Urban Drainage Modelling

TANAKA & CRUSE
Boundary Element Methods in Applied Mechanics

USCOLD (US Committee on Large Dams)
Development of Dam Engineering in the United States

WILLIAMS & ELDER
Fluid Physics for Oceanographers and Physicists

Related Pergamon Journals

(free specimen copy gladly sent on request)

China Ocean Engineering
Computers and Fluids
International Journal of Applied Engineering Education
International Journal of Engineering Science
International Journal of Rock Mechanics and Mining Sciences
International Journal of Solids and Structures
Journal of Terramechanics
Minerals Engineering
Ocean Engineering
Tunnelling and Underground Space Technology

Loose Boundary Hydraulics

THIRD EDITION

by

A. J. RAUDKIVI

Professor of Civil Engineering
University of Auckland, New Zealand

PERGAMON PRESS

Member of the Maxwell Macmillan Pergamon Publishing Corporation

OXFORD · NEW YORK · BEIJING · FRANKFURT
SÃO PAULO · SYDNEY · TOKYO · TORONTO

U.K.	Pergamon Press plc, Headington Hill Hall, Oxford OX3 0BW, England
U.S.A.	Pergamon Press, Inc., Maxwell House, Fairview Park, Elmsford, New York 10523, U.S.A.
PEOPLE'S REPUBLIC OF CHINA	Pergamon Press, Room 4037, Qianmen Hotel, Beijing, People's Republic of China
FEDERAL REPUBLIC OF GERMANY	Pergamon Press GmbH, Hammerweg 6, D-6242 Kronberg, Federal Republic of Germany
BRAZIL	Pergamon Editora Ltda, Rua Eça de Queiros, 346, CEP04011, Paraiso, São Paulo, Brazil
AUSTRALIA	Pergamon Press Australia Pty Ltd., P.O. Box 544, Potts Point, N.S.W. 2011, Australia
JAPAN	Pergamon Press, 5th Floor, Matsuoka Central Building, 1-7-1 Nishishinjuku, Shinjuku-ku, Tokyo 160, Japan
CANADA	Pergamon Press Canada Ltd., Suite No. 271, 253 College Street, Toronto, Ontario, Canada M5T 1R5

First edition 1967

Second edition 1976

Third edition 1990

Library of Congress Cataloging-in-Publication Data

Raudkivi, A. J.
Loose boundary hydraulics.
Includes bibliographies.
1. Sediment transport. I. Title.
TC175.2.R3 1989 627'.122 88-32920

British Library Cataloguing in Publication Data

Raudkivi, A J (Arved Jaan)
Loose boundary hydraulics. 3rd ed.
1. Sediments. Hydraulic transport
I. Title
627

ISBN 0-08-034074-1 (Hardcover)
ISBN 0-08-034073-3 (Flexicover)

Printed in Great Britain by BPCC Wheatons Ltd, Exeter

Contents

Preface

The subject of Loose Boundary Hydraulics extends from soil erosion and rivers to coastal processes, even snow drifts could be included, and the basic concepts also apply to numerous industrial processes. It involves not only the erosion of soils, transport of erosion products and the movement of beach materials, but also hydro- and aerodynamics. All the processes of erosion and transport are inseparably linked with fluid dynamics, from raindrops and stream flows to breakers on the beach. The fluid-solid interactions are complex and involve aspects of fluid mechanics which themselves have unsolved problems.

The literature on the subject is vast and scattered in publications and reports in which the emphasis varies according to the area of interest, e.g., geology, sedimentology, paleonthology, mineralogy, oceanography, coastal sedimentary environments, geography, agriculture, civil engineering, mining, etc. Apart from emphasis differences in interpretation occur. The whole subject of Loose Boundary Hydraulics could be likened to a huge zig-saw puzzle from which many pieces are still missing. My aim in writing was to give a summary, which would convey as good a description of the whole picture as possible, despite the missing pieces. In a field as complex as Loose Boundary Hydraulics the overall picture, a conceptual understanding or feel, is very important. Only a few simple problems are subject to rigorous analysis, the vast majority require a good deal of judgement before a solution to a practical problem is obtained. But judgement has to be based on understanding of the behavioural features. The presentation has an engineering bias, but is not written to be a recipe book on how to solve specific problems. It is hoped, however, that other readers will find the book of some value.

The reader will notice that numerical methods have been omitted, except for a few references. Numerical methods and numerical modelling are a subject in their own right. As stated above, the object here was to convey a picture of the relationships which describe the processes of loose boundary flows, that is, the analytical formulations on which numerical models have to be based. If these relationships give a poor description of the actual process then the results by the numerical model will be equally poor. Hopefully, the discussions and descriptions in the following pages will assist users with the evaluation of numerical models for practical applications.

It has been assumed that the reader has a working knowledge of fluid mechanics. Correspondingly, revelant relationships from fluid mechanics have been introduced without explanations.

I have drawn from the work of many people and I would like to thank everybody who has contributed to the study of this fascinating subject. I have attempted to give references to the sources, but if some have been missed, I offer my apologies. I would also like to thank the Leichtweiss Institute of the University of Braunschweig for their generous support during preparation of most of this third edition.

Auckland, New Zealand A.J.R.

List of Symbols

a - decay parameter for sediment concentration profile with elevation, an inverse mixing length.

$a = H/2$ - wave amplitude, amplitude.

a - the longest of the three mutually perpendicular dimensions of a particle; factor.

A - area; horizontal half axis of orbital movement; factor; constant; a function in cnoidal wave theory; scaling parameter.

A_c - cross-sectional area of tidal channels.

b - width of pier normal to flow; particle axis $a > b > c$.

B - width; bifurcation ratio; constant.

c - sediment concentration; wave form celerity, speed of translation; fraction of clay; the shortest of the three mutually perpendicular dimensions of a particle.

c_w - concentration of wash load; concentration by weight.

c_f - coefficient of friction or drag coefficient.

c_o - deep water wave celerity.

c_s - shallow water wave celerity.

c^* - wave group celerity.

C - Chézy coefficient; constant; cropping management factor.

$C' = 18 \log (12\ R_b/3\ d_{90})$.

C_D - drag coefficient.

C_s - Egiazaroff mobility number; coefficient of sorting.

C_u - uniformity coefficient.

$C_* = C/\sqrt{g}$

d_a - average particle size; characteristic armour grain size

d_g - geometric mean diameter

d_m - mean particle size

d_o = 2A - double amplitude of orbital movement
d_r - characteristic size of riprap
d_s - scour depth
$\bar{d}$ - mean particle size
D - diffusion coefficient; diameter; mean depth of flow (also $\bar{D}$); sediment delivery ratio; constant.
D_t - dissipation of energy by turbulence
D_s - scour depth measured from water surface
D_* = $(\Delta\, gd^3/\nu^2)^{\frac{1}{2}}$ - dimensionless grain size
$\dot{e}$ - erosion rate
E - energy; complete elliptic integral of the second kind; elevation difference; average annual soil loss; trap efficiency.
$\dot{E}$ - dimensionless erosion rate
ΔE - activation energy
f - Darcy-Weisbach friction factor; Lacey silt factor; force per bond in cohesive soils; $f = 1/T = \omega/2\pi$ - frequency.
f_w - wave friction factor
f', f'_w - friction factors based on grain roughness of bed.
F_b - bed factor in regime formula
F_D - drag force
$F_d = U/(\Delta\, gd)^{\frac{1}{2}}$ - densimetric Froude number
F_{gr} - mobility number
F_L - dimensionless limit velocity
$Fr = V/(gy_o)^{\frac{1}{2}}$ - Froude number
F_s - factor of safety; side factor in regime method
F_x - energy flux towards shore
g - gravitational acceleration
g_B - bed load transport rate weight or mass per unit width.
g_s - weight rate of sediment transport per unit width
G_{Ts} - total bed material discharge
h - still water depth; height of bed features
h_b - water depth at breaker line
h_i - depth to which intense sediment movement due to wave action occurs

h_o - seaward limit of sediment movement

h' - wave set up

Δh - height of mean energy level above MSL

H - head; elevation difference of water surfaces; wave height

$\bar{H}$ - mean wave height

H_o - deep water wave height

H_r - relative humidity

H_{rms} - root mean square wave height

H_s - significant wave height ($H_{1/3}$)

H_b - breaker height

i - headloss (energy gradient) in pipeline flow

i_b - immersed weight rate of sediment transport per unit width

i_m - energy gradient in terms of head of mixture water/sediment

J - function

k - roughness height; wave number $2\pi/L$; Heywood volume constant

k_s - equivalent (sand grain) roughness height

K - erosion coefficient or erodibility factor; complete elliptic integral of the first kind; constant

K_r - refraction coefficient

K_s - shoaling coefficient

^oK - absolute temperature in degree Kelvin

l - mixing length

L - length; wave length; Avogadro number; Monin-Obukhov length

L_o - deep water wave length

L^* - wave group length

m - mass per unit volume; a parameter in cnoidal wave theory; beach slope ($1:m = \tan\beta$)

M - mass; moment; proportionality factor; factor; bridge opening ratio

n - Manning coefficient; porosity; number of bonds per

m^2 in cohesive soil; number of class intervals; exponent; $n = \frac{1}{2}(1 + 2\ kh/\sinh 2\ kh)$

N - exponent, sample size; constant

p - pressure; symbol for $(\frac{1}{2} f_w)^{\frac{1}{2}}/\kappa$

p_i - percentage by weight of sediment size fraction i

Δp - dynamic pressure

P - conservation practice factor; length of wetted perimeter; protrusion; mean dynamic pressure on bed; sinuosity λ_p/λ; $P = \tau_o U$ - stream power; power; volume of tidal prism; probability

q - flow rate per unit width; characteristic velocity of turbulence

q_B - bed load transport rate; volume per unit width

q_s - suspended sediment transport rate; volume per unit width

q_T - total sediment transport rate; volume per unit width

Q - water discharge

r - radius; ripple factor; reflection coefficient

R - erosion potential; universal gas constant; resultant force

R_b - bed hydraulic radius (side wall correction applied)

$Re_* = u_* d/\nu$ - particle Reynolds number

Ri - Richardson number

R_u - wave run up

s - factor; distance of centre of a sphere from a boundary

S - Slope; radiation stress

SF - shape factor

S_c - sorting coefficient

S_i - sorting index

S_k - quartile arithmetic skewness

S_{kg} - geometric skewness

$S_{H^2}(f)$ - wave energy spectrum function

$S_s = \rho_s/\rho$ - specific gravity

S ‰ - salinity in parts per thousand

t - time

T - time; wave period; turbulence factor; total sediment eroded from catchment; Reynolds stress tensor; absolute temperature; $T = [(u'_*)^2 - u_{*c}^2]/u_{*c}^2$ - transport parameter

T^* - wave group period

u - horizontal component of velocity

u_m - maximum value of orbital velocity at bed

u' - component of turbulence velocity

$u_* = (\tau_o/\rho)^{1/2} = (gy_oS)^{1/2}$ - shear velocity

$u'_* = U\sqrt{g}/C'$

U - velocity

$\bar{U}_y$ - net mass transport velocity at elevation y

v - vertical component of velocity; void ratio

v_b - component of orbital velocity at bed

v_r - resultant velocity

V - volume

V_o - stable channel flow velocity

w - moisture content; percentage of water in soil by weight

W - work done; weight force

x_b - distance of breaker line from water's edge

x_p - distance between wave breaking and plunge lines

X - characteristic grain size in Einstein's formulae

y_o - flow depth

y_s - scour depth

y' - elevation of zero velocity according to logarithmic distribution

Y - sediment yield tonne/km^2; factor

z' - elevation of zero velocity above bed by logarithmic distribution

Z - partition function in rate process theory

α - constant; angle; reflected wave angle; concentration of roughness elements

β - constant; angle; bed slope; aspect ratio

$\gamma = H_b/h_b$ - breaker index

$\gamma_s^* = g(\rho_s - \rho)$

Γ - circulation

δ - boundary layer thickness; lag distance

δ' - laminar sublayer thickness

$\Delta = (\rho_s - \rho)/\rho$

∂ - partial derivative; mean rate of turbulence energy dissipation per unit mass

ε - momentum diffusion coefficient (eddy viscosity)

ε_s - sediment diffusion coefficient

ζ - theta potential

η - height of bed features; vertical ordinate of wave form or bed features; $\eta = y/h$ - relative depth; efficiency

$\bar{\eta}$ - height of mean water surface above SWL

θ - wave angle to contours; $\theta = \tau_o/(\rho_s - \rho)gd$ - Shields parameter

θ' - Shields parameter based on grain roughness

κ - Karman constant

λ - wave length of bed features; displacement distance; wave length of meander; distance between inflexion points; linear concentration

λ_1 - streamwise spacing of turbulence bursts

λ_3 - lateral spacing of turbulence butsts

λ_p - arc length of meander

μ - dynamic viscosity

ν - kinematic viscosity

ξ - Iribarren number or surf similarity parameter; sheltering function; $\xi = C(\frac{1}{2}f_w)^{\frac{1}{2}}$

ξ_1 - streawise dimension of turbulence bursts

ξ_3 - lateral dimension of turbulence bursts

ρ - density of water

ρ_s - density of solids

σ - standard deviation

σ_g - geometric standard deviation

$\sigma_t = \rho - 1000$ - sea water density

τ_o - bed shear stress

τ_c - critical bed shear stress; bed shear stress due to currents

τ_{cw} - bed shear stress due to waves and currents

τ' - shear stress based on grain roughness

ϕ - angle of friction or repose; $\tan\phi$ - friction factor; relative celerty; dimensionless sediment transport function; a function; bed angle; phi-index scale of grain size distribution; Galvani potential

ϕ_D - transport function in solids conveyance in pipilines

χ - potential

ψ - a transport parameter; Volta potential; $\psi = \rho u_m^2/(\rho_s - \rho)gd$ - mobility number

$\psi' = \Delta d_{35}/R'S = 1/\theta$

$\omega = 2\pi/T$ - phase velocity

Ω - wake coefficient

Chapter 1

Introduction

Sediment transport, two-phase flow, loose boundary hydraulics are some of the names used to identify problems of interaction between a fluid flow (water or air) and a granular material mixed into the flow and forming some of the boundaries of the flow which usually change with flow conditions. The problems occurring in the nature fall almost exclusively in the domain of loose boundary hydraulics. The transport may be that of the material of which the flow boundaries are composed or of material added to the flow. An exception to the loose boundary case is the conveyance of some granular material in pipelines where the boundaries are fixed, although deposits may form in the pipe. The fluid induced movement of granular material occurs in a multitude of natural and industrial processes; from soil erosion to transport by the streams and rivers, or materials handling, mineral dressing to a multitude of industrial processes. The water-sediment interaction affects water supply for all purposes and its quality. It affects recreation, fishing, wildlife, agriculture, forestry, navigation, flood control, hydropower and all forms of coastal processes. Specific facets of this interaction are sciences in their own right, e.g., sedimentology, geomorphology.

The crucial characteristic of all loose boundary problems is the *interaction* between the *fluid* and *sediments*, that is, *the erosion and sediment transport problems cannot be treated in isolation from hydro- or aerodynamics*. Indeed, the interactions involve areas of fluid dynamics which have many unsolved problems of their own, i.e. turbulence, boundary layer, diffusion and wave motion.

Soil is one of the most valuable natural resources and its erosion means a loss of a valuable asset. The aim must therefore be to reduce the rate of soil erosion by proper land management techniques. Soil conservation not only reduces problems of siltation, river control and water quality downstream but also creates an asset. At the same time the need for costly river improvement works downstream is reduced. Soil erosion is strongly linked to local climatic conditions and vegetal cover. The most dominant single factor is land management. The erosion rates from the same type of land in the same area can easily differ a thousand times depending on land use and management. The process of soil erosion could be subdivided as shown diagrammatically in Fig. 1.1. The noncohesive soils consist of individual grains, like sand, where the weight of the grains is a dominant force. In cohesive soils the grain size, if it can be defined at all, and its weight are usually insignificant compared to the electrochemical forces acting between the plates of crystals of the minerals involved. The particles bond together and form a coherent, cohesive mass.

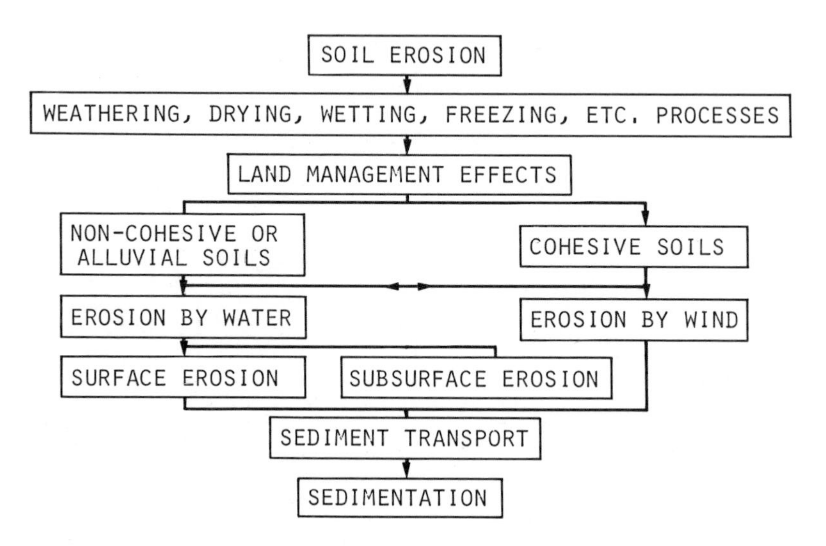

Fig. 1.1. Diagramatic subdivision of soil erosion.

Sedimentary particles arise from breakdown of older deposits and from volcanic eruptions, chemical precipitation and biological secretion. Of short term interest, are mainly particles eroded as solids from land. These become detached from bedrock by weathering and glacial activity. Estimates from the sediment content of Antarctic glaciers delivered to the sea suggest that continental glaciers are the world's greatest producers of sediments. Weathering is the other major producing agent which also acts to further break down the rock fragments. On the global scale, the amount of solids delivered to the sea is estimated to be of the order of 14×10^9 tonne per annum, plus an additional 4×10^9 tonne per annum of dissolved solids. This averages to about 100 tonne/km^2 per annum. Not all of the eroded material is transported by the river system directly to the sea. The rate is usually only about a quarter and often much less. Most of the sediment is deposited at intermediate locations, river valleys, lakes, etc. where it may rest for very long periods. The ratio of sediment yield at a measuring point, Y tonne/km^2 per annum, to total sediment eroded from the catchment upstream, T, is known as the delivery ratio D% = 100Y/T. Erosion is always accompanied by deposition. These, together with orogenic processes, shape the earth. The trend is for all soil and water to move to the lowest possible elevation, which, for soils ultimately means the ocean deep. There is, however, no likelihood for reduction of the height of mountains since indications are that, at present, the rates of orogeny are many times greater than the average rate of erosion.

Wind erosion can be serious in arid and semiarid regions and from cultivated land in dry seasons, particularly when coupled with over exploitation or unsuitable land management practices.

Surface erosion by water starts with raindrops. Rain is a major eroding agent, particularly of cultivated farmlands not yet protected by vegetation. A raindrop on hitting bare soil can by impact form a small crater and scatter soil particles, Fig. 1.2. The impacts can also lead to consolidation of the surface layer and a reduction of the infiltration capacity. The scattering of soil particles leads to a net transport of soil down the slope, where slope is measured relative to the direction of the path of raindrops. When the rainfall intensity exceeds the infiltration rate, overland flow develops which, with increasing depth, absorbs more and more of the energy of the raindrops. After a certain depth, direct impacts do not affect the soil surface but the extra input of energy leads to an overland flow with a higher intensity of turbulence than a sheet flow of the same depth without rainfall. Thus, rainfall on an idealised plane surface displays two distinct regimes, the impact regime and the energy regime (erosion by highly turbulent overland flow) with a transition from one to the other, Fig. 1.3. The impact regime dominates at light rainfalls, at the beginning of the rain, and it persists at the top of the slope for a distance depending on the rainfall intensity. On a natural soil surface the runoff water, whenever the rainfall intensity exceeds the infiltration rate, concentrates into rills and grooves. Thus the erosion process could also be divided into *rill erosion* and *inter-rills erosion*.

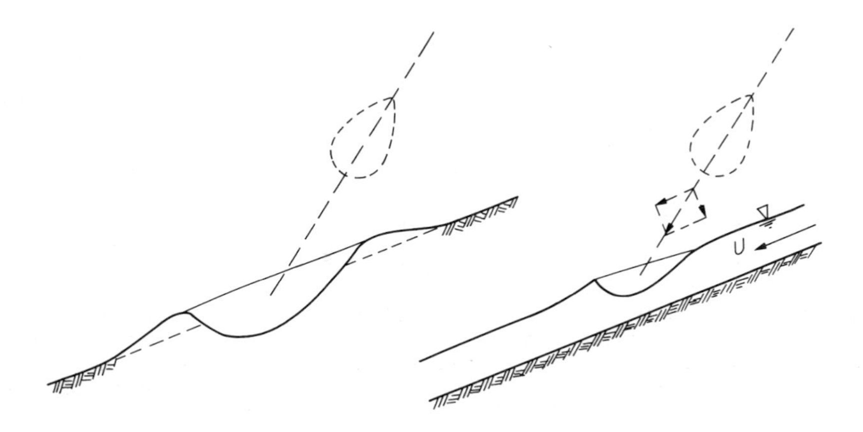

Fig. 1.2. Illustration of raindrop impact on dry surface and one covered with a layer of water.

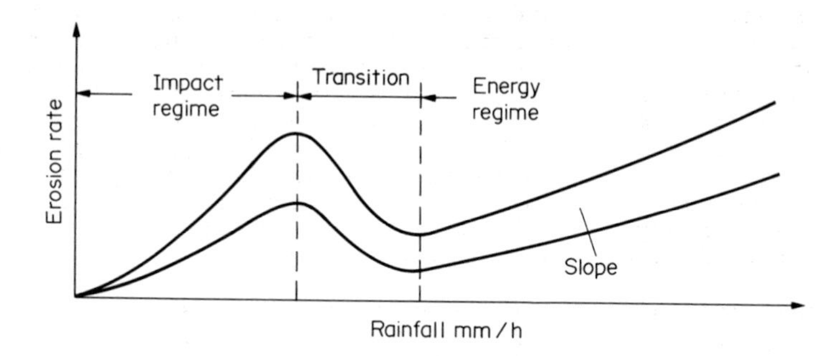

Fig. 1.3. Diagrammatic illustration of soil erosion by raindrops.

In the inter-rills area the soil surface is usually either exposed or covered only with a shallow film of water and the impacts are the main detachment mechanism of soil particles. The rills concentrate flow and the flow in the rills is deeper than the overland flow in the inter-rills areas. The effects of raindrop impacts on the rill flow may be absorbed by the flow before reaching the soil surface underneath, however, the raindrops do make the rill flow more turbulent than an equal open channel flow would be and more efficient in transporting and detaching soil particles.

Water which seeps into the ground can contribute to landslides and be the agent for subsurface erosion. Landslides are a major source of sediment supply to streams and can in some regions overshadow the supply from other sources. Water in the ground moves slowly and does not usually on its own transport granular material through the soil, with the exception of some freshly ploughed soils, some dispersive soils and some volcanic deposits. The subsurface erosion could be a chemical erosion, a dissolving of the soluable constituents of the soil, which can lead to tunnel erosion. The best known examples of the latter are the caves and rivers in limestone country. Tunnels can also develop in clay and loamy soils and in volcanic deposits. Deposits which are layered can concentrate seepage flows at certain elevations and if the deposit also has a gap grading, that is, if the smaller fractions can pass through the soil matrix formed by the coarser fractions of the deposit, then the soil will gradually "open up", giving rise to tunnels. Many of these tunnels collapse and start the development of erosion gulleys, particularly where the tunnels exit on the slope.

All the various products of erosion are transported by water. These transport processes together with coastal processes constitute loose boundary hydraulics.

In principle, the sediment transport problems are governed by the equations of motion and continuity, for both fluid and sediment, and boundary conditions. However, the interactions of the various causative and resisting forces are so manyfold and complex that a comprehensive analytical description of the processes of soil erosion, transport and geometry of streams or coastlines is even hard to imagine. Analytical models exist only for individual facets of the problem. The major stumbling block is the equations of motion. Not even the motion of a single grain in turbulent flow can be described adequately. The movement of a group of particles or of a granular boundary is still beyond analytical description. The problems become more complex still when the erodible material is cohesive. It is therefore not surprising that loose boundary hydraulics still relies heavily on experimental and field data and on *dimensional analysis*.

Dimensional reasoning is extensively used in discussion of sediment transport problems. The introduction of dimensional analysis to the study of these problems in the early part of this century gave a great boost to the development of sediment transport theories but gradually a situation was reached where it probably hindered rather than aided the advance of understanding. An assembly of variables can be arranged into dimensionless groups in very many different ways and plotting the same data over and over again, two parameters at a time, will not contribute much to understanding of the mechanics of sediment transport. In particular, plots of the form of yx^3 versus x on logarithmic paper are of very little value.

Mathematically, a physical phenomenon is described by a set of n independent, necessary and sufficient parameters $a_1, \ldots, a_n$. The n parameters a_i are independent if none of them is expressible as a function of the remaining n - 1 parameters. Stated differently, an independent parameter (variable) is one whose value is imposed externally on the system and as a corollary, a dependent variable is one whose value is determined by the values specified for the independent parameters. Any property A of the phenomenon must be definable by the a_i parameters as

$$A = f_A(a_i, \ldots, a_n). \tag{1.1}$$

The independent or characteristic parameters of the *fluid* are its density ρ and its viscosity μ. These two numerical values define the clear fluid for studies of mechanics. If the fluid contains suspended sediment, other than bed material entrained in the reach under consideration, then the initial concentration c_o has to be treated as a further independent parameter.

The *cohesionless granular material* is defined by its density ρ_s, a size characteristic d, and by the shape of the grain and grain-size distribution. The theory of dimensions is not suited for description of geometry and shape, and these are therefore omitted. This means that a particular functional relationship will be valid only for a given shape and grading properties of the sediment. When the shape or grading changes, the function f_A changes. It is also clear that because of incomplete understanding of the physics of cohesive materials an assembly of the independent parameters is not possible.

In a combined flow of fluid and sediment in an open channel, assuming two-dimensional steady uniform flow, the flow is determined by its depth y_o, slope S and gravity g, which provides the driving force. Hence, all variables of a two-dimensional steady uniform flow with sediment of a given shape and grading are defined by

$$\rho, \mu, \rho_s, d, y_o, S, g \tag{1.2}$$

Some of these can be replaced, if desired, with others but the quantities used for substitution must be dependent on the parameter substituted for. For example, $u_* = \sqrt{(g y_o S)}$ can be used to replace either g, y_o or S; $y_s{}^* = g(\rho_s - \rho)$ could be used to replace either g, ρ_s or ρ and we could write instead of equation (1.2)

$$\rho, \mu, \rho_s, d, y_o, y_s{}^*, u_* \tag{1.3}$$

Any flow property can then be expressed as a function of these independent parameters. For example, the flow rate

$$q = f_q(\rho, \mu, \rho_s, d, y_o, y_s{}^*, u_*) \tag{1.4}$$

and if the fluid and sediment have been specified then ρ, μ, d, ρ_s and $y_s{}^*$ are constants and

$$q = f_q(y_o, gS) \tag{1.5}$$

If the flow had acquired a concentration of sediment before entering the channel then

$$q = F_q(y_o gS, c_o) \tag{1.6}$$

A phenomenon of mechanics is defined by mass M, length L, and time T. Hence. equation (1.3) yields four dimensionless numbers.

The mechanics of dimensionless analysis is extensively covered in textbooks, e.g. Langhaar (1962) and, in particular, as applied to sediment transport, by Yalin (1965,1972). A very thorough treatment of the theory of dimensional analysis and similarity is given by Sedov (1959). Using d, ρ and u_* to define M, L and T yields

$$\frac{u_* d}{\nu}, \frac{\rho u_*^2}{y_s{}^* d}, \frac{y_o}{d}, \frac{\rho_s}{\rho} \tag{1.7}$$

and a dimensionless property X of the flow with sediment is defined by

$$X = f_x \left[\frac{u_* d}{\nu}, \frac{\rho u_*^2}{y_s{}^* d}, \frac{y_o}{d}, \frac{\rho_s}{\rho} \right] \tag{1.8}$$

These dimensionless terms are the well-known Reynolds number Re_*, the Shields' parameter θ, the relative roughness and relative density. These can be modified by substitution, by taking ratios or by raising to some power without changing the general nature of the statement. For example, in Yalin's sediment transport formula $g_s/y_s{}^*d = f(\mu u_*/_{y_s}{}^* d^2, \rho u_*^2/y_s{}^*d, y_o/d, \rho_s u_*^2/y_s d)$ where

$\mu u_*/\gamma_s^* d = [\rho u_*^2/\gamma_s^* d]/[u_* d/\nu]$ and $\rho_s u_*^2/\gamma_s^* d_* = [\rho_s/\rho][\rho u_*^2/\gamma_s^* d]$. Likewise we can find the parameter $\gamma_s^* d^3/\rho\nu^2$ which is $(u_* d/\nu)^2/[\rho u_*^2/\gamma_s^* d]$.

In principle, the flow behaviour in an alluvial channel should be definable in terms of these dimensionless numbers of equation (1.8). The function f will depend on the geometry of the channel and sand, and a change of sand or flume width will affect the result; a different function will pertain to each particular geometric condition.

The discussion relates to a basic simple situation. River systems and river processes are so complex that there is not even general agreement on which aspects are causes and which are effects. Clearly the inputs to a river reach are the water and sediment discharges and these determine the width, depth, velocity and sediment discharge through the reach, and rates of change of water and sediment storage in the reach. However, the bed roughness (friction factor) is interrelated with depth and velocity, sediment transport rate and probably also with the rate of scour or deposition.

In addition, many of the functions are multi-valued; for example, three different values of U are possible at a given slope. In a laboratory flume with a bed of fine sand and constant depth of flow, the mean velocity and sediment concentration cannot be expressed as a single-valued function of depth and slope; that is

$$\begin{aligned} f(U, y_o S) &\neq 0 \\ f(c, y_o, S) &\neq 0. \end{aligned} \tag{1.9}$$

This should not be interpreted to mean that the function does not exist, but rather that there are multiple solutions to $U = f(y_o S)$ and $c = f(y_o, S)$, and equation (1.9) cannot be used to predict mean velocity and sediment concentration without additional constraints.

In the simple case of an alluvial stream flowing at steady uniform depth and velocity, one has to consider (Task Committee, 1971): Q = water discharge, G_{Ts} = bed material discharge, c_w = concentration of wash load, B = width, y_o = mean depth, R = hydraulic radius, U = mean velocity, S = energy gradient, f = friction factor, ν = kinematic viscosity of water, ρ = density of water, ρ_s = density of sediment material, d_g = geometric mean particle size, σ = standard deviation of particle size grading, w = mean setting velocity, g = gravitational acceleration, and dimensions defining the plan form. Here seventeen parameters have been listed.

The relationships known or assumed are: continuity, $Q = UBy_o$; friction factor from $U = \sqrt{(8g/f)}\sqrt{(RS)}$; f = function of roughness; a sediment transport relationship; a width-depth relationship; a hydraulic radius-depth relationship; a relationship between plan-form and flow and sediment properties. These are seven dependencies. Hence, there are ten independent parameters.

On top of all this, it is necessary to distinguish between short-term and long-term behaviour of alluvial rivers and canals. Streams evolve with time so that they can transport sediment with the available water runoff. If too much sediment comes in, deposition takes place, the stream will steepen and the velocity increases until equilibrium is established.

It was shown by Birkhoff (1950) and Kline (1965) that normalizing of the governing equations of a system is the most rigorous method of deriving similitude parameters. The ideas can be illustrated by normalizing the equations of continuity and of motion. In a cartesian coordinate system with the x-axis parallel to the direction of flow, the time-averaged form of the equation of motion for a flow with variable density and viscosity is

$$\frac{d}{dt}(\rho u) + \left[\frac{\partial \overline{\rho u'^2}}{\partial x} + \frac{\partial \overline{\rho u'v'}}{\partial y} + \frac{\partial \overline{\rho u'w'}}{\partial z}\right] = -\frac{\partial}{\partial x}(\rho g h) + \nabla\mu\cdot\nabla u + \mu\nabla^2 u \qquad (1.10)$$

Similarly in y and z directions, but these will yield no additional similitude information.

The continuity equation is

$$\frac{\partial u}{\partial t} + \frac{\partial \rho u}{\partial y} + \frac{\partial \rho v}{\partial z} + \frac{\partial \rho w}{\partial z} = 0 \qquad (1.11)$$

In sediment-laden flow the density is

$$\rho = \rho_w + (\rho_s - \rho_w)c \qquad (1.12)$$

and for low concentration of sediment the viscosity could be expressed as

$$\mu = \mu_w(1 + kc) \qquad (1.13)$$

where $k \cong 2.5$.

In normalizing the equations each of the unknowns is made dimensionless with the aid of variables of the flow, preferably those which are readily measured. Using the following transformations:

$$u^* = u/U \qquad \rho^* = \rho/\rho_w \qquad x^* = x/y_o \qquad c^* = c/\bar{c}.$$
$$v^* = v/V \qquad t^* = tU/y_o \qquad y^* = y/y_o \qquad \mu^* = \mu/\mu_w,$$
$$w^* = w/W \qquad h^* = h/y_o \qquad z^* = z/y_o \qquad u'^* = u'/U, \text{ etc}$$

where y_o = total depth, h = piezometric head and

$$\bar{c} = q_s/q \qquad (1.14)$$

With these substitutions the equations can be transcribed in the normalized form

$$\frac{d}{dt^*}\rho^*u^* + \left[\frac{\partial\overline{\rho^*u'^{*2}}}{\partial x^*} + \frac{\partial\overline{\rho^*u'^*v'^*}}{\partial y^*} + \ldots\right] =$$

$$= -\frac{gy_o}{U^2}\frac{\partial\rho^*h^*}{\partial x^*} + \frac{\mu w}{\rho_w y_o U}(\nabla^*\mu^*\cdot\nabla^*u^* + \mu^*\nabla^2u^*) \qquad (1.15)$$

$$\frac{\partial\rho^*}{\partial t^*} + \frac{\partial\rho^*u^*}{\partial x^*} + \frac{\partial\rho^*v^*}{\partial y^*} + \frac{\partial\rho^*w^*}{\partial z^*} = 0 \qquad (1.16)$$

$$\rho^* = 1 + \frac{\rho_s - \rho_w}{\rho_w}\frac{q_s}{q}c^* \qquad (1.17)$$

$$\mu^* = 1 + k\frac{q_s}{q}c^* \qquad (1.18)$$

The similitude coefficients now are

$\frac{\rho_s - \rho_w}{\rho_w}$ differential density ratio, (1.19)

q_s/q ratio of the volumetric sediment discharge to the water discharge, (1.20)

$$gy_o/U^2 = 1/F_r^2 \qquad (1.21)$$

$$\mu_w/\rho_w y_o U = 1/Re \qquad (1.22)$$

The experimental, or analytical, solutions to the above equations depend on boundary conditions, which describe the properties of the physical system and do not normally appear explicitly in the equations derived. The lack of explicit functional relationships for the boundary conditions makes it necessary to use less rigorous means for prescription of the similitude coefficients.

When boundary conditions are adequately described in terms of grain size, this length scale could be made consistent with scales of the set of similitude coefficients obtained by normalizing the original equations. Thus, for d, this yields a dimensionless "length scale" of the form

$$D = \frac{\rho_w^{2/3} g^{1/3} d}{\mu_w^{2/3}} \tag{1.24}$$

since, Re $\rightarrow L^2 = T\nu$, Fr $\rightarrow T = L^{1/2}/g^{1/2}$; $L = \nu^{2/3}/g^{1/3}$ and using $\nu = \mu/\rho$. The particle diameter d alone is not adequate to describe the bed configurations. The same relationship could be written with channel width B instead of d. Ratios such like gd/U^2, d/y_o are of little help since their values depend on the flow variables U and y_o.

Willis and Coleman (1969) published an excellent discussion of presentation of experimental data in terms of similitude parameters

The concepts of similarity and dimensionless reasoning also form the foundations for modelling of loose boundary problems. For these the reader is referred to Yalin (1971) , Kobus (1980) and Novak and Cabelka (1981).

Chapter 2

Sediment Properties

2.1 Sediment Characteristics

For convenience the sediments forming the boundaries of a flow are subdivided into cohesive and non-cohesive sediments, although there is a fairly broad transition range. In alluvial or non-cohesive sediments the particle or grain size and weight, are the dominant parameters for sediment movement and transport. Non-cohesive sediments have a granular structure and do not form a coherent mass. However, the properties of alluvial soils change drastically with increasing clay content (fraction of soil composed of particles smaller than 2 μm). In most soils clay assumes control of soil properties at less than 10 per cent clay content. In cohesive soils the electro-chemical interactions dominate and the size and weight of an individual particle may be of little importance. Cohesive soils form a coherent mass. Its properties are discussed in the chapter on erosion of cohesive soils.

Soils are classified according to particle size, for example British Standard BS1377: 1975, as follows:

Very fine clay	0.24 - 0.5 μm	Very fine gravel	2 - 4 mm
Fine clay	0.5 - 1.0 μm	Fine gravel	4 - 8 mm
Medium clay	1.0 - 2.0 μm	Medium gravel	8 - 16 mm
Coarse clay	2.0 - 4.0 μm	Coarse gravel	16 - 32 mm
Very fine silt	4 - 8 μm	Very coarse gravel	32 - 64 mm
Fine silt	8 - 16 μm	Small cobbles	64 - 128 mm
Medium silt	16 - 31 μm	Large cobbles	128 - 256 mm
Coarse silt	31 - 62 μm	Small boulders	256 - 512 mm
Very fine sand	62 - 125 μm	Medium boulders	512 - 1024 mm
Fine sand	125 - 250 μm	Large boulders	1024 - 2048 mm
Medium sand	250 - 500 μm	Very large boulders	2048 - 4096 mm
Coarse sand	0.5 - 1.0 mm		
Very coarse sand	1.0 - 2.0 mm		

A very similar classification was proposed by the sub-committee on Sediment Terminology of the American Geophysical Union (Vanoni, 1975, page 20 Table 2.1).

The common definitions for particle size are;

sieve diameter - opening size of mesh (may include oblong particles)
sedimentation diameter - diameter of a sphere of the same density and the same fall velocity in the same fluid at the same temperature as the given particle
nominal diameter - diameter of a sphere of equal volume
triaxial dimensions a, b and c, where c is the shortest of the three perpendicular axes of the particle.

The sediment normally consists of a distribution of particle sizes. Figure 2.1 shows the two common presentations of the particle size distribution, the cumulative distribution curve as the log-normal and the log-probability plot. The 50% diameter on the log-probability plot is called the geometric mean diameter d_g and the geometric standard deviation is defined as

$$\sigma_g = (d_{84.1}/d_{15.9})^{\frac{1}{2}} \text{ or } \sigma_g = \frac{d_{84.1}}{d_{50}} = \frac{d_{50}}{d_{15.9}} \qquad (2.1)$$

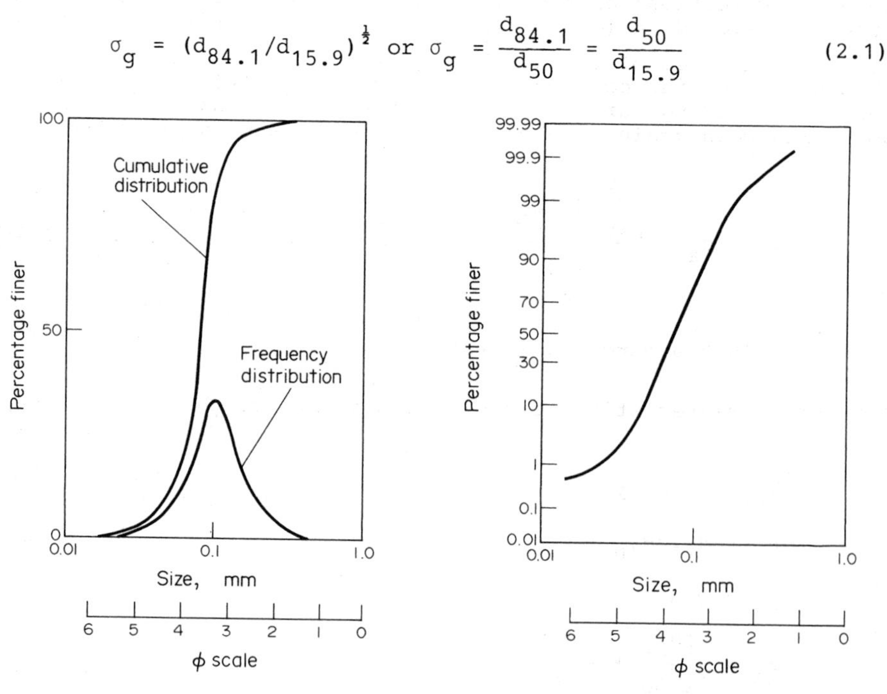

Fig. 2.1. Ways of presenting grain size distributions.

Many natural deposits plot as one or more straight line segments on log-probability paper. A two segment distribution can be shown to arise from superposition of two particle size distributions.

The natural mean diameter, $\bar{d}$, is given by the arithmetic mean of the size distribution and the standard deviation is calculated from

$$\sigma = \left\{ \frac{\sum_{i=1}^{n} f_i d_i^2 - \left[\left(\sum_{i=1}^{n} f_i d_i \right)_i \sum_{i=1}^{n} f_i \right]}{\sum_{i=1}^{n} f_i} \right\}^{\frac{1}{2}} \qquad (2.2)$$

where n is the number of class intervals in the grain-size distribution and f_i is generally taken as the *percent* frequency in each class. The values of $\bar{d}$ and d_g are related by

$$\bar{d} = d_g \exp [0.5 \ln^2 \sigma_g] \qquad (2.3)$$

Standard deviation of grain-size distributions which follow normal distribution is given by eqn (2.1).

Several definitions of the "effective" grain size, d_e, have been proposed, e.g. $d_e = (d_{10} + d_{20} + d_{30} + \ldots + d_{90})/9$, or from equal areas on the plot of 1/d as the ordinate and %-passing as the abscissa, i.e. the horizontal line from $1/d_e$ divides the area under the grading curve to equal parts. In the same sense a governing mean grain size or average size has been proposed

$$d_m = d_a = \frac{\sum_{p=0}^{100\%} \bar{d}_i \Delta p_i}{\sum_{p=0}^{100\%} \Delta p_i} \quad ; \quad \bar{d}_i = \frac{1}{2}(d_i + d_{i+1}) \qquad (2.4)$$

where Δp_i is the percentage weight of size d_i.

Further characteristics of sediment grading are the *sorting coefficient*

$$C_s = (d_{75}/d_{25})^{\frac{1}{2}} \qquad (2.5a)$$

or

$$S_c = d_{90}/d_{10} \qquad (2.5b)$$

or *sorting index*

$$S_i = \frac{1}{2}\left[\frac{d_{84}}{d_{50}} + \frac{d_{50}}{d_{10}}\right] \qquad (2.5c)$$

The ratio d_{60}/d_{10} was used by Hazen as the *uniformity coefficient*, whereas Kramer (1935) defined the *uniformity coefficient* as

$$C_u = \frac{\sum_{0}^{50} \Delta p_i \, d_i}{\sum_{50}^{100} \Delta p_i \, d_i} \qquad (2.5d)$$

The quartile *arithmetic skewness*

$$S_k = \frac{1}{2}(d_{75} + d_{25} - 2d_{50}) \quad (2.6a)$$

and the *geometric skewness*

$$S_{kg} = (d_{25}d_{75}/d^2{}_{50})^{\frac{1}{2}} \quad (2.6b)$$

Generally, the grain-size distributions near the source, like debris slides, are strongly skew towards the larger fractions (percentage by weight versus size). During the transport processes the distributions gradually transform and reach after a sufficient distance a normal (gaussian) form, i.e., the cumulative distribution plots as a straight line on log-normal paper.

The phi-index scale is by definition

$$\phi = \log_2 d_{(mm)} = -\log d/\log 2 \quad (2.7)$$

and it divides the sediment into arbitrary class intervals as follows:

d_{mm}:	8	4	2	1	0.5	0.25	0.125	0.0625
ϕ :	-3	-2	-1	0	1	2	3	4

The size of the sample for analysis depends on the size of the large particles present. The ASTM recommends a sample size M in kg

$$M = 0.082\ b^{1.5} \quad (2.8)$$

where b is the maximum intermediate triaxial dimension in mm.

There are two simple criteria by which it may be assessed if a sediment mixture can be treated as though it were uniform. If $d_{95}/d_5 < 4$ or 5 the sediment is uniform from the hydraulic point of view, and similarly if $\sigma_g < 1.35$ the sediment may be considered uniform. For mixtures which do not meet these criteria, the non-uniformity of grain size reduces the resistance to flow. The coarser grains tend to armour the crests of the bed features and thus reduce the effective roughness of the bed. The erosion of non-uniform sediments is a more complex problem than that of sediments of uniform grain size.

A further important parameter is the particle shape. One of the definitions in use is the shape factor

$$SF = c/(ab)^{\frac{1}{2}} \quad (2.9)$$

Another shape definition is the Heywood (1938) volume coefficient k, such that volume equals kd^3 where d is the diameter of the circle having the projected area of the particle in its most stable position, i.e. projected on the plane parallel to a-b axes. For natural grains $k \cong 0.3$ is a common value.

Density. Density of the particles is important and must be known. Where the sediment is composed of a variety of minerals the proportions and sizes need to be determined. The average density of the sample may change little but the variation from, say, pumice sands to iron sands or magnetite may be appreciable. Such variations in density affect sediment transport by segregation, e.g. the armouring effect of the heavy minerals on dune crests.

In problems involving erosion and deposition bulk density and porosity of the sediment have to be considered, e.g. newly formed fluvial deposits in reservoirs may have very high porosity and low weight per unit volume.

2.2 The Fall Velocity

The fall or settling velocity figures prominently in all sediment transport problems and although the concept is straightforward its precise evaluation or calculation is not. The literature dealing with the motion of particles of various shapes in ideal and in viscous fluids is extensive. A comprehensive summary was presented by Torobin and Gauvin (1959). The slow speed problem is treated by Happel and Brenner (1965). Reference is also made to the survey by Graf (1971).

The fall velocity is a function of size, shape, density and viscosity. In addition it depends on the extent of the fluid in which it falls, on the number of particles falling and on the level of turbulence intensity. Turbulent conditions occur when settling takes place in flowing fluid and can also occur when a cluster of particles is settling.

Falling under the influence of gravity the particle will reach a constant velocity - the terminal velocity - when the drag equals the submerged weight of the particle. For a spherical particle in stationary fluid the fall velocity w is

$$C_D \pi \frac{d^2}{4} \frac{\rho w^2}{2} = \pi \frac{d^3}{6} g(\rho_s - \rho)$$

$$w = \left[\frac{4}{3} \frac{1}{C_D} gd(S_s - 1)\right]^{\frac{1}{2}} \tag{2.10}$$

where S_S is the specific gravity of the grain and C_D is the drag coefficient which is a function of the Reynolds number $Re = wd/\nu$ where ν is the kinematic viscosity. The fall velocity depends on the shape of the particle and the Reynolds number.

For spherical particles of diameter d in a viscous fluid of infinite extent the drag coefficient is fairly well defined for laminar flow. The Stokes' solution of $F_D = 3\pi\mu dw$ yields

$$C_D = \frac{24}{Re} \tag{2.11}$$

for $Re < 0.5$ and approximately for up to 1.0, where $Re = wd/\nu$. The complete solution as provided by Goldstein gives

$$C_D = \frac{24}{Re} [1 + \frac{3}{16} Re - \frac{19}{1280} Re^2 + \frac{71}{20\ 480} Re^3 \ldots] \quad (2.12)$$

for $Re \leq 2$, where the first two terms in brackets are known as the Oseen solution.

For higher Reynold's numbers the theoretical treatments have as yet not succeeded in accurately predicting the value of the drag coefficient. The difficulties arise mainly from interaction of the turbulence with the particle. Here the effects of inertia and virtual mass have to be accounted for. The value of the drag coefficient depends strongly on the level of free stream turbulence, apart from turbulence caused by the particle itself. It also depends on whether or not the surface of the sphere is hydraulically smooth or rough.

Experimental data for $Re < 800$ appear to fit reasonably well

$$C_D = \frac{24}{Re} (1 + 0.150\ Re^{0.687}) \quad (2.13)$$

an expression given by Schiller and Naumann (1933). Kazanskij (1981) fitted experimental data for $10^{-3} \leq Re_w \leq 3 \times 10^4$ at 15°C with

$$C_D = \frac{24}{Re} + A + B - C \quad (2.14)$$

$$\log A = -1.01 + 2.77(\log Re + 1.2)\exp[-0.71(\log Re + 1.2)]$$
$$B = 6.726 \times 10^{-4}\ Re^{0.605}$$
$$C = 8.796 \times 10^{-15}\ Re^{2.86}$$

The trial and error procedure of estimation of the fall velocity can be avoided by the use of the dimensionless particle size

$$C_D Re^2 = \frac{4}{3} g \frac{\rho_s - \rho}{\rho} \frac{d^3}{\nu^2} \quad (2.15)$$

This yields $C_D Re^2$ and Fig. 2.2 the Reynolds number and hence C_D and w.

So far we have considered a single spherical particle in a fluid of infinite extent. If the fluid is externally bounded, as, for example, when testing for the fall velocity in a glass cylinder then the value of the drag coefficient will depend on the distance between the sphere and the boundaries.

Brenner (1961) showed that when the sphere is approaching the bottom of the container the Stokes' relationship for drag force has to be multiplied by a factor

$$K = 1 + (9r/8s) \quad (2.16)$$

for $r/s < 0.06$ where s is the distance of the centre of the sphere from the boundary at any given instant. If the plane is not solid but a free surface, like an interface between two liquids

$$K = 1 + (3r/4s) \quad (2.17)$$

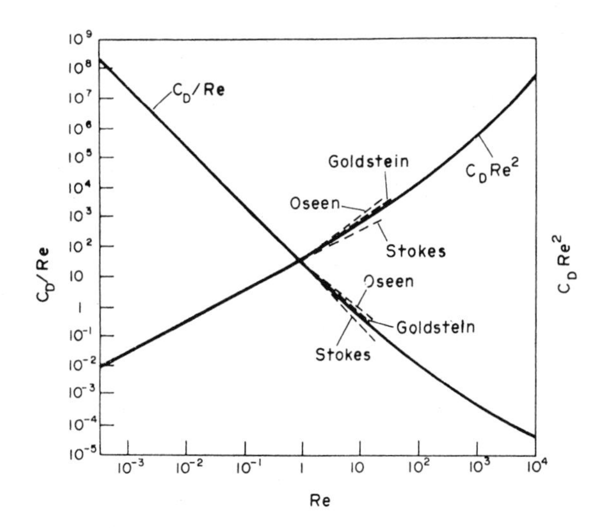

Fig. 2.2. C_D/Re and C_D/Re^2 vs. Reynolds number for sphere; for computation of the settling velocity and the particle diameter.

If the centre of the sphere is still 10r away from the solid floor K = 1.125, i.e., timing should not continue to the bottom.

There appear to be no analytical or experimental results for Re greater than the Stokes' range.

For a sphere falling near a single vertical wall, in Stokes' range, McNown (1951) gives

$$K = 1 + (18r/32s) \qquad (2.18)$$

and half-way between two plane walls

$$K = 1 + 1.006r/s. \qquad (2.19)$$

Slow-motion settling in a circular cylinder is discussed by Happel and Brenner (1965). For spherical particles falling on the axis of the cylinder

$$K = 1 + 2.1\, r/R \qquad (2.20)$$

where R is the radius of the cylinder. This relationship is supported by experimental results for $r/R < 1.0$. For $r/R > 0.1$ McNown et al (1948) give

$$K + \frac{3\pi\sqrt{2}}{8}\left[1 - \frac{r}{R}\right]^{-5/2} \qquad (2.21)$$

and at higher Re (McNown et al, 1951) when r/R is nearly equal to unity

$$K = \frac{r/R}{1-(r/R)^2} \qquad (2.22)$$

These are shown on Fig. 2.3 where the experimental data is by Fidleris and Whitmore (1961).

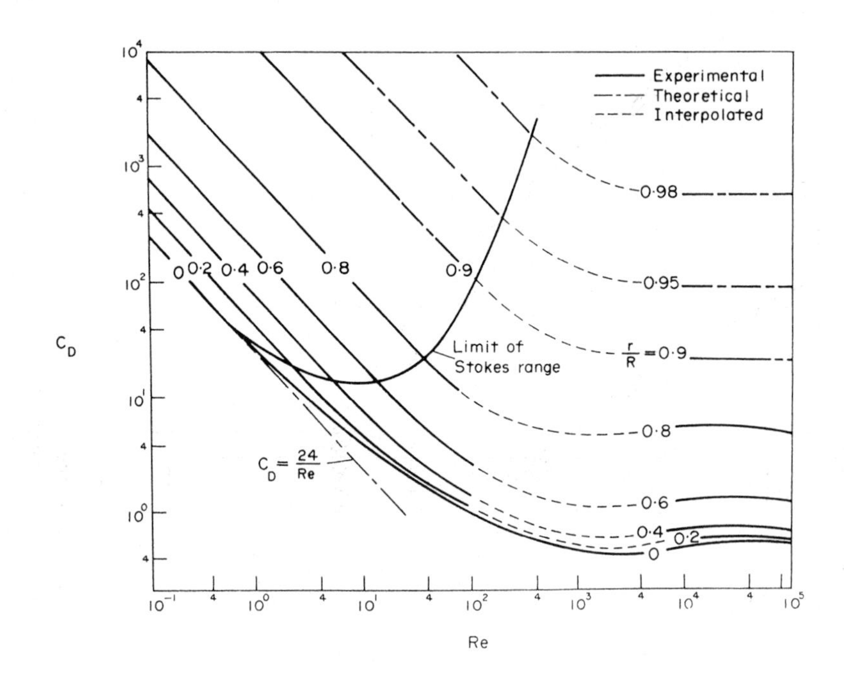

Fig. 2.3. Drag coefficient vs. Reynolds number, with r/R as parameter, in cylindrical tubes.

The fall velocity of naturally worn quartz grains in terms of nominal diameter (which is readily determined from weight and density), according to US Inter-Agency Committee (1957) is shown in Fig. 2.4 and Table 2.1. The relationship between the nominal and fall diameters and the fall and sieve diameters as a function of the shape factor is shown in Fig. 2.5. For naturally worn sediments over the range of about 0.2 to 20 mm the sieve diameter is approximately 0.9 times the nominal diameter.

Average values for quartz sands in water at 20°C appear to fit reasonably well to

$$w(\text{mm/s}) = 663\ d^2\ (\text{mm});\quad d < 0.15\ \text{mm} \tag{2.23}$$

$$w(\text{mm/s}) = 134.5\ d^{0.52} \simeq 134.5\ \sqrt{d}\ (\text{mm});\quad d > 1.5\ \text{mm} \tag{2.24}$$

with a transition region between $0.15 \leq d \leq 1.5$ mm as shown in Table 2.2, where d is the nominal diameter.

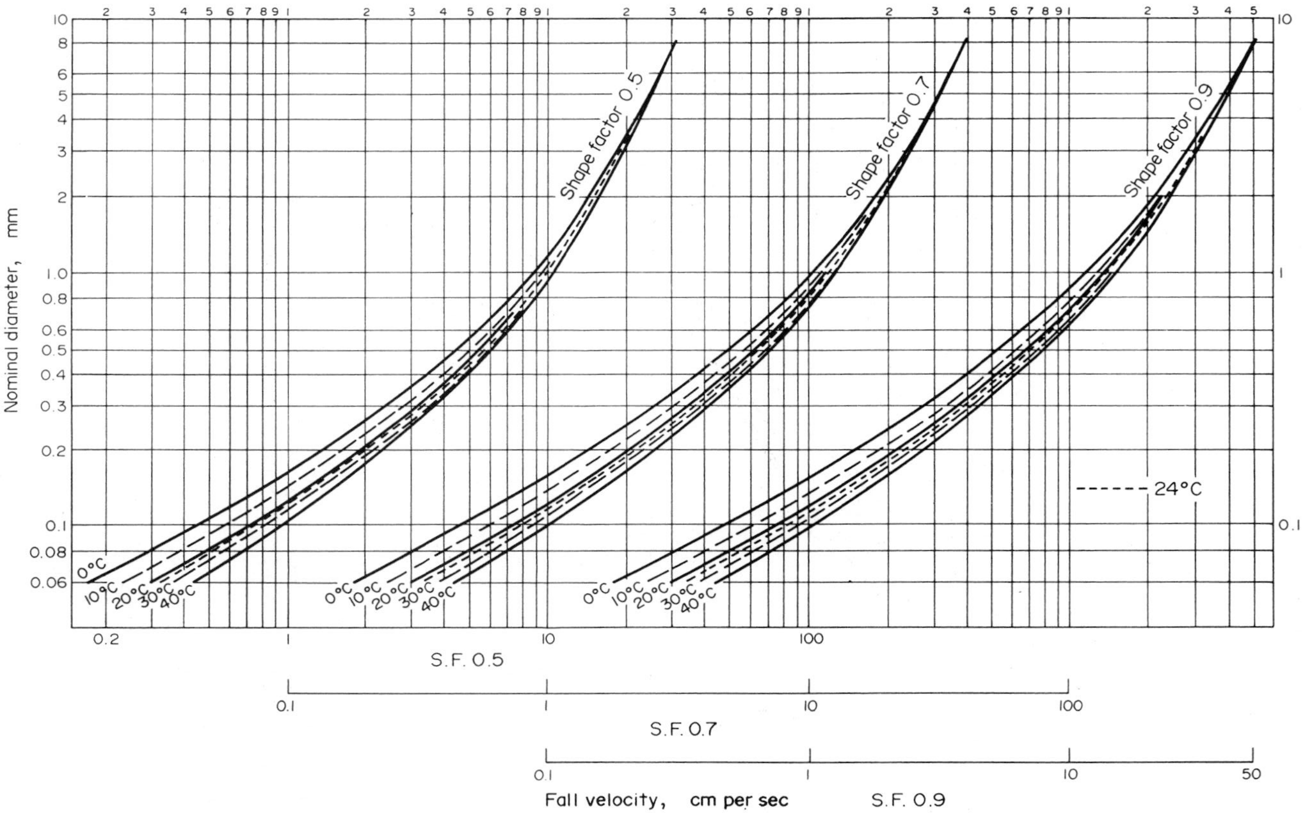

Fig. 2.4. Relation of nominal diameter and fall velocity for naturally worn quartz particles with shape factors (S.F.) of 0.5, 0.7, and 0.9 according to U.S. Inter-Agency Committee on Water Resources, Subcommittee on Sedimentation, 1957.

TABLE 2.1. Fall velocities of sediment particles in mm s^{-1}

	Specific gravity 2.00					Specific gravity 2.65					Specific gravity 4.30				
Temp.	Shape factor					Shape factor					Shape factor				
(°C)	0.3	0.5	0.7	0.9	S	0.3	0.5	0.7	0.9	S	0.3	0.5	0.7	0.9	S
Nominal diameter = 0.20 mm															
0	8.4	9.0	9.5	10.0	10.6	12.9	13.8	14.8	15.7	16.6	22.1	24.2	26.2	27.6	29.4
10	10.4	11.2	12.0	12.6	13.3	15.6	16.8	18.1	19.2	20.5	26.0	28.7	31.4	33.6	36.1
20	12.1	13.2	14.2	15.1	16.0	17.8	19.4	21.1	22.6	24.3	29.5	32.7	36.2	39.0	42.2
24	12.7	14.0	15.1	16.1	17.2	18.6	20.4	22.3	24.0	25.8	30.9	34.2	38.0	41.1	44.5
30	13.6	15.0	16.3	17.5	18.7	19.9	21.8	24.0	25.9	28.0	32.7	36.4	40.6	44.0	47.8
40	15.1	16.7	18.3	19.8	21.3	21.8	24.1	26.8	29.0	31.6	35.7	39.8	44.8	48.7	52.9
Nominal diameter = 0.50 mm															
0	27.9	31.4	34.7	37.9	40.4	40.1	44.7	50.2	54.8	59.2	64.1	72.6	81.9	90.6	99.0
10	31.9	36.1	40.2	44.1	47.3	45.0	51.2	57.2	63.0	68.8	71.0	81.5	92.2	103.0	113.0
20	35.3	39.9	44.7	49.5	53.5	49.0	56.3	63.1	70.2	76.8	76.2	87.9	101.0	113.0	124.0
24	36.3	41.3	46.4	51.6	55.8	50.3	57.9	65.3	73.0	79.7	77.9	90.4	104.0	117.0	128.0
30	38.0	43.2	48.8	54.3	59.0	52.4	60.3	68.4	76.6	83.8	79.9	93.2	108.0	122.0	134.0
40	40.2	46.2	52.5	58.7	64.0	55.2	63.8	73.0	82.4	90.5	82.4	97.6	114.0	130.0	143.0
Nominal diameter = 1.00 mm															
0	57.6	65.9	74.7	85.0	92.0	78.3	90.4	104.0	118.0	128.0	117.0	138.0	162.0	185.0	204.0
10	61.6	71.6	82.3	93.6	103.0	82.1	96.6	114.0	130.0	143.0	121.0	144.0	173.0	202.0	225.0
20	63.9	75.8	88.6	102.0	112.0	84.9	101.0	121.0	140.0	156.0	123.0	148.0	181.0	215.0	243.0
24	64.5	77.0	91.0	105.0	116.0	85.7	102.0	123.0	143.0	160.0	124.0	149.0	184.0	219.0	250.0
30	65.4	78.8	93.8	109.0	121.0	86.6	104.0	126.0	148.0	166.0	124.0	151.0	187.0	225.0	258.0
40	66.5	80.9	98.0	114.0	129.0	87.7	106.0	130.0	156.0	175.0	125.0	153.0	191.0	232.0	271.0
Nominal diameter = 2.00 mm															
0	95.0	114.0	138.0	163.0	181.0	124.0	149.0	184.0	221.0	252.0	177.0	217.0	269.0	328.0	383.0
10	96.6	117.0	144.0	174.0	198.0	125.0	153.0	190.0	231.0	273.0	177.0	220.0	274.0	339.0	412.0
20	97.3	119.0	148.0	181.0	211.0	125.0	155.0	193.0	239.0	289.0	177.0	222.0	277.0	346.0	434.0
24	97.6	120.0	149.0	183.0	216.0	126.0	156.0	194.0	240.0	294.0	177.0	223.0	278.0	348.0	442.0
30	97.9	121.0	151.0	187.0	222.0	126.0	157.0	195.0	243.0	301.0	178.0	224.0	279.0	351.0	451.0
40	98.3	123.0	153.0	190.0	231.0	126.0	158.0	197.0	247.0	310.0	178.0	225.0	280.0	354.0	464.0
Nominal diameter = 4.00 mm															
0	138.0	172.0	214.0	268.0	329.0	177.0	223.0	278.0	349.0	438.0	251.0	317.0	395.0	499.0	655.0
10	138.0	173.0	216.0	273.0	346.0	177.0	224.0	279.0	353.0	458.0	251.0	317.0	396.0	502.0	668.0
20	138.0	174.0	218.0	275.0	359.0	178.0	224.0	280.0	356.0	469.0	251.0	317.0	397.0	504.0	677.0
24	138.0	175.0	219.0	276.0	363.0	178.0	224.0	281.0	357.0	472.0	251.0	317.0	397.0	505.0	680.0
30	139.0	175.0	219.0	276.0	363.0	178.0	225.0	281.0	358.0	476.0	251.0	318.0	397.0	505.0	683.0
40	139.0	176.0	220.0	280.0	374.0	178.0	225.0	282.0	359.0	481.0	252.0	318.0	398.0	506.0	686.0
Nominal diameter = 8.00 mm															
0	195.0	247.0	309.0	392.0	524.0	251.0	317.0	396.0	504.0	675.0	354.0	448.0	560.0	713.0	963.0
10	195.0	247.0	309.0	392.0	530.0	251.0	317.0	396.0	504.0	675.0	354.0	448.0	560.0	713.0	954.0
20	195.0	248.0	309.0	393.0	531.0	251.0	317.0	397.0	505.0	675.0	355.0	448.0	561.0	714.0	945.0
24	195.0	248.0	309.0	393.0	531.0	251.0	317.0	397.0	505.0	675.0	355.0	449.0	561.0	714.0	941.0
30	196.0	248.0	310.0	394.0	530.0	252.0	318.0	398.0	506.0	675.0	355.0	449.0	562.0	715.0	936.0
40	197.0	249.0	311.0	395.0	529.0	252.0	319.0	399.0	507.0	675.0	356.0	450.0	563.0	716.0	928.0

S = Spheres

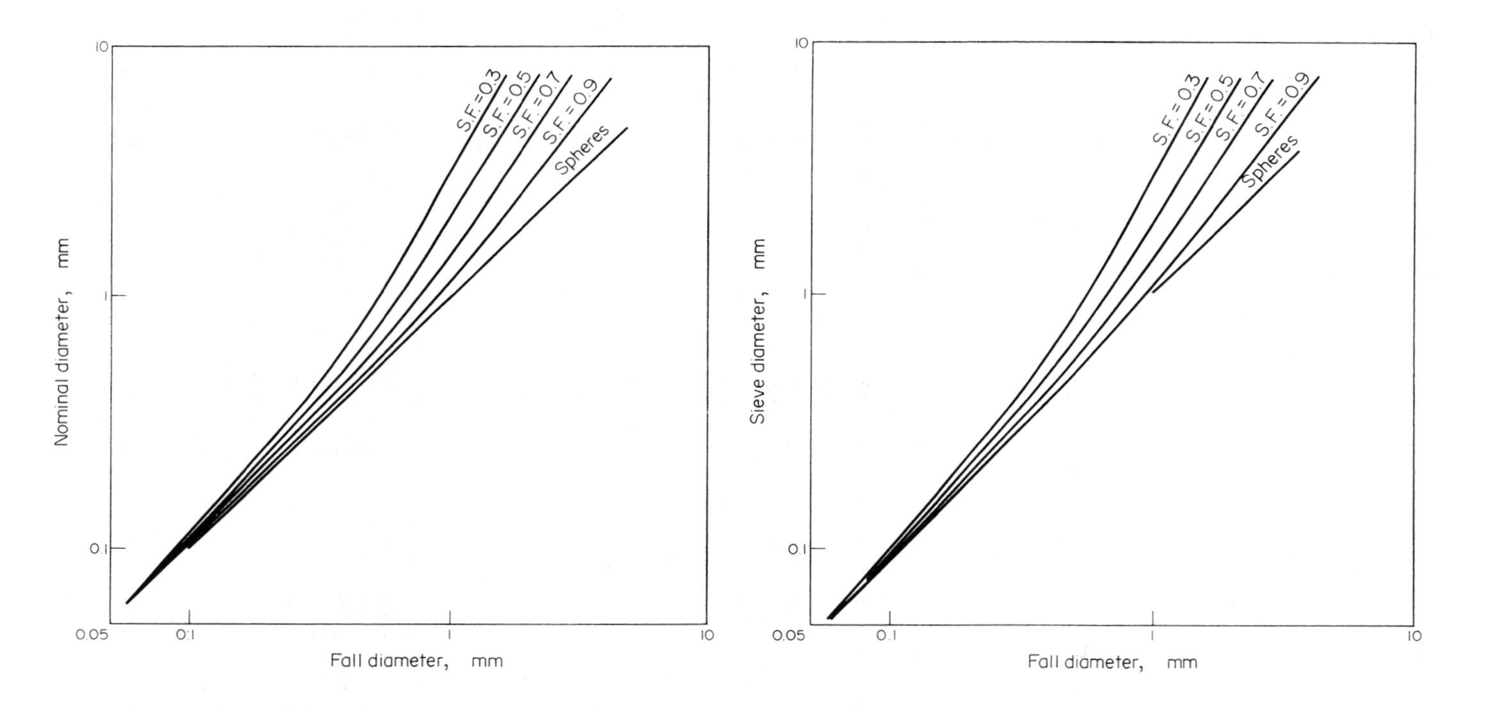

Fig. 2.5. Relationships between nominal diameter and fall diameter and sieve diameter and fall diameter for naturally worn quartz grains in terms of shape factor.

TABLE 2.2 Fall velocities of quartz sand of $0.15 \leq d \leq 1.5$ mm

d mm	0.15	0.2	0.3	0.4	0.5	0.6	0.7	0.8	0.9	1.0	1.2	1.5
w mm/s	14.8	21.1	36.1	50.0	64.0	76.4	88.6	99.0	110.0	121.0	137.3	166.0

Although the problem of the settling velocity of a single grain of arbitrary shape is far from being completely solved there is a fair amount of analytical and experimental information to guide application.

Unfortunately, not the single grain but a cloud of grains is the problem encountered in application. When it comes to settling, or to motion in general, of dispersed particles the theoretical and experimental information available becomes very meagre. The analytical work is essentially confined to the Stokes' range and is discussed by Happel adn Brenner (1965).

It can be readily observed that a few closely spaced particles settle much faster than the same individual particles. However, the fall velocity decreases when the same particles are dispersed throughout the fluid in quantity. This feature accounts for widely varying results in determination of fall velocities. Particles released as a group affect the flow pattern as illustrated in Fig. 2.6. However a single small particle, for example, falling in a glass cylinder is not easy to observe, but if a group of such particles is released these spread out into a dispersed cloud where some particles are close to each other, while at the same time these groups are widely separated longitudinally. The result is that the entire cloud may be observed to have a higher fall velocity than the single particles. The determination of the fall velocity of a dispersion of, say, 1.5% concentration by volume in still water is very difficult. Techniques where upward fluid velocity is used in the observation cylinder introduce an effect which has not yet been mentioned, and that is that the particles will assume preferred locations in cross-section. It has been observed that when the particles move downstream faster than the fluid they tend to migrate towards the pipe wall and vice versa. Neutrally buoyant particles migrate towards an equilibrium position at about 0.5 - 0.7 of the radius. This arises from the velocity distribution over the cross-section. A particle will have a velocity difference across it in radial direction which will give rise to a so-called shear lift and will also set the particle in rotation. The latter will give rise to a spin lift, Magnus effect. Lawler and Lu (1971) discuss the migration of particles in a pipe flow and offer an analytical model capable of explaining the observed results. Very briefly, a particle moving slower will have the Magnus effect acting towards the centre (Fig. 2.7). The relative velocity above the particle is greater than that nearer the wall and the circulation acts to reduce this further, hence the lift will be towards the centre. A particle moving faster has the direction of relative velocity reversed and the relative velocity nearer the wall will be greater. This is further increased by the circulation and the lift will be towards the wall.

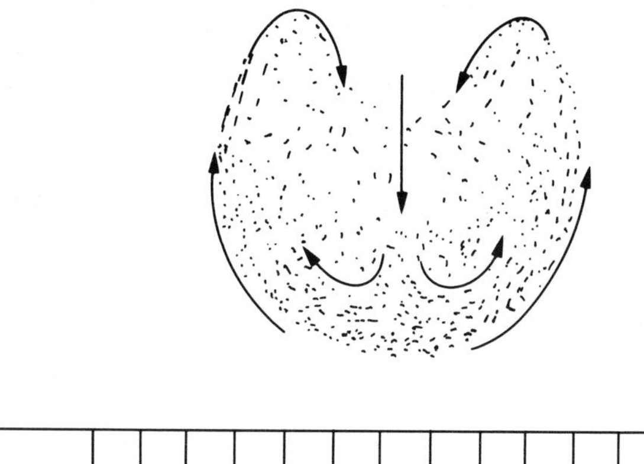

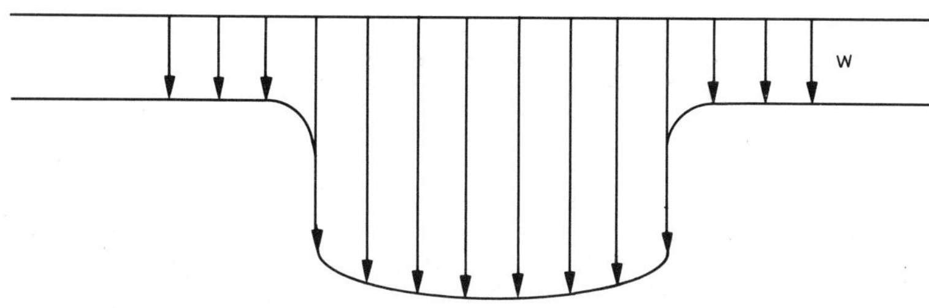

Fig. 2.6. Schematic flow pattern around a closely spaced group of grains.

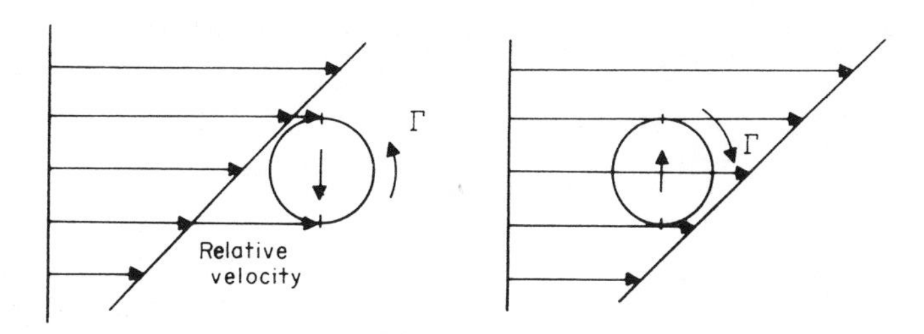

Fig. 2.7. Diagrammatic illustration of shear drift of a particle moving faster and a particle moving slower than the mean local fluid velocity.

The influence of particle concentration on the fall velocity can be expressed as

$$w = w_o(1-c)^{\beta} \tag{2.25}$$

where c is volumetric concentration, w_o is the fall velocity of a single grain and β is a function of Re = wd/ν and particle shape. It can also be expressed as a function of the nondimensional grain size

$$D_* = (\Delta g d^3/\nu^2)^{1/3} \tag{2.26}$$

where $\Delta = (\rho_s - \rho)/\rho$. For common natural grains (SF $\sim$ 0.7) the value of $\beta = 4.65$ for $D_* < 40$ and $\beta = 2.35$ for $D_* > \sim 800$, with a transition defined by $\beta = 7.478\ D_*^{-0.129}$.

The volumetric and weight concentration, c_w, are related to each other through

$$c = \frac{c_w/S_s}{(1-c_w) + c_w/S_s} \tag{2.27}$$

where $S_s = \rho_s/\rho$.

The *effective fall velocity* of a sediment mixture can be expressed as

$$w = \frac{\Sigma\ p_i w_{pi}}{\Sigma\ p_i} \tag{2.28}$$

where p_i and w_{pi} are the weight and fall velocity of the grains in the size range i. This fall velocity may be very different from that corresponding to d_{50}, as can be seen from the examples taken from U.S. Geological Survey analyses,† Table 2.3. When the grain size distribution is narrow and does not extend to Stokes range these fall velocities can be approximately equal.

†Professional Paper 422-1, 1966.

TABLE 2.3 Rio Puerco near Bernardo, New Mex. (a) (8.10.59)

Size limit (mm)	Percent finer	p	d_p (mm)	p_d (mm)	w_p (cm/s)	pw_p (cm/s)
1.0	100					
		0.002	0.71	0.00142	10	0.02
.5	99.8					
		.006	.35	.00210	5	.03
.25	99.2					
		.041	.176	.00720	1.9	.078
.125	95.1					
		.083	.088	.00555	.6	.049
.0625	86.8					
		.153	.044	.00670	.16	.0244
.0312	71.5					
		.101	.022	.00220	.04	.0040
.0156	61.4					
		–	.011	.0011	.01	.001
.0078	51.4					
		.014	.0055	.00007	.0025	.000035
.0039	50.0					
		.50 (b)				
Total		1.00		d_a = 0.026 mm	w = 0.206 cm/s	

d_{50} = 0.0039 mm and w = 0.0013 cm/s. Calculated w is 160 times this and corresponds to d = 0.016 mm or d_{75}. The four bigger grades contribute 5% of suspended sediment but 60% of w.

Elkhorn River at Waterloo Nebr. (a) (3.26.52)

Size Limit (mm)	Percent finer	P	d_p (mm)	pd_p (mm)	w_p (cm/s)	pw_p (cm/s)
0.5	100					
		0.04	0.35	0.014	5	0.2
.25	96					
		.19	.176	.0335	1.9	.36
.125	77					
		.15	.088	.0132	.6	.09
.0625	62					
		.13	.044	.0057	.16	.0208
.0312	49					
		.21	.022	.0046	.04	.0084
.0156	28					
		.04	.011	.00044	.01	.0004
.0078	24					
		.06	.0055	.00033	.0025	.00015
.0039	18					
		.18 (b)				
Total		1.00		d_a = 0.0718	w = 0.68 cm/s	

d_{50} = 0.033 mm and w = 0.09 cm/s. Calculated w is 7.6 times bigger corresponding to d = 0.092 mm.

A further complication arises from the particle shape. The behaviour of various geometrical shapes at low Reynolds numbers is discussed by Happel and Brenner (1965). Here quite a number of analytical solutions exist. However, outside the Stokes' flow region, and for irregular shapes, only experimental information is available. One of the significant features of the nonsymmetrical shapes is that they will develop a lift force, i.e. a force perpendicular to the direction of fall and their path will no longer be straight. This can under particular circumstances also happen with symmetrical bodies, for example, when spin develops on a sphere. In general, within Stokes' range a particle with unequal axes is stable in any orientation, but at higher Reynolds numbers it will tend to set itself broadside to the relative motion. At Re of the order of 1000 there is a tendency for an oscillatory motion to develop (Marchillon et al,1964), which is superimposed on the translation and has a strong effect on the drag coefficient. Graf (1971) shows graphs of the correction coefficient K versus shape factor and Re for various regular shapes. An extensive study of drag coefficients of isometric particles was carried out by Pettyjohn and Christiansen (1948).

Bagnold (1956) evaluated the drag coefficient as a function of the Reynolds number and volumetric concentration, Fig. 2.8. The results are based on experiments in a tube with upward velocity v without grains. The velocity used to calculate C_D from $F = \frac{1}{2}C_D\rho V^2$ and the Reynolds number is $V' = V/(1-c)$, where c is the volumetric concentration.

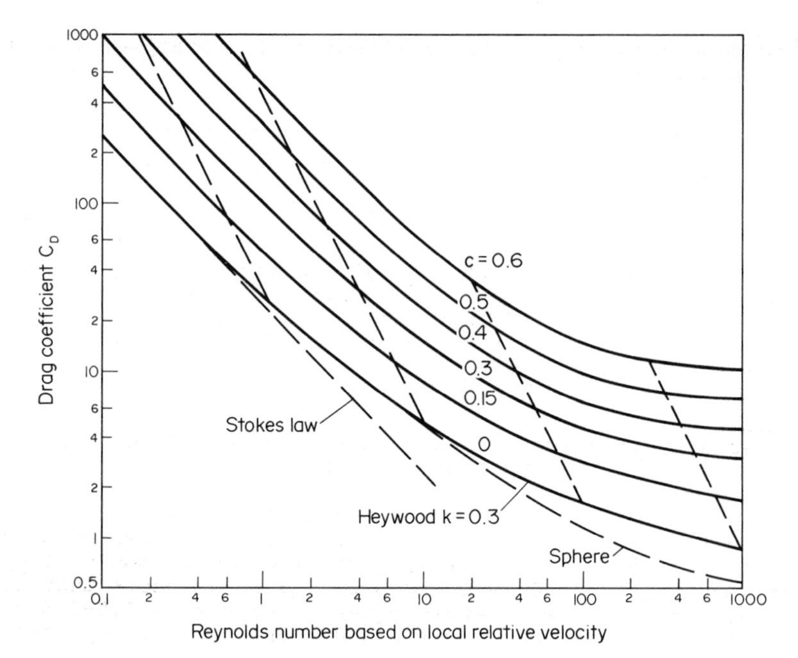

Fig. 2.8. Variation of drag coefficient with grain concentration. (By permission of the Royal Society, London).

2.3 Effect on Viscosity

The effect of fluid viscosity on the fall velocity enters through the Reynolds number and drag coefficient. However, suspended matter also affects the behaviour of the fluid. For example at high concentrations of colloidal matter water behaves distinctly as a non-newtonian fluid.

The viscosity of dilute suspensions is given by the Einstein equation

$$\frac{\mu_{susp}}{\mu} = 1 + k_1 c \qquad (2.29)$$

where c is the volumetric concentration and $k_1 = 2.5$ for $c < 2\text{-}3\%$. For greater concentrations higher order terms are necessary but difficulties are encountered with values of k_2, k_3, etc. For spherical particles Ward (1955) proposed

$$\frac{\mu_{susp}}{\mu} = 1 + k_1 c + k^2{}_1 c^2 + k^3{}_1 c^3 + \qquad (2.30)$$

which gives acceptable results for c up to 35% approximately. However, many formulae for μ_{susp} can be found in the literature concerned with suspensions. The results are generally independent of the size of the spheres. For non-spherical particles the k-values depend on shape. For ellipsoids, for example, k varies as shown below:

Axial ratio x/y	Minimum value of k	Maximum value of k
1	2.50	2.50
2	2.17	2.82
5	2.04	3.55
10	2.01	4.49

The viscosity of a suspension of irregular particles also depends on the particle size, Whitmore (1957).

The above expressions all refer to stable suspensions. When the soild particles are moving through the fluid as, for example, in a suspension which is settling, further complications arise. Experimental work by Oliver and Ward (1959) suggests that for low concentration the Einstein equation with k_1 = 3.0-3.6 gives acceptable results. For $0.1 < c < 0.3$ the data is represented by

$$\frac{\mu_{susp}}{\mu} = (1 + k) + (1 + 2k)k_1c + (1 + 3k)k^2{}_1c^2 + \qquad (2.31)$$

where $k = 0.33 \dfrac{\rho_s - \rho}{\mu}$.

The dynamic or absolute viscosity of water can be estimated from

$$\mu(\text{kgm}^{-1}\text{s}^{-1}) = \frac{1.79 \times 10^{-3}}{1 + 33.68 \times 10^{-3}T + 2.21 \times 10^{-4}T^2} \qquad (2.32)$$

and of air from

$$\mu(\text{kgm}^{-1}\text{s}^{-1}) = (17040 + 56.02T - 0.1189T^2) \times 10^{-9} \qquad (2.33)$$

where a simple approximation for density is

$$\rho_a \cong \rho_o e^{-0.125h} \qquad (2.34)$$

the temperature T is in °C, and h is elevation in km.

Chapter 3

Threshold of Particle Movement

3.1 Concepts of Threshold

The initiation or threshold of movement of a particle due to the action of fluid flow is defined as the instant when the applied forces due to fluid drag and lift, causing the particle to move, exceed the stabilising force due to weight force.

A grain on the surface is subject to a *weight force* and the *fluid forces*. These may further give rise to *shear stresses between the grains* in motion and those forming the stationary boundary, with the fluid between them taking part in this shearing. For analysis all forces are resolved into normal and tangential components. The tangential components maintain the forward motion. The tangential force may be transmitted *entirely by fluid* (transport by fluid over a horizontal bed) or at the other extreme it may be due *entirely to* the component of *weight force* as when granular material is sliding down a steep slope, but usually both contribute.

Generally near the bed there is a mean velocity profile $\bar{u} = f(y)$ and superimposed on it may be turbulent velocity fluctuations $u = \bar{u} + u'$ (Fig. 3.1).

If near the boundary the flow is laminar or if the laminar sublayer thickness is $\delta' \geq 5d$ then the individual grains will not shed eddies and the drag is due to viscous shear and is carried by the whole surface and not by a few more exposed grains. Here surface roughness should not influence the drag force. However, particle shape would influence the magnitude of the lift and drag components of the fluid force. These would also depend on the particle surface area as well as on piling position.

As the velocity increases the more exposed grains shed eddies and a wake is formed downstream. The size of the wake depends upon the size and shape of the particle and on the point of separation of the boundary layer formed on the particle. The point of separation in turn is a function of the shape of the particle

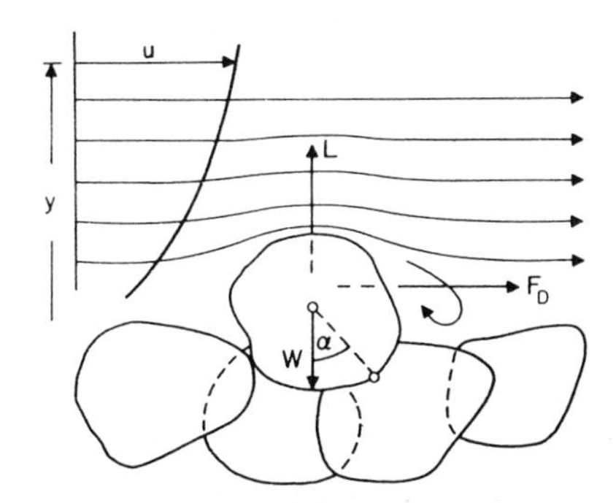

Fig. 3.1. Diagrammatic picture of an exposed grain subject to fluid force.

and the local Reynolds number. The drag force is now the resultant of the *surface drag* (viscous skin friction) and the *form drag* due to pressure differences in front and behind the particle. The point of application of the drag force depends on the magnitudes of *lift* and *drag* components. These in turn are functions of the shape and location of the particle and the local Reynolds number.

The elementary analytical picture by White (1940) (Fig. 3.1) considers the balance of moments only, i.e.

$$M_{weight} = \frac{\pi d^3}{6} g(\rho_s - \rho).\frac{d}{2}.\sin\alpha$$

and

$$M_{drag} = \beta\rho u_*^2 \frac{\pi d^2}{4} \frac{d}{2} \cos\alpha$$

where

$$F_D = \pi A \propto \rho u_*^2 \frac{\pi d^2}{4},$$

and β accounts for turbulence, the point of application of drag force, etc., i.e., all the effects besides weight and drag. Note that the model postulates a definite pivotal point for the two moments. Thus for equilibrium

$$u_{*_c} = \sqrt{\left[\frac{2}{3}\frac{\tan\alpha}{\beta}\right]^{\frac{1}{2}}}\sqrt{\left\{\left[\frac{\rho_s - \rho}{\rho}\right] gd\right\}^{\frac{1}{2}}} = B\sqrt{\left[\frac{\rho_s - \rho}{\rho} gd\right]^{\frac{1}{2}}} \quad (3.1)$$

and

$$u_c = 5.75B\sqrt{\left[\frac{\rho_s - \rho}{\rho} gd\right]^{\frac{1}{2}}} \log\frac{y}{y'} \quad (3.2)$$

Experimental data on threshold conditions in air plotted as u_{*_c} versus $\sqrt{d}$ may be interpreted to show that u_{*_c} is proportional to $\sqrt{d}$ for d greater than a certain diameter (Fig. 3.2). Such a plot

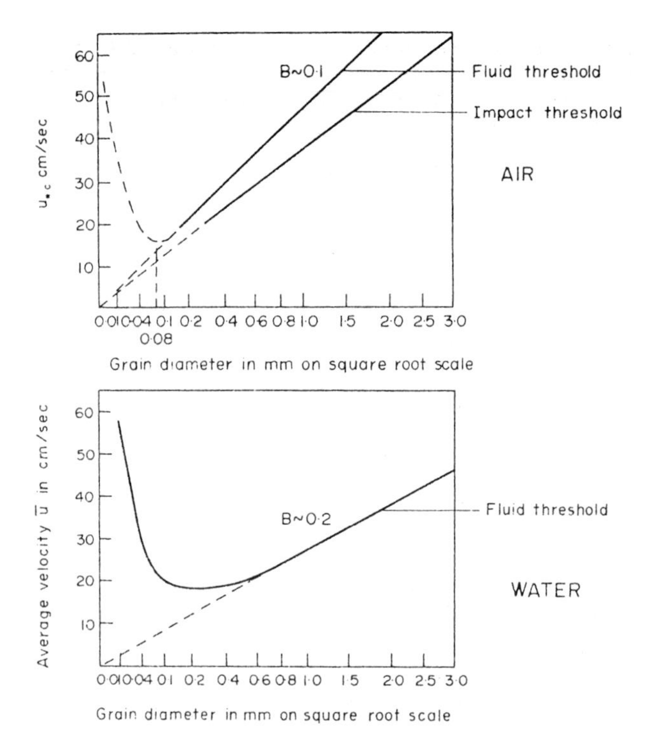

Fig. 3.2. Variation of threshold velocity with grain size, according to Bagnold (1941).

may be interpreted as dividing the data into a laminar and turbulent flow region near the boundary. For example, to the right of $d = 0.25$ mm the function is a straight line and in air for static threshold $B = 0.1$. The Reynolds number $Re_* = u_*d/\nu$ at $d = 0.25$ mm is 3.57. Colebrook and White (1937) showed that when $u_*d/\nu > 3.5$ the grain behaves as an isolated obstacle and sheds eddies. A prominent grain lying above the surface can carry the whole of the drag on the area it occupies and some of the drag on the area covered by the wake. It was argued that when $Re < 3.5$ the surface becomes hydraulically smooth, i.e. covered by a laminar sublayer which is thick relative to the grain size. The particles cease to shed eddies and the drag becomes a viscous one which is carried by the entire surface and not just by the exposed grains. This results in greater drag required to set the grains in motion and the proportionality of $\sqrt{d}$ does not hold. The minimum value of u_{*c} corresponds to approximately 0.08 mm particle size (particles finer than 0.08 mm feel soft and those coarser feel gritty). This increase in u_{*c} helps to explain soft dust on the ground in wind. Bagnold quotes that u_{*c} for fresh Portland cement powder equals that for $d = 5$ mm gravel.

In air the motion, once started, can be maintained at a lower value of u_* than required for initiation, referred to as the impact threshold. The terms static and impact threshold are analogous to static and dynamic friction.

In water only the static threshold value can be discerned ($B = 0.2$). The laminar sublayer model is tenable only when the sublayer is very thick relative to the grain size and more so in the air because of its much higher kinematic viscosity.

Shields (1936) approached the threshold problem by dimensional reasoning. The drag force

$$F_D = C_D A \frac{\rho u^2}{2} = f_1\left[a_1, \frac{ud}{\nu}\right]\rho d^2 u^2 \tag{3.3}$$

where u is the velocity at elevation $y = a_2 d$ and a_1 is the grain-shape factor. The velocity distribution for rough and smooth boundaries can be expressed as

$$\frac{u}{u_*} = 5.75 \log \frac{y}{k} + f\left[\frac{yu_*}{\nu}\right]$$

$$= 5.75 \log a_2 + f\left[\frac{du_*}{\nu}\right] \tag{3.4}$$

since $y = a_2 d$ and $k \propto d$.

Thus

$$F_D = \tau_0 d^2 f_3\left[a_1, a_2, \frac{du_*}{\nu}\right] \tag{3.5}$$

Resistance to motion was assumed to depend only upon the form of the bed and the immersed weight of particles, i.e.

$$R = a_3(S_s - 1)\gamma d^3 \tag{3.6}$$

where a_3 is a factor accounting for bed form and $\gamma = \rho g$.

Equating these, replacing τ_0 by τ_c, assuming a level bed and uniform particle size yields

$$\frac{\rho u_{*c}^2}{\gamma(S_s - 1)d} = \frac{\tau_c}{\gamma(S_s - 1)d} = f\left[\frac{du_*}{\nu}\right] \tag{3.7}$$

or

$$\theta_c = f(Re_*) \tag{3.8}$$

where the Shields' parameter

$$\theta = \frac{\tau_0}{(S_s - 1)\rho g d} \tag{3.9}$$

is a dimensionless shear stress, shear stress divided by the immersed weight of one layer of grains. Although τ_0 can be replaced by ρu_*^2 and θ expressed as $u_*^2/\Delta gd$, this does not change its physical meaning to a Froude number.

Shields plotted his experimental data as θ versus Re_* and identified areas where different kinds of bed features developed. His sediments included amber (ρ_s = 1.06) brown coal (ρ_s = 1.27), granit (ρ_s = 2.7) and barite (ρ_s = 4.25) and the Re_*-values covered 6 to 220. For fine sediments a 45° slope was postulated for the θ-Re_* function. The line defining the θ-Re_* function seen in most representations today was not drawn by Shields but added later. Likewise, many more data points have been added since. A form of Shields diagram is shown in Fig. 3.3. It defines the threshold of movement of uniform grains under unidirectional steady flow.

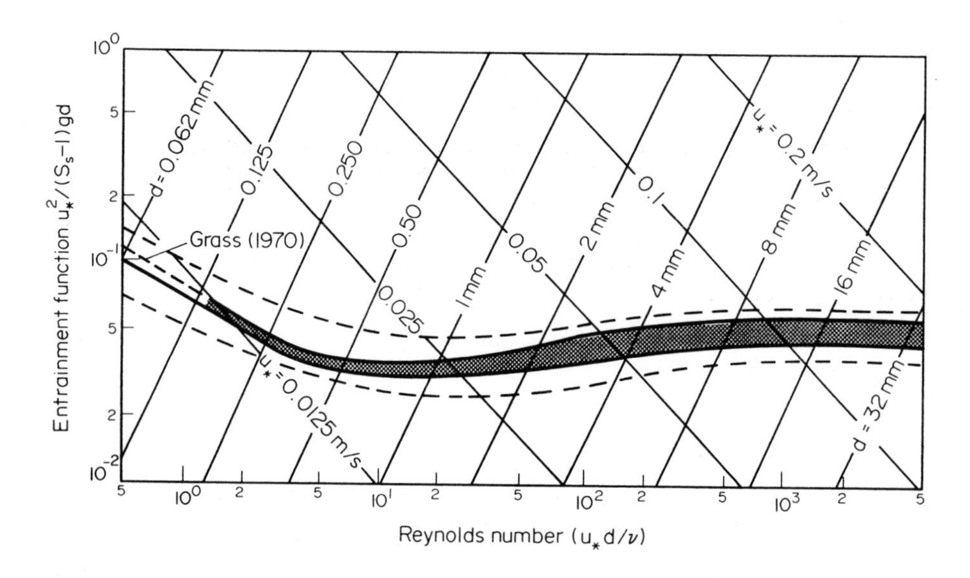

Fig. 3.3. Threshold of sediment entrainment as a function of Reynolds number. The shaded band indicates the spread of data by Shields, the dashed lines the envelope to most of the published data.

The studies by Grass (1970) indicated for small grains, $Re_* < 2$, a flatter than 45° slope and those by White (1970) even flatter, Fig. 3.4. Mantz (1973) used small flakes and found that when the flake length was used for diameter the results were similar in trend to those by White.

Unsöld (1982) combined previous data and presented his own, Fig. 3.5, which indicates that the θ_c-value increases much more slowly with decreasing Re_* than was assumed by Shields. Unsöld presented this data in terms of equal transport intensities $q^+_s = q_s/\rho_s du_*$. This in terms of the dimensionless transport rate ϕ (see Sediment

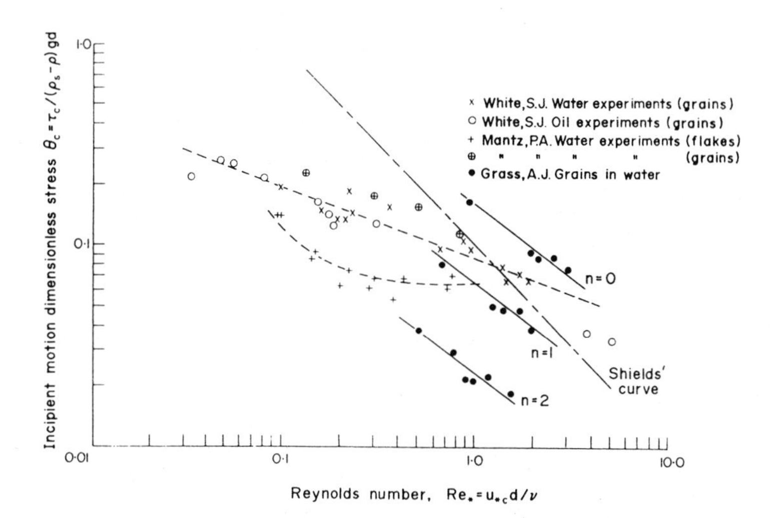

Fig. 3.4. Critical shear stress versus particle Reynolds number for fine grains and flakes. (Mantz, 1973).

Load) is $q^+_s = \phi/\sqrt{\theta}$. The data points are on +2 slope for d = const, ρ_s = const and ν = const with increasing τ.

Bagnold reasoned that the entrainment function has a maximum of the order of 0.4. The shear stress on the top layer could be written as

$$\tau = c(\rho_s - \rho)gd \tan \alpha$$

where c is the concentration and tan α is the friction factor of the top layer with respect to the stationary grains below. Thus

$$\frac{\tau}{(\rho_s - \rho)gd} = c \tan \alpha \cong 0.63 \times 0.63 = 0.4 \qquad (3.10)$$

A drawback of the Shields diagram is that the shear velocity appears on both axes. In order to avoid trial and error solutions u_* and d contours have been added in Fig. 3.3. Bagnold (1963) also plotted θ_c against d but such a graph is limited to given grains and fluid.

Bonnefille (1963) was one of the first to present the threshold in terms of the dimensionless particle size $D_* = (\Delta gd^3/\nu^2)^{1/3}$ and thus avoid the trial and error estimation of u_*. He showed that the data on threshold is described by $Re_* = f(D_*)$ as

$$D_* = 2.15\ Re_* \quad ; \qquad Re_* < 1 \qquad (3.11a)$$

$$D_* = 2.5\ Re_*^{4/5}; \quad 1 < Re_* < 10 \qquad (3.11b)$$

$$D_* = 3.8\ Re_*^{5/8}; \qquad Re_* > 10 \qquad (3.11c)$$

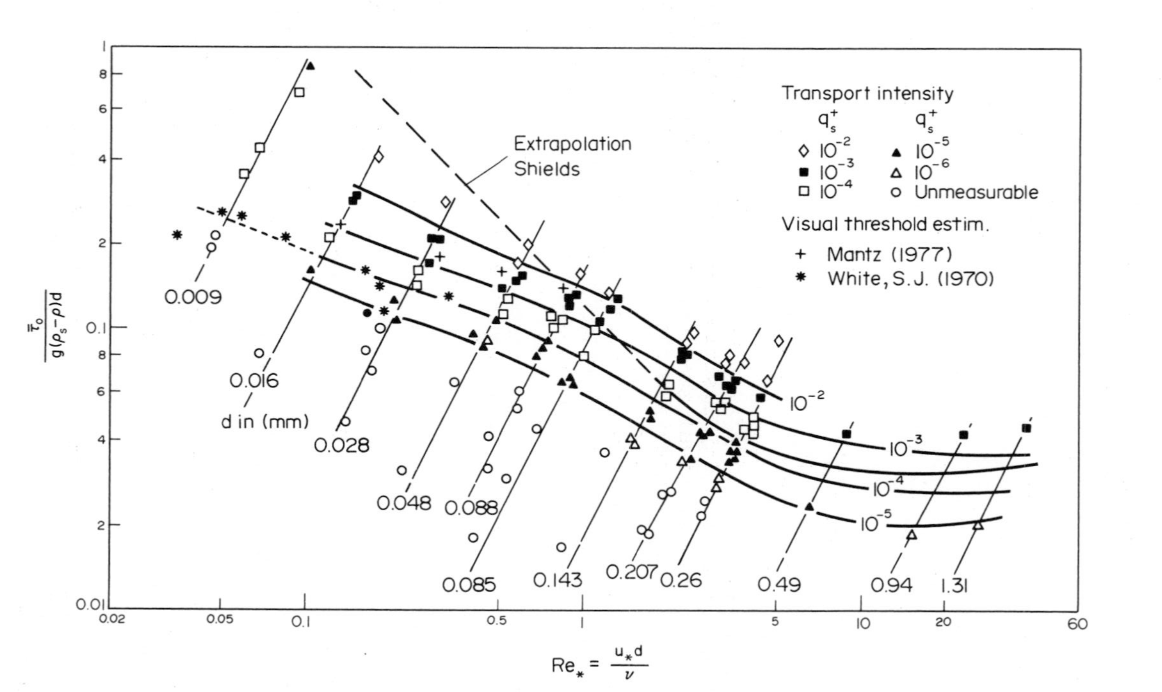

Fig. 3.5. Threshold of fine-grained noncohesive uniform sediments with transport intensity $q^+_s = q_s/\rho_s du_*$ as parameter, according to Unsöld (1982).

Vollmers and Pernecker (1967) showed that for very fine sediments ($Re_* < 1$) which are subject to some cohesion

$$D^* = Re_*^2 \tag{3.11d}$$

Yalin (1972) presented the data as θ_c versus D_*-diagram. A detailed discussion of the threshold problem and the Shields diagram was presented by Miller et al. (1977).

Numerous formulae exist for the "critical velocity" for particle movement. The critical velocity can also be calculated for given conditions from theoretical shear stress or shear velocity. Thus, the mean critical velocity is

$$U_c = u_{*_c} \, C/\sqrt{g} \tag{3.12}$$

or at a given elevation

$$u_c = 5.75 \, u_{*c} \log y/y' \tag{3.13}$$

where C is the Chézy coefficient and y' is the elevation at which the logarithmic velocity distribution has zero velocity. The problem with the critical mean velocity is that the bed shear stress for the same mean velocity decreases with increasing depth of flow. The use of a critical bed velocity u_b suffers from the difficulty of definition of the elevation where it is to be measured and its relationship to the mean velocity. Probabaly the best known of the early contributions on critical velocity is the Hjulström (1935) curve, Fig. 3.6. Neill (1967,1968) specified a "conservative design curve" for uniform coarse material (gravel) in terms of critical mean velocity V_c as

$$\frac{V_c^2}{(\rho_s/\rho - 1)gd} = 2.0 \left[\frac{d}{y_0}\right]^{-1/3} \tag{3.14}$$

where y_0 is the depth of uniform flow.

Yang (1973) used the conventional drag and lift concepts combined with the logarithmic velocity distribution and arrived at

$$\frac{V_c}{w} = \frac{2.5}{\log \dfrac{u_* d}{\nu} - 0.06} + 0.66; \quad 1.15 < \frac{u_* d}{\nu} < 70 \tag{3.15}$$

where the numerical constants are from empirical curve fitting. At values of $Re_* > 70$, V_c/w is no longer a function of Re_* and the formula reduces to

$$V_c/w = 2.05 \tag{3.16}$$

Zanke (1977) proposed

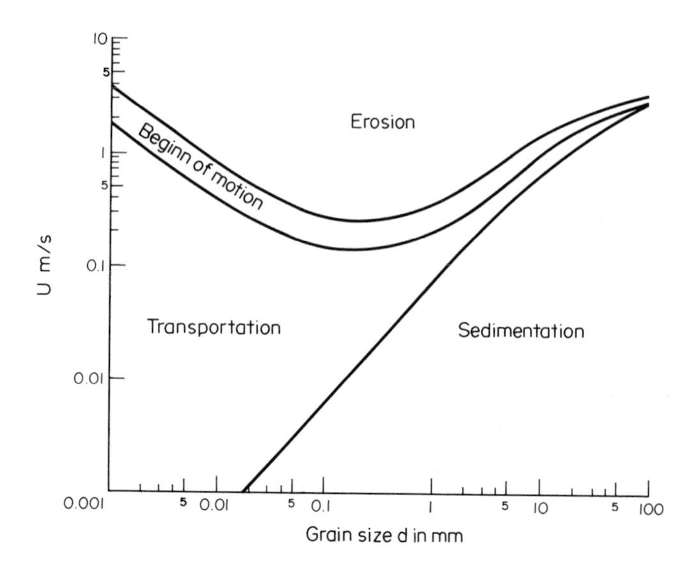

Fig. 3.6. Erosion and deposition criteria for uniform particles according to Hjulström.

$$V_c = 2.8(\Delta gd)^{\frac{1}{2}} + 14.7\,\frac{\nu}{d}\,c \qquad (3.17)$$

where V is in m/s, d in m and c is a coefficient for cohesiveness varying from c = 1 for noncohesive to 0.1.

A major problem with threshold is the experimental definition of it because the initiation of motion in a turbulent flow is a probabilistic process. Shields measured small rates of sediment transport and extrapolated these to zero rate but, as will be shown later, the transport rates at low values of shear stress excess ($\tau_0 - \tau_c$) are very sensitive to this value. The plot of transport rate against ($\tau_0 - \tau_c$) or ($\theta - \theta_c$) on logarithmic paper does not have a constant slope. Indeed, the results by Unsöld, Fig. 3.5, show a different threshold for each transport rate. Thus, an extrapolation of data will introduce errors. Neill and Yalin (1969) proposed

$$N = nd^3/u_* = f(Re_*, \theta) \qquad (3.18)$$

and set $N = 1 \times 10^{-6}$ as the practical lower limit that can be reasonably observed in open channel experiments. In it n is the number of grains displaced per unit area of bed per unit time. Yalin (1972) modified the expression to $\varepsilon = (m/AT)(\rho d^5/\rho_s g)^{\frac{1}{2}} > 0$ where m is number of grains, A area and T time. Neill and Yalin also make the point that for equal values of θ the shear velocity u_* "must be approximately 30 times greater in air" than in water and n will be greater. Thus, if the same number of grains are observed an apparently lower value of θ in air is recorded. This could account, for example, for the difference of B = 0.08 to 0.10 in air and B ≅ 0.20 in water, Fig. 3.2.

3.2 Effects of Flow Characteristics and Bed Surface Structure

It has to be borne in mind that in turbulent flow both lift and drag are fluctuating quantities in magnitude, point of application and in direction and not even the viscous sub-layer can be looked upon as steady two-dimensional laminar flow. Studies of the laminar sub-layer (Kline et al. 1967; Corino and Brodkey, 1969; Grass, 1971) have revealed a complex flow structure which, although dominated by viscosity, consists of large three-dimensional high- and low-speed velocity streaks. When a high-speed eddy from the flow hits the sub-layer it penetrates to the boundary and ejects low momentum fluid on both sides into the outer region. This low momentum fluid retards the local velocity and gives rise to another strong eddy which leads to repetition of the process. These high- and low-velocity regions alternate laterally across the flow and appear as stream-wise streaks on the surface of a flat bed covered by very fine sand (Fig. 3.7).

The wall region of the boundary layer flow has an extremely complex structure and most of the turbulence is produced there. The researchers at Stanford introduced the concept of turbulent bursts produced by emission of fluid from the low-speed streaks. The sequence is described as *lifting-up of low-speed fluid tongues*, *vortex production* downstream of the tongue and *break-up* and diffusion as turbulence into the flow.

On a plane boundary these bursts lead to strong fluctuation of pressure, Fig. 3.8. The average streamwise spacing of the bursts is

$$\lambda^+{}_1 = \lambda \frac{u_*}{\nu} = 500 \qquad (3.19)$$

and lateral

$$\lambda^+{}_3 = 100 \qquad (3.20)$$

The average frequency of bursts is

$$\frac{U_0 \, \overline{T}_B}{\delta} = 5 \qquad (3.21)$$

where δ is the boundary layer (flow) thickness and U_0 is the surface velocity of flow.

These turbulence bursts impose a rapid and significant pressure fluctuation on the bed surface which have an important effect on entrainment of sediment. The flow disturbances at the boundary diffuse into the flow as turbulence. The size of the individual eddies decreases in this diffusion process without significant loss of energy, the so-called cascade of turbulence, until the eddies become small enough for the viscous stresses to become dominant and these dissipate the energy into heat.

Turbulence can effect entrainment in a number of ways:

(i) The particle may be moved by the drag exerted by a passing eddy, i.e. an impulse force on the particle.

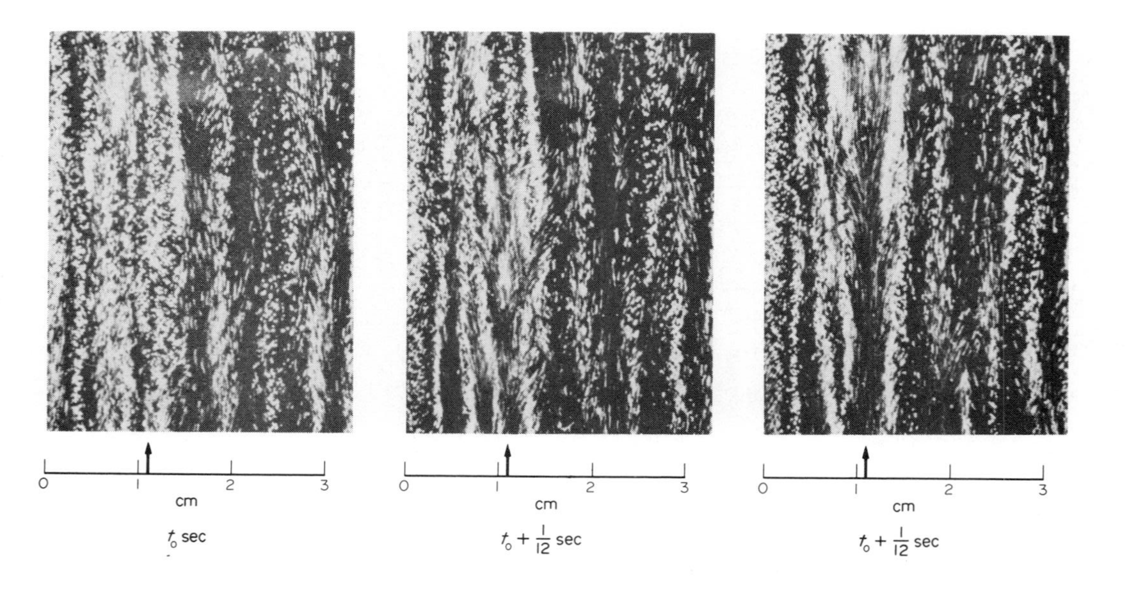

Fig. 3.7. Viscous sublayer flow structure visualized by means of fine 0.1-mm diameter sand moving over a smooth black boundary. The sequence of three motion film frames, separated in time by 1/12 sec with a 1/30-sec exposure, illustrates the development of a typical fluid inrush phase, indicated by the arrow. $\nu = 0.0125$ cm²/s, $u_* = 2.13$ cm/s, $\nu/u_* = 0.006$ cm. (Grass, 1971).

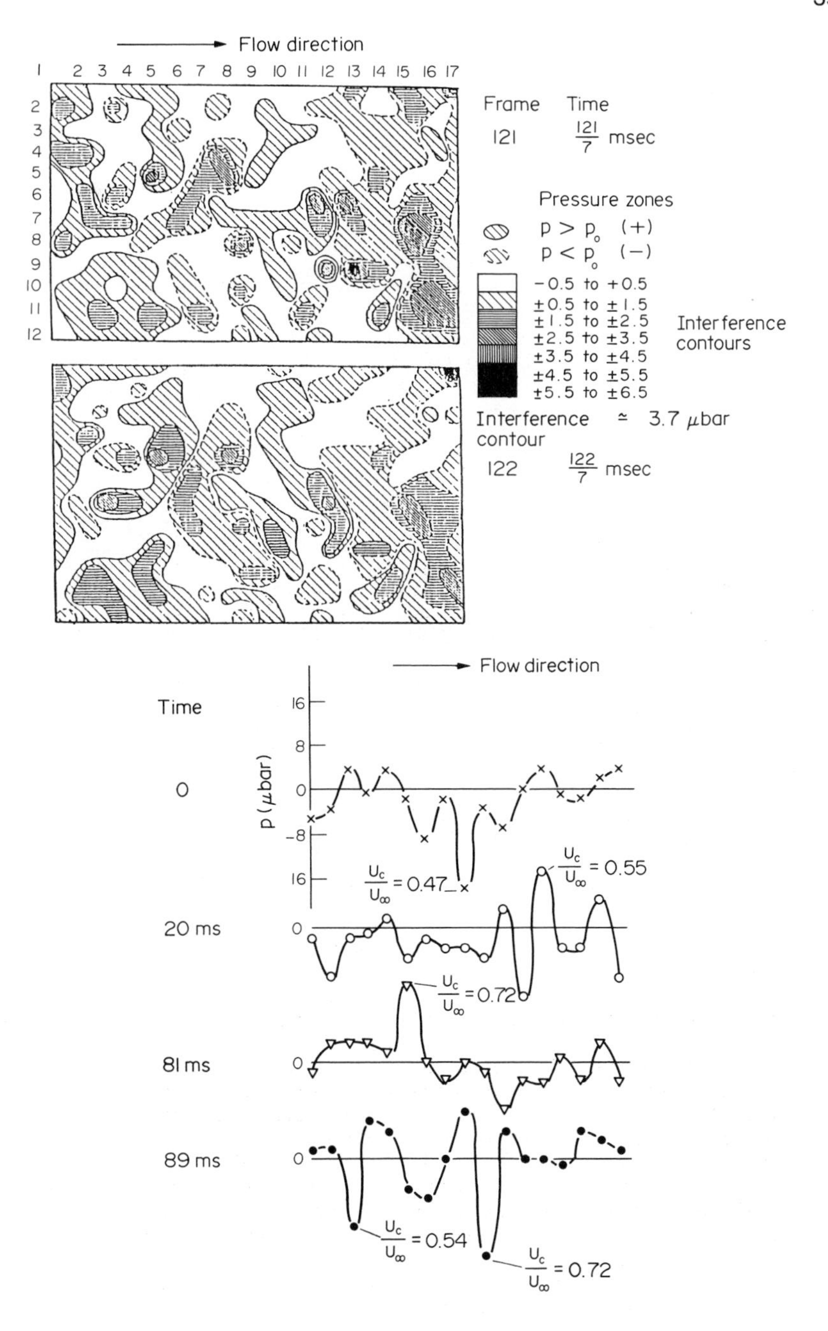

Fig. 3.8. Instantaneous pressure distributions on a boundary and pressure fluctuations at a point in turbulent flow, according to Emmerling (1973).

(ii) The eddy may lower the local pressure and the particle may be ejected from the bed by hydrostatic pressure. By this mechanism even sheltered particles can be entrained.

(iii) Entrained particles may be put into suspension rather than moved along the bed.

An important consequence is also that the size of the channel will have a bearing on entrainment. In large channels larger eddies can occur and these contain more energy. This aspect will be discussed under suspended sediment transport. In general, the sediment entrainment is a function of both the temporal mean drag, $\bar{\tau}_0$, on the bed and the intensity of turbulence over it. Here $\bar{\tau}_0$ and the agitation by turbulence may be looked upon as the two asymptotes to entrainment. At one limit the material may be moved along by $\bar{\tau}_0$ as in laminar viscous flow and at the other limit material may be suspended by random agitation alone while $\bar{\tau}_0 = 0$. The temporal mean drag carries the material downstream while turbulent agitation makes the particles more mobile, that is it lowers the critical shear value at which they will move. Measurements of turbulence and of temporal mean shear along the ripple form by Raudkivi (1963) and Sheen (1964) provide clear experimental evidence for the above argument (Figs. 3.9 and 3.10). It is seen that $\bar{\tau}_0$ is zero at the reattachment point and increases to a maximum over the crest, while turbulence intensity is a maximum at the reattachment point and decreases downstream. For more than half the distance along the ripple face the shear stress is seen to be less than τ_c. Since the sediment is kept in motion it follows that the transport is maintained by the combined action of the turbulent agitation and temporal mean shear stress. Literature contains numerous papers on measurement and calculation of lift and drag on individual particles. These are valuable as aids to understanding of the entrainment process but as yet have only a limited value for practical application.

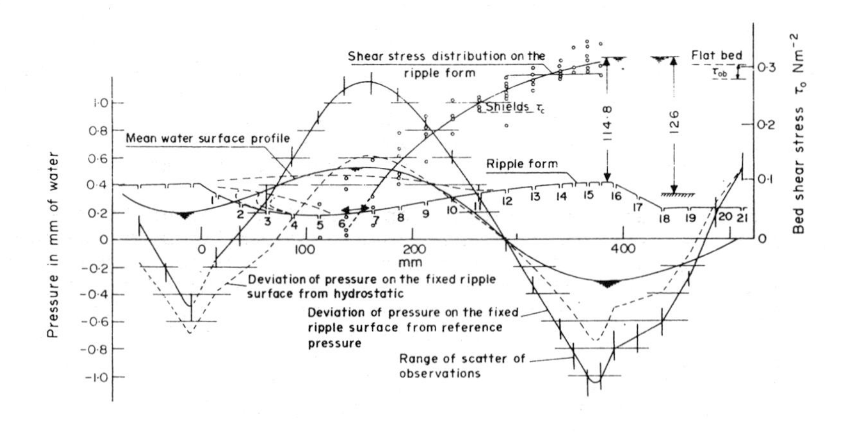

Fig. 3.9. Ripple form, water surface profile, distribution of pressure deviations and bed-shear stress.

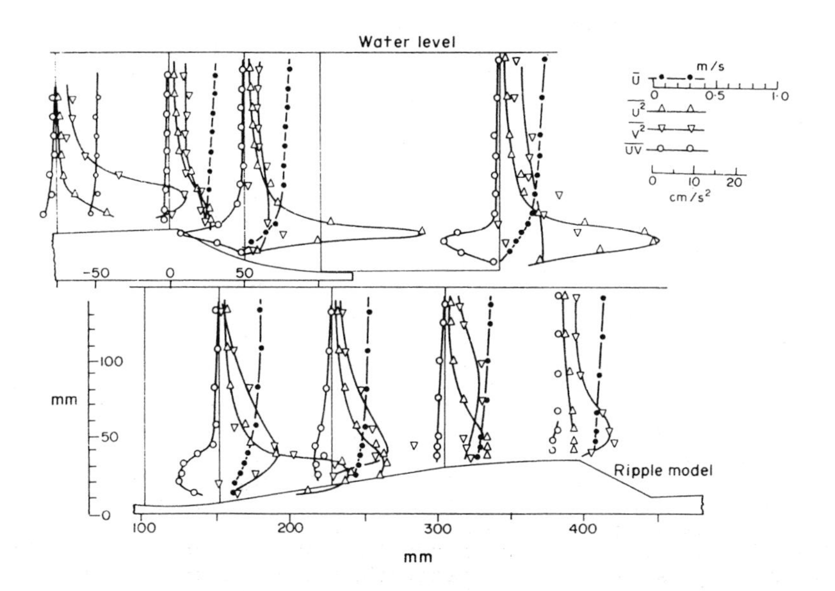

Fig. 3.10. Profiles of mean velocity, longitudinal and transverse components of velocity fluctuations, and turbulent shear measured on the ripple form shown in Fig. 3.9.

A further feature of the threshold condition is that the applied shear stress of a turbulent flow is probabilistic, that is, it has frequency distribution as illustrated in Fig. 3.11. However, this distribution is bounded. Energy considerations will show that in a given steady flow the extreme bursts of shear stress of low probability must be limited in the magnitude. Likewise, the critical shear stresses required to move a grain on a surface of unit area are randomly distributed due to the positions of the grains. When these two distributions start to overlap the grains which have the lowest τ_c will start to move. Grass (1970) related the distance between the peaks of the two distributions to Shields diagram by $n({}^{\sigma}\tau_0 + {}^{\sigma}\tau_c)$, where σ is the standard deviation, and showed that with n = 0.625 the data merged with that by Shields.

In nature the grains are neither spherical nor of uniform size. This means that the flow may be capable of moving some but not all of the bed material. The grains in the surface layers become coarser and we speak of armouring. The armouring effect causes the critical particle movement stress distribution to be time dependent. Thus an eroding surface can become immobile again. However, if the τ_0 distribution covers the τ_c distribution completely all grains will be moved and no armouring will take place. Such a probability reasoning, although not using the same terminology, is used by Gessler (1965, 1970, 1971, 1973) in his development of design criteria and critical shear for mixtures. The question of armouring will be discussed in more detail later.

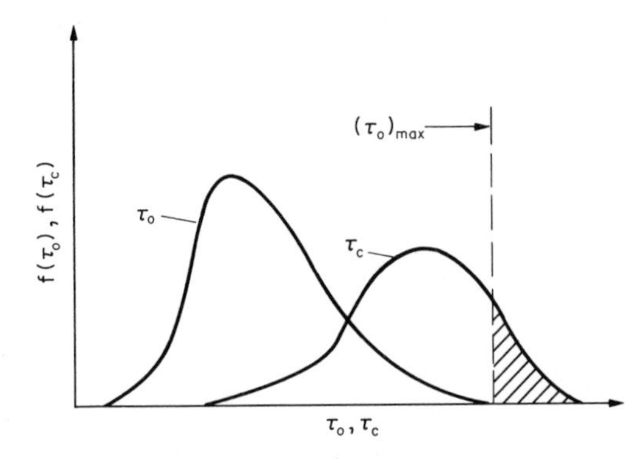

Fig. 3.11. Illustration of frequency distributions of boundary shear stress τ_0 and the critical shear stress τ_c of surface grains.

The threshold of movement of a grain is also strongly influenced by its position in the surface. A grain from a co-planar layer is obviously more difficult to move than one laying on top of the others. The conventional threshold analysis considers the balance between the overturning moment by the fluid forces acting on the particle and the restoring moment arising from its weight. This analysis incorporates the assumption that the bed provides one or more pivotal points for this balance of moments to be realised. If the fluid forces makes the bed grains on which, for example, a larger particle rests mobile then these cannot provide effective pivotal support for this large grain. Under these conditions the larger particle either gets embedded into the bed material or it slides over it. Coleman (1967) discussed threshold of spherical particles resting on top of similar particles and Fenton and Abbott (1977) carried out experiments in which a grain was given a different preset amount of protrusion, P, above the bed of similar grains. Their results, augmented by data from Chin (1985), are shown in Fig. 3.12. It is seen that for spherical grains the critical value of θ for entrainment from a co-planar bed is $\theta_c = 0.13$ whereas $\theta_c = 0.02$ at only half diameter protrusion and will become almost zero on full protrusion on a flat surface (equal to rolling resistance). Thus, an exposed particle can be transported over the bed without disturbing it, by a bed shear stress, τ_0 equal or less than the critical shear stress, τ_c, of the bed material in general. This transport can be of coarser grains as well as finer ones.

When the time average of the applied bed shear is τ_0, the critical or threshold shear stress of the bed material is τ_c and τ_i is the minimum shear stress required to move a particular grain over the bed surface then one can identify a number of special cases:

(a) $\tau_i > \tau_c > \tau_0$ and $\tau_c > \tau_i > \tau_0$, no grain movement at all;

(b) $\tau_0 > \tau_c > \tau_i$ and $\tau_0 > \tau_i > \tau_0$, all grains are in motion;

(c) $\tau_i > \tau_0 > \tau_c$. Here the bed material moves but not the overlying exposed grains which will become embedded;

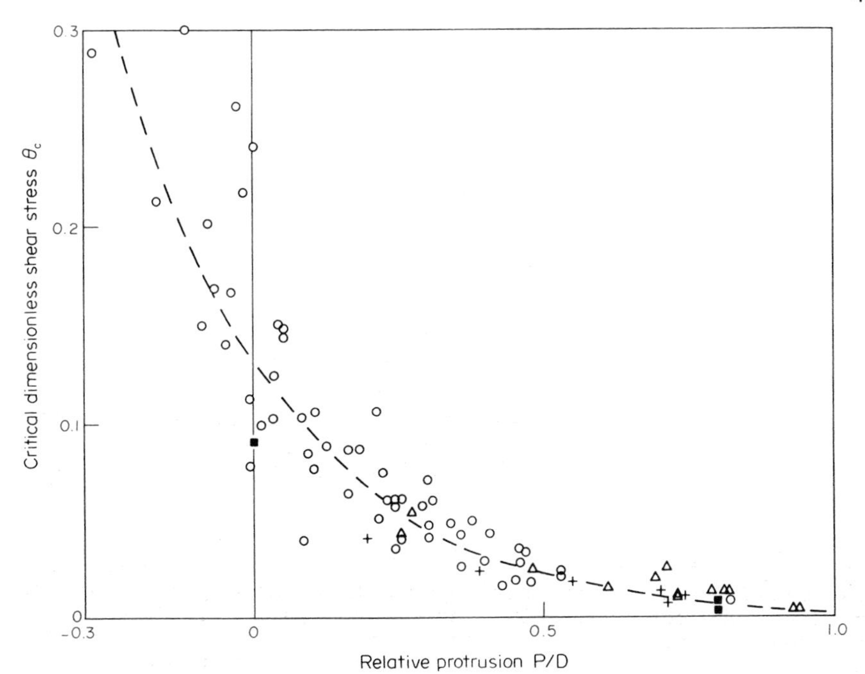

Fig. 3.12. Threshold shear stress θ_c versus relative protrusion above the bed. O - Data by Fenton and Abbott (1977); Δ, ■, ♦, ●, - Data by Chin (1985).

(d) $\tau_c > \tau_0 > \tau_i$. The exposed particles move over the bed which itself remains undisturbed. This form of movement has been called overpassing. Larger particles overpassing a bed composed of smaller particles, roll and slide, whereas smaller overpassing particles show a saltating or bouncing form of movement.

The critical shear stress of a size fraction d_i in the sediment mixture cannot be determined from the Shields diagram. The threshold shear of the fraction d_i depends on the grading of the mixture and surface structure. It is also a time dependent value because the surface armours (as will be discussed later). Some indications may be obtained from the relationships proposed by Hayashi et al. (1980):

$$\theta_{ci} = \theta_{ca} d_i/d_a \; : \; d_i/d_a < 1.0$$

$$\theta_{ci} = \theta_{ca}\left[\frac{\log 8}{\log 8d_i/d_a}\right]^2 \; ; \; d_i/d_a > 1.0 \qquad (3.22)$$

where d_a is the average grain size of the mixture and θ_{ca} the critical dimensionless shear stress for d_a size.

3.3 Threshold in Air

The Shields diagram is based on threshold in liquids. The threshold in air has not been researched as extensively. Bagnold introduced the concept of fluid threshold and impact threshold and concluded that for coarse grains the factor B in water was about twice that in air. Neill and Yalin argued that this difference might arise from visual assessment of threshold. However, the tendency for lower values of τ_c with increasing ρ_s/ρ is also apparent in Shields data who didn't address this problem. The differences between air and water are further magnified because in practice fluid threshold in water is compared with impact threshold in air. Bagnold noted that fluid threshold conditions in wind tunnel, with rolling motion of grains, could be observed at the upstream end for a distance of only 30 to 50 cm, that is, the motion rapidly changes into saltation as soon as a few grains start to move, a kind of chain reaction of impacts. Figure 3.13 shows a superposition of data in water, as a function of transport intensity, with Bagnold's data in air and with all available data on fluid threshold in air. A similar plot was presented by Miller et al. (1977), Fig. 3.14, which covers particle sizes from 0.78 μm to 1.34 mm and densities from 1300 to 11 350 kg/m³. In both graphs the rapid rise of θ_c at $Re_* < 1$ is in keeping with Bagnold's explanation as explained in conjunction with Fig. 3.2. Bagnold expressed the threshold shear velocity of grains $d > 0.1$ mm in air as

$$u_{*c} = A\sqrt{\Delta g d} \;;\quad A = 0.08 - 0.12 \tag{3.23a}$$

or

$$u_c = 5.75\, A\sqrt{\Delta g d}\, \log y/y' \tag{3.23b}$$

where y' is the height of the "focal point", Fig, 4.1, of $y' \cong$ 10 mm. Zingg (1953) related τ_0 simply to grain diameter.

$$\tau_c = 0.335\, d_{mm} \tag{3.24}$$

where τ_0 is in Nm^{-2}. (For lb/ft² the factor is 0.007.) Chepil (1959) proposed for turbulent wind of uniform velocity

$$\bar{\tau}_0 = \frac{0.66\, gd(\rho_s-\rho)\eta \tan\phi}{(1 + 0.85 \tan\phi)T} \tag{3.25}$$

where η is the ratio of drag and lift on the whole bed to that on topmost grains ($\eta \cong 0.2$), T is a turbulence factor ($T \cong 2.5$) and ϕ is the angle of repose.

The threshold in air is dependent on the moisture of air. The grains absorb moisture as surface films and these films give the sand mass a degree of cohesion. Of course, the moisture can be derived from other sources, like rain, but even if the source was moisture in air the moisture films may not be in equilibrium with air. For equilibrium conditions, studies reported by Belly (1964) indicate a moisture content w (difference in moist and

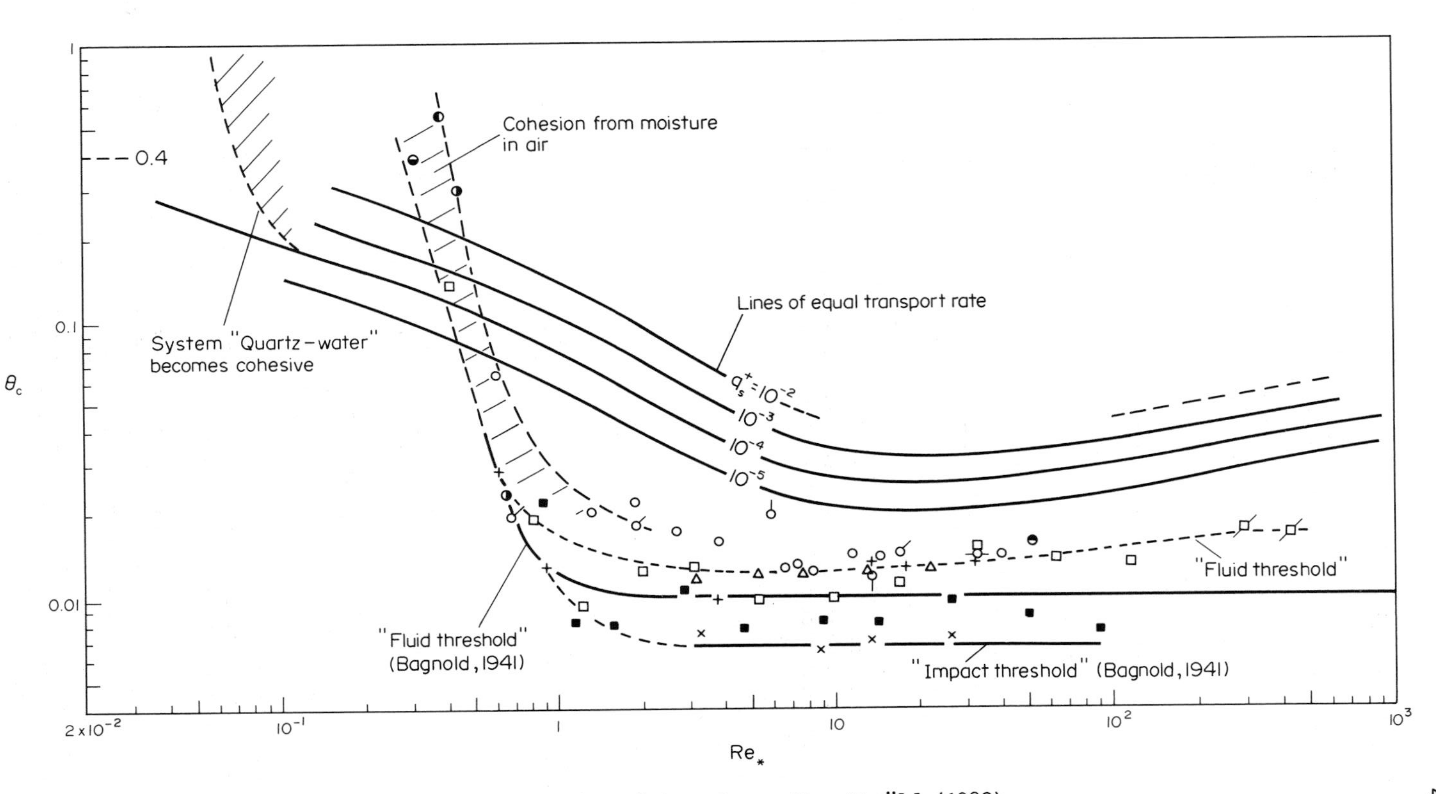

Fig. 3.13. Threshold in air and in water, after Unsöld (1982).

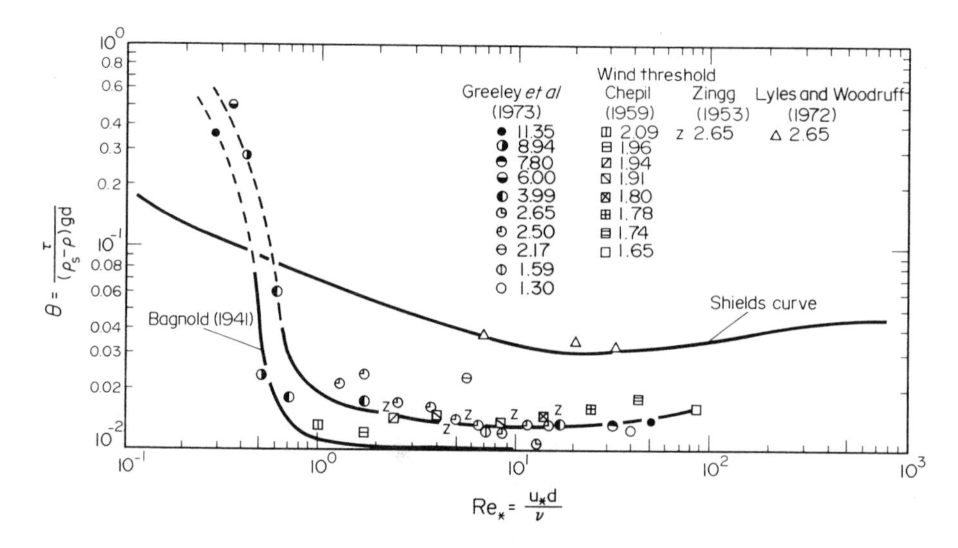

Fig. 3.14. Threshold of grains in the wind, according to Miller et al. (1977). Data symbols are identified by source and grain density (g/cm³).

dry weights of sample divided by dry weight) as a function of relative humidity H_r

$$w = 2.6 \times 10^{-3} H_r \tag{3.26}$$

where both w and H_r are in percent. The result is based on experiments with relatively coarse sand where even at 100% humidity the moisture content was less than ½%. The relationship between moisture content and critical velocity was for $0.1 < w < \sim 4\%$ given as

$$\log w = -3.12 + 6.18\, u_{*c} \tag{3.27a}$$

or

$$u_{*c} = 0.162 \log w + 0.505 \tag{3.27b}$$

Zanke (1982) suggested that the cohesion which arises from moisture could be treated as an increase in the apparent density of grains, expressed as

$$\rho_{sm} = f(w)/d^2 \tag{3.28}$$

where the function f(w) was empirically related to the humidity of air for $w > 0.008$ as

$$f(w) = 0.0266 + 3.696 \times 10^{-3} \ln w \tag{3.29a}$$

or

$$f(w) = 8 \times 10^{-3} \ln (H_r + 1) \qquad (3.29b)$$

where $0 < H_r \leq 1$ and f(w) has units of g/cm and d in eqn (3.28) is in cm. In SI system eqn (3.29b) would have a factor 8×10^{-4} kg/m and d in eqn (3.28) would be in m. Zanke expressed the threshold for $Re_* > 1$ as

$$\frac{u_{*c}^2}{\Delta\, gd} = \frac{0.01}{Re_{*c}^2} + 0.01 \left[1 - \frac{1}{Re_{*c}}\right] \qquad (3.30)$$

which for $Re_* > 5$ is approximately equal to

$$u_{*c} = 0.05(\Delta gd)^{\frac{1}{2}} + [\Delta gd(0.0025 - 0.05D_*^{-1.5} + 0.214D_*^{-2})]^{\frac{1}{2}} \qquad (3.31)$$

where now $\Delta = (\rho_s - \rho)/\rho + \rho_{sm}/\rho$ and this value is also used to calculate the dimensionless grain size D_*. Figure 3.15 shows the effect of moisture on the threshold shear velocity.

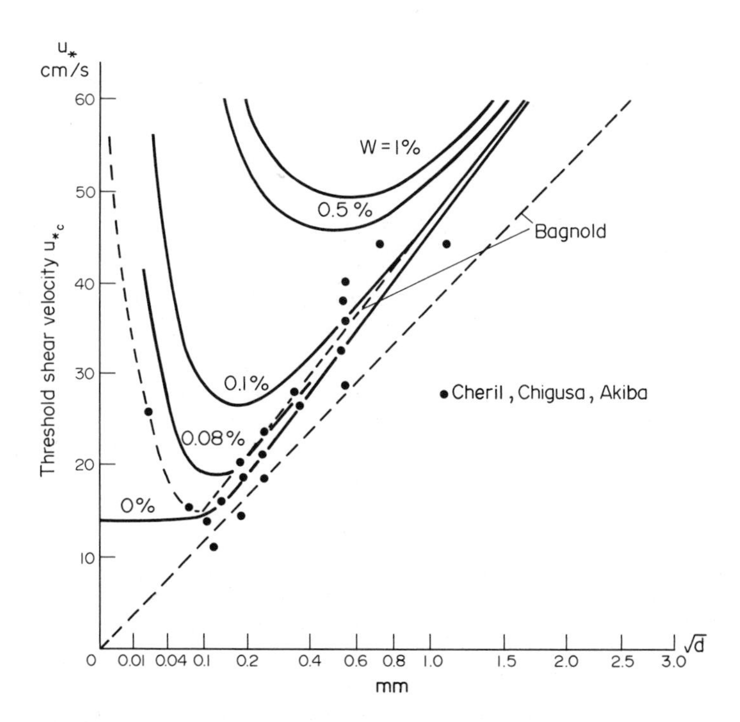

Fig. 3.15. Critical shear velocity for the threshold of grain movement in air as a function of moisture content of sand, according to Zanke (1982).

3.4 Dilatation Stress in Grain-Fluid Mixtures

Grains set in motion and dispersed into the flow are supported either by

(i) The upward diffusion of turbulence from the bed boundary, i.e., the weight of the grain is balanced by the upward component of fluid momentum transferred to it

or by

(ii) The vertical component of forces arising from transfer of momentum from grain and grain to bed.

The case of suspension by diffusion of turbulence will be discussed under the heading of suspension of sediment.

The transfer of momentum from grain to grain and grain to bed is clearly important near the bed when the concentrations are high. In order for two layers of spherical particles to move relative to each other due to an applied shear stress the layers have to move apart if the particles are not to be crushed. This normal stress arising from the applied shear was called *dilatation stress* by Reynolds. On a bed of sediment the top layer must move upwards against the weight force. Bagnold (1954,1956,1966) showed that this *dispersive* or dilatation stress does not disappear when the grains lose continuous direct contact. That grains dispersed in fluid give rise to stress fields, even at very low concentrations, was shown by Einstein in 1906. These stresses are usually accounted for by the use of the apparent viscosity. The collisions and momentum transfer between grains and grains and the bed lead to both a dispersive stress, p_y, in vertical direction and an increase in the total shear stress. For the total shear stress Bagnold wrote

$$\tau = \tau' + \tau^* \tag{3.32}$$

where τ' is due to intergranular fluid and τ^* is the component due to collisions of grains.

Bagnold developed analytical models for the cases when grain inertia and viscosity effects, respectively, dominate. The dispersive pressure in the case when inertia dominates was expressed in the form

$$p_y \propto f(\lambda)\rho_s d^2 \left[\frac{dU}{dy}\right]^2 \cos\alpha_i \tag{3.33}$$

where α_i is the angle between the resultant force and its vertical component, and λ is a linear concentration. When distance between the centres is $s + d = bd$ then $b = (1/\lambda) + 1$ or in terms of volumetric concentration

$$c = \frac{c_0}{[(1/\lambda) + 1]^3} \tag{3.34}$$

where c_0 is the maximum possible concentration ($\lambda = \infty$, $s = 0$) equal to $\sqrt{2}\,\pi/3 = 0.74$. The associated grain shear stress is

$$\tau^*_{xy} = p_y \tan\alpha_i \tag{3.35}$$

which is additive to τ' and goes to zero as $\lambda \to 0$.

For the combined grain and fluid shear case Bagnold obtained

$$\tau = \mu(1 + \lambda) \quad 1 + \left[\frac{f_1(\lambda)}{2}\right] \frac{dU}{dy} \tag{3.36}$$

where μ is the dynamic viscosity of fluid. These expressions are in form similar for those derived for flow of dense gases.

The experimental data on τ^*/λ and p_y/λ versus $\rho_s\lambda d^2(dU/dy)^2$ show both τ^* and p_y becoming proportional to $(dU/dy)^2$ at higher speeds. All data for $\lambda^2 < 14$ plot on one line. The ratio $\tau^*/p_y = \tan\alpha_i$ for $\lambda < 14$ approaches 0.32 and approximately 0.4 for $\lambda > 14$. The most significant observation is that the dispersive pressure persists even in the absence of grain inertia effects. This means that there is a second mechanism for support of grains against gravity besides the diffusion of turbulence from boundary.

Bagnold presented the results, Fig. 3.16, as

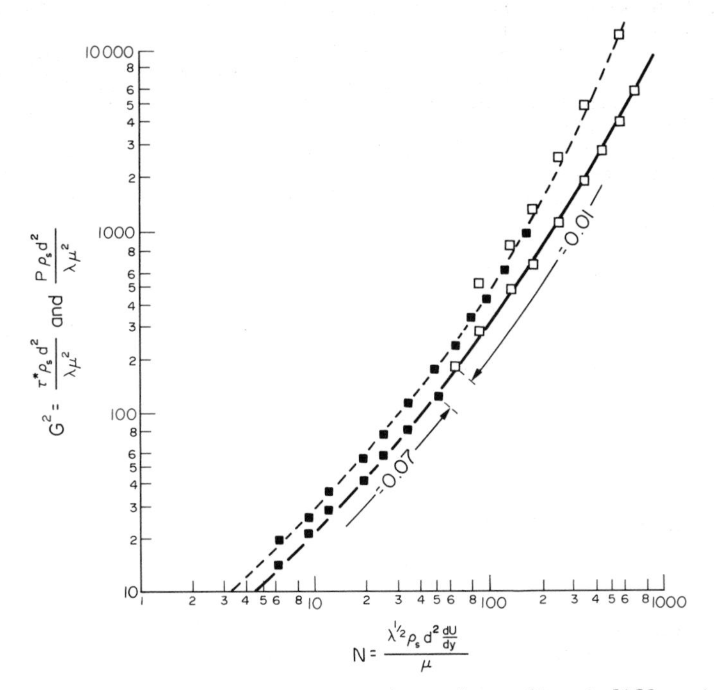

Fig. 3.16. Conformity of experimental results at different viscosities. Complete line shows shear stress; broken line pressure, λ = 11. , μ=0.01 (water); ,μ=0.07. (By permission of the Royal Society, London).

$$G^2 = \frac{\tau^* \rho_s d^2}{\lambda \mu^2} \text{ and } \frac{p_y \rho_s d^2}{\lambda \mu^2} \tag{3.37}$$

versus

$$N = \frac{\lambda^2 \rho_s d^2 (dU/dy)^2}{\lambda^{3/2} \mu (dU/dy)} = \frac{\lambda^{\frac{1}{2}} \rho_s d^2 dU/dy}{\mu} \tag{3.38}$$

where G has the form of a Reynolds number and N is the ratio of shear stress from grain inertia to viscous shear, with the proportionality to $\lambda^{3/2}$ inserted from observations. The ratio τ^*/p_y was found to range from 0.32 to about 0.75, Fig. 3.17. The transition region covers approximately $450 > N > 40$ or $3000 > G^2 > 100$.

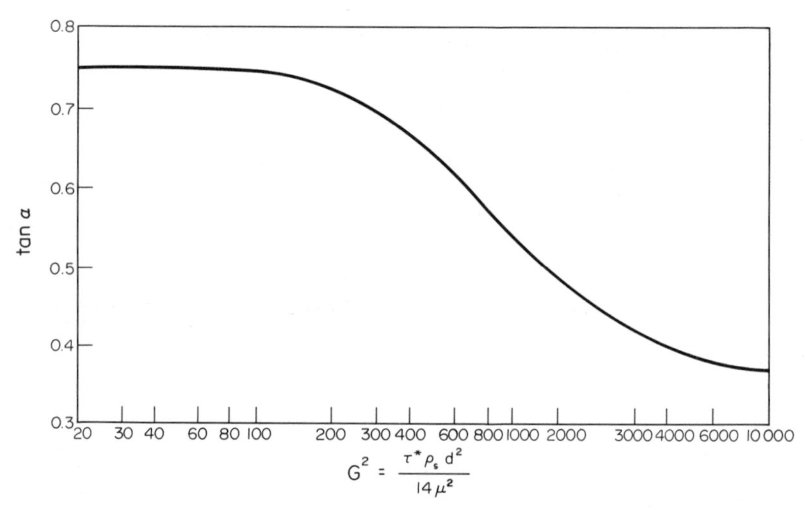

Fig. 3.17. Values of the solid-friction coefficient tan α in terms of the Reynolds-number criterion G for granular shear. (Bagnold, 1966).

Chapter 4

Sand Transport by Air

Once particles have been set in motion it is necessary to distinguish between dust storms and sand storms. The small particles of dust are kept in suspension by upward currents of the air movement. It is a suspension phenomenon and the dust cloud may rise to a height of 1000 m or more. In an erosion desert, where there is very little dust, the initial dust cloud is blown away and, soon after the wind exceeds the threshold strength, one can observe a thick low-lying sand cloud driving across the country. It has a clearly marked upper surface about 0.9 - 1.2 m above the ground.

The problem is to explain how large grains get up to heights of 0.9 - 1.2 m and what keeps them there. Measurements show that the upward eddy currents near the surface are not sufficiently strong to keep these grains in suspension or to lift them up, particularly since the vertical component of turbulence is much reduced by the presence of the solid boundary, the ground.

Bagnold and others have shown that the grains move like ping-pong balls. Once the grains are in the air, they receive a supply of energy from the forward drag of the wind. The wind drives them onwards until on falling to the ground they bounce up again and so on. Immediate observational support is furnished when the sand cloud moves over a stretch of hard ground or one covered with pebbles; then the sand cloud thickness increases appreciably. When the sand cloud moves over a surface composed of loose grains similar to those in movement the falling grains splash into the loose mass and frequently become buried but in the process they eject a number of other particles into the air. These particles do not rise as high as the particle which ejected them did because some energy has been lost in disturbing the surface.

Theoretical considerations (Bagnold, 1936), as well as experiments, show that the downward velocity of the grain is very near to the terminal velocity when it hits the ground. By this time the forward velocity of the grain is nearly that of the wind. Hence the angle at which the grain hits the ground is given by

$$\tan \beta = \frac{\text{terminal velocity of grain}}{\text{wind velocity}}$$

Observations show that the angle of impact remains remarkably constant over a wide range of conditions at $10° < \beta < 16°$. For example, if the grain does not rise high enough to reach full wind velocity, it will also not reach the full terminal velocity of fall and the angle β will not be affected much.

This bouncing movement of sand grains has been called "saltation".

4.1 The Surface Creep

The grain in saltation strikes the surface at a flat angle. A portion of its energy is passed on to the grains which are ejected upwards as it hits the loose surface. The rest of the energy is transmitted to the loose surface in disturbing many grains. This continued bombardment results in a slow forward creep of the grains composing the surface, i.e. the grains in the surface creep are not affected directly by wind and receive their momentum by impact from saltating grains. The surface creep amounts to about one-quarter to one-fifth of the total transport, but it plays a very important part in sand movement by wind. It is the means of transport of grains whose weight is far too great to be shifted by the wind drag alone. A high speed grain in saltation can by impact move a surface grain of about 200 times its own weight or 6 times its diameter.

Owing to the great difference in speed between the grains in saltation and in surface creep, surface creep is responsible for the changes in the size grading of sand deposits.

4.2 Effects of Sand Movement of Wind

When velocity measurements, taken over a wind-blown sand surface fixed by moistening, are plotted on log-normal paper (Fig. 4.1) they all converge at a point y', which is somewhat greater than $k/30.2$ if k is taken to be the mean grain diameter. The equivalent roughness height k corresponds better to the depth of the little bombardment craters. If the measurements are repeated over the same sand surface but now dried out, the wind can be increased in strength up to a threshold velocity without any movement taking place and without change in velocity distribution. But once movement has started at some point the velocity of the wind close to the surface downstream drops to a lower value - the impact threshold - due to the extra drag of the saltating grains.

The impact threshold is the limiting value at which particle motion can be maintained once it has been started. It was pointed out that these two separate threshold values, that is fluid (or static) and impact threshold, can be distinguished in transport by wind only.

The drag on the fluid by the saltating grains causes a radical

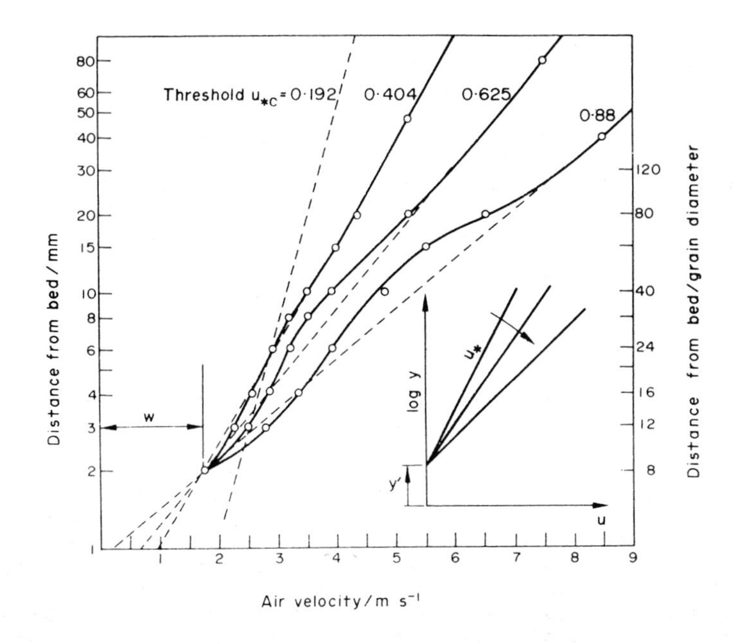

Fig. 4.1. Measured wind velocity profiles, as modified by the saltation of sand grains of diameter 0.25 mm. (Bagnold, *Proc. Roy. Soc.* A, 332, 1973).

change in the velocity distribution near the boundary. Above the layer of saltating grains the logarithmic distribution of velocity is re-established. If these rays on the log-normal plot of velocity, corresponding to particular values of drag on the flow (u_*), are extended downwards they cross each other at a nearly constant distance from the boundary and at a nearly constant value of velocity (Fig. 4.1). This constant velocity is approximately equal to the fall velocity of the grains, w. It is seen that the velocity profiles, measured in a laboratory wind tunnel, depart from straight lines within the lower 30 mm. In the upper part of this zone the changes in the velocity profiles are in keeping with the transfer of momentum from air to the slower moving grains and vice versa in the lower part. The virtually constant velocity near the boundary Bagnold (1973) explains as follows: "The answer appears to lie in the physical mechanism by which, under any constant shear stress, the transport rate of solids remains stable at a certain definite value. If the fluid velocity at the base of the saltation were to increase to any higher value than that at the threshold of bed movement, the number of solids in motion would increase indefinitely as time went on. It follows therefore that the maintenance of a steady transport rate by any fluid, air or water, requires the resistance to the fluid flow exerted by the saltation above the bed surface to be always such that the fluid velocity at the base of the saltation never increases beyond its threshold value. In air the constant fluid velocity is in fact less than that at the threshold of motion, owing to the effect of the chain-reaction."

The mean height of saltation paths is given as

$$y(\text{mm}) = 74.35\ d_{mm}^{3/2}\ \tau_0^{1/4} \tag{4.1}$$

where τ_0 is in Nm^{-2}.

Bagnold further suggests that the straight lines of the logarithmic distribution can be represented by a variable roughness height k in the logarithmic formula as

$$u = 5.75\ u_* \log\left[y/d\left[0.03 + b\ \frac{u_* - u_{*c}}{u_{*c}}\right]\right] \tag{4.2}$$

in which for the above experiments d = 0.25 mm, b = 1 and u_{*c} = 0.192 ms^{-1}.

Assuming that the velocity at the base of the saltating layer remains at the threshold value, an analogous situation would exist in water, except that y/d is reduced from about 8 to about 0.5 which is effected by putting b = 0.08 instead of unity. Note that the numerical values of Fig. 4.1 refer to laboratory measurements where the thickness of the layer of saltating grains is limited to a much smaller value than that in the open air.

It is important to observe that irrespective of the wind speed the velocity near the ground remains constant, at about the fall velocity, and that the saltating grains exert a strong drag on the flow. According to Bagnold the conditions near the ground at u_* = 0.9 ms^{-1} with saltating grains of d = 0.25 mm are the same as over a fixed surface of gravel of d = 35 mm.

Theory as well as observation show that the kink in the velocity distribution curve corresponds to the height of the statistical mean path of the saltating grains. This in turn corresponds with the mean height of the sand grains above the surface as obtained from concentration measurements (Bagnold, 1941, 1973; Belly, 1964).

4.3 Instability of a Flat Sand Surface

If it is assumed that the distribution of descending grains is uniform over the whole area, then they can be represented diagrammatically by parallel equi-distant lines (Fig. 4.2).

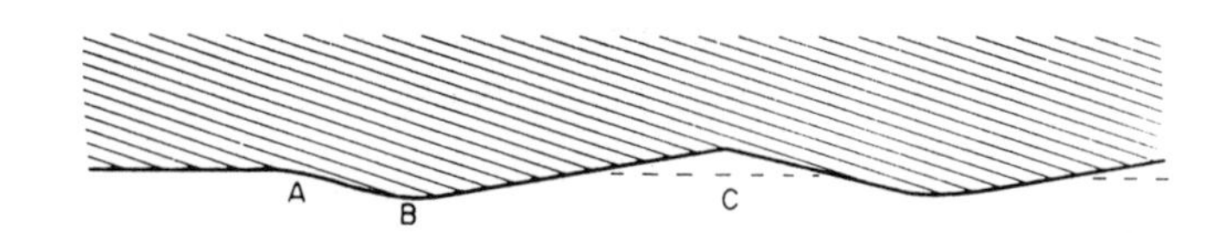

Fig. 4.2. Diagrammatic distribution of falling sand grains.

If is further assumed that the mass flow in surface creep q_s at any point is proportional to the number of forward impulses which each unit area at that point receives per second, it follows that q_s at that point is proportional to the closeness to one another of the points of impact in the diagram. If now by one cause or another a tiny hollow is formed, then on the lee side AB these points are further apart and on the windward side they are closer together. Hence many more grains are driven up the slope BC than down the reverse slope AB. This means that the original hollow will get bigger. Grains excavated from the hollow will accumulate at C because they are not removed downwind as quickly as they are arriving, thus forming a second lee slope beyond C.

Each grain in saltation describes a certain path through the air from the place of its ejection from the surface to the point where it strikes the surface and ejects another grain. From grain to grain the paths may be very different, but for any given strength of wind there exists in a statistical sense a mean grain path. In other words, of all the grains which are ejected from any given small area of the surface a greater number will fall on a second small area a distance of one characteristic path length L downwind than on any other area.

If grains were ejected in equal numbers from all over the surface, the distribution of their impacts would still be uniform; but if owing to a local tilting of the surface a greater number of ejections occurs at one point, then the effect is most markedly felt at another point distance L downstream. This is the principle of the mechanism of ripple formation. The initial irregularity is exaggerated by the surface creep up the downwind slope of the hollow and the mean path length of saltating grains leads to the systematic repetition (Fig. 4.3).

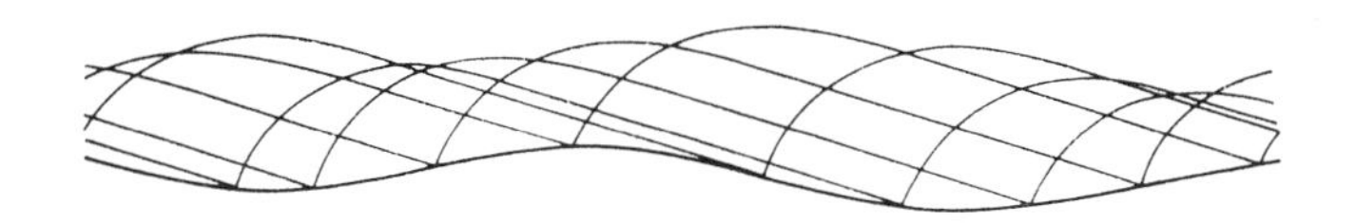

Fig. 4.3. Schematic ripple pattern and characteristic path of saltating grains.

Ripple length and the mean path length are in close agreement. However, for a uniform sand, ripples disappear when u_* exceeds about 3 times the corresponding threshold value. The problem gets more complicated when the sands are non-uniform. A knowledge of the relationships between ripple length and wind velocity for natural non-uniform sands would be extremely useful. The mean wind velocity at any place could be determined with a tape measure.

The explanation as it stands would make the ripple crests continuously higher and the hollows deeper, but as the height of the crests increases, so does the wind velocity. Over the crests the wind velocity increases with the height at a greater rate than over the hollows, due to convergence of streamlines. Therefore, as the crests rise, more and more grains are pushed over and deposited in the hollows. This soon stabilizes the shape.

With uniform sands the ripple height to length ratio is usually about 1:20 to 1:30. With non-uniform sands this ratio may be 1:15 to 1:10 because coarser grains collect and pave the crests. Ripples of the latter type become decidedly asymmetrical and the cross-section is controlled by the proportion of coarse grains in the sand.

4.4 Ridges and Dunes

If there is over a surface of non-uniform sand a wind of lower strength than the ultimate threshold needed to set the largest grains in motion, the sand movement lasts for a limited time only, until the surface becomes paved by coarse grains. Ripples produced under these conditions have crests paved by coarse grains. If, however, there is a continuous supply of saltating grains coming into the area, the picture changes. By impact the saltating grains can dislocate grains more than 6 times their own diameter, so that coarse grains are still kept in motion. The falling grains can penetrate into the hollows, which were sheltered, and expose more bed material. The finer grains of this bed material go into saltation and are carried off, so that more and more coarse grains are excavated and driven up the windward slope. The grains moving over the crest come into the complete shelter of the lee slope (Fig. 4.4).

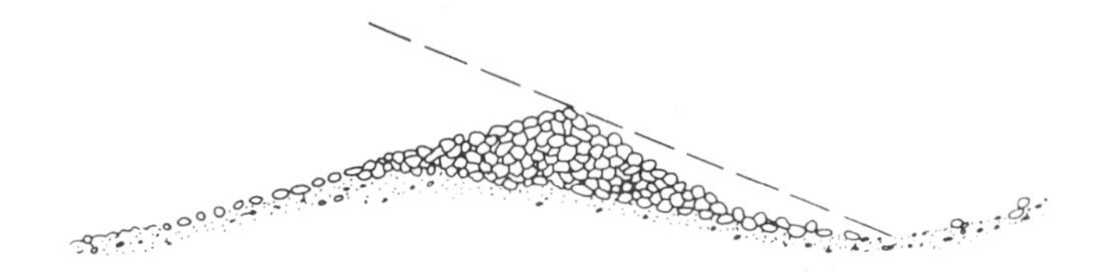

Fig. 4.4. Ridge formation.

Since new grains continue to arrive and no corresponding grains leave the crest (they arrive by creep) the characteristic path length ceases to be relevant. The ripples will go on growing and become ridges, some of which are about 60 cm high and 20 m in wavelength.

The essential difference between ripples and ridges lies in the relative magnitudes of the wind strength and the dimensions of the crest grains. The wind conditions favourable for ridge formation are between the fluid or static threshold and the impact threshold. For the ripple, the wind is strong enough to carry away the topmost crest grains when the crest rises above a limiting height, whereas in the case of a ridge the wind is not strong enough to do this.

This leads to the basic features of dune formation. Imagine a stretch of sand which has reached a non-erosion stage in a given wind, i.e. covered by pebbles larger than those movable by this wind. These pebbles shelter the finer particles. Let this area be followed by a patch of fine sand.

If now the wind increases, more sand goes into motion from between the pebbles over the pebble-covered area. But for any given wind the rate of sand flow over a surface containing pebbles, that is a harder surface, is greater than the rate possible over the surface of plain soft sand, where much of the energy in saltation goes into friction and surface creep. Hence the new supply of sand, with the increased wind, passes downwind over the pebble area at a greater rate than it can pass over the sand patch. Consequently, a strong wind causes accretion of sand on the soft sand patch. The slowing down of sand transport over the sand patch causes the air to become overloaded with sand and this has the effect that accretion starts upwind of the upwind border of the soft sand patch and the sand patch extends upwind as well as increases in thickness. This action lasts only as long as there is a plentiful supply of sand stored on the pebbly upwind surface and stops when equilibrium is reached, i.e. when the pebbles shelter the sand so that no more erosion at the given wind strength occurs. After that the movement on the sand patch still continues and the patch becomes longer downwind and thinner.

If the wind slackens, then any sand coming from a sand source will be stored in the area dotted with pebbles. The pebbles shelter the sand until a new equilibrium condition is reached.

Hence, the pebble area acts as an accumulator of sand during slight winds, which may be from any direction, and this "stored" sand is used by the strong dune-building wind. For coastal dunes the sand supply is also independent of dune-building wind. The sand arrives on a beach by the action of the sea and becomes available for dune-building by wind when it dries out.

Excellent descriptions and explanations on the formation of various forms of dunes, on grading of sand deposits, etc., are given by R.A. Bagnold (1941).

4.5 Transport by Wind

Wind erosion is a very complex and important problem in relation to soil conservation and agriculture. The factors influencing wind erosion are numerous but the most important ones may be grouped as follows:

I. *Air*	II. *Ground*	III. *Soil*
Velocity	Roughness	Soil structure
Turbulence	Obstruction	Texture
Humidity	Topographic features	Content of organic matter
Viscosity	Cover	Lime
Pressure	Temperature	Specific gravity
		Moisture content

Wind erosion depends on the mutual relationship of these factors and the effect of any one may be to aggravate or to reduce the erosion. A simple example is that of wind turbulence which increases erosion. But although surface roughness increases the turbulence, the erosion decreases because the wind velocity near the surface is reduced. A detailed treatment of wind erosion particularly of agricultural soils, is given in the literature by Chepil and Woodruff (1963,1965).

Bagnold (1938,1941) expressed the sand transport rate (kg m^{-1} s^{-1}) by surface creep and saltation by

$$g_s = C \left[\frac{\bar{d}}{d_s}\right]^{\frac{1}{2}} \frac{\rho}{g} u_*^3$$

$$= \alpha\ C \left[\frac{\bar{d}}{d_s}\right]^{\frac{1}{2}} \frac{\rho}{g} (u_1 - u_*) \qquad (4.3)$$

where d_s is reference sand size d = 0.25 mm, $\rho \cong 1.24$ kg/m³, u_* is the velocity gradient (shear velocity) of wind above the sand cloud, and C = 1.5; 1.8 and 2.8 for uniform sand, naturally graded sand and for a sand with a broad range of grain sizes, respectively. Kawamura (1951, in Horikawa (1978, p.285) expressed the transport as

$$g_s = K \frac{\rho}{g} - (u_* - u_{*c})(u_* + u_{*c})^2 \qquad (4.4)$$

where K ≅ 1.0.

The shear velocity is related to wind velocity with the aid of the logarithmic velocity distribution

$$u = 5.75\ u_* \log y/y' + u_t \qquad (4.5)$$

where y' is the height of the "focal" point for the velocity distribution in air with sand transport and u_* is the corresponding threshold velocity, Fig. 4.1. Bagnold put for desert sands (d ≅ 0.25 mm) y' ≅ 0.01 m so that in terms of wind velocity at 1 m elevation $u_* = (1/11.5)(u_1 - u_t)$. Thus, $\alpha = (1/11.5)^3 = 6.58 \times 10^{-4}$. Zingg (1953) related the height of the focal point and the threshold velocity to grain size

$$y'_{mm} = 10\ d_{mm} \qquad (4.6)$$

$$u_t = 20\ d_{mm}\ \text{(mph)} = 8.94\ d_{mm}\ \text{(m/s)} \qquad (4.7)$$

Horikawa and Shen (1960) used Zingg's expressions and d = 0.3 mm and expressed u_* and u at 1 m and 4.465 m elevations as

$$u_* = 0.0690\ u_1 - 18.4\ \text{cm/s} \qquad (4.8a)$$

$$u_* = 0.0548\ u_{4.465} - 14.7\ \text{cm/s} \qquad (4.8b)$$

From wind measurements over sand (surface d_{50} = 0.4 mm, blown sand d_{50} = 0.3 mm) Kubota et al. (1982) by inserting average measured values of u_t and y' obtained

$$u_* = 0.0644\, u_1 - 12.9 \text{ cm/s} \tag{4.9a}$$

$$u_* = 0.0511\, u_5 - 10.2 \text{ cm/s} \tag{4.9b}$$

Their field data showed that the measured and computed sand transport rates compared well when C = K = 1.5 was used. However, laboratory data gave a better fit with the factor equal to 1.0. According to Bagnold the surface creep accounts for about 20 to 25% of the transport rate.

Zanke (1982) proposed transport equations for "near the surface" transport rate, g_s, and the suspended load (flying) transport rate g_{ss} which yield comparable results to Bagnold:

$$g_s = 0.02 \left[\frac{u_*^2 - u_{*c}^2}{w^2}\right]^{1.5} D_*^3 \frac{\nu}{1-n} \tag{4.10}$$

and

$$g_{ss} = 0.02 \left[\frac{(u_*^2 - u_{*c}^2)(u_*^2 - u_{*s}^2)}{w^4} D_*^4\right]^{0.75} \frac{\nu}{1-n} \tag{4.11}$$

where n is the porosity of deposit and u_{*s} is a critical shear velocity for suspension. According to Bagnold $u_{*s}/w = (0.4\Delta)^{\frac{1}{2}} \cong$ 1.2 for sand in water and $u_{*s}/w \cong 29.4$ in air or about 15 if the u_{*c}-values in air are half of those in water, i.e., for d = 0.25 mm and w ≅ 2 m/s, the $u_{*s} \cong 30$ m/s which is clearly too high. Engelund (1965) derived $u_{*s} \cong 0.25$ w. Zanke (1976) proposed from laboratory data

$$u_{*s} \cong 0.4\, w \text{ and } u_{*s} \cong 0.2\, w \tag{4.12}$$

for water and air, respectively. The $u_{*s} \cong u_{*c}$ in air for d = 0.12 mm and d = 0.22 mm in water.

Apart from desert conditions, sand transport by wind is important at sand beaches. Onshore winds, particularly at low tide, pick up sand from the surface, as it dries out, and carry it landwards and build the coastal dunes. The sand volume of the dune provides a natural safety barrier against extreme storm wave attacks. Where the dunes are small their growth can be assisted with the aid of dune fences and planting. The effect of plants is illustrated in the photo, Fig. 4.5.

Fig. 4.5. Illustration of the effectiveness of plants in trapping sand (ammophilia arenaria and desmoschoenisis spiralis). Note the wind swept surface outside the planted area.

Chapter 5

Geometry of Fluvial Channels

The studies of geometry of rivers form a branch of science known as geomorphology or river morphology, and has a literature in its own right. The treatments are mainly by geologists, mostly in qualitative terms and in time scale of thousands of years. For engineering purposes quantitative information is required and for relatively short periods, time in decades rather than millenniums. The information available is almost exclusively empirical and for predictive purposes of fairly low confidence levels. The reasons for this empiricism are easy to see; river morphology involves the more complex aspects of hydrodynamics simultaneously with erodible boundaries and transport of sediment. The sediment deposits are usually extremely inhomogeneous, ranging from rocks, alluvial deposits to the various clays. Alluvial deposits may have complex grain size mixtures and may be stratified. Correspondingly, any analytical model of the problem faces enormous difficulties.

It is convenient for description of rivers to separate the various characteristic features according to their order of size. The first subdivision is according to slope from mountain streams, steep rivers to rivers of the plains. These segments in turn can be associated with large scale geometric features in plan, such as braiding and meandering. Superimposed are smaller features known collectively as bed features. Apart from changes of slope and geometry the flow rate increases and the average sediment size decreases with distance downstream. In earlier literature the decreasing grain size was attributed to wear. However, the major cause is the sorting of grains due to preferential transport of certain sizes under given flow conditions. The average longitudinal profile of the river bed, as well as of water surface, can be approximated by an exponential curve. Superimposed on such a bed profile will be deeps at bends and shallows at inflexion points. With the increasing flow and decreasing slope the cross-sectional area of flow increases, etc. The interaction of this multitude of factors creates the form of the particular river.

As a generalisation, a river could be looked upon as consisting of an upper, a middle and a lower reach which correspond to erosion, regime and aggradation states, respectively. Over the steep upper reach the sediment transport capacity is usually greater than supply, leading to erosion of the stream bed. The middle reach is in a latent erosion state or in regime. Here the sediment transport rate is less than the stream's transport capacity because transport for substantial periods of time is greatly reduced or stopped by the armouring of the stream bed. In the delta and lower reaches of the river the transport capacity decreases due to decreasing slope. From where the capacity gets less than sediment supply from upstream the river bed starts to aggrade. Each of the reaches of the river is characterised by plan and cross-sectional form of the river and by the bed features.

It should also be noted that rivers as "simple" open channels exist only over short reaches. The river, in general, is an intricate system of joining channels with spatially and in time varying flows. The parts of the river system are usually identified by the *stream order*, values. The smallest channel without side branches is the first order channel. Downstream of the confluence of two first order channels is a second order channel, two second order channels lead to a third order, etc but the joining of a first order in second or second order in third does not change the order rating downstream. Horton (1945) formulated the law of *stream numbers* which states that "the number of streams of different orders in a given drainage basin tends closely to approximate an inverse geometric series in which the first order term is unity and the ratio is the bifurcation ratio". (The ratio of the number of streams segments of a given order to the number of segments of the next higher order.)

5.1 Large Scale Features

In most mountain streams, or torrents, the flow is totally controlled by topographical features where changes in geometry are very slow. However, where the river has any freedom to move laterally it will constantly attempt to attain a state of dynamic equilibrium, a regime, in response to the varying flow and sediment load by adjusting its channel pattern, cross-sectional shape, channel roughness and slope. The resulting river channel patterns may be subdivided into three types: straight, meandering and braided channels, although further subdivisions have been proposed. For detailed discussion of these aspects reference is made to Leopold et al. (1964), Schumm (1977) and Richards (1982).

In the steep upper reaches where the discharge and sediment load fluctuate violently and the winding pattern of the stream is broken up by spill-overs, i.e. as the water level rises it takes the shortest course possible and runs over the inside banks of bends eroding channels in these, the rivers are usually *braided*, Fig. 5.1. In the lower reaches of the river, through alluvial plains, a single winding channel develops, called *meandering*, Fig. 5.2. Leopold and Wolman (1957) related the slope, S, to the bank full discharge, Q_b which separates the braiding and meandering regimes as

Fig. 5.1. Aerial view of a section of braiding gravel river.

Fig. 5.2. An oblique aerial view of a meandering river.

$$S = 0.012\, Q_b^{-0.44} \tag{5.1}$$

Henderson (1961) showed that the data also satisfies

$$S = 0.0002\, d_{mm}^{1.15}\, Q^{-0.46} \tag{5.2}$$

which brings the grain size d in mm into the expression. Plotted with slope as the ordinate and discharge as the abscissa, the area above the line, defined by the above equation, is the domain of braiding. It is of interest that eqn (5.2) closely resembles

$$S = 0.335\, d_m^{1.15}\, Q^{-0.46} \tag{5.3}$$

derived for a threshold stable channel, see Stable Channel Design. The effect of varying flows on meander geometry is discussed by Ackers and Charlton (1970). Straight natural channels of substantial length are rare. They do occur on very flat slopes, and on very steep slopes as broad and very shallow channels. However, there are numerous reaches of canals which have remained straight, although the flow in these oscillates from side to side as manifested by the existence of alternate bars on the bed. The three regimes in a laboratory study; straight, meandering and braided, were presented by Schumm and Khan (1972) in terms of sinuosity versus valley slope, Fig.5.3. *Sinuosity* is defined as the ratio of the arc length λp (distance along the deepest flow path) to the wave length λ which refers to the distance between the inflexion points

$$P = \lambda_p/\lambda \tag{5.4}$$

Schumm (1977) shows that the shape of Fig. 5.3 remains the same when slope is replaced by stream power $\tau_0 V$. Generally the slope

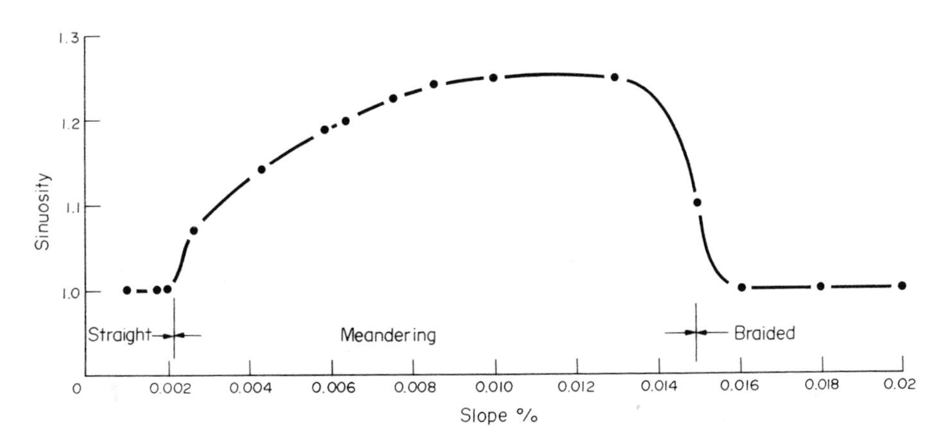

Fig. 5.3. Relationship between sinuosity and flume slope, Schumm and Khan (1972).

and sediment transport rate increase as the channel changes from straight into meandering, into braiding.

The methods of description of channel geometry may be divided into empirical descriptions, with or without theoretical justification, and analytical models. These mostly utilize methods of stability analysis, but thermodynamic analogy, probabilistic models and energy models exist, of which the last are very promising. A straight channel has the independent parameters of water flow and sediment discharge to which flow depth, width and channel slope adjust themselves. The meandering stream has additionally its plan geometry as a dependent variable.

The stability analyses start with the original work on river meanders by Callander (1969), Adachi (1967), Sukegawa (1970), Engelund and Skovgaard (1973), Parker (1976) and Fredsøe (1978). These models apply linearised stability analysis of flow to straight non-deformable channels and predict the onset of formation of alternate bars. The case of self-formed straight channels was treated by Parker (1978).

Reviews of the empirical and linearised stability analysis methods are presented by Callander (1978) and Richards (1982).

It was demonstrated in laboratory by Friedkin (1945) and Callander (1969) that at certain slopes a straight channel develops first alternate bars and gradually a stable "sinusoidal" meander form evolves.

Ikeda et al. (1981) and Parker et al. (1982) developed a stability model which allows for bank erosion and development of a truly meandering channel. They used the dynamic description of flow in bends, arising from Engelund (1974) second approximation and a kinematic description of bank erosion. The analysis shows that at wave lengths shorter than a critical wave length the bends stabilise, and grow at wave lengths longer than the critical value. The physical reason for stability or otherwise is expressed in terms of velocity distributions at crossings (fords) and apex of the bend as illustrated in Fig. 5.4.

In nature the meander patterns show an unmistakable regularity, yet their wave length and amplitude can be described only in statistical sense. The random element arises mainly from the inhomogeneity of the terrain through which the river flows. Many relationships have been proposed for predictions of the meander wave length, λ, and its double amplitude, a_m. Wave length λ, refers to the distance between inflexion points. The earlier relationships, e.g., Inglis (1949) gave λ as proportional to root of discharge, with the constant of proportionality in SI-units of 53.6 for meanders in flood plain and 46.0 for incised rivers. Dury (1965) proposed

$$\lambda = 54.3\ Q_b^{0.5} \tag{5.5}$$

and Anderson (1967) transcribed the statement to

$$\lambda / A^{0.5} = 72\ F_r^{0.5} \tag{5.6}$$

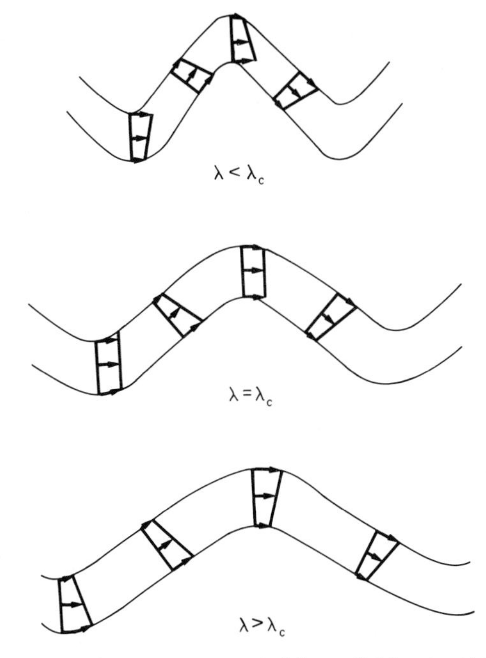

Fig. 5.4. Physical interpretation of linear stability criterion, according to Parker et al. (1983).

where A is the cross-sectional area of flow and $F_r = \overline{V}/(gy_0)^{0.5}$, which, with depth assumed proportional to $Q^{0.43}$, yields

$$\lambda = 47\ Q^{0.39} \tag{5.7}$$

However, much stronger correlation has been shown to exist with the average channel width, B. A few of these relationships are shown in SI-units in Table 1.

TABLE 1: Examples of relationships between meander dimensions and channel width

	Meander length	Double amplitude	Source
1)	$\lambda = 6.5\ B^{0.99} = 6.6\ B$	$a_m = 18.4\ B^{0.99} = 17.38\ B$	Inglis
2)	$\lambda = 11.45\ B$	$a_m = 27.3\ B$	"
3)		$a_m = 10.6\ B^{1.04} = 14\ B$	"
4)		$a_m = 30.8\ B$	"
	$\lambda = 11.0\ B^{1.01}$ $\lambda = 4.6\ r_c^{0.98}$	$a_m = 3.0\ B^{1.1}$	Leopold and Wolman (1957, 1960)
	$\lambda = 10.0\ B^{1.025}$	$a_m = 4.5\ B^{1.0}$	Zeller (1967)

1) Rivers in flood plains
2) Incised rivers
3) In flood plains (Wisconsin), Bates data
4) Incised rivers, Bates data

In Table 1 the radius r_c of the meander bend is included, a parameter which is also linked with the water depth in the bend. Since meander bends are generally non-circular the radius is estimated by dividing the meander path into segments of equal length and drawing perpendiculars to the segments. The perpendiculars describe a polygon and the distance from its centre to the maximum amplitude point is used to define the radius, which is usually two or three times the average channel width.

The channel width and depth are important parameters themselves. Lacey (1929) deduced for the length of the wetted perimeter, P, which usually approximates closely to the stream width, a relationship independent of sediment size

$$P = 4.8\ Q^{\frac{1}{2}} \tag{5.8}$$

In general, width, B, and depth, y_0, of a given river system are approximately proportional to the mean annual discharge, Q_m, taken to the power of 0.5 and 0.4, respectively, where the constant of proportionality is a function of the sediment. For example, Kellerhals (1967) proposed for streams with non-cohesive bed material

$$y_0 = 0.42\ Q_d^{\ 0.40}\ d_{90}^{\ -0.12} \tag{5.9}$$

where Q_d is the "dominant discharge", one which is assumed to produce the same geometry as the fluctuating discharge of the river ($Q_d \cong Q_T \cong 1.58$) and d_{90} is the sediment size of which 90% is finer. The width and depth are also functions of the type of sediment and the sediment load. Coarse sediments lead to broader and shallower rivers and fine sediments to deeper and narrower streams. Schumm (1960) correlated the width to depth ratio with the percentage of silt-clay material in the sediment composing the perimeter of the channel using

$$M = \frac{S_c B + 2\ S_b y_0}{B + 2\ y_0} \tag{5.10}$$

where S_c and S_b are the percentages of silt and clay in channel bed and channel banks, respectively, where the percentage of silt and clay is the material passing 75 μm sieve. The correlation obtained is

$$B/y_0 = 255\ M^{-1.08} \tag{5.11}$$

Schumm also suggests that aggrading channels have a larger width-depth ratio than indicated by eqn (5.11) and degrading channels have a lower ratio. Equation 5.11 has been criticised because the correlation includes both width and depth in its derivation and that width should be a function of bank-silt content. Ferguson (1973) showed that Schumm's data yield

$$B = 33.1\ Q_{ma}^{\ 0.58}\ S_b^{\ -0.66} \tag{5.12}$$

where Q_{ma} is the mean annual flood (return period 2.33). Schumm

(1968) also developed the meander wave length expressions

$$\lambda = 1935\ Q_m^{0.34}\ M^{-0.74}; \quad r^2 = 0.89$$

$$\lambda = 618\ Q_b^{0.43}\ M^{-0.74}; \quad r^2 = 0.88 \qquad (5.13)$$

$$\lambda = 395\ Q_{ma}^{0.48} M^{-0.74}; \quad r^2 = 0.86$$

where Q_m is mean annual flow rate. From both sets of Schumm's data Richards (1982) proposed

$$B = 25.5\ Q_{ma}^{0.58}\ S_b^{-0.6} \qquad (5.14a)$$

$$y_0 = 0.03\ Q_{ma}^{0.35}\ S_b^{0.6} \qquad (5.14b)$$

or since $Q = By_0V$

$$V_{ma} = 1.23\ Q_{ma}^{0.07} \qquad (5.15)$$

The river width and cross-section are also a function of sediment load. If a tributary, for example, brings in large quantities of suspended sediment, the width of the river downstream of the confluence may decrease and depth increase. Large bedload input leads to an increased width and decreased depth downstream.

The inability of the analytical models to predict the channel geometry under the extremely complex conditions in nature has encouraged the study of alternate methods.

Leopold and Langbein (1962) introduced the thermodynamic concept of entropy as an analogy into stream morphology and this has led to some very promising results. The analogy is discussed in some detail by Yang (1971). The main conclusion is that a stream channel which approaches equilibrium conditions develops a course of flow which makes the rate of expenditure of potential energy per unit mass of water flowing a minimum.

The variation of kinetic energy and of friction losses in a stream channel is governed by complex relationships but the sum of these in any reach must equal the loss of potential energy in the same reach. The analogy leads to the result that each stream order has an equal amount of fall, i.e., $Y_1 = Y_2 = Y_3 = \dots Y_n$. Thus, the relationship

$$\frac{Y_i}{Y_{i+1}} + \frac{Y_i}{y_{i+2}} + \dots = 1 \qquad (5.16)$$

could be used to test the river system for dynamic equilibrium. In analogy to thermodynamics where the rate of production of entropy per unit mass goes to a minimum on approaching a steady state Yang minimized the stream power per unit width, VS.

Chang (1979) extended this argument, using the unit stream power integrated over the cross-sectional area of flow, $\rho g\ QS$, and proposed that the "condition of equilibrium occurs when the stream power per unit channel length ρg QS is a minimum subject to given constraints", i.e. since Q is given this means minimum slope. Two minima are possible, one at lower and the other at the upper flow regime (see Chapter 6). The channel slope for regime channels cannot exceed the valley slope, S_V, and generally, the minimum slope for the upper flow regime, S_u, is greater than that for the lower regime, S_L. Both must be equal or greater than S_V. When the minimum equals the valley slope the channel ought to be straight, or more strictly when $S_L = S_V$ because when $S_L > S_V$, then $S_L > S_V$, which is also a possible condition, and the channel would lengthen, i.e., meander. Chang (1979) used the Engelund-Hansen, Einstein-Brown and Du Boys sediment transport formulae and the Engelund method of flow resistance estimation and carried out numerical experiments. These led to the diagram shown in Fig. 5.5, reproduced from Chang (1985,1986). The diagram is divided into four regions, each separated by a threshold or delineation criterion. The first threshold, I, is that for bed load movement defined by

$$\frac{S_c}{\sqrt{d}} = 0.000386\ Q^{-0.51} \tag{5.17}$$

where d is in mm and Q in $m^3 s^{-1}$. The line II is defined by

$$\frac{S}{\sqrt{d}} = 0.00704\ Q^{-0.55} \tag{5.18}$$

Lines I and II define Region I. According to Chang "A regime channel in this region is characterized by a flat slope, low velocity, small bed load, and low flow resistance in the lower regime of ripples and dunes. The channels are relatively deep. Man-made canals are in this region. Natural channels in this region tend to meander." The channel width B in m for Region 1 is given as

$$B = 5.68 \left[\frac{S}{\sqrt{d}} - \frac{S_c}{\sqrt{d}}\right] Q^{0.47} \tag{5.19}$$

and depth D in m

$$D = 0.83\ Q^{0.47} \exp\left[-\ 0.38 \left[\frac{S}{S_c} - 1\right]^{0.4}\right] \tag{5.20}$$

in terms of discharge in m^3/s. For Region 3

$$B = 278\ Q^{0.93} (S/\sqrt{d})^{0.84} \tag{5.21}$$

and

$$D = (-0.112 - 0.0379 \ln Q - 0.0743 \ln S/\sqrt{d})\, Q^{0.45} \tag{5.22}$$

Line III

$$\frac{S}{\sqrt{d}} = 0.00763\ Q^{-0.51} \tag{5.23}$$

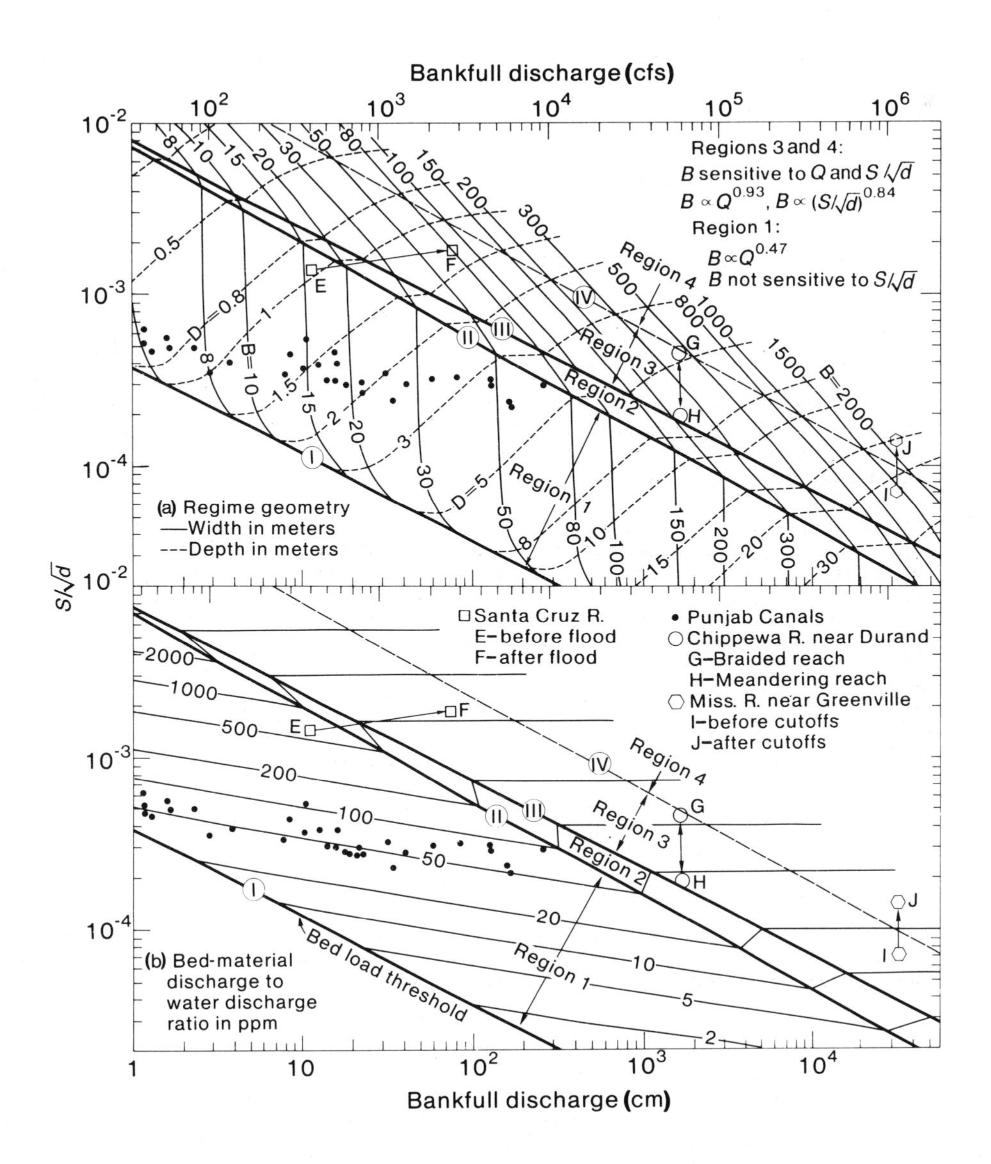

Fig. 5.5. Regime relationships for sand-bed rivers, according to Chang (1986).

encloses the small Region 2. Rivers in this region "are more likely to assume a unique channel geometry" corresponding to the lower regime minima. The width to depth ratio is large and rivers here tend to be braided.

In Region 3 both pool and rapid (riffle) sections of river may coexist. "Both channel width and depth are sensitive to the slope". The "rivers in this region may be braided". The width

Region 4 is similar to 3 but slopes are steeper. The diagram is based on formulae which are at best approximations of the processes in nature. In particular, the problems with energy losses in the turning and twisting, bifurcating and joining of flows in braiding rivers are as yet not subject to an analytical prediction. Nevertheless, the diagram (Fig. 5.5) presents a valuable addition to the methods available and is helpful for understanding the processes involved.

The plan geometry of river affects not only the energy loss of the flow but introduces also the so-called secondary currents and leads to significant variation of water depth along the "talweg", the deepest points of the cross-sections.

A bend in the river leads to the water banking-up due to centrifugal effect. The banking-up is approximately given by

$$z_o - z_i = \frac{V_s{}^2}{g} \ln \frac{r_o}{r_i} \tag{5.24}$$

where V_s is the surface velocity and subscripts o and i refer to outer and inner radii and elevations, respectively. Since the velocity drops to zero at the bed the magnitude of the centrifugal force diminishes with depth. This leads to a flow downwards at the outer bank (down the dynamic pressure gradient) and by continuity, a surface flow towards the bend. Thus, a spiral current is superimposed on the flow which affects both the flow patterns and the sediment movement. In the bend, the point of maximum velocity moves close to the outer bank and also downward, i.e. the maximum velocity streamline is meandering not only in plan but also in elevation. The flow in a meandering river converges in flowing into the bend and diverges as it flows out of the bend towards the inflexion point, Fig. 5.6. The problem of flow in open channel bends is discussed in detail by Rozovskii (1961), and in a recent review by Falcón (1984). Its exact analytical solution is still one of the challenges to the researchers. Rozovskii shows that the near bed components of velocity and shear stress are related by

$$\frac{\tau_{or}}{\tau} = - \frac{u_r}{u} \tag{5.25}$$

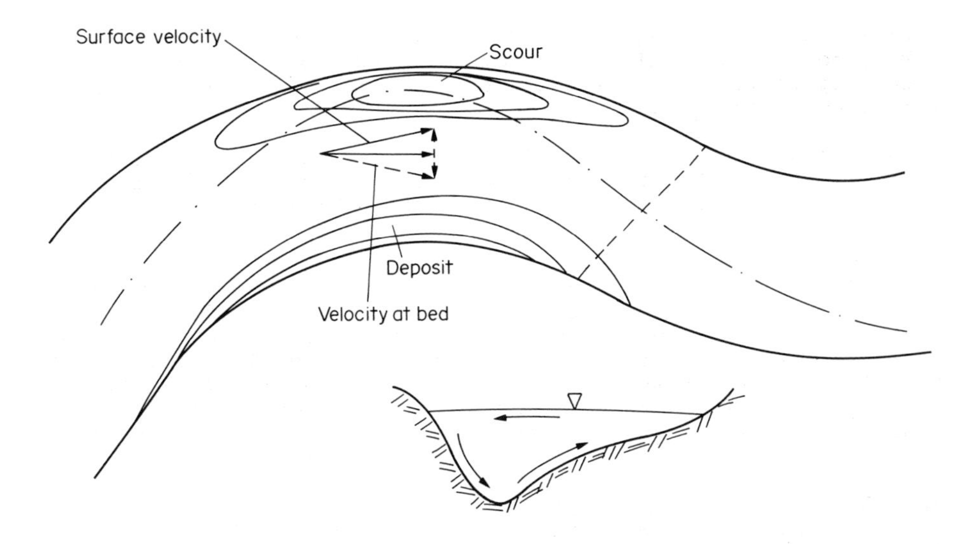

Fig. 5.6. Diagrammatic picture of flow through a bend.

where the subscripts r denotes radial components, his eqn 1.91. It also follows from his analysis and data for channels with rough beds that the radial components of velocity at the bed and surface are equal in magnitude. The development of this secondary circulation in a fixed boundary bend is described by

$$u_{rs} = u_{rs\infty}\left[1 - \exp\left[- \frac{2\kappa^2}{n}\frac{r}{y_o}\phi\right]\right] \tag{5.26}$$

where the subscripts s and ∞ refer to surface and fully developed, respectively, κ is Karman constant, $n = \kappa\sqrt{8/f} = \kappa C/\sqrt{g}$ is the exponent in the power-law velocity distribution, f is Darcy-Weizbach friction factor, C the Chezy coefficient, r radius, y_o water depth at r and ϕ the bend angle. For $u_{rs} \cong 0.9\ u_{rs\infty}$

$$\phi = 1.2\ \frac{n}{\kappa^2}\ \frac{y_o}{r} \tag{5.27}$$

which is typically of the order of 0.2 radians or 10 to 15 degrees.

The distribution of vertically averaged mean velocities in the cross-section is given by

$$\frac{U}{U_c} = \left[\frac{y_o}{y_{oc}}\right]^{C+\frac{1}{2}} \left[\frac{r_c}{r}\right]^{\frac{1}{2}} \tag{5.28}$$

where the subscript c refers to centreline values and the exponent is from the Chezy $C \propto (y_o/k_s)^c$, see eqn 6.21, and is usually of the order of 1/6.

A number of analytical models have been proposed for calculation of the transverse bed slope in a river bed. The simple free vortex model also leads to

$$\frac{dy}{dr} = -\frac{dz}{dr}\frac{1}{1 - Fr^2} \tag{5.29}$$

which means that, for constant total energy, the bed of a rectangular channel bend should fall towards the outside when $Fr < 1$ and inside when $Fr > 1$. Since in the free vortex $dy/dr = y/r$

$$\frac{dz}{dr} = (Fr^2 - 1)\frac{y}{r}$$

or

$$z + \frac{gr}{u_o r_o} + \frac{u_o r_o^2}{2g\ r^2} = H_o = \frac{U^2}{2g} + y + z = y_o + \frac{U_o^2}{2g} \tag{5.30}$$

All the analytical models are restricted to bends with fixed banks, i.e., changes in bed elevation only, and constant discharge. Examples are van Bendegom (1947) and its modified form in Jansen et al. (1979), Allen (1970), Engelund (1974), Kikkawa et al. (1976), Zimmermann and Kennedy (1978), Zimmermann and Naudasher (1979), Falcon and Kennedy (1983) and Odgaard (1981,1982,1984). The slope β of the curved plane in radial direction to horizontal is determined in all these models from the balance of the radial component of fluid drag, F_{Dr}, parallel to the bed and the immersed weight component of grain, $W_i = (\rho_s - \rho)\ g\forall$, where $\forall$ is the volume of grain, as

$$F_{Dr} = W_i \sin\beta \tag{5.31}$$

The mean velocity is assumed to be normal to the radius. Odgaard (1981) shows that the results reduce to

$$\sin\beta = K\frac{y_o}{r} \tag{5.32}$$

where y_o is the depth of flow at radius r. The individual derivations vary in the form and definition of the term K. Odgaard assumes that the particle in the bed surface is at the threshold for longitudinal movement. This means that the excess shear stress, $\tau_o - \tau_c$, is dissipated by the sediment in motion, a concept similar to the Bagnold's energy model.
It leaves open the question of the role of any form drag which may arise when ripples are present. The radial force balance for a particle at threshold is given by

$$A\ \tau_{c,r} = W_i\ \sin\beta \tag{5.33}$$

where A is the projected surface area of the grain and $\tau_{c,r}$ is the radial component of the critical shear stress acting on the grain. Assuming that

$$\frac{\tau_o}{\tau_{or}} = \frac{\tau_c}{\tau_{c,r}} = -\frac{u_r}{u} \tag{5.34}$$

according to eqn 5.25, and with $u = \beta u_*$, where β is a constant of proportionality

$$\frac{\tau_{c,r}}{\tau_{or}} = \frac{u_{*c,r}}{u_*} \tag{5.35}$$

Using $u_* = \kappa U/n$, where $n = \kappa\sqrt{8/f}$ is the exponent in the power law velocity distribution (not Manning n), and u_{*c} from the Shields parameter θ, Odgaard obtains for the transverse bed slope $\beta = dy_o/dr$

$$\beta = \frac{3\alpha}{2\kappa} \sqrt{\theta} \, \frac{n + 1}{n + 2} \, F_d \, \frac{y_o}{r} \tag{5.36}$$

where α is the ratio of projected surface area to volume for a nonspherical particle divided by that for a sphere ($\alpha = 1.27$ for particles with a shape factor SF = 0.7) and F_d is in form similar to a densimetric Froude number

$$F_d = \frac{U}{\sqrt{\Delta g d}} \tag{5.37}$$

Zimmermann and Kennedy (1978) put $n = 1.13/\sqrt{f}$. For the streamwise development of the bed slope Odgaard uses eqn 5.26, i.e., eqn 5.26 is multiplied by $\{1-\exp[-(2\kappa^2/n)(r_c/y_{oc})\phi]\}$ to yield β as a function of bend angle ϕ. A first order estimate of β is obtained with average velocity Q/A, average depth, centre line radius and an estimated κ. The most difficult parameter is the grain size and its variation with radius. The d_{50} of surface layer is suggested but the surface layer samples are difficult to obtain in the field.

For the development of streamwise velocity distribution Odgaard (1984) derived

$$\frac{U}{U_c} = \left\{ \left[\frac{r_c}{r}\right]^{\frac{1}{2}} \exp\,(-Bs) + E[1-\exp(-Bs)] \right\}^{\frac{1}{2}} \tag{5.38}$$

where $s = r_c\,\phi$ is the distance along the centre line

$$B = \left[\frac{n}{n+1}\right]^2 \left[\frac{\kappa^2}{n^2 y_o} + 10\,\frac{(n+1)^2 y_o}{(n(n+2)r^2}\,G\right]$$

$$G = \frac{1}{2}\,\frac{m(1+2c) - 1}{1 + m(1+2c)[(r_c/r)^{\frac{1}{2}} - 1]} + 1$$

$$m = 4.8[U_c/(\Delta g d_c)^{\frac{1}{2}}]\sqrt{\theta}$$

$$E = H/(BU_c^2)$$

$$H = 2\left[\frac{n}{n+1}\right]^2 gS$$

where S is the streamwise slope, d_C is the grain size at the centre line and c = 1/6. The factor 4.8 for m arises from $3\alpha/2\kappa$ using α = 1.27 and κ = 0.40. The G-values for fully developed bend flow are typically in the range of 3.5 to 6.7. In the field where

$$\frac{y_o}{r} \leq \frac{0.138\kappa}{n+1}\left[\frac{n+2}{nG}\right]^{\frac{1}{2}} \tag{5.39}$$

the predictions for velocity development are said to come within 10% of fully developed values using

$$B = \frac{2\kappa^2}{y_o(n+1)^2}$$

$$E = \left[\frac{n}{n_c}\right]^{\frac{1}{2}}\left[\frac{y_o}{y_{oc}}\right]\left[\frac{r_c}{r}\right]$$

and

$$\frac{n}{n_c} = \left[\frac{y_o}{y_{oc}}\right]^c$$

The length of the bend required for the development of flow is given by

$$\phi = -\frac{(n+1)^2}{2\kappa^2}\frac{y_o}{r}\ln\frac{0.19}{1-1/E} \tag{5.40}$$

which is typically about 0.7 radians or 40 to 45 degrees.

The distribution of depth in the cross section is predicted by

$$\frac{y_o}{y_{oc}} = \left\{1 + m(1+2c)\left[(r_c/r)^{\frac{1}{2}} - 1\right]\left[1 - \exp\left[-\frac{2\kappa^2}{n}\frac{r_c}{y_{oc}}\phi\right]\right]\right\}^{-\frac{1}{c+\frac{1}{2}}} \tag{5.41}$$

At this stage the above equations by Odgaard, or any of the others, can only be used as indicators. The modified van Bendegom solution yields

$$\sin\beta = \frac{3}{2}\frac{y_o u_*^2}{gr\Delta d}\alpha \tag{5.42}$$

where

$$\alpha = \frac{2n^2}{\kappa^2(n+2)(n+3)}$$

$$\Delta = (\rho_s - \rho)/\rho$$

yo is depth and d is particle size. Assuming that $\sin\beta \cong dy_o/dr$ and substituting $u_*^2 = gy_oS$

$$\frac{dy_o}{y_o^2} = \frac{3}{2}\frac{\alpha S}{d\Delta}\frac{dr}{r}$$

and by putting $Sr = S_o r$, the values at the outer bank, yields

$$\frac{1}{y_o} - \frac{1}{(y_o)_o} = \left[\frac{1}{r} - \frac{1}{r_o}\right] \frac{3\alpha S_o r_o}{2d\Delta} \quad (5.43)$$

where $6.25 \leq \alpha \leq 7.67$ for $6 \leq n \leq 9$ and in most cases $1.5\,\alpha\, S_o r_o / d\Delta > 1$, i.e. the lateral bed profile of the river bend is upwards convex. The depth y_o has to be estimated for given constant flow. The derivation assumes uniform bed material and fixed banks.

Equation 5.43 is similar to a relationship developed by Engelünd and Fredsøe (1982), i.e.

$$\frac{y}{y_o} = \left[1 + \frac{n}{R} \cos\left[\frac{2\pi s}{L}\right]\right]^{b \tan\phi} \quad (5.44)$$

where y_o is the mean depth of flow, n is the radial distance measured from inner bank, s is the distance along the centre line of the bend, R is the minimum radius of centre line, L is the meander length measured along the centre line, $\tan\phi$ is angle of friction and b is a factor approximately equal to 10.

The river bends are seldom with non-erodible banks. In a sinuous stream the path of the maximum velocity is not parallel to the banks. It moves closer to the outer bank in the bends. The maximum velocity path also oscillates in elevation, moving closer to bed in the bends. This leads to increased velocity gradients, i.e., increased boundary shear τ_o at the outer boundary of the bend, and to an increased erosion. The material eroded from the outer bank is carried by the bottom current towards the downstream side of the inner bank of the bend where some of it is deposited. In many rivers, extensive sediment banks are found at the downstream side of the inner bank of bends. The radial grain size distribution in the surface layer of the bend varies substantially.

The actual longitudinal profile of the river is a succession of deeps at the bends and shallows at the inflexion points. The formulae, such as eqns (5.9) or (5.14) yield an estimate of average flow depth but little has been reported in the literature on how to determine the maximum depth in bends. In general, observations show it is inversely proportional to the radius of the bend but the actual value depends on the sediment and flow. In principle, the cross-sectional shape reflects the balance between fluid forces and sediment movement, i.e. the radial component of shear stress on the bed would be in balance with the weight component of the sediment grain on the cross-sectional profile if the width of the channel remained fixed. If the banks are eroding, then that rate of erosion has to be accounted for in the balance equation. The analytical models are not yet capable of predicting the cross-sectional shape of a river bend or even the maximum depth with a degree of reliability required in practice. The designer still has to rely heavily on field observations on a particular stream.

Some useful empirical quidance is given by the relationship presented by Ripley (1927), based on field observations. Accordingly, the depth

$$y = \bar{D}\left[1 - \frac{x^2}{W^2}\right] + \bar{D}\,\frac{5.34}{r}\left[1 - \frac{x^2}{W^2}\right]x \qquad (5.45)$$

where $\bar{D}$ is the mean depth (area/surface width) in metres, multiplied by 1.445, W is the half-width of channel (B/2) in metres, r is the radius of curvature of the outer bank in metres and x and y are coordinates with the origin in water surface at mid-width, x positive outwards and y-downwards. (Notice that x changes sign.) The maximum of y occurs at

$$x = \left[\frac{W^2}{3} + \left[\frac{r}{16.02}\right]^2\right]^{\frac{1}{2}} - \frac{r}{16.02} \qquad (5.46)$$

An interesting deduction by Ripley is that when r is less than 40 times the square root of the channel cross-section no further deepening results from increased curvature. For these cases $r = 40\sqrt{A}$. Bends in streams, "the radii of curvature of which are greater than 40 times the square root of the area of the cross-section, are constructive bends and tend to stability of channel both in position and depth; whereas "bends with $r < 40\sqrt{A}$" are destructive bends, tending to cause shifting of the position of the channel and to form cut-offs and crevasses". The formula is said to be inapplicable when r exceeds about $110\sqrt{A}$.

In the closure of discussion the author gave a modified form of eqn (5.45) in which both of the terms in brackets were replaced by $\{[4.8(1-x^2/W^2) + 3.61]^{0.25} - 1.9\}$.

For streams which do not occupy the full width of the channel in the flood plain the mean depth in eqn (5.45) is multiplied by 1.65 (instead of 1.445) and 5.34 is replaced by 8.01. Consequential change in eqn (5.46) is 24.03 for 16.02.

The empirical rules of the regime method recommend multiplying factors to the "stable channel depth" of K = 1.5 for moderate bends, K = 1.75 for severe bends, K = 2.0 for abrupt right angle bends and K = 2.25 - 2.5 alongside cliffs or walls. The stable channel depth of an unregulated alluvial channel could be estimated from the Lacey (1929) expression for hydraulic mean radius R, which in SI-units is

$$R = 0.4725(Q/f)^{1/3} \cong 0.5(Q/f)^{1/3} \qquad (5.47)$$

where f is the silt factor for which Lacey proposed an approximate formula

$$f = 1.6\,\sqrt{d_{mm}} \qquad (5.48)$$

and gave a table of values. (Reproduced in Chapter 8).

For the case of fixed or firm cohesive banks

$$R = 1.34(q^2/f)^{1/3} \qquad (5.49)$$

were q is the flow rate per unit width. Blench (1969) for the same conditions gives

$$y_o = (q^2/F_b)^{1/3} \qquad (5.50)$$

where the "bed factor" $F_b = V^2/y_o$ for sand beds is

$$F_b = 1.9\sqrt{d_m}\ (1 + 0.012\ c) \qquad (5.51)$$

where c is the volumetric concentration of sediment in ppm, i.e. c = 0 leads to F_{bo}. Eqn 5.51 applies for $d < 2_{mm}$; for larger grains $F_{bo} = 1.75\ d_{mm}^{1/4}\ (\nu_{20}/\nu)^{1/6}$ where ν_{20} is the kinematic viscosity at 20°C. Blench (1969) also gives

$$F_{bo} = 0.58\ w_{cm/s}^{11/24}\ (\nu_{20}/\nu)^{11/72} \qquad (5.52)$$

which generally yields larger values than the expressions above.

It should be noted that the regime approach is applicable to about bank full flow and not to low flow stages of the river.

The meander pattern slowly translates, usually the bends become more and more exaggerated until a natural cut-off occurs, leaving the old bend isolated, the oxbow lakes. There are no reliable methods, other than observations on the given river, to estimate the occurrence of the cut-offs.

5.1.1 Scouring of the river bed

The behaviour of the river bed is of great importance to the designer. Some of the questions are: is there a significant lowering of the river bed during a flood, a general scour; does a localised channel develop into the river bed; is the bed fluidised so that its bearing capacity is reduced over a certain depth? As yet, there are no conclusive answers to these questions. On the question of general scour the literature contains only vague comments. Leopold et al. (1964) states that "Some channels change their beds but little during a flood. Presumably this variation in bed stability is caused by differences in the type of material comprising the bed and its degree of consolidation, imbrication, packing, or cohesiveness". However, they also produced data from Colorado River, which show a progressive erosion at rising stages and fill again during falling stages of a flood produced by snow-melt, and state that "this phenomenon is characteristic of ephemeral streams and apparently large rivers in semiarid climates". Frequently, local scouring is followed by a fill, like at bends and fords (crossings, inflexion points), but there are measurements indicating longer reaches of cut or fill than just a bend-ford length. The amounts of sediment which accumulate on deltas do not indicate large overall scour depths but this does not exclude substantial cuts and fills over certain reaches of the river.

The general scour features are greatly influenced by the cross-sectional shape of the river valley. If the flood flows are confined, so as to lead to large changes in water surface elevation, the scouring power of the streams is greatly increased. If on the other hand the flood flows spill out onto berms with only minor rise in the water level, the conditions in river may change very little. Thus, one approach is to calculate for a given reach of the river the equilibrium depth of flow and relate this to the observed water levels. The difference would indicate the depth of general scour. It also becomes evident that the cuts will be associated with narrow valleys and fills with broad valleys or flood plains. The whole process is strongly influenced by the sediment supply.

Apart from changes in the average bed level during flood, many rivers, and in particular, wide gravel rivers, have a tendency to develop quite deep localised stream channels relative to mean flow depth within the cross-section. Field measurements at Ohau River in New Zealand (Thompson and Davoren, 1983) showed local channels of about 3 m depth in a braod gravel river with an average depth of the order of 0.75 m and mean velocity 2.38 m/s (bed material d_{50} = 20 mm, d_{65} = 35 mm and d_{84}/d_{50} = 3.5). These channels frequently change their positions during floods. Another feature of gravel rivers is the large moving gravel banks which lead to partial or at times even total bifurcation of flow at their upstream ends and confluence downstream. The joining of the two currents, angled to each other, usually leads to a strong spiralling current downstream of the confluence which scours a deep channel for a limited distance, like the deeps in river bends. Again the bars and the associated channels move. There are no reliable methods for prediction of the depth or location of such channels or when they occur. Mosley (1976) and Ashmore and Parker (1983) showed that the maximum scour depth, D_s, at the confluence depended strongly on the initial confluence angle θ for $15° < \theta < 90°$, where $\theta = \theta_1 + \theta_2$, both measured from the downstream direction. Amoafo (1985) fitted $D_s/D = 2.24 + 0.031\theta$ to the data, where D_s is measured from water surface and D is the mean depth of flow. The influence of confluence angle decreases with increasing asymmetry of flow in the branches. From field measurements, Mosley (1982) expressed the depth at the confluence as

$$y_o = 0.531\ Q^{0.343} \tag{5.53}$$

where Q is the combined discharge of the two channels. Where the channels are ill-defined, the estimation of Q is a problem. Likewise, there is at present, not enough data to ascertain the range of validity of the above expression. In general, the designer has to rely on the observations of the river in question.

5.2 Bed Features

Whenever the velocity of flow over an alluvial bed exceeds the threshold value, bed features develop. These are likely to interact to some extent with the larger features like meanders but their major effect is on resistance to flow, on sediment transport and turbulence of flow. The bed features of sandy beds are conventionally subdivided into *ripples*, *dunes* and *antidunes* but no clear-cut definition of ripple and dune exists. Therefore, in some non-English language literature only the term bed feature is used without distinction of individual forms. There are numerous presentations which attempt to delineate conditions leading to a particular form of bed features, for example, Fig. 5.7 by Simons et al. (1964) in terms of stream power $\tau_o U$ and grain size. Znamenskaya (1969) proposed a plot of Froude number against the ratio of mean velocity to fall velocity with the steepness η/λ as a parameter and incorporating the translation velocity c, Fig. 5.8. Bogardi (1966) classified bed features on gd/u_*^2 - d diagram and on $u_*[1.65\rho/(\rho_s - \rho)]^{\frac{1}{2}}$ - d diagram. Liu (1957) used a u_*/w - u_*d/ν plane to delineate the zones of individual types of bed features, Fig. 5.9. Hill et al. (1969) used a $u_{*c}d/\nu$ - gd^3/ν^2 plane, Fig. 5.10, where for constant $(\rho_s - \rho)/\rho$ the gd^3/ν^2 term is the dimensionless grain size D_*. The dimensionless grain size was also used by van Rijn (1984) Fig. 5.11 together with his transport parameter

$$T = [(u_*')^2 - u_{*c}^2]/u_{*c}^2 \qquad (5.54)$$

where $u_*' = U\sqrt{g}/C'$ and the Chezy C is based on grain roughness, $C' = 18 \log(12R_b/3d_{90})$, using the hydraulic radius relating to bed (side wall correction) and $3d_{90}$ as the roughness height k_s. The u_{*c} value is obtained from the Shields threshold criterion.

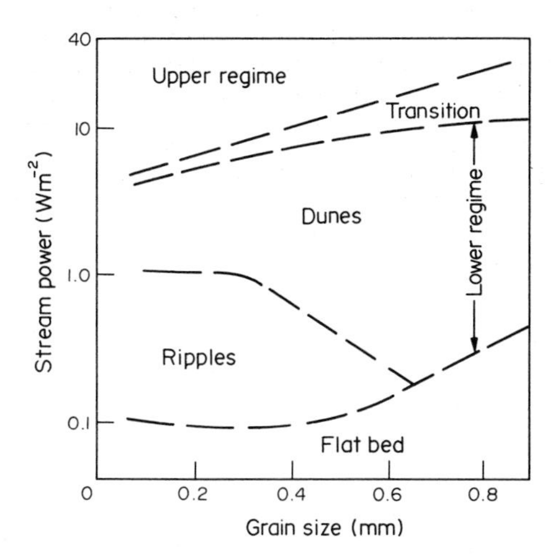

Fig. 5.7. Classification of bed features according to Simons et al. (1964).

There are numerous other presentations. All of which present some observed data, almost all of it from laboratories. Frequently, grain size grading or temperature are not given and all the data carry some effects of the particular equipment and experimental technique used. Most of the data relate to sand beds, i.e. d_{50} < 2 mm assuming uniform grain size. The bed features in gravel rivers differ substantially, as will be discussed later, and much of this difference is due to differences in grain size distribution. Sands are usually almost uniform whereas gravels may have geometric standard deviations in excess of σ_g = 4 and also a much greater variety of grain shapes. An extensive

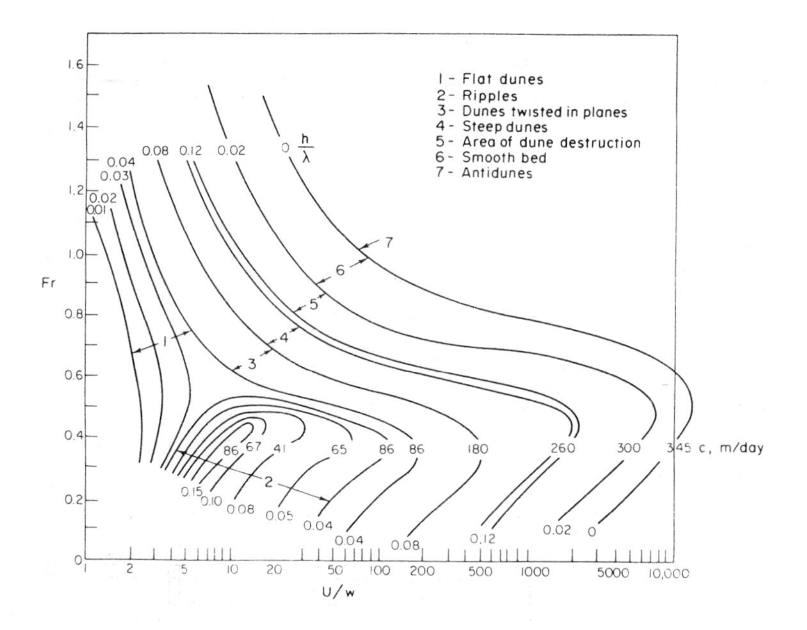

Fig. 5.8. Relationship of Fr vs. U/w with c and η/λ as parameters (After Znamenskaya, 1969).

discussion of the multitude of bed features was presented by Allen (1968).

A prominent characteristic of these bed features on a loose alluvial bed, more or less transverse to the direction of flow, is the presence of features of different size and the occurrence of two separate modes of growth of the features on a flat bed at subcritical flow conditions. These two modes have been labelled *ripples* and *dunes*.

Ripples develop at small values of shear stress excess $(\tau_o - \tau_c)$ or at low values of θ/θ_c, at Reynolds numbers $Re_* = u_* d/\nu < 22$ to 27, and are said to occur only with fine grained sediments, $d < 0.7$ to 0.9 mm. The size of the ripples is assumed to be independent of flow depth. An empirical formula for average ripple length (in laboratory flumes) is

$$\lambda = 1000\ d \tag{5.55}$$

According to Richards (1980) ripples form when

$$0.0007 < kk_s < 0.16 \tag{5.56}$$

where $k = 2\pi/\lambda$ is the wave number and k_s is the effective roughness height for which he used the results by Owen (1964)

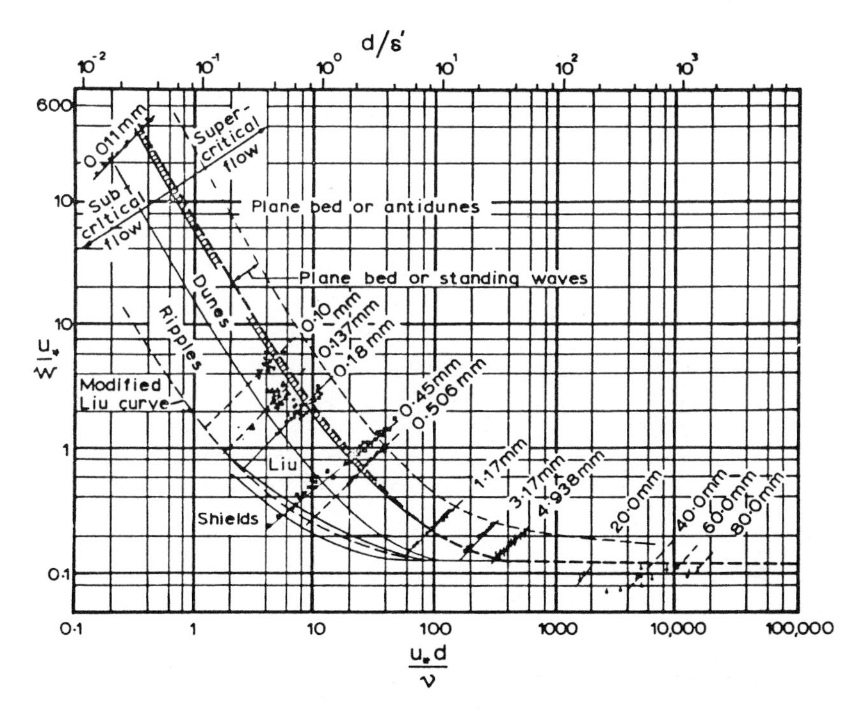

Fig. 5.9. Plot of experimental data as u_*/w versus Reynolds number, according to Liu (1957).

$$k_s = \alpha_o \frac{\tau_o - \tau_c}{(\rho_s - \rho)g} + k_n \; ; \quad \tau_o > \tau_c \tag{5.57}$$

where k_n is the Nikuradse sand grain roughness height and α_o = 26.3 based on data from Columbia River (median grain size 0.33 mm). His analysis led to

$$203 \; d < \lambda < 4050 \; d \tag{5.58}$$

Yalin (1977) argued that the ripple length was a function of the particle Reynolds number, Re_*, to which he added (1985) the dimensionless particle size, D_*, i.e., $\lambda/d = f(Re_*, D_*)$. He incorporated the D_* effect in Re_* as $\lambda/\alpha\ d = f(\alpha\ Re_*)$ and expressed α empirically as $\alpha = 3.38\ D_*^{-0.25}$. This condensed the data on ripple length into more or less a single function with a $(\lambda/\alpha\ d)_{min}$ of about 600 (± 150) at $\alpha\ Re_* \cong 5.5$. Thus, $600\ d < \lambda < \sim 2000\ d$ which is a similar but narrower range than indicated by eqn (5.58). Lau (1987) in discussion presented data which show lower α-values at smaller and larger D_* ($D_* = 8$, $\alpha \cong 1.6$; $D_* = 1000$, $\alpha \cong 0.45$) and drew a convex function on double logarithmic paper which more or less agrees with Yalin's power law, $\alpha = 3.38\ D_*^{-0.25}$, at $D_* \cong 100$ but presented no data for this region.

For the steepness of bed features Yalin (1972) deduced by dimensional reasoning that

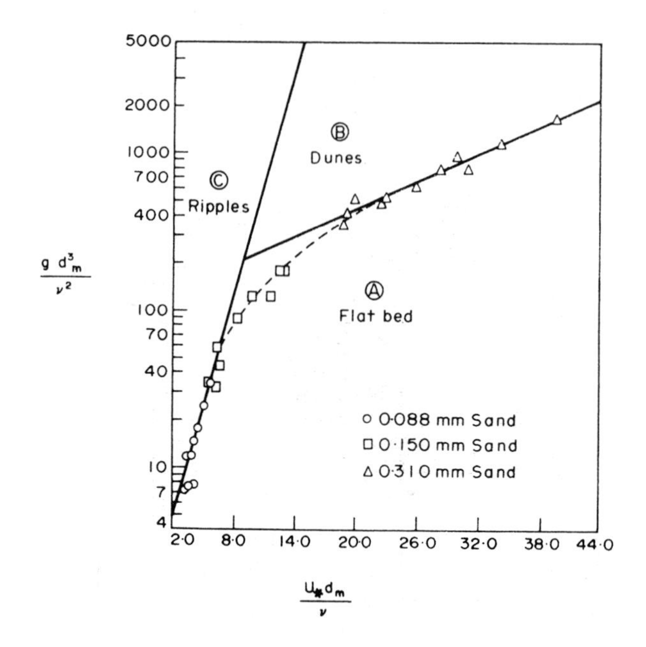

Fig. 5.10. Classification of bed features according to Hill et al. (1967).

$$\frac{\eta}{\lambda} = f\left[\frac{\theta}{\theta_c}, \ Re_*, \ \frac{y_o}{d}\right] \tag{5.59}$$

where η and λ are the height and length of the bed feature and y_o is flow depth. Thus, if for ripples η/λ is independent of flow depth and they occur at small near constant Re number then

$$\frac{\eta}{\lambda} = f\left[\frac{\theta}{\theta_c}\right] \tag{5.60}$$

Experimental data fall within the band indicated in Fig. 5.12. In his subsequent paper (Yalin, 1985) included D_* in his ripple height description, $\lambda/d = f(U/\theta_c, D_*)$. The data on η_{max}/d versus D_* for $y_o/d > 600$ suggest a decreasing maximum value of η/d with increasing D_* ($D_* = 10$, $\eta/d \cong 150$; $D_* = 100$, $\eta/d \cong 92$; $D_* = 400$, $\eta/d \cong 40$).

The ripples, in principle, are the boundary interaction with flow in the lower constant shear stress layer (Raudkivi, 1983). Ripples are seldom two-dimensional, except in narrow laboratory flumes and wave induced ripples. Generally, the pattern is three-dimensional and at times shows remarkable regularity. The three-dimensional pattern can be related to the concentrated vortices (Raudkivi, 1965). The ground roller in the lee of the ripple can be looked upon as a vortex filament. These in frictionless fluid must extend from infinity to infinity, form a loop or bind

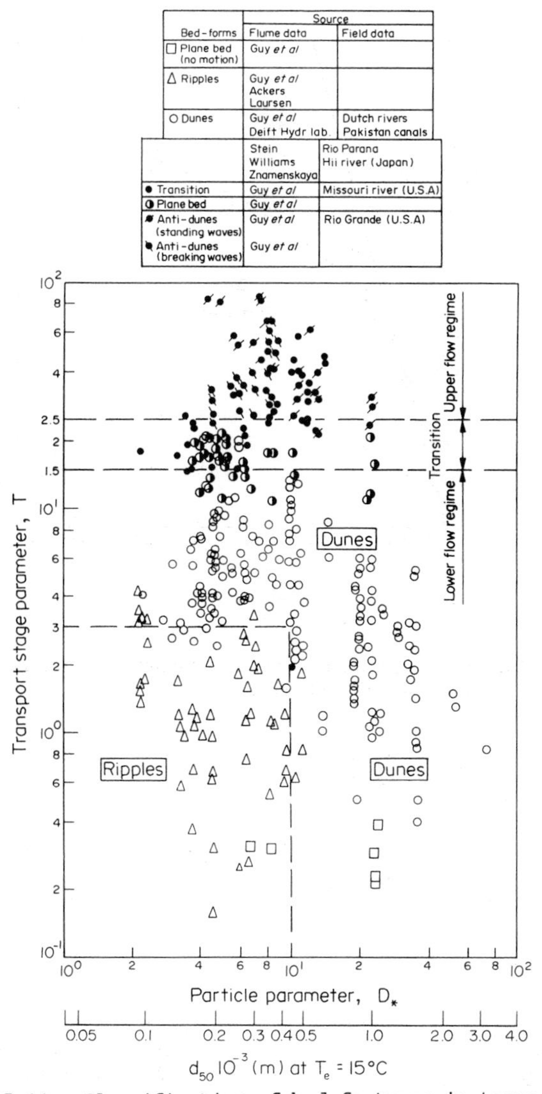

Fig. 5.11. Classification of bed features in terms of transport parameter T and dimensionless particle size D_*, according to van Rijn (1984).

to a surface. Since ripples form at relatively low velocities the vortices are weak and hence easily disturbed. Hence the straight vortex filament in the lee of a ripple cannot survive as a straight filament, it either has to attach to the bed as a horseshoe, close into a ring in the lee of the ripple or form a net of vortex filaments covering the surface as illustrated in Fig. 5.13. The vortex induced velocities indicate a tetrahedral pattern of bed features, similar to what can be observed in nature.

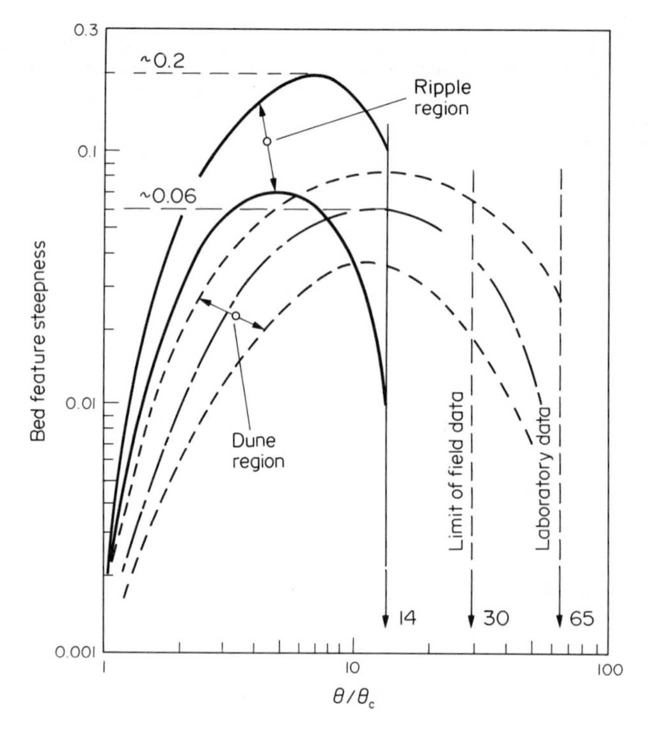

Fig. 5.12. Steepness of ripples and dunes according to experimental data. Dune data for $Re_* > 32$ and $y_0/d > 100$, after Yalin (1972).

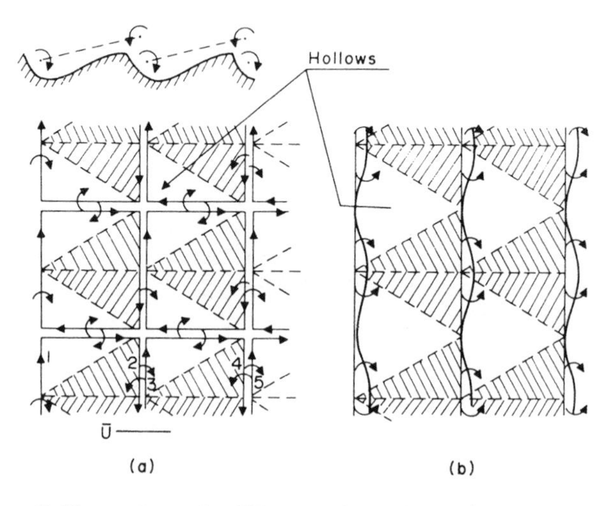

Fig. 5.13. Schematic illustration of vortices over a rippled bed, (a) as a net and (b) as horseshoe vortices.

Dunes are generally larger bed features and develop at higher bed shear stresses than ripples. They are the result of the boundary interaction with the total flow depth. Their wave length is assumed to be proportional to water depth. Yalin (1972) argues that

$$\lambda = 2\pi y_o \tag{5.61}$$

whereas according to Allen (1970) in deep flows (y_o > 10 m) a better fit is given by

$$\lambda = 1.16\ y_o^{1.55} \tag{5.62}$$

The variation of steepness of dunes in terms of θ/θ_c is shown in Fig. 5.12. Yalin and Karahan (1979) also investigated $\eta/\lambda = \theta/\theta_c$, y_o/d) and concluded that $\eta/\lambda\delta \rightarrow f(\theta/\theta_c)$ for $y_o/d > 100$. For y_o/d < 100 the results were expressed by

$$\frac{\delta}{\delta_{max}} = \xi \exp(1 - \xi) \tag{5.63}$$

where $\xi = X/\overline{X}$, $X = \theta/\theta_c - 1$ and $\overline{X}$ is the X-value for $\delta = \delta_{max}$. The function for δ_{max} and X was defined by four points:

$20 \leq y_o/d \leq 30$	$\delta_{max} = 0.0095$	$\overline{X} = 2.03$
$40 \leq y_o/d \leq 50$	$\delta_{max} = 0.018$	$\overline{X} = 3.85$
$65 \leq y_o/d \leq 75$	$\delta_{max} = 0.027$	$\overline{X} = 5.78$
$100 \leq y_o/d$	$\delta_{max} = 0.06$	$\overline{X} = 12.84$

A similar expression for dune steepness was also proposed by Fredsøe (1975).

$$\frac{\eta}{\lambda} = \frac{1}{8.4}\left[1 - \frac{0.06}{\theta} - 0.4\ \theta\right]^2 \tag{5.64}$$

Führböter (1979) derived for equilibrium conditions dune height

$$\frac{\eta}{y_o} = \frac{2}{2n + 1} \tag{5.65}$$

where n is the exponent in a relationship for sediment transport q_s as a function of mean velocity, $q_s \propto U^n$, where n appears to be within 3 < n < 6. Orgis (1974) derived for the dune height

$$\frac{\eta}{y_o} = \frac{1}{\nabla}\left[\frac{y_c - y_o}{3y_c - 2y_o}\right](1 - Fr^2) = \frac{1}{\nabla}\left[\frac{y_o^{2/3} - y_c^{2/3}}{3y_o^{2/3} - 2y_c^{2/3}}\right](1 - Fr^2) \tag{5.66}$$

with the maximum value given by

$$\frac{\eta_{max}}{y_o} = \frac{1}{1 - \nabla}\left[1 - \frac{Fr^2}{2} - \frac{3}{2}\sqrt[3]{Fr^2}\right] \tag{5.67}$$

where ∇ is a dune parameter equal to $^1/_2$ for a triangular and $^2/_3$ for a parabolic form, y_c is the critical depth and $Fr^2 = U^2/gy_o$. The translation of the dune form was expressed as

$$\frac{cd}{\nu} = Ad^{2/3}\frac{Fr^3}{1 - Fr^2} \tag{5.68}$$

where c is the celerity of the wave form and $A = 0.4 \times 10^6$ for the minimum, $A = 1.7 \times 10^6$ for the mean and $A = 5.1 \times 10^6$ for the maximum celerity of the dune form. This relationship agrees well with the data presented by Kondratev (1962).

Yalin (1972) plotted data on length of bed features as $\lambda/y_o = f(y_o/d, Re_*)$. The result is schematically shown in Fig. 5.14. On this diagram the ripple data cluster about a line $\lambda/d \sim 1000$ and show little dependence on Reynolds number. Yalin concludes that ripple data is limited to $Re_* < 24$.

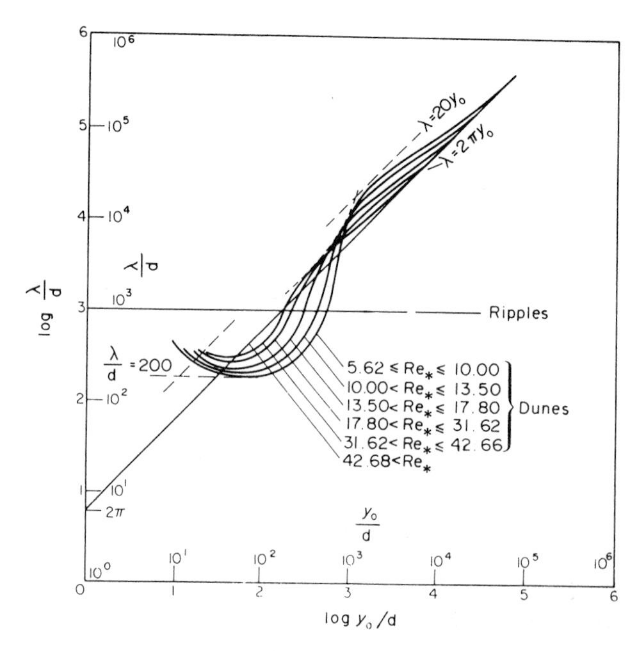

Fig. 5.14. Yalin's plot of λ/d versus y_o/d with Re_* as parameter.

The dune data tend to form a family of curves with Re_* as the parameter. At high values of λ/d all these curves become asymptotic to $\lambda = 2\pi y_o$. At $Re_* > 30$ the points scatter around the line $\lambda = 2\pi y_o$ and at $Re_* > 43$ this straight line is a good fit to the points. Although plots of this kind are deceptive because a scatter of points which covers an order of magnitude (tenfold) appears as a fairly narrow band, they are valuable as indicators of trend. It is seen that when y_o/d drops below a certain value

($\sim$ 700 for laboratory data used) λ/d becomes less than 1000 and has a lower limit of the order of 200. Thus $\lambda_d/\lambda_r \sim (200/1000)$ d_d/d_r and unless $d_d > 5d$, the dune length becomes less that that of the ripple. Here d_d and d_r refer to the particle size in bed covered by dunes and ripples, respectively. At large values of y_o/d the dune lengths are about a thousand times greater than the ripple lengths and both ripples and dunes could be present at sufficiently small values of Re_*. At small values of y_o/d $< \sim 700$ dunes or ripples could be present depending on the value of Re_*, but not both. Put in other words, a bed feature of a given size could be either a dune or a ripple depending on the value of Re_*.

Dunes at small values of Re_* and large y_o/d appear to touch a tangent of the order of $\lambda = 20\ y_o$. These dunes must be very flat and Yalin calls these the sand bars.

The overlap of the ripples and dunes can also be displayed on the Shields' diagram (Fig. 5.15). For a discussion reference is made to Jensen (1970).

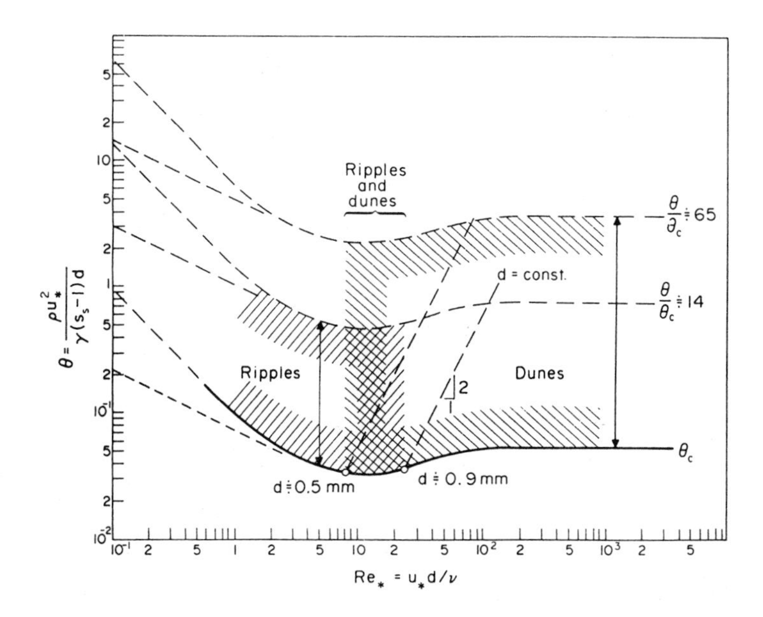

Fig. 5.15. Ripple and dune domains on the Shields' plane.

As seen from Fig. 5.12 bed features first increase in size and steepness and then flatten with increasing bed shear stress, i.e., their height decreases while length remains essentially constant. The wave length grows during the initial stages of development and again when the dune height becomes very low. At these low dune heights an increase in bed shear stress rapidly leads to a flat bed, the so called *transition flat bed* condition.

The great interest in the geometry of bed features is due to their effect on flow. They are roughness elements. The problem is further complicated by the observation that the dune surface can be covered by ripples, for example, when the flow decreases rapidly and leaves residual dunes. The flow over the dunes could then generate ripples. In nature flow rates and flow depths vary continuously and the bed features take time to develop. A small ripple pattern may stabilise in 10 minutes whereas a "mega-ripple" or a large dune pattern may take several days to reach equilibrium. Consequently, with changing flow new regimes can develop on relics of bed features. The flat bed at threshold of motion is essentially a conceptual idea which can be established in laboratory but hardly exist to a significant extent in nature.

Van Rijn (1984) analysed experimental dune data from 84 flume experiments covering the grain sizes $0.19 \leq d_{50} \leq 2.3$ mm and some field data ($0.49 \leq d_{50} \leq 3.6$ mm). The dune height was related to the dimensionless grain size D_*, and a transport parameter (eqn 5.54). From dimensional reasoning van Rijn expressed both η/y_o and η/λ as $f(d_{50}/y_o, D_*, T)$. The best fit function obtained for dune heights was

$$\frac{\eta}{y_o} = 0.11 \left[\frac{d_{50}}{y_o}\right]^{0.3} [1 - e^{-0.5T}][25 - T] \qquad (5.69)$$

and dune steepness

$$\frac{\eta}{\lambda} = 0.015 \left[\frac{d_{50}}{y_o}\right]^{0.3} [1 - e^{-0.5T}][25 - T] \qquad (5.70)$$

Flat bed conditions were assumed for both $T \leq 0$ and $T \leq 25$. The results are shown in Fig. 5.16a and b. The scatter is seen to be of the order of ± 200%. Part of the scatter could be due to lumping data over too wide a grain size range. Fine sands are essentially uniform in grain-size distribution but grains up to 3.6 mm occur generally only in mixtures. The shape and size is also a function of the grain size distribution.

Wijbenga and Klassen (1981) and Klassen et al. (1986) have reported on studies of bed form dimensions for unsteady flow conditions. These show a transition time for dune height in laboratory flume from flat bed to equilibrium and vice versa of the order of 1 hour, and about 2 hours for dune length. From the measurements on lower Rhine van Urk (1982) reports an about four day time lag in dune height development. This has an appreciable effect on water levels. The time lag for dune lengths was considerably longer than for dune heights.

A method of calculation of the dune shape was proposed by Fredsøe (1982). The method incorporates a sediment transport relationship and a shear stress variation from point of reattachment of flow (see Fig. 3.9) to crest of dune. The latter is based on data downstream a negative step and his own measurements on a triangular bed form downstream a step. A function is fitted to yield the variation of $\theta^*/\theta^*_{crest}$ in terms of dune height, flow depth and distance, where $\theta^* = \theta - \mu dh/dr$ is an adjusted Shields

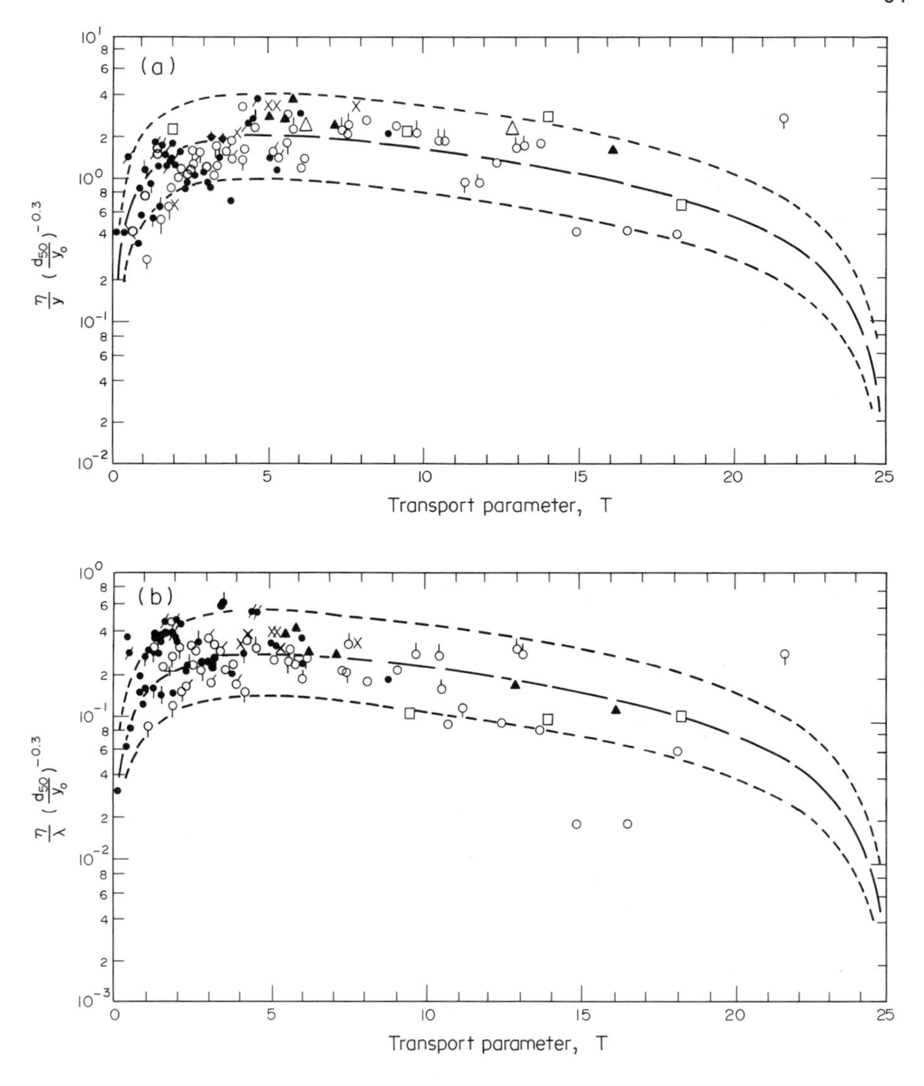

Fig. 5.16. (a) Height of bed features as a function of particle size, flow depth and transport stage.
(b) Steepness of bed features as a function of particle size, flow depth and transport stage.

parameter accounting for the adverse bed slope on the dune and $\mu \cong 0.1$ is a factor. The stress variation used differs from that measured by Raudkivi (1963) over a dune (Fig. 3.9). However, the method is capable (with any stress distribution) of predicting η/λ and η/y_o as a function of θ' as well as the form of the dune.

The functional forms of steepness and height relationships are shown in Fig. 5.17. The infinitely coarse sediment means that the transport is as bed load only with increasing amount of suspension as η/λ and η/y_o decrease.

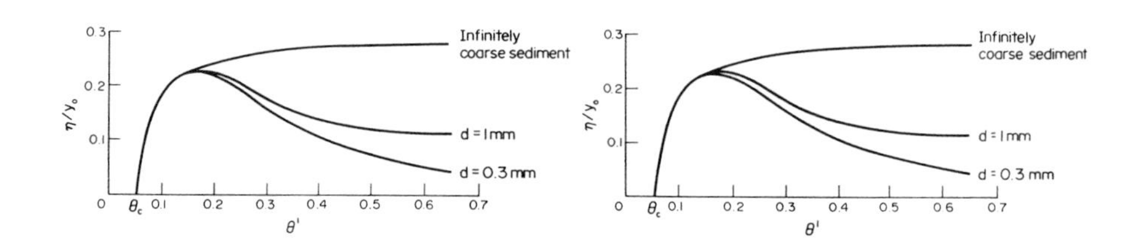

Fig. 5.17. Variation in dune height/water depth ratio with bed shear stress for different grain sizes.

An intriguing feature is the oblique dunes observed in some relatively straight reaches of rivers with a transverse slope of the river bed. Engelund (1974) suggests that these oblique dunes are the result of the varying depth and discharge across the cross-section. Figure 5.18 shows the bottom streamlines mapped on a plaster of paris model of an idealized oblique dune. It is seen that the velocities near the sand surface will have a net uphill component which will carry bed load as well as suspended particles towards the shallow side. Streamlines in the water surface are deflected away from the dune. The oblique dune conditions were created and maintained in a laboratory flume. However, experiments with fine sand and ripple conditions showed no tendency to this obliquity.

Rivers meander and the curvature or bend effect leads to secondary currents and sloping bed surfaces which are frequently covered by oblique dunes. Engelund (1973,1974) discussed the motion of a particle on a sloping plane subject to a forward force proportional to $\tau_o d^2$ and a weight component down the slope $(S_s - 1)$ $d^3 \sin\alpha$, leading to a resultant at angle β to the forward direction. He concluded that $\tan\beta = 0.04 \sin\alpha/\theta$. The object of the discussion was to show that a sloping bed can be maintained with the aid of oblique dunes. Allen (1968) demonstrated with his plaster models that the lee eddy of such an oblique dune has a net uphill component which carries sediment up the slope, Fig. 5.18. Engelund's experimental results showed a curved dune face, but some of this would be due to the boundary layers from side walls. The deflection of the current over the oblique step leads to a curvature in flow with the associated centripedal accelerations and the secondary currents observed. The angle of the dunes observed was of the order of 35°.

The sloping bed itself implies oblique dunes because, as was shown by Exner (1931), the speed of translation of a bed feature

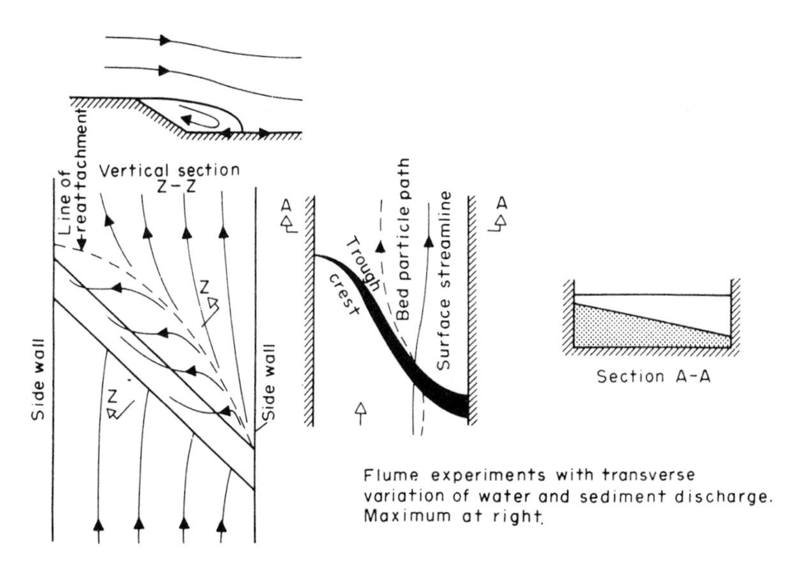

Fig. 5.18. Bottom streamlines mapped on plaster model of (idealized) oblique dune. (Engelund, 1974).

$$c = \frac{q_s}{\eta} \tag{5.71}$$

is inversely proportional to its height, η, where q_s is the rate of sediment transport over its crest. For a given slope the bed shear stress varies directly with depth and dune height increases with bed shear stress, except when the transition flat bed conditions are approached. Thus, if $\tau_o \propto y_o \propto \eta$ and $q_s \propto \tau_o^{3/2}$ the speed $c \propto y_o^{\frac{1}{2}}$ and the dune moves faster on the upslope side leading to an oblique crest.

With increasing bed shear stress, the dunes become longer and flatter, leading to what is, at times, described as an *undulating bed*, and this gives place to *transition flat bed*. At still higher shear stresses, *antidunes* develop. *Antidunes* are bed features which are uniquely associated with free surface waves, and do not form in air or in closed conduits. Antidunes are more symmetrical than dunes, except under the extreme conditions where the surface waves break, and the form of antidunes can translate both upstream and downstream. Engelund (1966) suggested the name of sinusbed.

Whenever the Froude number, Fr, is greater than half, a small discontinuity in the bed, or a change in bed roughness, will cause stationary surface waves in a channel with otherwise smooth walls. The amplitude of these waves increases as the Froude number approaches unity and decreases thereafter but waves can still be observed at about Fr = 1.25. In a fixed flat-bed channel the amplitude of these waves decreases with distance downstream, but if the bed can conform a steady-state condition can be created. The variation of the wave length with Froude number is shown in Fig. 5.19.

It should be noted that the data on bed features are essentially

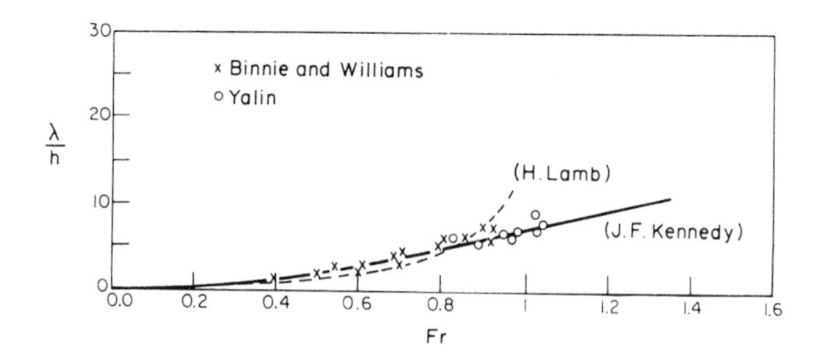

Fig. 5.19. Variation of wave length with Froude number. (Binnie and Williams, 1966).

confined to uniform bed material. However, dunes formed from a uniform gravel are very different from those on a bed composed of a broad grading of sediment sizes, e.g. $\sigma_g = 3.5$. Also where flood flows bring in heavy loads of colloidal suspended sediment the bed features are affected. Their development is delayed, they are lower and the transition flat bed appears sooner than in a flow without suspended clay.

5.2.1 Analytical models

The bed features have been the centre of attention for researchers for a long time. References appear already in ancient literature but Du Buat (1786) is probably the first technical reference, followed by Hübbe (1861), Darwin (1883), Deakon (1894), Blasius (1912), Gilbert (1914), etc. The literature on purely analytical models could be dated to the work by Exner (1920). In the following only a very brief summary of the analytical models is presented. For detailed reviews reference is made to Kennedy (1969), Reynolds (1976), Richards (1980) and Engelund and Fredsøe, (1982).

None of the models explain the cause of the bed forms. However, a certain amount of success has been achieved with description of the geometry and movement. The modelling of dunes and antidunes has been more successful than that of ripples.

The models may be subdivided into five types: models based on classical mechanics, the Helmholtz instability model, kinematic models, synergetic models and statistical models. Statistical because wave lengths and amplitudes of ripples and dunes can be defined in statistical sense only. Truly regular bed features in unidirectional flow, even in laboratory, are extremely rare.

Models based on classical mechanics. The studies by Exner (1920, 1925, 1931) are based on concepts of classical mechanics and mark the beginning of the current era of modelling. He started from the assumption that the capacity of the flow to transport sediment depends on its velocity. Then, when the flow is saturated with sediment, that is, when it is carrying the maximum amount of sediment for the given velocity, an acceleration of flow implies erosion and a deceleration implies deposition. Thus scour and deposition depend on $\partial u/\partial x$. If y and η denote the water surface and sand bed ordinates respectively, both measured above a common datum, then $\partial\eta/\partial t$ will represent the rate of scour, leading to Exner's first equation

$$\frac{\partial\eta}{\partial t} = - K \frac{\partial u}{\partial x} \tag{5.72}$$

where u is the velocity of flow near the bed and is a function of the distance x along the bed. K is the "erosion coefficient", a factor relating sediment discharge to flow velocity.

The continuity requirement with u = U yields

$$uB(y - \eta) = Q = \text{const.} \tag{5.73}$$

where B is the width. If b and y are assumed constant then eqns (5.72) and (5.73) yield

$$\frac{\partial\eta}{\partial t} + \frac{Kq}{(y - \eta)^2} \frac{\partial\eta}{\partial x} = 0 \tag{5.74}$$

in which K, the discharge per unit width q, and y are constants. The bed at time t = 0 is assumed to be given by a cosine form of wave length λ and amplitude a

$$\eta = a_o + a \cos \frac{2\pi x}{\lambda} \tag{5.75}$$

and the bed form at any time t is then given by

$$\eta = a_o + a \cos \frac{2\pi}{\lambda} \left[x - \frac{qK}{(y - \eta)^2} t \right] \tag{5.76}$$

The wave amplitude a remains constant and the wave propagates downstream at a celerity given by

$$c_r = \frac{qK}{(y - \eta)^2} \tag{5.77}$$

This expression shows that the crest travels faster than the trough and consequently the sinusoidal bed will transform into an asymmetric wave with a gentle upstream slope. The theoretical downstream slope will develop with time an overhanging portion which in nature is impossible.

The analysis was extended to variable width, and to flow with friction at constant width and variable width. The theory does not explain the initial sinusoidal bed form. Exner's results are summarized in English by Leliawsky (1955).

The Exner-type model was developed further by Ertel (1966) and Führböter (1967,1969). Führböter showed that for steady state conditions and bed load transport only, the speed of propagation

$$c = q_{max}/\eta \tag{5.78}$$

where q_{max} is the transport rate over the crest in terms of volume of deposit, i.e. divided by (1-n), and assumed to be a f(U_{max}) of of the form $q_{s\ max} = aU^n_{max}$. This led to eqn 5.65. Note that low rates of sediment transport mean low heights of bed features.

Velikanov (1936,1955,1958) and his co-workers introduced the idea of turbulence into the Exner type model. From studies of turbulent open-channel flow they found that large-scale periodic eddies occur in the flow with transverse dimensions comparable with the depth of flow. Similarly, measurable variations in the suspended sediment concentration were found along the stream. It was concluded that these large-scale perturbations possess the greatest amount of fluctuating energy and are directly related to the formation of ridge-type bottom deposits. Velikanov put forward a mathematical model showing that turbulence could cause erosion and deposition along the bed. Velikanov started from Exner's treatment, which did not show how the initially flat sand bed becomes wavy, and proceeded to show that turbulence could lead to this deformation of the flat surface.

Helmholtz instability models. Liu (1957) proposed a pure instability model of the Helmholtz type. He assumed that the bed was a fluid of very much higher viscosity than the water or air flowing over it. In reality a granular bed does not deform as a result of the pressure distribution in the flow. An extension of the Kelvin-Helmholtz instability analysis to sand waves is the work by Moshagen (1984).

Kinematic models. The kinematic instability models are the most extensively developed. Anderson (1953) proposed a potential flow model and obtained a relationship between the Froude number and depth-wave length ratio. However, the major contribution is due to Kennedy (1961,1963,1969). A two-dimensional potential flow with an erodible boundary is given a sinusoidal perturbation in the boundary, and the equations are combined with a transport equation of the power law form. A quantity δ is introduced which is supposed to be the distance by which the local sediment transport rate lags behind the local velocity at the mean bed level. The equations are linearized and lead to expressions for two limiting cases. One is the case where the lag distance δ is a phase shift between the local flow properties and the bed waves, and is taken to be a multiple of the wavelength, i.e. $\delta = a\lambda$. This leads to

$$Fr^2 = \frac{\cosh^2 ky_o}{ky_o(\sinh 2ky_o + ky_o)} \tag{5.79}$$

where $k = 2\pi/\lambda$.

The other limiting condition is that where the transport lag distance δ is prominent and independent of the wave length. In this case δ is expressed in terms of flow depth as $\delta = jy_o$. The resulting equation for the dominant values of ky_o is

$$Fr^2 = \frac{U^2}{gy_o} = \frac{1 + ky_o \tanh ky_o + jky_o \cot jky_o}{(ky_o)^2 + (2 + jky_o \cot jky_o)ky_o \tanh ky_o} \tag{5.80}$$

These relationships are shown in Figs. 5.20 and 5.21. The area

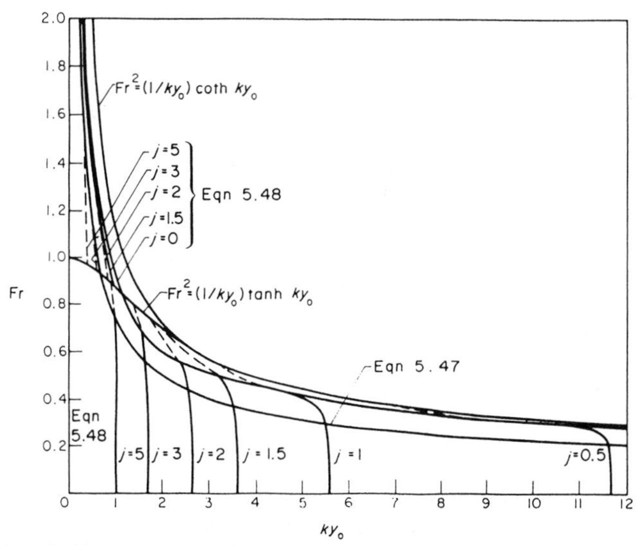

Fig. 5.20. Dominant ky_O given by eqns. (5.79) and (5.80) and regions of occurrence of various bed forms. (The various configurations are idendified by the character of the lines representing eqn. (5.80), except $j \to 0$, as follows. $Fr^2 ky_O$ < $\tanh ky_O$; solid lines correspond to ripples and dunes, dashed lines to transition. $Fr^2 ky_O$ > $\tanh ky_O$: solid lines correspond to antidunes moving upstream, dashed lines to antidunes moving downstream). (Kennedy, 1969).

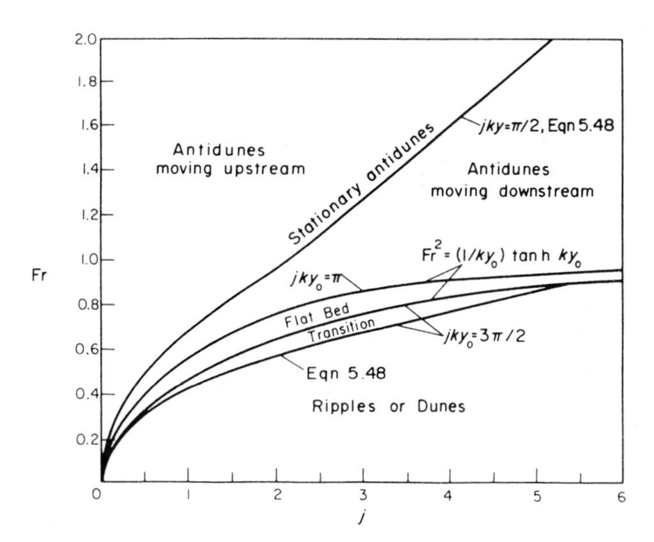

Fig. 5.21. Conditions for occurrence of various bed forms. (Kennedy, 1969.)

between the $F^2 = (1/ky_o)\coth ky_o$ and $F^2 = (1/ky_o)\tanh ky_o$ lines and the joint line to the right in Fig. 5.20 is the region of antidunes, and the area below is the region of dunes (and ripples?). The area above the line $F^2 = (1/ky_o)\coth ky_o$ is more likely not to be part of the solution plane, as far as sand waves are concerned, than the area for ripples or three-dimensional (short-crested) features. Figure 5 by Kennedy (1969) shows that this area is empty of experimental points.

Equation (5.80) is of the form of

$$\lambda/y_o = f(Fr, j) \tag{5.81}$$

and from Fig. 5.20 we see that for large values of Fr the value of λ/y_o (= $2\pi/ky_o$) becomes independent of j. When Fr becomes sufficiently small then λ/y_o becomes a function of j only. Kennedy reasoned that the analysis applies to ripples but that the ripple wave length is proportional to the particle size and does not depend on the depth of flow.

Although Kennedy (1969) wrote that "The evidence now appears overwhelming that δ is not merely an artifice introduced to achieve the desired prediction, but a real physical entity", the evidence for this is lacking. At lower transport rates the particles move intermittently and a physcial meaning of "transport relaxation time" is hard to conceive. As an analytical artifice the j is very effective. In general, the analysis is very successful in describing the observed data. Very similar results were obtained by Shirasuna (1973). He assumed that the initial disturbance of the flat bed is caused by turbulence (as did Velikanov) and derived a relationship between the space correlation function of bed undulation and the time - space correlation of turbulence. The stability of the undulation was then discussed as a two-layer potential flow of different densities (cf. Helmholtz instability models). The sand wave travels at a much lower velocity than the fluid. Between these is a fluidized sand layer which flows as a fluid. The sand waves were treated as internal waves and their stability was analysed for open channel and pipe flow.

Engelund (1966) related this "delay distance" to the phase shift between surface and bed waves. The phase-lag problem was also analysed by Hayashi (1970). In addition Engelund made a significant extension by developing a model which includes the shear stress on the bed. Varying shear stress along the bed form was also considered by Raudkivi (1966). Engelund (1970) assumed that the stability of an erodible bed depends on the phase shift between the bed form and the local rate of sediment transport. This phase shift arises from a phase shift between the bed shear and the bed form and from the variation of the rate of deposition of suspended particles. The bed load is assumed to respond immediately to shear stress and this leads to the instability of the bed at small Froude numbers. The model was formulated in terms of vorticity transport, using a constant eddy viscosity ε, and includes sediment concentration. It was deduced that if all the sediment moves in suspension the stability boundaries are as shown in Fig. 5.22. When bed load dominates, stable-unstable regions reverse. The usual situation of the sediment transport leads in both modes to stability boundaries which vary

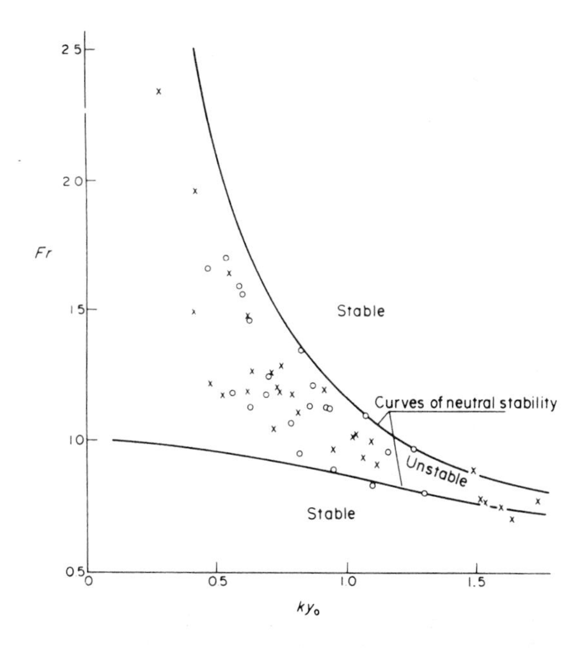

Fig. 5.22. Asymptotic stability boundaries. Experiments by Guy et al. (1966): O,d = 0.19 - 0.47 mm; and by Kennedy (1961): X,d = 0.23 - 0.55 mm. (Engelund, 1970).

with grain size, see Fig. 5.23. The diagrams are for the same depth of flow, but the amount of suspended sediment is larger with fine sediment at a given flow. The dotted lines refer to the largest growth rate of perturbations. The author concludes that the transition between the dunes and the plane bed is a function of the ratio of suspended to bed load.

The influence of gravity on the local bed load transport rate and the growth of the dunes was evaluated by Engelund and Fredsøe (1974) by using a second-order approximation. The results are summarized in Fig. 5.24. The transition between dunes and the transition flat bed varies with the mean fall diameter of the grain. For grain size stand. dev. $\sigma = 0.4$ and $d_m > 0.5$ mm, Fr $\cong 0.85$ is constant. Dunes form at Fr < 0.85. For $d_m = 0.2$ mm, Fr $\cong 0.53$, $d_m = 0.3$ mm, Fr $\cong 0.6$ and $d_m = 0.45$ mm. Fr $\cong 0.84$. These Fr values are slightly higher for $\sigma = 0.35$ and lower for $\sigma = 0.45$, e.g. at $d_m = 0.3$ mm, the change in Fr is $\pm$ 0.03, respectively. Figure 5.24 also offers an explanation to the increase in wave length of dunes as temperature decreases. A temperature drop increases the kinematic viscosity of water which in turn increases the number of suspended particles.

All the above analyses were essentially related to dunes and indicate the conditions of maximum growth rate of a bed disturbance.

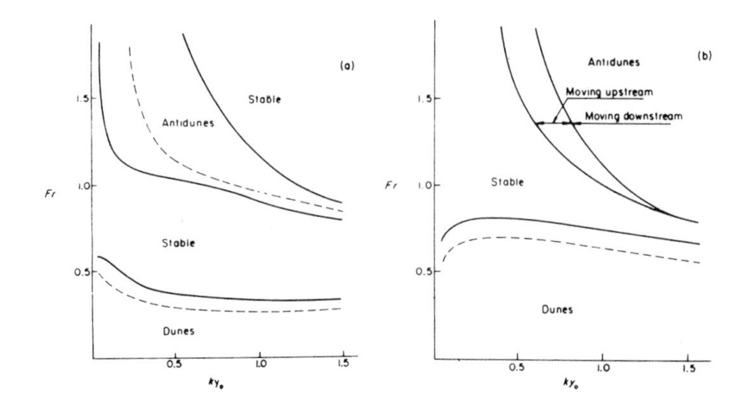

Fig. 5.23. Stability diagram for the complete solution. The parameters are: (a) $U/u_* = 21$ and $u_*/(wFr) = 2$, and (b) $U/u_* = 21$ and $u_*/(wFr) = 1$. (Engelund, 1970).

Richards (1980) included the effective bed roughness together with a one-equation turbulence model in his stability analysis and predicted that the bed is also unstable at low shear stresses, i.e. there is not one but two local maxima of growth rate of a disturbance, Fig. 5.25. The analysis showed that the dune maximum was strongly dependent on flow depth whereas the ripple maximum was independent.

The instability analyses indicate the wave length at which an initial perturbation will grow under given conditions, not the equilibrium form or wave length. The theory also indicates a given wave length but in laboratory or nature the wave length can only be expressed in statistical terms.

Raudkivi (1981,1983) examined the argument that ripples are associated with bed load transport and dunes with suspended transport by using the empirical rule that

$$w/u_* < 0.85 \quad \text{- suspension}$$

$$0.6 < w/u_* < 2 \quad \text{- saltation}$$

$$2 < w/u_* \quad \text{- bed load}$$

where w is the fall velocity of grains. Figure 5.12, for example, indicates a maximum ripple steepness when $\theta/\theta_c = 5$. For this condition w/u_* varies form 0.24 to 1.63 for 0.1 to 0.6 mm sand, respectively, and clearly does not satisfy the proposition.

The reason for the two kinds of bed features is most probably associated with the velocity distribution of flow. The velocity distribution over an initially flat bed could be approximated by two zones, Fig. 5.26, one of constant shear (du/dy = const.) and variable velocity and the other of constant velocity and zero shear. In the constant shear zone the momentum of flow increases

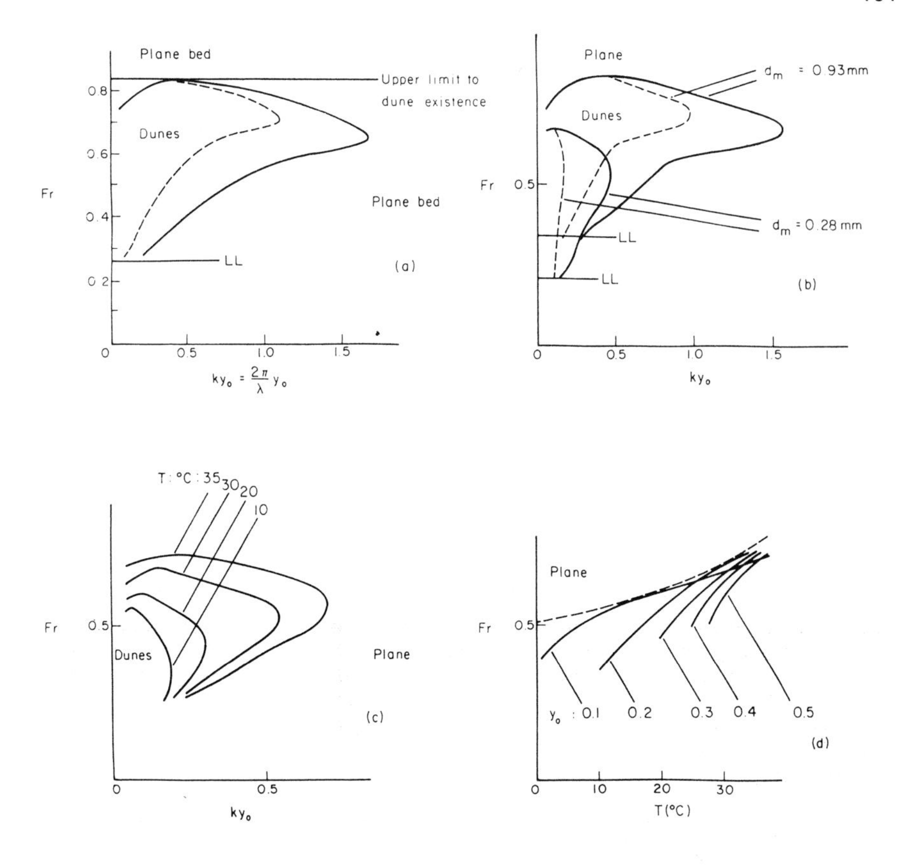

Fig. 5.24. (a) Stability limits to the formation of dunes, y_o = 0.2 m, d_m = 0.45 mm with σ = 0.4, T = 20°C. LL indicates the lower limit of bed-load transport. The dotted curve corresponds to maximum amplification. (b) Variation of the dune instability region with the mean sediment size d_m; y_o = 0.2 m, σ = 0.4, T = 20°C. (c) Variation of the instability region with temperature, y_o = 0.1 m, d_m = 0.2 mm, σ = 0.3. (d) Variation with temperature of the upper limit to existence of dunes at given values of ky_o. The dotted line indicates the absolute upper limit to existence of dunes. Flow conditions are as for (c). (Engelund and Fredsøe, 1974).

rapidly with distance from bed and consequently the development of a disturbance at the boundary is retarded. A disturbance in the boundary could set up an oscillation of streamlines in the constant shear zone. The hypothesis is that ripples are bed features associated with flow oscillations confined to the constant shear zone. When the disturbance is too large to be

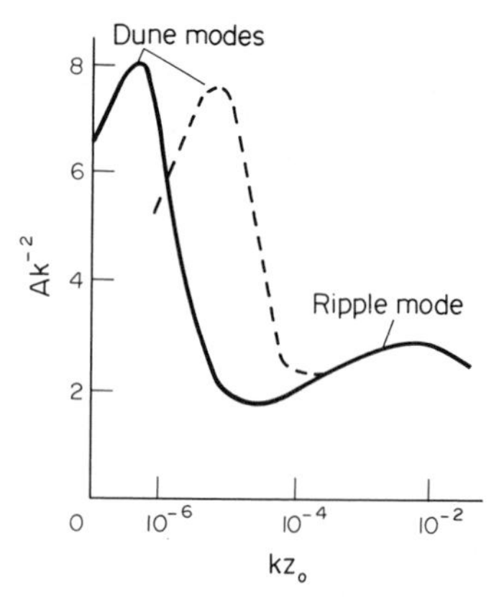

Fig. 5.25. Variation in amplification factor A with wave number k, after Richards (1980) z_o = bed roughness, —— $y_o/z_o = 10^5$, --- $y_o/z_o = 10^4$.

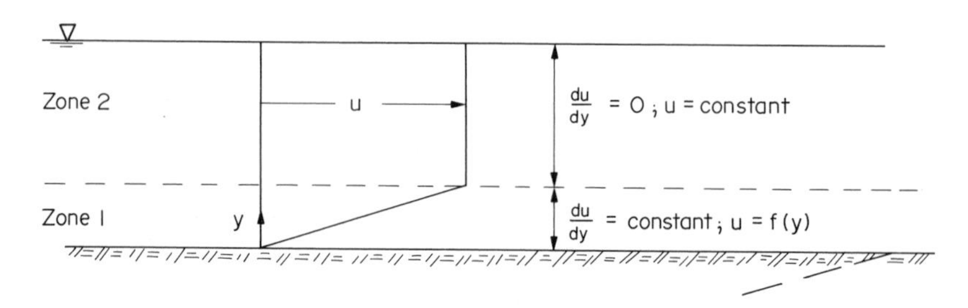

Fig. 5.26. Diagrammatic presentation of velocity distribution.

confined to the constant shear zone the constant velocity zone will be brought into action, i.e. the depth of flow enters into the oscillatory picture leading to dunes. Clearly the natural frequency characteristics of the constant shear and constant velocity layers are quite different.

Furthermore, the interaction between flow and the boundary involves energy transfer. If one looks at the flow and boundary as an oscillating system which reaches through energy transfer a state of equilibrium, then it is also known that the energy transfer is generally not restricted to the "resonant" frequency (wave length). For example, the studies of interaction of flow with an oscillating cylinder (Raudkivi and Small, 1974) showed that the energy transfer occurred over the range $0.6 < f_r < 1.8$

for $3 \times 10^4 < Re < 8 \times 10^4$, where f_r is the ratio of natural frequency of the cylinder to the shedding frequencies of vortices. Since the two zones of flow have their own different "natural" frequencies, one could visualise an energy transfer as illustrated diagrammatically in Fig. 5.27. The main feature of this argument is that it allows the co-existence of both the deterministic and random features as are observed in nature.

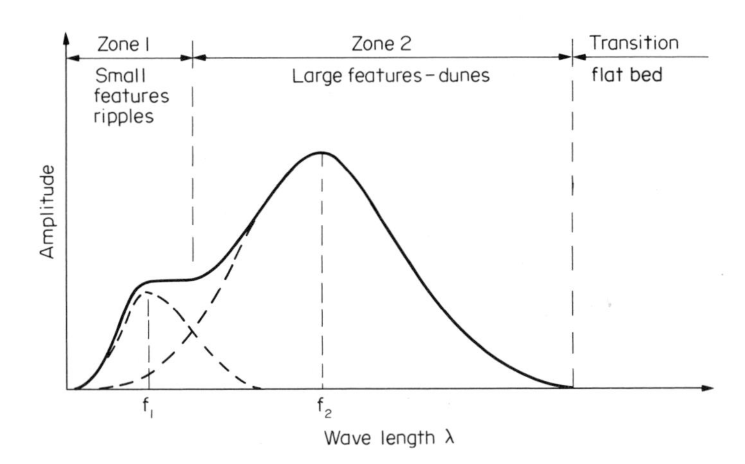

Fig. 5.27. Diagrammatic presentation of distribution of wave lengths of bed features.

In general terms, this means that the total freedom of development of the bed features (randomness) is curtailed and that a degree of order is imposed (the deterministic part). This kind of a compromise is a feature of many natural phenomena as well as social systems. A well-known analogy is the entropy from thermodynamics.

The development of bed features could also be viewed as the reverse of the branching process of the biological sciences. This methodology is known as synergetics. Führböter (1983) showed that since the speed of translation of bed features is inversely proportional to their height, eqn 5.78, the lower features travel faster and catch up with larger ones. This leads to coalescence and reduction in number. If the sediment transport is as bed features only then the combined height will be

$$h = (h_1^2 + h_2^2)^{\frac{1}{2}}$$

and as a first order approximation

$$u_1 = q_s/h_1 \ , \ u_2 = q_s/h_2 \ , \ u = q_s/(h_1^2 + h_2^2)^{\frac{1}{2}}$$

or

$$u = \frac{1}{\sqrt{1/u_1^2 + 1/u_2^2}}$$

or

$$u(n) = \frac{1}{(1/u_1{}^2 + 1/u_2{}^2 + \dots 1/u_n{}^2)^{\frac{1}{2}}} \tag{5.82}$$

A numerical experiment shows that starting with N random disturbances of bed the pattern of bed features rapidly settles down as illustrated in Fig. 5.28, where a indicates the spacing of the bed features.

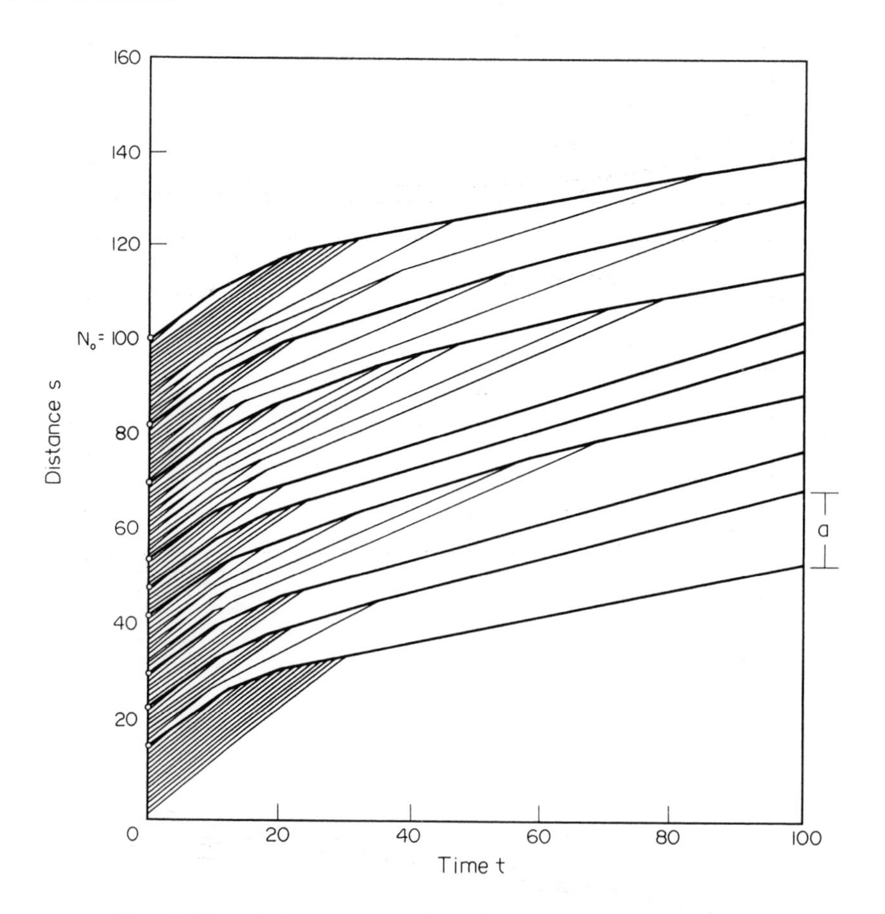

Fig. 5.28. Coalescence of bed features on the distance - time plane.

Figure 5.29 summarizes some of the numerical results for N = 1000 initial randomly distributed bed features when the range of translation velocities $u_{max}/u_{min} = \alpha = 2$ and the number of velocity increments z is large. The development with time shows variations with z only at the beginning, e.g., $z \cong 2$ and t less than 50 units but at t = 500 units the difference between z = 2 and $z = \infty$ is almos undetectable. N = 1000 at t = 500 leads to about 30 ± 5.

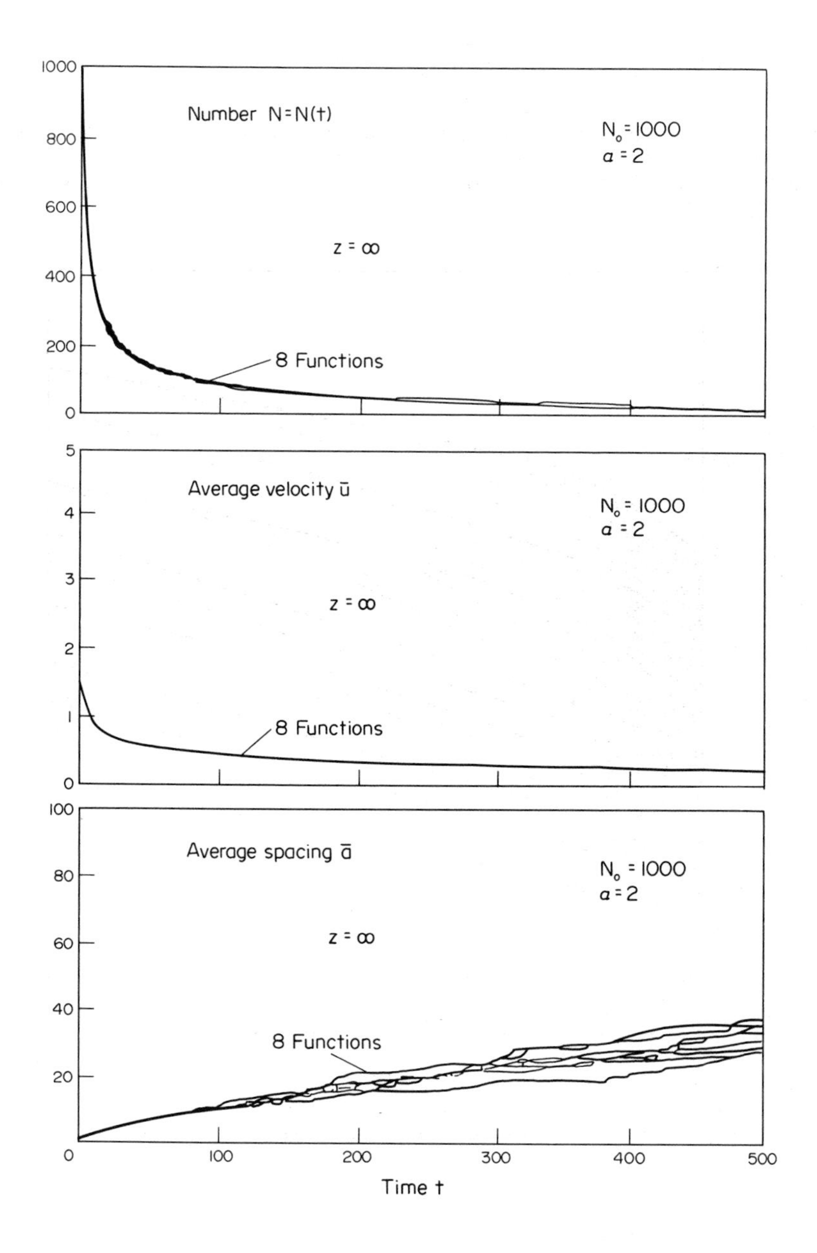

Fig. 5.29. The variations with time of number of bed features, their average velocity and spacing starting with 1000 randomly distributed bed features.

This type of semi-random order or ordered chaos is also in keeping with the ideas put forward above.

Statistical models. The bed features which form are never identical, not even under laboratory conditions, and the same pattern of features cannot be reproduced in the next experiment. Hence, there is a distinct feature of randomness. If so, what are the statistical properties of the bed features? Observations show that a wide range of wave lengths is present, even when the pattern appears to be fairly regular. It is the randomness of the features of the pattern which has prompted the study of these by statistical methods. The question of which of these methods, the statistical or the deterministic, is the better cannot be answered at present, and it appears that neither is the complete answer. Even to a casual observer there is enough regularity in the features to recognize a distinct pattern but when it comes to measurement we have to resort to averages. It seems therefore that the bed features can be described neither by the deterministic analytical models, described before, not by the purely probabilistic models alone. The "final" model may have to be a combination of both.

For the probabilistic models the bed elevation Z is assumed to be the realization of a random process,

$$Z = f(x) \tag{5.83}$$

In this formulation the observer moves at the celerity c of translation of the bed features. The features are two-dimensional and stationary in time. For three-dimensional features $Z = f(x,y)$. The theories concentrate on the definition of this function which would then define the geometry of the bed. Any realization of this function would give a possible bed configuration. All realizations would have the same values of the statistical parameters but could differ appreciably in detail. There is a sizeable amount of literature dealing with random processes and no attempt will be made here to go into the details of the theory.

The random process here is the bed elevation as a function of distance x along the channel, $Z(x)$. At any given location x_i the quantity $Z(x_i)$ is a random variable, part of a family known as a stochastic process. From observed records we can calculate the mean amplitude $\overline{Z}(x_i) = E[Z] = \mu$ (for convenience this is usually set equal to zero) the variance $\sigma^2 = E[(Z - \mu)^2] = E[Z^2] - \mu^2$, the autocovariance function and the autocorrelation function. The autocovariance function for a stationary process with zero mean is related to the spectral density function $S(k)$ of the process $Z(x)$, where $k = 2\pi/\lambda$ is the wave number. It appears from laboratory data that the frequency distributions of bed elevations approximate a truncated Gaussian form.

Instead of the bed elevation Z, the wave lengths of the features could be treated as a stochastic process. A variation on this technique is to consider the zero crossings, that is the intervals formed by the profile crossing the mean bed elevation. Frequency histograms can be prepared of the zero crossing intervals and it appears that a number of these can be fitted by a two-parameter gamma distribution.

By analysing observed records we may plot the autocovariance function versus lag or its Fourier transform, the spectral density function, versus wave number. From the shape of these functions we may attempt to draw conclusions about the form of the function describing the random process. However, it is clear that the statistical parameters which specify this function $f_Z(x)$ must vary with the variation of the bed features. For example, the function must yield $\sigma = 0$ at the threshold of motion and at conditions which yield the transition flat bed. Likewise, because the formation of ripples and dunes is governed by different parameters the form of the function $f_Z(x)$ is also likely to be different for these features. At present little is known about the properties of this random function.

This method of study was initiated by Nordin and Algert (1966) and has since then been added to by a number of researchers, e.g. Nordin (1969), Engelund (1966), Hino (1968), Jain and Kennedy (1971), Ashida and Tanaka (1967), Fukaoka (1968), Annambhotla et al. (1972). Nordin and Algert proposed a second-order Markov model to describe dune profiles. The results have been promising in the high-frequency range. Hino deduced the "-3 power law" for sand waves, but his results are based on dune data only. It can be readily shown that the spectral density function in this context has the dimension of length cubed and if the random wave length is the only influence on S(k) then S(k) must be proportional to k^{-3} as derived by Hino. However, as seen from Fig. 5.30, the -3 slope is realized only for large values of k; that is, for relatively short waves. The bed features analysed in Fig. 5.30 are classified by Nordin and Algert as dunes. The "-3 power law" for the high wave number range by Hino is given by $f(\phi)k^{-3}$, where $f(\phi)$ is a function of the angle of repose of the sand. However, it appears from spectral data that $f(\phi)$ is also a weak function of k and of hydraulic variables. The "-3 power law" in the high wave number range is apparently unaffected by the grain size and gradation of the sediment. Hino also suggested that for frequencies near the spectral peak the slope changes to a "-2 power law". This, however, does not appear to be supported by data. It rather appears that the exponent is a function of the Froude number. At low Froude numbers the slope is essentially constant at approximately -3. It decreases with increasing Froude numbers to a value of 2 when $Fr \to 0.7$ (when surface waves become prominent).

The sand wave spectra do not appear to centre around any dominant wave number, but cover a broad range of wave numbers. There are frequently secondary peaks which indicate second- and higher-order harmonics.

At present, mainly due to lack of suitable data, only vague deductions have been possible. However, the statistical studies hold promise for additional insight into the phenomenon of bed features or resistance to flow depending on the object of study. For study of resistance to flow the lower part of the frequency spectrum is likely to be the significant part, just as in turbulence where the lower frequencies contain most of the energy.

It needs to be understood that when we talk of spectral analysis of sand waves, we are applying techniques of analysis which refer

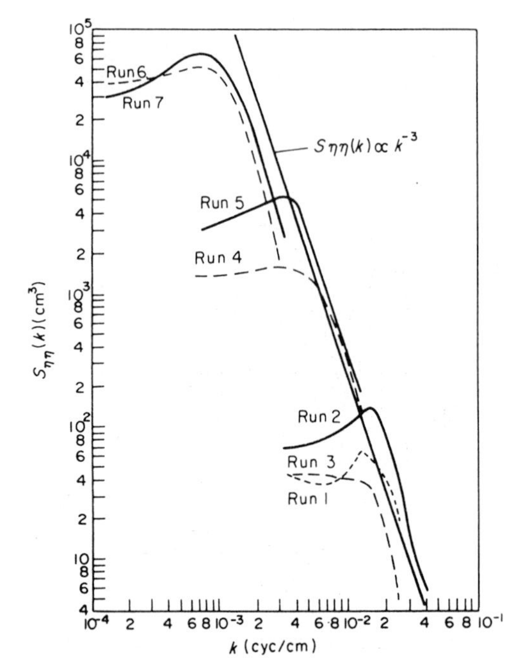

Fig. 5.30. Wave-number spectra and sand waves compared with the "-3 power law". (Curves are re-plotted from the graphs by Nordin and Algert (1966), Hino (1968).)

to continuous media. Unlike deformation waves of a continuous medium, the sand waves do not pass through each other but are merged into the upstream faces of the slower moving larger bed forms.

If a frequency spectrum of the movement of the bed forms is inferred from the wave-number spectrum through use of a relationship between celerity and wave number, c(k) = f(k), then a continuous sand wave medium has been assumed in which each component travels independently at a celerity corresponding to its wave number (Ashida and Tanaka, 1967; Hino, 1968). Ashida and Tanaka arrived at

$$c(k) \propto \sqrt{k}; \quad k = \frac{1}{\lambda} \tag{5.84}$$

Hino used Kennedy's

$$c(k) \propto k \coth 2\pi k y_o$$

and assumed that $c(k) \propto k$, which corresponds to conditions when $y_o >> \lambda$. If $\lambda > y_o$, the above relationship indicates that c(k) is inversely proportional to the depth of flow. One of the major problems facing these methods of analysis is the collection of experimental data suitable for checking of the various statistical theories. For example, consider eqn 5.59. It implies a frequency distribution for four variables. Even a study with Re_* or y_o/d constant involves a family of surfaces as illustrated in Fig. 5.31.

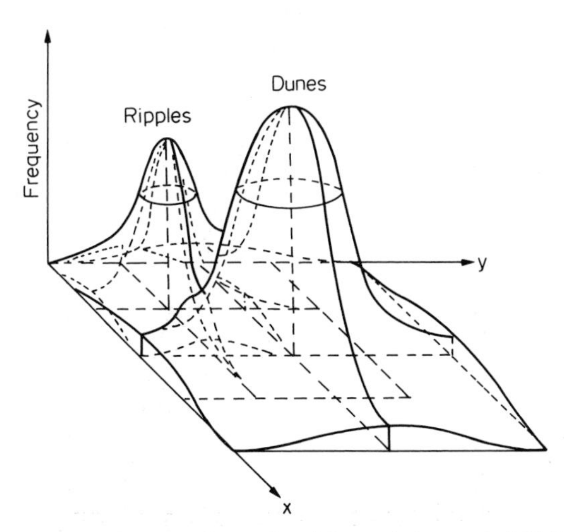

Fig. 5.31. Illustration of a bi-variate probability function.

5.3 Bed Features of Gravel Rivers

The problem of bed features in *gravel rivers* is even more complex than in sandy streams. A major factor for this additional complexity is the difference in the structure of sediment. Sands in general form beds of almost uniform grain size whereas gravel beds usually contain a broad range of grain sizes, at times from fine sand to large boulders. As a consequence there is a preferential transport as well as armouring of the bed surface. The different grain sizes also lead to the applied shear stress excess, $(\tau_o - \tau_c)/\tau_c$ in gravel rivers being less than over sandy beds. The equivalent of transition flat bed condition in sandy rivers, is in gravel streams, if at all, a very rare event. The movement of grains of a gravel bed is much more selective and associated with the formation and breakdown of grain clusters on the surface. In these clusters grains support each other mutually and armour the surface. An indiscriminate transport occurs only at shear stresses greater than that for the limiting armour layer, i.e., at stresses at which no armour layer can reform. In principle then, the complexities arise from the grading of the sediment and relative flow strength. Quite regular two-dimensional bed features can be generated in laboratory with uniform gravels, i.e., $\sigma_g < 1.3$ to 1.5.

Already the formulation of the mutually supporting clusters of grains leads to local three-dimensionality of flow. The cluster protects the surface locally but the bed erodes at the periphery of the cluster until it fails through undermining and the formation of a new cluster begins (Chin, 1985). The armouring process also affects the large scale bed features in gravel-bed rivers which are relatively flat, compared to sand dunes, and are referred to as *bars*. The bars have a multitude of forms and their lengths are usually of the order of channel width or more.

Gravel rivers with well-defined source of debris input tend to develop initially a complex braided pattern which becomes more stable downstream and eventually turns into a "meandering" pattern. In the process the skewness of the size distribution of the bed material decreases. Parker (1976) related the stable configuration to a parameter $E = S_e/\pi(Fry_o/B)$, where S_e is the energy slope, Fr the Froude number, y_o the flow depth and B the width of channel. The channel meanders when $E < 1$ and braids when $E > 1$. The value of E also indicates the complexity of braiding. Gravel rivers also tend to exhibit large scale "waves" in bed slope, much longer than the length of bed features. These probably arise from massive point inputs of debris which are then slowly spread out into long delta-like features on the river profile.

The nomenclature for the features and types of bars in gravel rivers is still a controversial topic and the descriptions have a massive literature in their own right. For more details and references to literature reference is made to Hey et al. (1982) and Richards (1982).

Bars are in principle an accumulation of grains and the flow must be capable of moving grains to the top of the bar. The limiting condition for this is given by the Shields criterion, $\theta_c = \tau_c/(\rho_s - \rho)gd_c \cong 0.05$, where d_{90} refers to a characteristic grain size for which d_{90} is fairly frequently used. Thus, the slope $S \cong 0.08\ d_{90}/y_o$ or if $d_{90}/y_o \cong 1/4$ the limiting slope would be of the order of $S = 0.02$. Characteristically, the slope on the bar tends to flatten and a lee face forms from the coarser material. With time, flow down this lee face may become a mini-rapid, in general the word *riffle* has become the synonym for this type of rapid flow. The term riffle is used in conjunction with a *depositional bar* and pool sequence. Non-depositional rapids formed by bed rock or landslides do not involve gravel deposition.

Numerous forms of bars can develop depending on conditions, three of which are illustrated in Fig. 5.32. The development usually starts with the central bar. The medial bars normally close to one bank and give rise to the diagonal bars. Common are also the point bars which develop at the convex banks of streams. In general, bars attached to the banks are more stable than central banks.

In well defined "straight" channels, such as have been created by river regulation works, a system of alternate banks develops. These are in principle, the beginning of the meandering. Bars which form in the middle of the stream are associated with the braiding tendency of the stream. Jaeggi (1984) proposed for the upper limit of alternate bar development

$$\theta/\theta_c = 2.93\ \ln(\theta_\beta/\theta_c) - 3.13(B/d_m)^{0.15} \qquad (5.85)$$

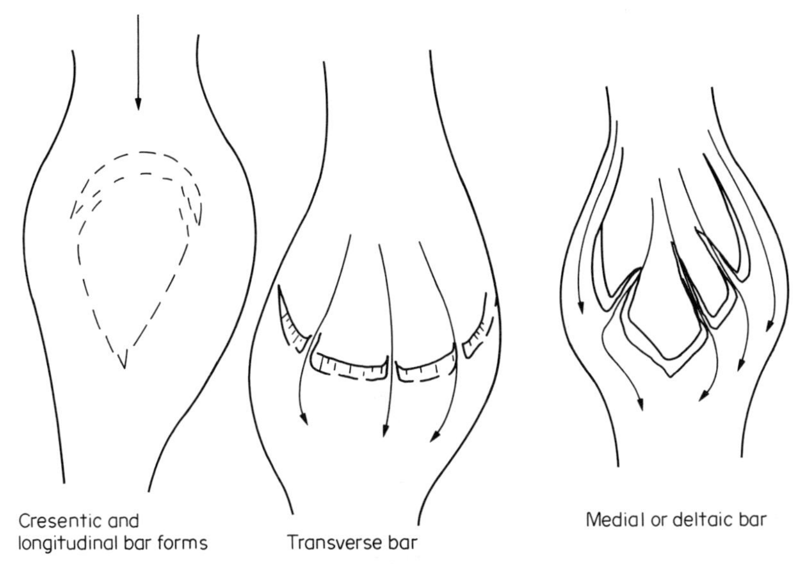

Fig. 5.32. Illustration of bar forms in gravel rivers.

where $\theta_\beta = BS/(S_s - 1)d_m$, B is the width of channel, S is slope and $d_m = \sum_{p=0}^{100\%} \bar{d}_i \Delta p / \sum_{p=0}^{100\%} \Delta p$, $\bar{d}_i = \frac{1}{2}(d_i + d_{i+1})$. The d_m is said to be a governing grain size. The lower limit for alternate bar development is the threshold condition. However, in gravel rivers this is affected by armouring. Jaeggi proposed for

$$\frac{\theta_a}{\theta_c} = \left[\frac{d_a}{d_m}\right]^{0.67} \cong \left[\frac{d_{90}}{d_m}\right]^{0.67} \tag{5.86}$$

Studies by Chin (1985) indicate that the limiting armour layer has a d_{50}-size, d_{50a}, approximately equal to about 0.55 of d_{max} of the grading of parent sediment, as indicated by extrapolation of the grading curve to 100% passing grain size and ignoring large randomly occurring stones. The maximum critical shear stress, τ_{ca}, was found to be about 75% of τ_c given by the Shields criterion for d_{50a}. According to Jaeggi the slope for alternate bar formation

$$S > \frac{\exp[1.07(B/d_m)^{0.15} + M]}{12.9\ B/d_m} \tag{5.87}$$

where $M \cong 0.34$ for uniform material and $M \cong 0.7$ for material with a broad grading. At slopes steeper than 1% the critical value of θ tends to increases and this can reduce the factor 12.9 to about 6. The scour depth associated with alternate bars (below the average bed level) was expressed by

$$y_s = 0.76\ h_{ab} = \frac{B}{6(B/d_m)^{0.15}} \tag{5.88}$$

where h_{ab} is the total height of alternate bars (sum of scour and bar height), for which Kishi (1980) proposed $h_{ab} = 0.05\ B$ for $B/d_m = 19\ 000$. From comparisons with field data Jaeggi replaced B in eqn 5.88 with $B-2my_s$, where 1:m is the slope of the bank. This, however, leads to an iterative solution of eqn 5.88. For resistance in these alternate bar channels

$$\frac{U}{u_*} = 2.5\left[1 - \exp\frac{\alpha y_o/d_{90}}{\sqrt{S}}\right]^{\frac{1}{2}} \ln 12.3\left(\frac{y_o/d_{90}}{\beta}\right) \qquad (5.89)$$

was proposed, where from experimental data $\alpha \cong 0.02$ and $\beta \cong 2$.

Equation 5.87 separates channels with a three-dimensional alternate bar system from channels with a pool-riffle system in which the riffles lie more or less perpendicular to the channel and the pools extend over the width of the channel.

Single channel gravel streams tend to develop diagonal riffles. Material is eroded where the flow converges towards the bank and is deposited in the inflexion (ford or crossover) region where the flow diverges and moves from one bank to the other. The deposition propagates upstream and reduces locally the slope. The leading edge of the deposit, the lee face, forms a riffle with a steeper gradient for the flow. This stretch of flow is frequently called the *chute*, Fig. 5.33. The chute draws some water off the bar and thus encourages deposition. During low flows the riffle sections dissipate most of the flow energy. At high flow stages the gradient through the pools increases and decreases over the riffles. This encourages scouring in pools and protects the riffles, i.e., the riffle-pool sequence has a self stabilizing characteristic. Typical slopes over riffles in "straight" channels are of the order of 1:50 and 1:80 to 1:100 in curved channels. The slopes over pools are of the order of 1:500 to 1:300 with curved channels showing steeper slopes for pools. The wave length of the riffle-pool sequence is usually in the range of $L = 2\pi B$ to 14B, where B is the channel width.

A very interesting study of the development of alternate gravel bars and return to meandering after the river was straightened by river control was presented by Lewin (1976).

5.4 Armouring

Most alluvial sediments, with the exception of fine sands, have a broad grain size distribution. The geometric standard deviation of river gravels is normally about 3.5 to 4. The erosion of such beds under certain conditions can lead to the formation of an armour layer on the bed surface, also known as paving, which protects the bed.

The development of a stable armour layer depends on the stability of individual large particles or stones and their numbers in the bed. An individual large particle on the surface tends to scour a hole around its front perimeter. The particle in time slides upstream into this hole. This reduces the particle exposure and

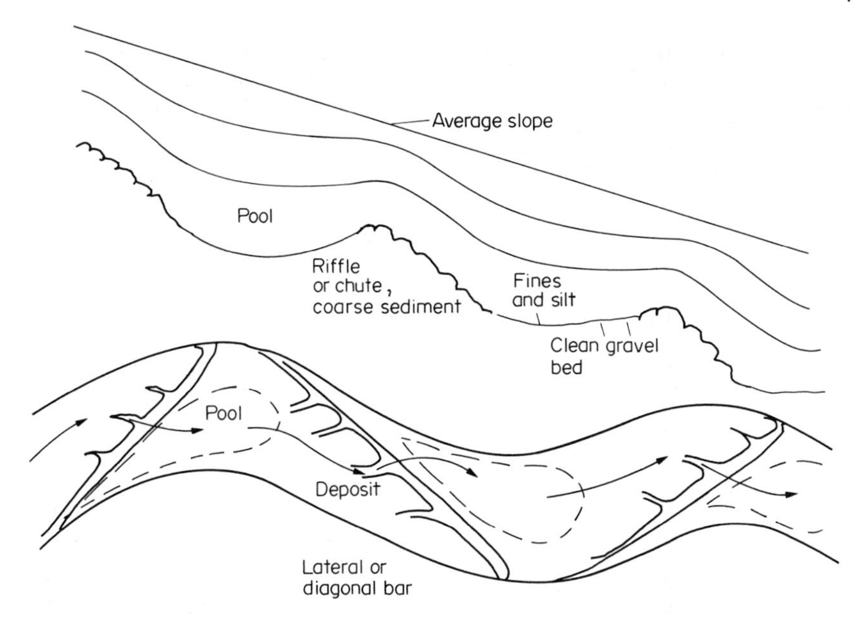

Fig. 5.33. Diagrammatic illustration of bar-riffle-pool sequence in a single channel gravel river.

increases it stability. Other coarser particles also tend to be trapped in the scour hole and gradually a cluster of particles develops. Finer particles are deposited in the shelter of larger ones and in the crevices between the larger stones. All these particles support each other with the large ones acting as anchor stones. By scouring and forming clusters the bed particles equalise their susceptibility to entrainment. The formation of clusters explains why particles of all sizes of the bed material are present in the armour layer. A cluster may fail by erosion of the bed around it, the anchor stone may then move downstream to a new more stable position and the cluster formation is started anew. The tendency for armouring and formation of clusters decreases with increasing uniformity of bed material.

The armouring is a function of the applied bed shear stress. Figure 5.34 illustrates the gradings of the armour layer at different applied shear stresses. As the shear stress is increased a condition will be reached where the surface does not armour any more and all particles are indiscriminately transported by the flow. Up to this limiting shear stress, or critical shear stress of the armour layer, the d_{50} particle size of the armour layer, d_{50a}, increases with shear stress. The development as illustrated in Fig. 5.34 is insensitive to the original grading. The process is controlled by the coarse fraction and the d_{max} size. The grading affects the time scale of development. Initially d_{50a} increases approximately in proportion to shear stress but becomes insensitive to shear stress near the limiting shear. Chin (1985) found that the ratio of d_{max}/d_{50a} approached a lower limit

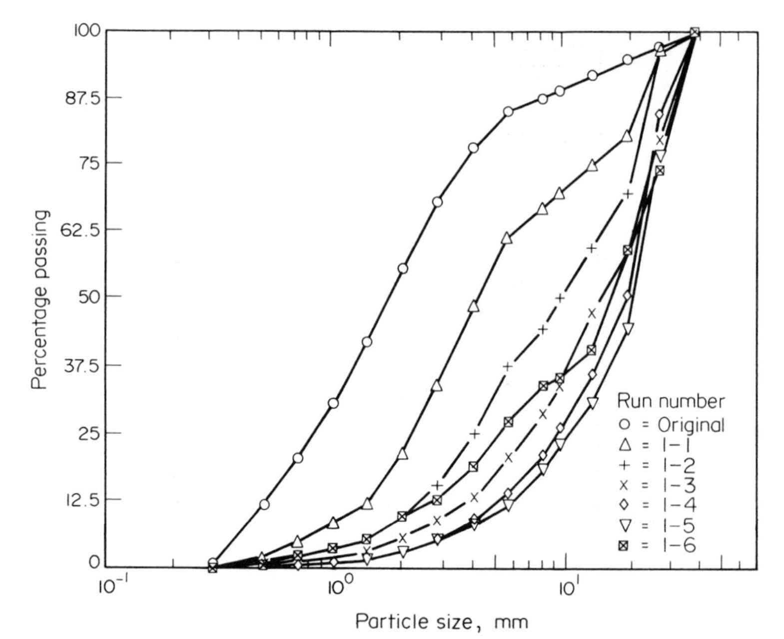

Fig. 5.34. Variation of equilibrium armour layer grain size distribution with increasing shear velocity. The shear velocities were 41, 53, 66, 76, 85 and 95 mm/s for runs 1 to 6, respectively, with d_{50a} of 4.2, 9.5, 14.5, 19.0, 20.5 and 16.0 mm.

$$\frac{d_{max}}{d_{50a}} \cong 1.8 \tag{5.90}$$

This limit is approached asymptotically and is reached when $u_{*a}/u_{*ac} \geq 0.9$ where u_{*ac} is the limiting critical armour layer shear velocity. Equation 5.90 implies a $d_{84}/d_{50} \cong 1.5$ or approximately uniform grading of the coarse fractions. The critical shear θ_{ca} was found to be a function of $(d_{50a})_{max}/d_{50}$. The equation proposed by White and Day (1982) for the stability of individual particle size fractions:

$$\frac{u_{*i}}{u_{*c}} = 0.4(d_i/d_a)^{-\frac{1}{2}} + 0.6 \tag{5.91}$$

where d_a is a characteristic size when $u_{*i} = u_{*c}$

$$\frac{d_a}{d_{50}} = 1.6 \left[\frac{d_{84}}{d_{16}}\right]^{-0.28} \tag{5.92}$$

fitted Chin's results quite well, i.e.:

$$\frac{\theta_{ca}}{\theta_c} = \left[0.4\left[\frac{(d_{50a})_{max}}{d_{50}}\right]^{-0.5} + 0.6\right]^2 \tag{5.93}$$

where the original bed material is characterised by d_{50} and θ_c = 0.05.

The characteristic size d_{max} of the bed material is the most difficult one to estimate and varies from sample to sample. The armour layer is clearly not affected by an occasionally occurring extra large stone. The grading is characterised by σ_g but the value of σ_g is based on the central part of the grading curve. The gradings of coarse material are more typically of the form of the original bed material shown in Fig. 5.34. Thus, extrapolation with σ_g (straight line) would seriously underestimate d_{max}. One method to obtain d_{max} is to extrapolate the grading curve on the basis of the last two or three data points to the 100% passing size. This approach relies also to some extent on judgement. The d_{max}-value could also be estimated by using two sieves, such that all particles just pass the coarser sieve. If the sieves are in the $2^{1/4}$-series the error involved by assuming d_{max} to be equal to the coarser sieve would be less than 20%, i.e. $(2^{1/4} - 1)$. If an observed grading of the limiting armour layer then extends to the same d_{max} size (see Fig. 5.34) then this would be a confirmation for the d_{max} estimate.

The major effect of the armouring is that at all shear stresses less than the critical armour layer shear stress the sediment transport rate initially decreases with time as the surface armouring develops. Usually these armour clusters here and there fail, giving rise to locally increased erosion until the armour there builds anew, and this maintains on average a steady bed load transport rate. However, when there is a decrease in flow rate the sediment transport rate can effectively go to zero. There is only an indiscriminate steady bed load transport when the bed shear stress exceeds the limiting armour layer shear stress.

The armouring effect present with broad grain-size distributions also affects the size and shape of bed features.

5.5 Secondary Currents

In connection with the discussion of the meanders it was emphasised that strong secondary currents are created by the bends. At the outside bank the secondary current is downwards and towards the inner bank at the bed, with the surface current towards the outer bank completing the circulation. Consequently, the sediment concentration at the outer bank is the lowest. This secondary current is also responsible for the large variation in depth in the bends, as was discussed earlier. However, secondary currents are present even in straight channels and play a significant part in the distribution of suspended sediment and where the sediments deposit or scour, but as yet no complete explanation for their existence or description is available. A major factor is the non-uniform distribution of shear stress over the periphery of the cross-section. A typical velocity distribution is illustrated in Fig. 5.35.

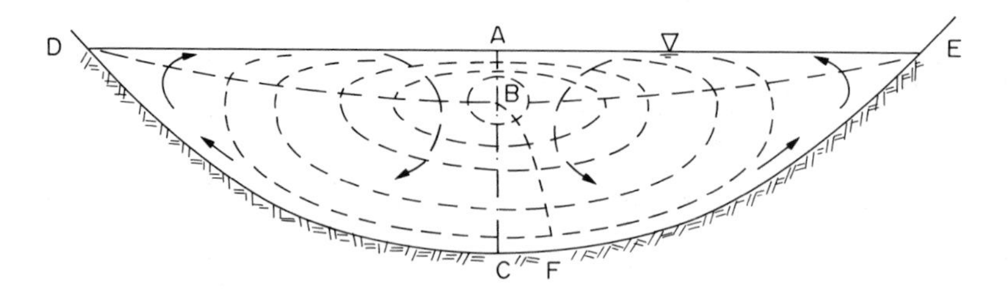

Fig. 5.35. Diagrammatic illustration of equal velocity contours and superimposed spiral motion in open channel.

It is usual to find that the maximum velocity does not occur at point A but at point B, a distance below the surface. Orthgonals to the isotachs separate the region DBE from the rest of the cross-section. The time avearage shear stress across the orthogonals is zero (since the mean local velocity gradient across the orthogonal is zero) and this leads to the apparent paradox that the weight component of the volume OBEA is not balanced by a shear stress. (It should be noted, however, that due to turbulent momentum exchange, there are always instantaneous apparent shear stresses acting across the orthogonals.) In terms of mean values, the shear stress on the boundary is, for example, for the segment FC

$$\bar{\tau}_o = \frac{\rho g \Delta A S_o}{FC} \tag{5.94}$$

where ΔA is the area FBC. It is also seen that the velocity is low at the edges A and E, and this means that the hydraulic grade line towards the edges rises due to the reduction of the velocity head, something like the line DBE. This leads to a flow in the cross-section from the edges towards the centre, secondary flows. These currents are always present as can be seen by the tendency of floating debris to move from the banks towards the middle. Continuity would require a return flow which leads to a pair of vortices superimposed on the flow as indicated in Fig. 5.35. In wide channels, more than one pair have been reported. One could thus argue that the weight component of volume DBEA goes into accelerating the fluid which enters the region DBEA at low velocities from near the banks.

Secondary circulation alters the distribution of shear stress acting on the bed. If the flow in the cells is at the bed towards the banks, as in Fig. 5.35, the scouring action on the bed, between the cells, is increased. If the circulation is in opposite sense the shear stress on the bed, over the area between the cells, is reduced and there is a tendency for a ridge to form. Thus, the circulation tends to stabilise the particular feature of the cross-section over a reach of the stream, for example the riffle and pool sequences. The shallow depth in the middle of the riffle reach induces flow towards the banks of the surface and towards the middle of the bed, and vice versa in the following pool section.

Apart from the helical secondary currents, natural rivers contain a multitude of flow features which are frequently referred to as macro-turbulence in order to differentiate these from the small scale features of the usual turbulence. The name is somewhat unfortunate because these large features do not satisfy the basic requirement of turbulence, that is, random character of properties. Many observers have described these large scale features and Matthes (1947) even proposed a classification. Although no rigorous classification is possible, the flow features do possess certain characteristics which allow some grouping:

(1) *Rhythmic and cyclic surges*. These consist of velocity and level fluctuations. Most level and velocity recordings contain such cyclic movements, the velocity records more so at lower river levels and near the bed whereas surface elevations fluctuate more during rising stages. Sudden forced changes of flow direction also lead to cyclic rise and fall of water level and eddying which, at times, reverses in direction.

(2) *Transverse oscillations* arise mainly from curvature effects. These may be associated with river bends and superelevation of water surface or with underwater contours.

(3) *Continuous rotating movements*, such as back eddies with vertical axis downstream of bends and where the channel has excess width. More concentrated eddies of this kind can occur, for example, at the lee side of bridge abutments.

(4) *Vortex action*. Concentrated vortices are common in open channel flow. These may be caused by submerged obstructions as horseshoe vortices, cast-off vortices, vortices in the surface of discontinuity, etc., and they are present in the lee of bed features. The vortices approximate free vortex motion and have a low pressure core which leads to the vortices with vertical axes acting like vacuum cleaners, i.e. lifting sediment from the bed and spreading it into the flow. A very intriguing and common feature on the surface of most rivers is the upwelling of water masses without an apparent rotation in horizontal plane, commonly called boils. The boils are likely to be vortex rings. From fluid dynamics, it is known that a vortex cannot start or end in flow. It has to stretch from infinity to infinity (the textbook case), from boundary to boundary where it dissipates itself in the boundary layer or the vortex has to close itself into a loop (smoke rings). A vortex, for example, at the lee of a three-dimensional bed feature tends to bind itself to the bed like a horseshoe. However, if a flow surge lifts such a vortex, it could form itself into a ring. Such a ring in a horizontal plane, propels itself due to the induced velocity field up or down, depending on the direction of rotation. The speed of the vortex ring in a direction normal to the plane containing it is given approximately by

$$V = \frac{\Gamma}{4\pi r} \ln \frac{8}{a/r} \qquad (5.95)$$

where Γ is the circulation of the vortex, r is the radius of the ring and a is the radius of the vortex (for derivation, see e.g. Raudkivi and Callander, 1975, page 110). Thus, the vortex may rise to the surface as a boil. It is significant that the boils show a strong central upward flow (as required), little whirling about a vertical axis and their diameter increases with depth of flow. It has also been reported that the longitudinal component of fluid velocity in a boil is significantly less than the mean stream velocity. This suggests that the fluid, with near zero forward velocity at the lee of a dune, has not been accelerated to stream velocity during its rise to the surface. The periodicity of "casting" of these boils to the surface could be linked with the wave length of the dunes, i.e. the periodicity of the dynamic system. Thus, to the first approximation the period could be expressed as the ratio of dune length to mean velocity $T = 2\pi y_o/U$ or $UT/y_o = 2\pi$, a value close to 5 as given by equation 2.110.

Jackson (1976) reported that the mean period of boils on the surface of lower Wabash and Polomet rivers was approximately $U_o\bar{T}_B/\delta = 7.6$ which could be compared to 2π quoted earlier. It should be noted that the streamwise adverse pressure gradients at the bed, lee side of the bed features, encourage the lifting of the tonques.

In summary, secondary currents arise from any flow curvature as well as from boundary drag. Turbulence bursts lead to rapid and significant pressure fluctuations on the bed surface and these have an important effect on entrainment of sediment. It may be that these bursts are strong enough to cause boils on the water surface, although it is more probable that these are associated with bed roughness and bed features. Whatever the mechanisms, the boils have an important effect on the suspended sediment load.

Laboratory studies with transport of fine sediment over a flat bed show a fairly regular pattern of small vertical "tornadoes" translating downstream which are not necessarily visible on the water surface. These tornadoes are most likely assocaited with the vorticity of flow due to side wall boundary layers, i.e. they do not appear to be periodic.

In general, it is important to remember that the flow in open channels is a boundary layer flow problem.

Chapter 6

Channel Roughness

6.1 General Relationships

The Prandtl mixing length concept for turbulent flow leads to the *law of the wall*

$$\frac{u}{u_*} = \frac{1}{\kappa} \ln \frac{y}{y'} = \frac{1}{\kappa} \ln \frac{y}{y_1} + \frac{u_1}{u_*} \qquad (6.1)$$

where u is the mean velocity at elevation y, $u_* = (\tau_o/\rho)^{\frac{1}{2}}$ is the shear velocity, κ is the Karman constant, y' is the elevation at which the logarithmic velocity is zero and y_1 where $u = u_1$. Setting $y_1 = \delta' = 11.6\ \nu/u_{*1}$ the laminar sublayer thickness, yields for turbulent flow over *smooth boundaries*

$$\frac{u}{u_*} = \frac{1}{\kappa} \ln \frac{yu_*}{\nu} + C \qquad (6.2)$$

which is independent of roughness height k when $C_1 = u_s/u_* =$ constant or empirically if $\delta'/k > 10$. If $\kappa = 0.40$ then $C_1 = 5.5$.

If the roughness height is large relative to δ', then empirically $y' = k/30.2$. This, with $\kappa = 0.40$, yields for fully turbulent flow over a rough boundary

$$\frac{u}{u_*} = \frac{1}{\kappa} \ln \frac{y}{k} + C_2 = 2.5 \ln \frac{y}{k} + 8.5 \qquad (6.3)$$

For an equivalent roughness k_s is used, the Nikuradse sand roughness, and

$$\frac{u}{u_*} = \frac{1}{\kappa} \ln \frac{y}{k_s} + B \qquad (6.4)$$

$$B = 2.5 \ln \frac{u_* k_s}{\nu} + 5.5 \ ; \quad \frac{u_* k_s}{\nu} < 5$$

$$B = 8.5 \qquad ; \quad \frac{u_* k_s}{\nu} > 70$$

The range $5 < u_* k_s/\nu < 70$ ia covered by a transition formula, e.g. the Colebrook-White formula.

The velocity *defect law* is valid for both the smooth and the rough boundaries

$$\frac{u_o - u}{u_*} = -\frac{1}{\kappa} \ln \frac{y}{y_o} \tag{6.5}$$

The logarithmic velocity distribution deviates from the observed one at distances $y/y_o > 0.15$ to 0.20, where y_o is the thickness of the boundary layer or depth of the fully developed turbulent flow. Thus, if the velocity defect is calculated with u_o an additional constant is required. Clauser (1956) showed that the smooth and rough boundary data over the range of $y/y_o < 0.15$ satisfy

$$\frac{u_o - u}{u_*} = -2.5 \ln \frac{y}{y_o} + 3.7 \tag{6.6}$$

As seen from eqn 6.4 the law of the wall for uniform rough and smooth surfaces differs only in the value of the constant which is a function of $u_* k_s/\nu$. The effect of the roughness it to shift the velocity distribution. Accordingly, Hama (1954) wrote eqn 6.3 as

$$\frac{u}{u_*} = A \ln \frac{u_* y}{\nu} + B - \left(A \ln \frac{u_* k_s}{\nu} + B - C_2\right) \tag{6.7}$$

where the bracket term represents the velocity shift. The constant $B-C_2$ depends on the type of roughness. The general velocity distribution (Cebeci and Smith, 1974) has the form

$$\frac{u}{u_*} = \left[\frac{1}{\kappa} \ln \frac{u_* y}{\nu} + B\right] - \frac{\Delta u}{u_*} + \frac{\Omega}{\kappa} f\left[\frac{y}{\delta}\right] \tag{6.8}$$

where the last term is "the wake region velocity augmentation function". Coles (1956) proposed empirically

$$f\left[\frac{y}{\delta}\right] = 2 \sin^2\left[\frac{\pi}{2} \frac{y}{\delta}\right] \tag{6.9}$$

where δ is the boundary layer thickness or flow depth. For $\delta = y_o$

$$\frac{u_o}{u_*} = \frac{1}{\kappa} \ln \frac{u_* y_o}{\nu} + B - \frac{\Delta u}{u_*} + 2 \frac{\Omega}{\kappa} \tag{6.10}$$

and the velocity deficit

$$\frac{u_* - u}{u_*} = \left[\frac{1}{\kappa} \ln \frac{y}{y_o}\right] + 2 \frac{\Omega}{\kappa}\Big] - \frac{\Omega}{\kappa} f\left[\frac{y}{y_o}\right] \tag{6.11}$$

From measured values the wake coefficient

$$\Omega = \frac{\kappa}{2}\left[\frac{u_o - u}{u_*}\right]_{y/y_o=1} \tag{6.12}$$

Figure 6.1 shows diagrammatically the law of the wall and the velocity deficit as defined by eqn 6.11. In terms of sediment transport it is important to note that there are two parameters, Ω and κ, which define the velocity distribution.

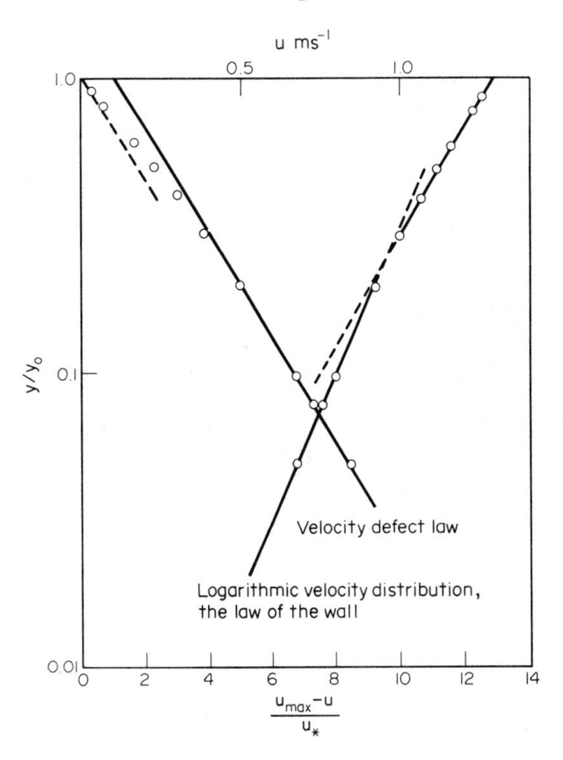

Fig. 6.1. Illustration of the velocity defect law and the law of the wall (upper scale) velocity distribution in open channel flow. Eqn 6.12 yields for this example with κ=0.40 a wake coefficient Ω = 0.2.

For roughness which consists of distinct elements O'Loughlin and MacDonald (1964) set $k_s = \alpha\beta k$, where α is concentration of roughness elements and β is an aspect ratio. Wooding et al. (1973) defined α as the ratio or the frontal area of element to wall area per element, and β as the ratio of roughness height to its streamwise length to a power of n, to be determined experimentally. They set

$$\beta = (k/L)^{0.4} \tag{6.13}$$

Thus, the law of the wall is

$$\frac{u}{u_*} = \frac{1}{\kappa} \ln \frac{y}{\alpha\beta k} + C \tag{6.14}$$

The data by O'Loughlin and Macdonald was shown to satisfy

$$\frac{u_o}{u_*} = 2.87 \ln \frac{y_o}{\alpha\beta k} - 2.05 \tag{6.15}$$

Perry et al. (1969) showed that data from distinct roughness patterns, for which the coordinate origin was at 0.25 k_s below the tops of the roughness elements, satisfy

$$\frac{u_o}{u_*} = 2.5 \ln \frac{y_o}{k_s} + 5.9 \tag{6.16}$$

i.e., $B-C_2 = -0.4$, whereas for sand roughness $B-C_2 = -3.5$.

Equation 6.3 integrated over depth of flow yields the flow rate per unit width, and divided by depth, the mean velocity U as

$$\frac{U}{u_*} = 5.75 \log \frac{y_o}{k_s} + 6 \tag{6.17}$$

and

$$\frac{u - U}{u_*} = 5.75 \log \frac{y}{y_o} + 2.5 \tag{6.18}$$

The mean velocity occurs at $y = 0.368\ y_o$. Hence, the approximation for one point velocity measurement at $0.4\ y_o$. Observing that $u_* = (gy_oS)^{1/2}$, where S is the slope, and comparing with the Chézy formula

$$V = C\sqrt{RS} = C\sqrt{y_oS} \tag{6.19}$$

where R is the hydraulic mean radius, shows that

$$C = 18 \log(y_o k) + 18.8 = 18 \log(11.1R/k) \tag{6.20}$$

Plotted on log-log scale this relationship for C has a slope of $\tan^{-1}(1/6)$ at about $y_o/k = 30$, i.e., at this relative roughness

$$C \propto (y_o/k)^{1/6} = K(y_o/k)^{1/6} \tag{6.21}$$

Substituting into the Chézy formula yields

$$V = \frac{K}{k^{1/6}} y_o^{2/3} S^{1/2} \tag{6.22}$$

which yields the Manning or Strickler formula

$$V = \frac{1}{n} R^{2/3} S^{1/2} \tag{6.23}$$

where n is the Manning coefficient (ln ft-lb system of units K = 1.486). The Chézy and Manning coefficients are related by

$$C = \frac{1}{n} y_o^{1/6} \tag{6.24}$$

Strickler (1923) expressed n for natural channels as

$$n = \frac{1}{21.1} d^{1/6} = 0.0474\ d^{1/6} \tag{6.25}$$

where d is the grain size. For this Lane proposed d_{75}-size. When

$$n = d_{(m)}^{1/6}/24 = 0.04168\ d_{(m)}^{1/6} = 0.01312\ d_{(mm)}^{1/6} \tag{6.26}$$

the Manning formula yields

$$\frac{U}{u_*} = 7.66 \left[\frac{R}{d}\right]^{1/6} \tag{6.27}$$

which is practically identical with the logarithmic formula.

Note that the Manning formula is consistent with the logarithmic velocity distribution in the neighbourhood of $y_o/k \cong 30$. In the region of $y_o/k = 10$ the Chézy coefficient $C \propto (y_o/k)^{1/4}$ yielding $V \propto R^{3/4}S^{1/2}$ which is the form proposed by Lacey (1929).

Expressing the shear stress τ_o as

$$\tau_o = c_f \frac{\rho U^2}{2} = \frac{f}{4} \frac{\rho U^2}{2} \tag{6.28}$$

where f is the Darcy-Weisbach friction factor, yields

$$u_* = \left[\frac{\tau_o}{\rho}\right]^{\frac{1}{2}} = U \left[\frac{f}{8}\right]^{\frac{1}{2}} \tag{6.29}$$

and

$$C_* = \frac{C}{\sqrt{g}} = \left[\frac{8}{f}\right]^{\frac{1}{2}} \tag{6.30}$$

Thus, replacing in eqn 6.18 u_* by $U(f/8)^{\frac{1}{2}}$ yields

$$\frac{U - U}{U\ \sqrt{f}} = 2.03 \log \frac{y}{y_o} + 0.88 = \frac{(u - U)C}{U\ \sqrt{8g}} \tag{6.31}$$

This leads to resistance equations for two-dimensional open channel flow, similar to those for pipe flow, as follows:

(i) for smooth boundary: $\left[\frac{u_* k}{\nu} < 3\right]$, $Re = 4RU/\nu$

$$\frac{1}{\sqrt{f}} = 2.03 \log (Re \sqrt{f}) - 0.47; \tag{6.32}$$

(ii) and for rough boundaries: $\frac{u_* k}{\nu} > 70$

$$\frac{1}{\sqrt{f}} = 2.03 \log \frac{4R}{k} + 0.91 \tag{6.33a}$$

or

$$\frac{1}{\sqrt{f}} = 2.03 \log \frac{y_o}{k} + 2.12 \tag{6.33b}$$

while a Colebrook-White type transition formula for $5 < \frac{u_* k}{\nu} < 70$.

A power law velocity distribution is frequently used, instead of the logarithmic form, for its advantages for numerical work. The distribution can be written as

$$\frac{u}{u_*} = \frac{n}{\kappa} \left[\frac{y}{y_o}\right]^{1/n} \tag{6.34}$$

and

$$U = \frac{n^2}{(n + 1)\kappa} u_* \tag{6.35}$$

or

$$\frac{u}{U} = \frac{n + 1}{n} \left[\frac{y}{y_o}\right]^{1/n} \tag{6.36}$$

Data on velocity distribution can be plotted on log-log paper as u/U versus y/y_o. The exponent n is then given by the slope of the line of best fit.

6.2 Resistance in Alluvial Channels

In an alluvial channel the resistance differs from the fixed-boundary relationships described above. In particular, the roughness of an alluvial channel varies with flow and is only indirectly related to the grain size and grain size distribution of the boundary material. No method exists to date by which one could reliably calculate the roughness and energy loss in an alluvial channel. The multitude of empirical methods illustrates that the solution has not yet been found.

In an alluvial channel the roughness depends on

(a) the nature of bed material, its grading and properties (particularly shape) and on the spatial variation of these in the channel, and
(b) the flow depth and velocity (or shear velocity) which determines the nature of the bed features for a given bed material.

The inherent difficulty with the estimation of the steady state flow depth in an alluvial channel arises from the variation of boundary roughness or shear stress with the velocity of flow as illustrated for a sand bed in laboratory flume in Fig. 6.2.

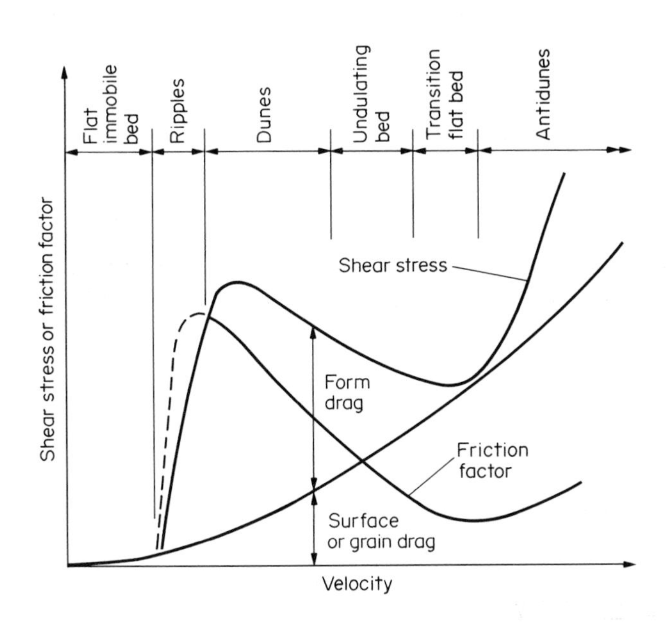

Fig. 6.2. Variations of bed shear stress τ_o and Darcy-Weisbach friction factor f with mean velocity U in flow over a fine sand bed.

The changes in roughness due to bed features can be substantial in flow over fine sand beds as illustrated by laboratory data in Fig. 6.3. The Darcy-Weisbach f-value can increase about tenfold from the initial flat bed value and it returns to almost the same value at the transition flat bed conditions. The grain roughness and that caused by the sediment in motion combined is seen to equate to approximately 6 d_{50}. This may well be a more representative value for the transition flat bed in general than the idealised flat bed of grain roughness only. The initial flat bed may be rare in nature but the transition flat bed is a common occurrence. The first observation is that for a given f-value there may be three solutions for flow velocity. In nature the slope of the stream can not change significantly with flow rates. Therefore, the effect of the changes in roughness is a substantial variation in flow depth. Since the bed shear stress for approximately constant channel slope is proportional to flow depth the stage-discharge curves too vary with the changing bed features and show a "discontinuity" where the flow passes through the transition flat bed conditions, from the *lower regime* to the *upper regime*. Examples can be found in the paper by Simons et al. al. (1962), Nordin (1964) and others.

The form drag over well-formed sand dunes and ripples is high because their steep lee faces (about 30°) approximate to a

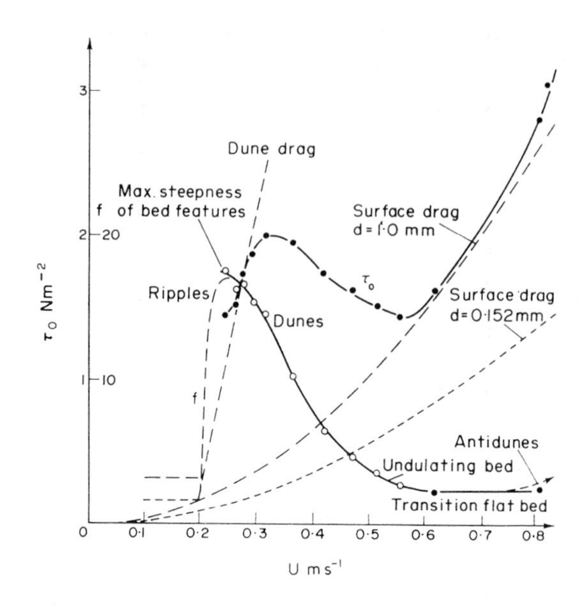

Fig. 6.3. Variation of bed shear stress τ_O and Darcy-Weisbach friction factor, f, with mean velocity, U, in flow over a fine sand bed.

negative step with a substantial lee eddy and expansion loss. In nature the range of f-values is likely to be smaller, because the bed features are more random in size and distribution, but the form drag over sand beds is still a very significant component of energy loss.

For a gravel bed the variation of the form drag component is not so pronounced and usually the slope of the total shear stress function remains positive with increasing velocity, i.e. no downward slope as shown in Fig. 6.2. In gravel rivers the bed features are much more rounded in form and their form drag contribution is usually small compared to energy loss due to surface drag and secondary currents. As a consequence roughness of gravel rivers is more readily described by formulae than of sandy streams. However, special difficulties arise with the description of roughness in very shallow gravel rivers where the roughness height is of the same order of size as the flow depth. The flow in such shallow streams, usually braided, is also twisting and turning, bifurcating and joining together and all this consumes energy as well.

The description of the resistance to flow as a function of sediment and flow parameters is one of the unsolved problems of loose boundary hydraulics. The methods in current use are essentially empirical and have a low level of accuracy for prediction purposes. The reasons for this are many. Not even in a steady two-dimensional flow can the roughness height and the dimensions of bed forms be predicted reliably. In nature, the flow rate varies and the bed is covered with residual and newly formed features. The features are usually strongly three-dimensional, the composition of the bed material is not uniform and frequently the material varies over cross-section as well as along the stream.

The usual approach is to separate the roughness of an alluvial bed into surface or grain roughness and roughness due to bed features, the form drag. Not even for the grain roughness is there unanimity on which size fraction is the representative one. The recommended values range from $k_s = 1.25\ d_{35}$ (Ackers and White 1973) to $k_s = 5.1\ d_{84}$ (Mahmood, 1971). Einstein and Barbarossa (1952) used $k_s = d_{65}$, Engelund and Hansen (1967) $k_s = 2\ d_{65}$, Hey (1979) $k_s = 3.5\ d_{84}$ etc.

A detailed discussion of the problem of resistance to flow in gravel rivers can be found in Hey (1979), including that of the effects of cross-sectional shape and different bank roughness, and in Limerinos (1970). Hey proposed

$$\frac{1}{\sqrt{f}} = 2.03 \log \frac{11.75\ R}{3.5\ d_{84}} \quad \text{or} \quad \left[\frac{8}{f}\right]^{\frac{1}{2}} = 5.62 \log \frac{a\ R}{3.5\ d_{84}} \qquad (6.37)$$

where a varies with slope $11.1 \leq a \leq 13.46$. This differs from the formula by Limerinos in only of the factor 3.5 instead of 3.16. The relationship yields reasonable values when y_o/d_{84} is greater than about six. The data base comes from rivers with slopes less than about 1%. Bathurst (1978) proposed an empirical relationship

$$\left[\frac{8}{f}\right]^{\frac{1}{2}} = \left[\frac{R}{0.365\ d_{84}}\right]^{2.34} \left[\frac{B}{y_o}\right]^{7(\lambda - 0.08)} \qquad (6.38)$$

where B is the width of stream of depth y_o and $\lambda = 0.139 \log (1.91\ d_{84}/R)$.

In steep mountain streams the individual roughness elements are large and may even protrude through the water surface. This makes the concept of roughness height at least questionable. The energy losses are mainly governed by the dominant roughness. With increasing size of roughness elements one ought to calculate the form drag of the individual elements but this is impracticable. Here, use could be made of the blockage concept, as described in texts on wind tunnel testing. The blockage factor accounts for the higher effective velocity due to the effect of fixed tunnel walls. Accordingly, the drag coefficient on the model of projected area a in a tunnel of cross-sectional area A is increased by a factor $(1 - a/A)^{-2}$. Thus, if the dominant roughness elements, for example, occupy one fifth of the channel width B and have a height k then $a/A = 0.2\ Bk/By_o = 0.2\ k/y_o$ or $(1 - a/A) = (1 - bk/R)$ and

$$\frac{1}{\sqrt{f}} = 2\left[1 - \frac{bk}{R}\right] \log \frac{12R}{k} \qquad (6.39)$$

Thompson and Campbell (1979) suggested from their field results $b = 0.1$ and $k = 4.5\ d$, where d is the mean diameter of the stones (median d of a counting sample).

Bathurst (1985) delineated with aid of data from rivers with slopes steeper than 0.4% the region of $(8/f)^{\frac{1}{2}}$ in terms of y_o/d_{84}, as shown in Fig. 6.4. Accordingly, the equation by Hey (eqn 6.37) effectively defines the lower limit of roughness of uniform flows but values of over 60% greater are possible. Bathurst proposed a simple empirical equation

$$\left[\frac{8}{f}\right]^{\frac{1}{2}} = 5.62 \log \left[\frac{y_o}{d_{84}}\right] + 4 \qquad (6.40)$$

which lies more or less in the middle of the area with accordingly smaller plus or minus variation.

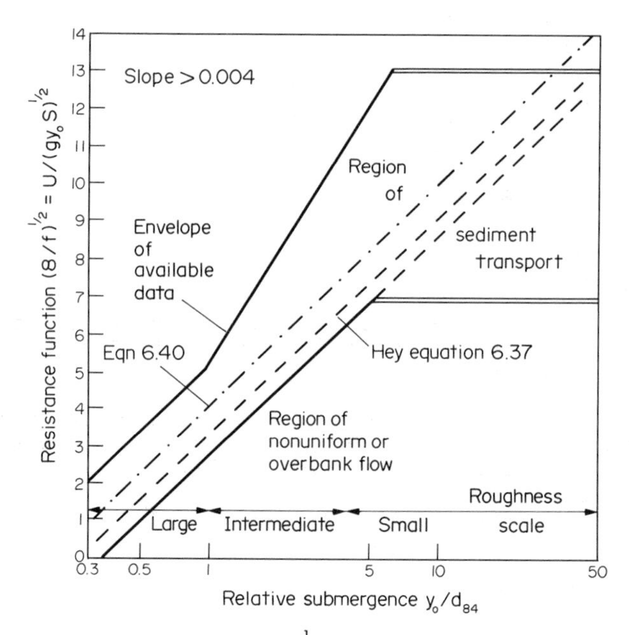

Fig. 6.4. Region of $(8/f)^{\frac{1}{2}}$ as a function of y_o/d_{84} for uniform flows according to Bathurst (1985).

Brownlie (1983) proposed from analysis of published data a similar "bulk" formula for sandy rivers, grain sizes $0.088 < d_{50} < 2.8$ mm. For wide channels ($y_o \equiv R$) the lower regime was fitted by

$$\frac{R}{d_{50}} = 0.3724\,[q/(gd_{50}{}^3)^{0\cdot5}]^{0\cdot6539} S^{-0\cdot2542} \sigma_g{}^{0\cdot0150} \qquad (6.41a)$$

and the upper regime (beyond transition flat bed conditions) by

$$\frac{R}{d_{50}} = 0.2836\,[q/(gd_{50}{}^3)^{0\cdot5}]^{0\cdot6248} S^{-0\cdot2877} \sigma_g{}^{0\cdot08013} \qquad (6.41b)$$

The maximum velocity for the lower regime for $S < 0.006$ was given by

$$0.8(1.74\ S^{-1/3}) = V/(\Delta g d_{50})^{0.5} \tag{6.42}$$

For upper regime the minimum velocity is given by using a factor of 1.25 instead of 0.8. The formulae for the essentially sandy sediments are insensitive of the mean particle size d_{50} and of the geometric standard deviation σ_g of the grain-size distribution.

Equations 6.41 are essentially the Manning formula. Bruschin (1985) rearranged the formulae and expressed the Manning n for the lower regime as

$$n_L = \frac{d_{50}^{1/6}}{12.38}\left[\frac{RS}{d_{50}}\right]^{1/7.3} \tag{6.43}$$

and the upper regime as

$$n_u = \frac{d_{50}^{1/6}}{20.38}\left[\frac{RS}{d_{50}}\right]^{1/9} \tag{6.44}$$

i.e. the Manning n is

$$n = \frac{k_s^{1/6}}{a\sqrt{g}} \tag{6.45}$$

where from above $a_L = 3.95$ and $a_u = 6.5$. Strickler's expression for n is equivalent to a = 6.74. According to Bruschin for gravel rivers $4.6 < a < 5.3$.

A very helpful collection of data on river geometry, the measured values of Manning's n and photographs of the rivers was published by the US Geological Survey (Barnes, 1967).

For flow over bed features the effective roughness from eqns 6.14 and 6.16 is

$$k_s = \alpha\beta k[\kappa(5.9 - C)] = K\eta\ h/\lambda \tag{6.46}$$

when $k = \eta$, $\alpha = \eta/\lambda$ and $\beta = 1$, or

$$\frac{k_s}{\eta} = f(\eta/\lambda) \tag{6.47}$$

where η/λ is the steepness of the bed features. Evaluating the constant K with the aid of eqn 6.15, using $\kappa = 0.40$ instead of 0.35 which leads to C = -0.5 to -1.0 depending on the range of $y/\alpha\beta k$-values, yields

$$k_s = (12.9 \text{ to } 15.8)\eta\ \eta/\lambda \tag{6.48}$$

Shinara and Tsubaki (1959) proposed

$$\frac{k_s}{\eta} = 7.5\left[\frac{\eta}{\lambda}\right]^{0.57} \tag{6.49}$$

and van Rijn (1982) fitted to data from flumes and field

$$\frac{k_s}{\eta} = 1.1(1 - e^{-25\ \eta/\lambda}) \qquad (6.50)$$

The data covered $0.08 \leq y_o \leq 0.75$ m, $0.25 \leq U \leq 1.1$ m/s and $0.1 \leq d_{50} \leq 2.4$ mm. Equation 6.50 appears to follow the centre of the band of data points, Fig. 6.5 and eqn 6.49 the upper limit. The total hydraulic roughness of an alluvial bed was expressed as

$$k_s = 3d_{90} + 1.1\ \eta(1 - e^{-25}\eta/\lambda) \qquad (6.51)$$

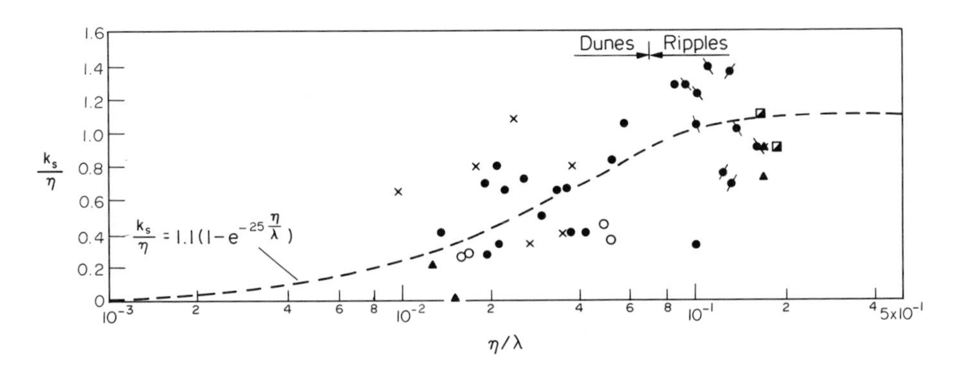

Fig. 6.5. Equivalent form drag roughness as a function of height and steepness of bed features, according to van Rijn (1982).

Furhter complications arise from the time lag in development of the bed features. Thus, if the flow conditions are such that the bed features increase with increasing flow rate, then with relatively rapidly increasing flow the roughness will be less than indicated by instantaneous flow rate. If, however, the flow approaches transition flat bed conditions then the roughness is greater than that for the given flow rate. For any given conditions the effect of the time lag can be assessed with the aid of Fig. 6.2.

The above arguments in general apply to two-dimensional situations. Complications arise already on the side slopes of the flow cross section. For the same bed material the side slope lowers the threshold of movement. Usually the particle size also varies over the periphery. The resistance, particularly over ripples and dunes, is strongly affected by heavy concentrations of colloidal wash load. The development of bed features starts at higher bed shear stresses than in clear water. The features which form are lower and smoother and they are washed out sooner, i.e., the transition flat bed condition occurs at lower shear than in clear water. Significant energy losses arise from momentum transfer in cross sections which have a deep channel flanked with shallow flow areas, for example, over bank flows on flood plains. For these problems reference is made to Knight and Demetriou (1983), Rajaratnam and Ahmadi (1979, 1981).

Natural streams are neither two-dimensional or straight. Frequently sandy streams have cohesive banks and it may be necessary to consider the bank resistance separately. A simple method is to consider the cross section as areas associated with bed and banks $A = A_b + A_w$. Then

$$A_b = \frac{P_b f_b U^2}{8gs} \qquad A_w = \frac{P_w f_w U^2}{8gs} \qquad A = \frac{PfU^2}{8gs} \tag{6.52}$$

where P is the length of the respective wetted perimeters, and

$$Pf = P_b f_b + P_w f_w \tag{6.53}$$

In meandering streams bend losses have to be added to friction loss. Rouse (1965) from laboratory tests with 90° bends gave the loss as $0.5\ U^2/2g$ for $B/y_o = 16$ and $0.2\ U^2/2g$ for $B/y_o = 8$. However, losses vary with curvature, length of arc and the Froude number. A very approximate average value in rivers is $0.2\ U^2/2g$. In laboratory channels an approximate relationship for the bend loss S" is

$$\frac{S''}{S} = 300 \left[\frac{y_o}{r_c}\right]^2 \tag{6.54}$$

where r_c is the centre line radius. In sharp bends lee eddies form downstream of the bend which reduce the effective cross section and consume flow energy. At times even small surface waves occur at the inner bank of the bend and add to the energy losses. Rozovskii (1957) expressed the energy gradient S" due to secondary currents in the flow through the bend as

$$S'' = (12\sqrt{g}/C + 30g/C^2)(D/r_c)^2 Fr^2 \tag{6.55}$$

where C is the Chézy coefficient, D is mean depth and $Fr^2 = V^2/gD$.

6.3 Methods of Estimation of Flow Depth

A description of the proposed methods of estimation of flow depths in alluvial channels has been given by Vanoni (1975) and numerical examples have been presented by Raudkivi (1982). In principle, the methods are of two kinds; a simultaneous solution of the continuity and Chézy or Manning equations, or separation of surface (grain roughness) drag and form drag due to bed features and use of empirical functions. The first group also includes the *regime method*, for example, the Lacey formula

$$U_o = 0.646\ \sqrt{fR} = 10.8\ R^{2/3}\ S^{1/3}\ ms^{-1} \tag{6.56}$$

where R is the hydraulic mean radius, f is the Lacey silt factor and U_o is the stable channel velocity. In a wide channel $R \cong y_o$

and $q = U_o y_o$, i.e., eqn 6.56 defines the y_o-U_o-function and its intersection with $y_o = q/U_o$ defines the flow depth for the given flow rate q. From $Q = U_o A = U_o RP$ the length of the wetted perimeter P is defined as a function of Q and if this does not adequately approximate to channel width, the shape of the channel has to be defined. The regime method is further discussed in the chapter on Stable Channel Design.

The *Einstein-Barbarossa* (1952) method of estimation of the depth of flow in alluvial channels is one of the most widely known techniques. In it the boundary shear stress τ_o is separated into two components relating to grain roughness of the flat fixed surface, τ_o', and form drag of the bed features, τ_o'', respectively, i.e. -

$$\tau_o = \tau_o' + \tau_o'' \tag{6.57}$$

The value of U/u_*'' was empirically related to

$$\psi' = \left[\frac{\rho_s - \rho}{\rho}\right] \frac{d_{35}}{R'S} \tag{6.58}$$

where U is the mean velocity of flow, u_*'' is the shear velocity corresponding to $\tau_o'' = \rho g R'' S = \rho (u_*'')^2$, d_{35} is the particle size of which 35% is finer, ρ_s and ρ are the densities of grains and water, respectively and S is the slope. This relationship for U/u_*'' is shown in Fig. 6.6. The relationship was based on data from streams and rivers carrying fine sediments and sands and has been found to be unsatisfactory for $\psi' > 10$. Shown are also modifications by Sentürk and by Cunha (1967). The function by Cunha is intended for gravel beds. The data relate to a river with $S = 0.8 \times 10^{-3}$ and bed material of $d_{90} = 6.1$ mm, $d_{35} = 1.7$ mm. Line 3 was recommended for use when $(wd_{35}/\nu)[U/gR)^{\frac{1}{2}}] > 70$, where w is the fall velocity of d_{35} size particles. The Einstein-Barbarossa method also leads to results which differ substantially from observations when the flow conditions are just above the threshold of sediment transport. The method implies that form drag commences when flow starts, i.e., zero threshold velocity. For fine sand beds this is approximately true but with coarser sediment the bed may be immobile at lower velocities.

Engelund (1966,1967) also divided the resistance into surface and form frag. His calculation of the depth-discharge relationship for sandy rivers proceeds as follows:

(1) Assume a y_o' (a depth corresponding to grain resistance) and calculate

$$\theta' = \frac{y_o' S}{(S_s - 1) d_{50}} \tag{6.59}$$

(Engelund used the mean fall diameter)

(2) From Fig. 6.7, or $\theta' = 0.06 + 0.4\theta^2$ for the ripple and dune region, find the dimensionless shear stress θ. For $0.4 < \theta' < 1.0$ two values of θ can be obtained, one for lower and one

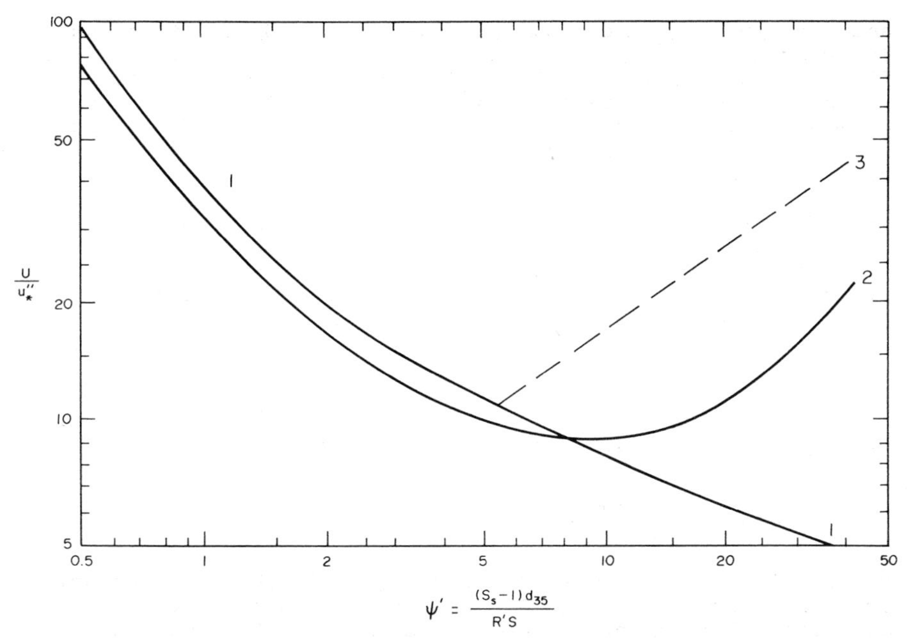

Fig. 6.6. Einstein-Barbarossa function for form drag: 1 - Einstein-Barbarossa curve; 2 - Modification by Sentrk; 3 - Cunha modification for gravel when $(wd/\nu)\ (U/\sqrt{gR}) > 70$.

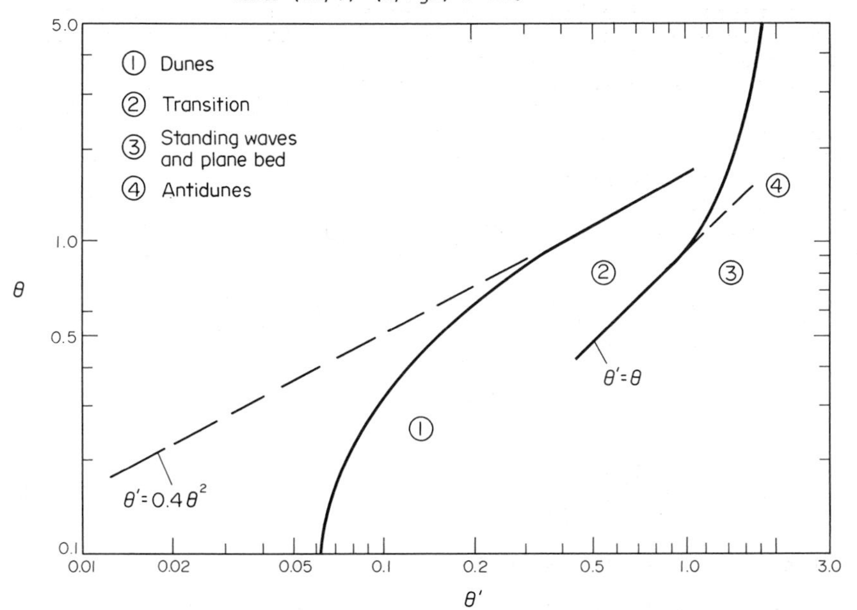

Fig. 6.7. Engelund's relationship between dimensionless shear stress θ and dimensionless grain shear stress θ'.

for upper regime of flow. Alternatively $\theta' = \theta_c + 0.3\theta^{3/2}$; $\theta \leq 0.6$.

(3) From θ calculate

$$y_o = \frac{\theta(S_s - 1)d}{S} \tag{6.60}$$

(4) Calculate mean velocity from

$$\frac{U}{\sqrt{g y_o' S}} = 5.75 \log \frac{y_o'}{2d_{65}} + 6 \tag{6.61}$$

(5) Calculate discharge $q = Uy_o$.

The methods which rely on separation of shear into components due to grain roughness and form drag assume in principle that a unique relationship exists between the grain roughness component and total shear. No allowance is made for changes in viscosity or turbulence intensity near the bed. The energy dissipated by form drag increases turbulence which affects sediment suspension, apparent viscosity and size of bed features.

White et al. (1979) argued that the $\tau_o' = f(\tau_o)$ assumption is insufficient and that at least the effect of the dimensionless grain size $D_* = [(S_s - 1)gd^3/\nu^2]^{1/3}$ should be accounted for. Data were analysed in terms of the "mobility parameter", F_{gr}, introduced by Ackers and White (1973), and D_*:

$$F_{gr} = \frac{u_*^{\,n}}{[(S_s - 1)gd]^{\frac{1}{2}}} \left[\frac{1}{\sqrt{32} \log 10\, y_o/d}\right]^{1-n} \tag{6.62}$$

which for fine sediments (n = 1) reduces to

$$F_{fg} = \frac{u_*}{[(S_s - 1)gd]^{\frac{1}{2}}} = \sqrt{\theta} \tag{6.63}$$

It was found that the data could be fitted by a function of the form

$$\frac{F_{gr} - A}{F_{fg} - A} = f(D_*) = 1\text{-}0.76\{1\text{-}\exp[-\log D_*^{\,1.7}]\} \tag{6.64}$$

for Froude numbers less than 0.8, i.e., the lower flow regime, using d_{35} of parent material, or

$$f(D_*) = 1\text{-}0.70\{1\text{-}\exp[-1.4 \log D_*^{\,2.65}]\} \tag{6.65}$$

in terms of d_{65} of surface material size distribution. In these

$$\left.\begin{array}{l} n = 0 \\ A = 0.17 \end{array}\right\} \text{for } D_* \geq 60$$

$$\left.\begin{array}{l} n = 1.0 - 0.56 \log D_* \\ A = 0.23/\sqrt{D_*} + 0.14 \end{array}\right\} \text{for } 1 \leq D_* < 60$$

Thus, with known $u_* = (gy_oS)^{\frac{1}{2}}$ and F_*, eqn 6.64 or 6.65 yields F_{gr} which in turn, eqn 6.62, yields the mean velocity V and the friction factor

$$f = 8(u_*/V)^2 \tag{6.66}$$

White et al. (1987) extended this analysis to the upper flow regime with the empirical relationship

$$\frac{(F_{gr} - A) + 0.07(F_{gr} - A)^4}{F_{f_g} - A} = 1.07 - 0.18 \log D_* \tag{6.67}$$

The flow regimes were defined in terms of dimensionless stream power

$$P_* = VS/[(g\nu)^{1/3} D_*] \tag{6.68}$$

where eqn 6.64 or 6.65 yields P_{*L} for the lower and eqn 6.67 the P_{*u} for the upper flow regime. The transition region was identified to be in the range of $0.011 < P_* < 0.02$ and the flat bed at $P_* > 0.02$, i.e., $P_{*L} < 0.011$. In the absence of guidance by field data the authors recommend

$$P_{*L} + P_{*u} \begin{cases} < 0.022 - \text{lower flow regime} \\ > 0.022 - \text{upper flow regime} \end{cases}$$

The predictions of roughness by the above method, by Engelund and by van Rijn methods were compared by van Rijn (1984). The results imply that predictions by the method proposed by van Rijn are closer to measured values from field data than the other two.

Lovera, Alam and Kennedy (Alam and Kennedy, 1969, Lovera and Kennedy, 1969) method is based on empirical design charts, Figures 6.8 and 6.9. The procedure of calculation is as follows:

(1) From known values of d_{50} and ν, a selected value of mean velocity U and an assumed value of R_b calculate $U/(gd_{50})^{\frac{1}{2}}$, R_b/d_{50} and UR_b/ν, where R_b is that part of hydraulic mean radius associated with the roughness of bed, i.e. laboratory data have been adjusted for side wall drag. The hydraulic mean radius R is used below implying wide channels in which the resistance caused by the bed is much larger than that by the banks.

(2) Read f' and f" values from Figures 6.8 and 6.9 respectively.

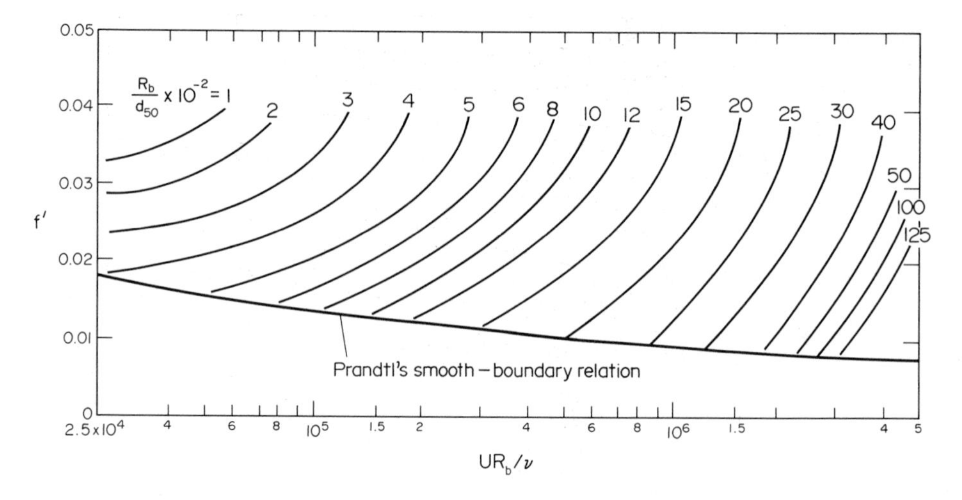

Fig. 6.8. Lovera and Kennedy chart for grain roughness friction factor f' as a function of Reynolds number UR_b/ν for constant values of relative depth R_b/d_{50}. (Data points shown on the original chart have been omitted.)

Use the value of f' corresponding to smooth boundaries if the UR/ν versus R/d_{50} intersection falls below the smooth boundary line.

(3) The friction factor is then $f = f' + f''$.

(4) Calculate R_b using f from step (3) and known S

$$R_b = fV^2/8g\,S \tag{6.69}$$

This has to equal the assumed R_b.

Figure 6.8 is in principle a Moody diagram and the relative roughness lines should become more or less horizontal with increasing Re, not vertical. The method requires a trial and error solution and frequently field conditions are not covered by Fig. 6.8 and 6.9.

Van Rijn (1984) proposed a trial and error solution for flow depth based on Fig. 6.5 and using the transport parameter T used to describe the height and length of bed features, Fig. 5.16. The procedure is as follows:

1. Compute dimensionless particle size $D_* = \left[\frac{(S_s - 1)gd_{50}^3}{\nu^2}\right]^{1/3}$

2. Compute u_{*c} from θ_c versus D_*

3. Compute transport stage $T = [(u_*')^2 - (u_{*c})^2]/(u_{*c})^2$

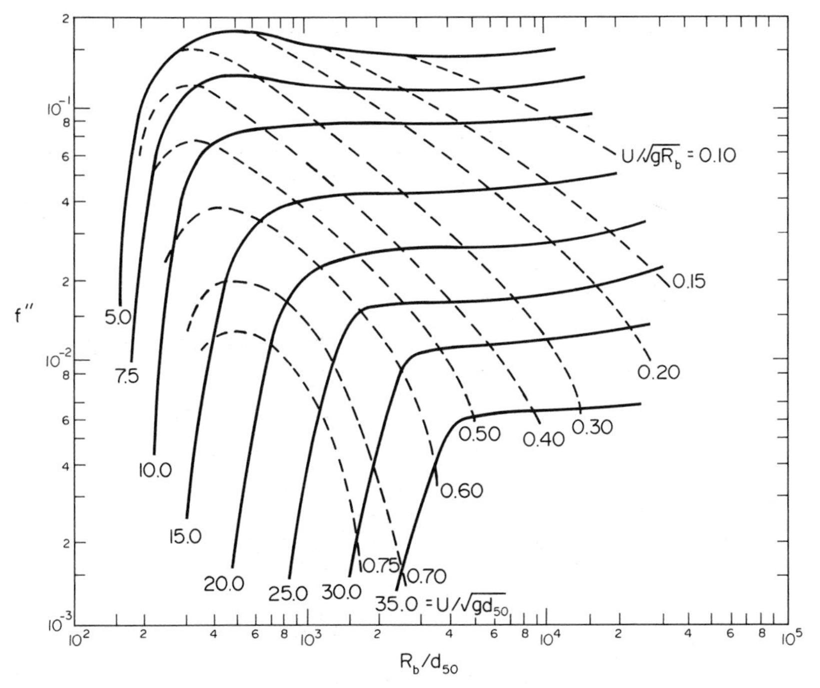

Fig. 6.9. Alam and Kennedy chart for form drag friction, f" as a function of relative depth, R_b/d_{50} for (i) constant values of grain Froude number, $U/(gd_{50})^{\frac{1}{2}}$ and (ii) constant values of Froude number, $U/(gR_b)^{\frac{1}{2}}$. (Data points shown on the original chart have been omitted.)

where $u_*' = (g^{0.5}/C')U$; $C' = 18 \log(12\ R_b/3d_{90})$

4. Compute bed form height η from eqn 5.69.
5. Compute bed form length λ from $\lambda = 7.3\ y_o$
6. Compute equivalent roughness $k_s = 3d_{90} + 1.1\eta(1 - e^{-25\eta/\lambda})$, eqn 6.51.
7. Compute $C = 18 \log(12\ R_b/k_s)$

The known values are mean velocity U, flow depth y_o, width, particle size d_{50} and d_{90} as well as densities and viscosity. The hydraulic mean radius R_b relates to bed, i.e., for flume data side wall correction procedure is applied (e.g. Vanoni and Brooks 1957, Williams, 1970).

Raudkivi combined laboratory and field data into a graph of $U/(u_*^2 - u_{*c}^2)^{\frac{1}{2}}$ versus $\rho u_*^2/\rho g(S_s - 1)d_{50}$ where u_{*c} is the shear velocity at the threshold of sediment movement. The functions interpolated to the data points are shown in Figure 6.10. The curves

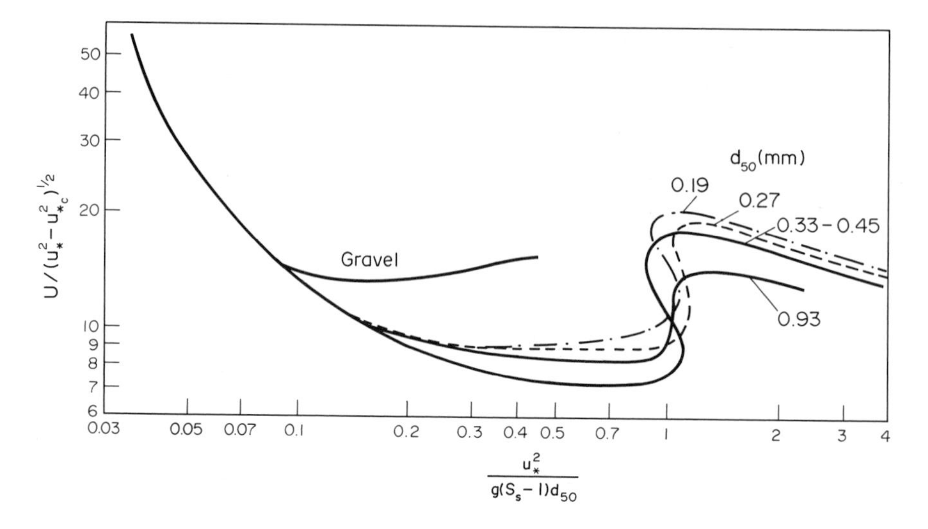

Fig. 6.10. Resistance functions for alluvial channels, Raudkivi (1967).

at any location represent more or less the modal value of $U/(u_*^2 - u_{*c}^2)^{\frac{1}{2}}$. On any perpendicular to a curve the data points indicate a skewed distribution of scatter with the long tail upwards or to the left in the vicinity of θ = 1 on Figure 6.10. The calculation of flow depth requires the selection of a typical curve. It appears that any particular velocity should be qualified by confidence limits, analogous to correlation not accounted for by regression analysis. However, data from which the spread of the distribution of data points could be estimated with some confidence are not available. The method tends to lead to large relative errors at low values of θ.

Some further methods are described by Garde and Ranga Raju (1985) and Simons and Sentürk (1977.

Example 6.1

Calculate the flow depths for flow rates of q = 0.4; 0.5; 0.6 and 1.0 m^3/s per metre width in a river with slope $S = 9 \times 10^{-4}$, bed material d_{35} = 0.30 mm, d_{50} = 0.35 mm, d_{65} = 0.41 mm, S_s = 2.65 and $\nu = 1.141 \times 10^{-6}$ $m^2 s^{-1}$.

The surface drag due to grain roughness from eqn 6.17 or 6.27 with, for example, $k_s = d_{65}$ is

y_o (m)	0.2	0.5	0.8	1.0	1.5
u_* (m/s)	0.042	0.066	0.084	0.094	0.115
U (m/s)	0.90	1.58	2.09	2.39	3.05

For illustration the Einstein-Barbarossa, Engelund and Regime methods are used.

Einstein-Barbarossa Method

The river is assumed to be wide and R $\equiv y_o$, with R' = y_o' being the part associated with grain roughness. The y_o values in the above calculation are for a fixed plane surface of grain roughness d_{65} and are therefore the y_o' values required in loose boundary calculations. Thus, for any selected value of U the y_o' value is known from the table of calculation above (or a plot of U vs y_o). The parameters used in conjunction with Fig. 6.6 are

$$\psi' = (S_s - 1)\frac{d_{35}}{y_o'S} = \frac{0.55}{y_o'}$$

$$u_*''^2 = gy''S = 0.0088\ y'' \text{ or } y'' = 113.26(u_*'')^2$$

U (m/s)	0.5	0.75	1.0	1.25
y_o' (m)	0.08	0.5	0.24	0.34
ψ'	6.9	4.4	2.3	1.72
U/u_*''	9.9	13.5	18.0	23.0
u" (m/s)	0.051	0.056	0.056	23.0
y" (m)	0.29	0.35	0.35	0.33
y_o (m)	0.37	0.50	0.59	0.67

U (m/s)	1.5	2.0	2.5
y_o' (m)	0.46	0.74	1.08
ψ'	1.2	0.73	0.51
U/u_*''	34.0	58.0	100.0
u" (m/s)	0.044	0.035	0.025
y" (m)	0.22	0.13	0.07
y_o (m)	0.68	0.87	1.15

Engelund Method

$\theta' = 1.558\ y'$ $\quad y_o = 0.642\ \theta$

y' (m)	θ'	θ	y_o (m)	U (m/s)	q (m^2/s)
0.2	0.312	0.793	0.51	0.83	0.422
0.3	0.468	1.009	0.65	1.07	0.691
0.4	0.628	1.19	0.77	1.28	0.975
		0.62	0.40		0.510
0.5	0.779	1.34	0.86	1.46	1.258
		0.80	0.51		0.746
0.6	0.935	1.48	0.95	1.64	1.552
		0.95	0.61		0.997
0.8	1.247	1.4	0.90	1.95	1.755
1.0	1.558	2.0	1.28	2.23	2.856

Regime Method

$U = 10.8\ (R^2 S)^{1/3} = 1.0427\ y^{2/3}$

y_o (m)	0.4	0.8	1.2
U (m/s)	0,57	0.90	1.18

Predicted depths

q (m^2/s)	0.4	0.5	0.6	1.0
Depth (m)	y_o	y_o	y_o	y_o
Einstein-Barbarossa	0.51	0.56	0.59	0.68
Engelund	0.49	0.55/0.40	0.60/0.44	0/77/90.61
Regime	0.56	0.64	0.71	0.97

The alternate values by the Engelund method relate to the lower and upper flow regimes.

The results are plotted in Fig. 6.11. In the same manner other methods can be applied and superimposed.

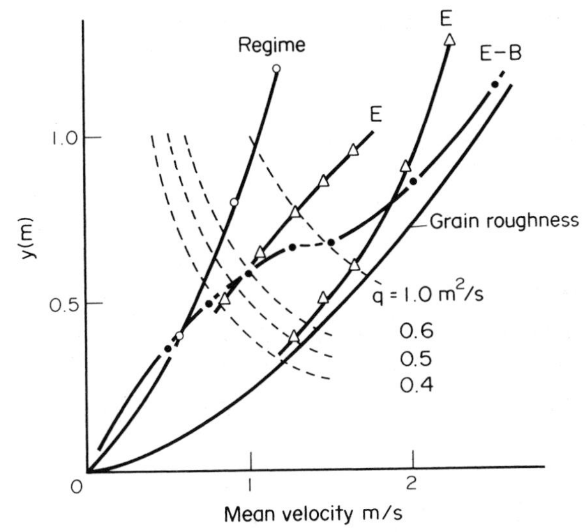

Fig. 6.11. Estimation of flow depths, Example 6.1.

Chapter 7

Sediment Transport

When the applied shear stress on the bed of an alluvial channel exceeds the threshold value for the given bed material the particles on the bed begin to move downstream. The transport is usually mainly for convenience of description, subdivided into *bed load*, *saltation* and *suspension*. A rule of thumb for this subdivision is

$$6 > w/u_* > 2 \qquad \text{bed load}$$

$$2 > w/u_* > 0.6 \qquad \text{saltation}$$

$$0.85 > w/u_* > 0 \qquad \text{suspension}$$

where w is the fall velocity of grains. The bed load is sediment transported in sliding and rolling mode along the bed, hence also known as contact load. This mode of transport occurs at relatively low shear stress excess, $\tau_o - \tau_c$, and the grains generally move intermittently. As the shear stress excess increases more and more particles are put into saltation, a hopping or bouncing mode of movement. The saltation mode is very dominant in transport by air. In water the saltation phase is narrow and the saltation height low because of the small differences in solid and water densities. The true saltation height in water, according to Bagnold (1973) is only of the order of a grain diameter, whereas it can be over a metre in air. However, in laboratory flumes grains can be seen to bounce higher and to remain in suspension for a while (Sumer and Deigard 1981, Sumer and Oguz, 1978). The rolling, sliding and saltation is called the bed load transport. At still higher shear stress excess *suspension* develops. In this mode the upward diffusion of turbulence from the boundary maintains the particles in suspension against gravity. The turbulence bursts, pressure fluctuations and saltation would be involved in entraining the particles from the bed. Once the suspension phase of transport has developed the subdivision is less meaningful, although there would still be particles which roll and slide and those which saltate.

Frequently, a further mode of transport is present, the *wash load*.

The wash load consists of very fine particles, usually in clay and fine silt range, which in general tend to be suspended in Brownian motion. These particles are brought into the stream by overland flow - dirt and dust - and are usually not present in quantities in the bed material. An additional source of wash load is the abrasion of gravel in transport.

The various sediment transport relationships could be divided according to their derivation into groups. The oldest of these are the relationships in terms of $(\tau_o - \tau_c)$ or $(\theta - \theta_c)$, then come formulae based on probabilistic arguments, formulae based on work done (stream power) and more recently, formulae based on computer optimisation of observed data using various dimensionless parameters and computer models. Since all the formulae are based on experimental data, the quality of predictions depends on whether the conditions correspond to those on which the formula was based. The assessment of the quality of any formula does not only depend on the formula but also on the fact that sediment transport in nature is extremely difficult to measure accurately.

7.1 Bed Load Transport

The definitions of bed load transport vary. Bagnold defined bed load as the particles which are in successive contacts with the bed and the processes are governed by gravity only, whereas suspension is carried by upward diffusion of turbulence. Einstein defined bed load as a thin layer of two particle diameters thick with the saltation being part of suspension. This is an arbitrary definition whereas that by Bagnold has a physical basis.

Most of the bed load transport formulae are of the form of the Du Boys (1879) formula which is reproduced here for historic interest.

7.1.1 Du Boys type formulae

Du Boys assumes the bed to move as a series of superimposed layers of thickness d', presumably of the same magnitude as the particle diameter, and the velocity of the layers to vary linearly, Fig. 7.1. If the nth layer from the top remains at rest the surface layer must have a velocity $(n - 1)\Delta V$ and

$$q_B = \frac{1}{2} nd' \ (n - 1)\Delta V$$

The longitudinal component of fluid weight, $\gamma y_o S$, is assumed to be balanced by the friction at the bed. The friction factor between successive layers is assumed constant, f_s, so that the force balance is

$$\gamma y_o S = f_s(\gamma_s - \gamma)nd' = \tau_o$$

The threshold conditions are given when the top layer just resists motion, i.e. when n = 1,

$$\tau_o = \tau_c = f_s(\gamma_s - \gamma)d'$$

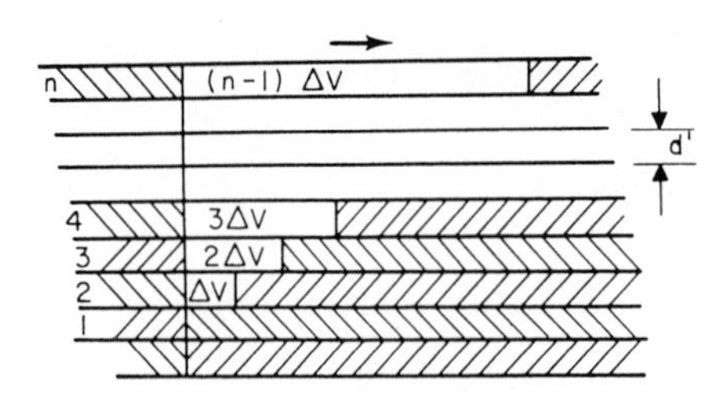

Fig. 7.1. Du Boys model for bed-load transport.

Hence

$$\tau_o = n\tau_o \text{ or } n = \tau_o/\tau_c$$

Substituting this value of n in q_B and rearranging

$$q_B = \frac{\Delta V d'}{2\tau_c^2}\tau_o(\tau_o - \tau_c) = C_s\tau_o(\tau_o - \tau_c) \tag{7.1}$$

Straub (1950) gave

d mm	1/8	1/4	1/2	1	2	4
C_s $(m^6/N^2s) \times 10^5$	3.282	1.945	1.175	0.689	0.405	0.243
C_s $(ft^6/lbf^2\ sec)$	0.81	0.48	0.29	0.17	0.10	0.06
τ_c (N/m^2)	0.766	0.814	1.054	1.533	2.443	4.311
τ_c (lb/ft^2)	0.016	0.017	0.022	0.032	0.051	0.09

or $$C_s = \frac{0.17}{d_{mm}^{3/4}} \tag{7.2}$$

and $$C_s = \frac{0.689 \times 10^{-5}}{d_{mm}^{3/4}}$$

in fps and SI units, respectively.

Introducing from Manning's formula

$$\tau_o/\gamma = yS = S^{7/10}(qn)^{3/5}$$

yields

$$q_B = C_s \frac{S^{1.4}\gamma^2}{(1/n)^{1.2}} q^{3/5} (q^{3/5} - q_c^{3/5}) \tag{7.3}$$

There are many formulae of basically the same form, e.g. Shields (1936):

$$\frac{g_B}{\gamma q} \frac{\gamma_s - \gamma}{\gamma S} = 10 \frac{\tau_o - \tau_c}{(\gamma_s - \gamma)d} \tag{7.4}$$

where g_B is rate of sediment flow by weight per unit width. Schoklitsch (1934) fitted

$$g_B = \frac{7000}{\sqrt{d}} S^{3/2} (q - q_c) \tag{7.5}$$

to data for uniform sediment from d = 0.305 to 7.02 mm where

$$q_c = 1.944d\ 10^{-5}/S^{4/3}\ (m^3/s\ m) \tag{7.6}$$

The formula was modified to

$$g_B = 2500\ S^{3/2} (q - q_c) \tag{7.7}$$

in which

$$q_c = \frac{1}{n} y_c^{5/3} S^{1/2} = 0.26\Delta^{5/3} d^{3/2} S^{-7/6} \tag{7.8}$$

and for $d \geq 0.006$ m

$$y_c = 0.076 \frac{\gamma_s - \gamma}{\gamma} \frac{d}{S} = 0.076\ \Delta d/S \tag{7.9}$$

and

$$n = 0.0525\ d^{1/6} \tag{7.10}$$

E.T.H. in Zurich, in the period 1930-50, proposed several formulae, e.g. the Meyer-Peter (1934) formula,

$$g_B^{2/3} = 2.5\ q^{2/3} S - 42.5\ d \tag{7.11}$$

or

$$0.4 \frac{g_B^{2/3}}{d} = \frac{q^{2/3} S}{d} - 17$$

The formula was fitted to laboratory data using gravel of d = 5.05 and 28.6 mm. An average grain size was used, expressed as

$$d = \frac{1}{100} \Sigma d_i \Delta p_i$$

where Δp_i is % of particle size fraction of mean diameter d_i.

This formula was extended by including data from experiments with material with a range of specific gravities:

$$\frac{q^{2/3}S}{d} - 9.57\,(\gamma_s - \gamma)^{10/9} = 0.462\,(\gamma_s - \gamma)^{1/3}\left[\frac{g_B(\gamma_s - \gamma)}{\gamma_s}\right]^{2/3} \tag{7.12}$$

A further modification, the Meyer-Peter and Müller formula, includes the separation of bed resistance into resistance due to grain roughness, S', and bed forms, S", as

$$\frac{\gamma R(k/k')^{3/2}S}{(\gamma_s - \gamma)d} - 0.047 = 0.25\,\frac{\rho^{1/3}}{(\gamma_s - \gamma)d}\left[\frac{g_B(\gamma_s - \gamma)}{\gamma_s}\right]^{2/3} \tag{7.13}$$

where R is the hydraulic mean radius (y_o for unit width). The k/k' term is frequently called the roughness factor or ripple factor. If one followed the partition

$$\tau_o = \tau_o' + \tau_o'' \text{ or } u_*^2 = u_*'^2 + u_*''^2$$

where u_*^2 is proportional to RS or y_oS then

$$\frac{k}{k'} = \frac{(y_oS)'}{y_oS} = \left[\frac{C}{C'}\right]^2 \tag{7.14}$$

where C is the Chézy coefficient. The ripple factor was empirically set equal to

$$\frac{S'}{S} = \left[\frac{k}{k'}\right]^{3/2} \tag{7.15}$$

instead of $(k/k')^2$ as per eqn 7.14, and $k' = 26/d_{90}^{1/6}$ was used. It was reported that k/k' varied within 0.5 to 1.0 which is in keeping with coarse sediment and small form drag contribution. For mixtures $d = d_m$ (eqn 2.4).

Kondrat'ev et al. (1962) give the formula by Shamov

$$g_B = 0.95\,d^{1/2}\,(U/U_c)^3(U - U_c)(d/y_o)^{1/4}\ \text{kg s}^{-1}\,\text{m}^{-1} \tag{7.16}$$

where

$$U_c = 3.83\,d^{1/3}\,y_o^{1/6}\ (\text{ms}^{-1}) \tag{7.17}$$

is the critical velocity for beginning of sediment movement.

Note that g_B is the weight rate of transport per unit width which in the formulae appear as kg and can be interpreted as kg mass. The immersed weight rate of transport is $[(\rho_s - \rho)/\rho_s]g_B$. The bulk volume of transport is given by $q_B/(1 - n)$ where q_B is m³ s^{-1} m^{-1} and n is the porosity of the deposit.

7.1.2 Formulae based on probability concepts

Einstein (1942,1950) departed from the mean tractive force concept. The starting-point of his argument is that in the turbulent flow the fluid forces acting on the particle vary with respect to both time and space, and therefore the movement of any particle depends upon the probability, p, that at a particular time and place the applied forces exceed the resisting forces.

The probability of movement of any one particle is expressed in terms of weight rate of sediment transport, the size and immersed weight of particles, and a characteristic time which is a function of particle size/fall velocity ratio. It is postulated that a given particle moves in a series of steps and that a given particle does not stay in motion continuously.

From these considerations a transport function is developed as

$$\phi = \frac{g_B}{g\rho_s}\left[\frac{\rho}{\rho_s - \rho}\,\frac{1}{gd^3}\right]^{\frac{1}{2}} = \frac{q_B}{\sqrt{\{(S_s - 1)gd^3\}}} \qquad (7.18)$$

Next, the probability is interpreted as the fraction of the bed on which, at any given time, the lift on a given particle is sufficient to cause motion. The flow parameters are based on the logarithmic velocity distribution. The expression derived is

$$\psi = \frac{\rho_s - \rho}{\rho}\,\frac{d}{R_B S} = \frac{1}{\theta} \qquad (7.19)$$

and is the reciprocal of the Shields' entrainment function. Einstein subdivides the total drag into form and surface drag and R_B is the fraction of the hydraulic mean radius appropriate to surface drag. The result is given as

$$\phi = f(\psi)$$

Generally, most formulae can be reduced to the form

$$\phi = f(\theta) \qquad (7.20)$$

Experimental data from Zurich and data published by Gilbert and Murphy (1914) plotted as ψ versus ϕ are shown in Fig. 7.2. Data for $\phi < 0.4$ are represented by

$$0.456\,\phi = e^{-0.391\psi}; \quad \phi < 0.4 \qquad (7.21)$$

For large rates of sediment transport the data deviates from this straight-line relationship, (1) on Fig. 7.2.

The probability-bed load intensity relationship is written in the form

$$\frac{p}{1 - p} = A_* \left[\frac{i_B}{i_b}\right] \phi = A_*\phi_* \qquad (7.22)$$

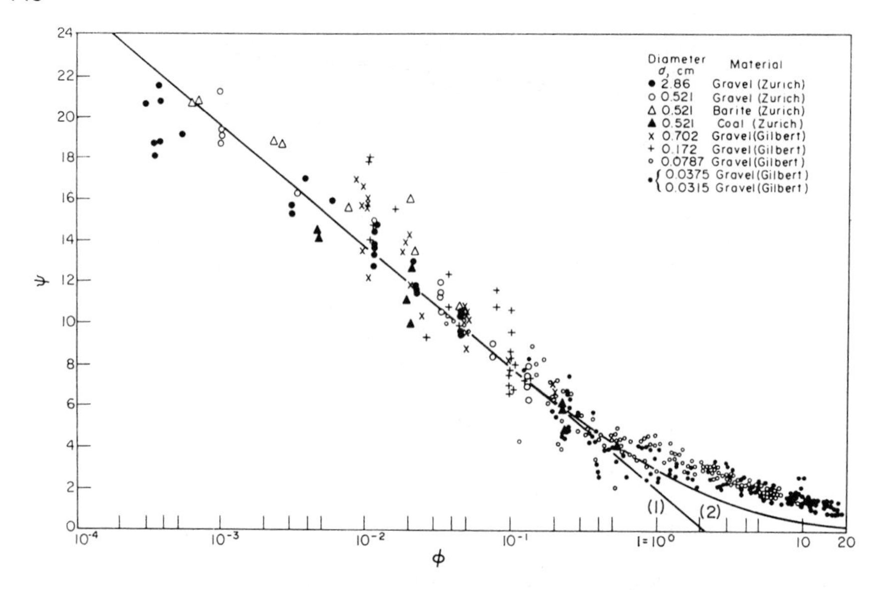

Fig. 7.2. Einstein's bed-load equations. (After Einstein, 1942).

where A_* is a constant determined by experiments (= 43.5), i_B is fraction of bed load in a given grain size, and i_b is fraction of bed material in the given grain size. The above relationship is shown as curve (2) in Fig. 7.2. The presence of suspended load is given as the cause for deviation of experimental points from this line. The probability of erosion p was assumed to follow the normal error distribution which combined with eqn 7.22 lead to

$$1 - \frac{1}{\sqrt{\pi}} \int_{-B_*\psi_* - 1/\eta_o}^{B_*\psi_* - 1/\eta_o} e^{-t^2}\, dt = p = \frac{A_*\phi_*}{1 + A_*\phi_*} \tag{7.23}$$

where B_* is another constant equal to 1/7, η_o = 0.5 and t is variable of integration only. This relationship ϕ_* versus ψ_* is shown in Fig. 7.3 where

$$\psi_* = \xi Y(\beta^2/\beta^2_x)\psi \tag{7.24}$$

in which ξ and Y are correction factors. The factor ξ accounts for sheltering of smaller particles and is shown as ξ versus d/X on Fig. 7.4a. In here

$$X = 0.77\ \Delta \text{ if } \Delta/\delta' > 1.80$$

$$X = 1.39\ \delta' \text{ if } \Delta/\delta' < 1.80$$

where X is a characteristic grain size of the mixture, $\Delta = k_s/N$ is the apparent roughness in the logarithmic velocity distribution

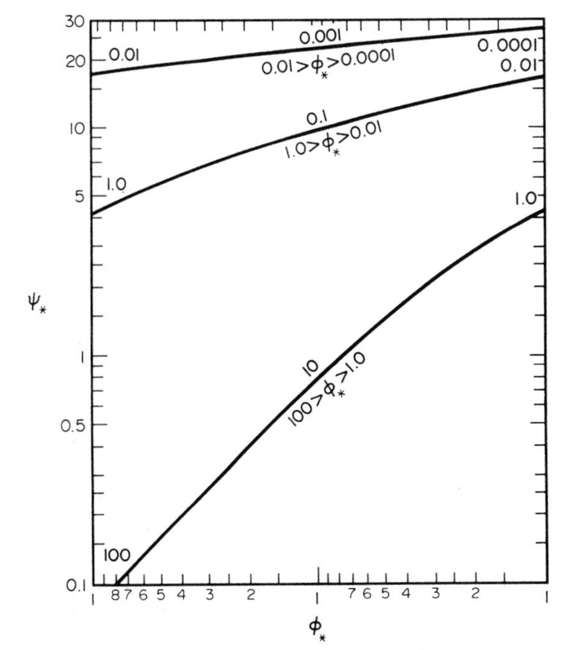

Fig. 7.3. Plot of Einstein function $\phi_* - \psi_*$.

$$u_b = 5.75\ u_* \log \left[\frac{(30.2)(0.35\ X)}{\Delta}\right] \tag{7.25}$$

which is the effective velocity at a distance 0.35 X from the theoretical bed. The correction factor N is a function of k_s/δ', Fig. 7.4b, where $\delta' = 11.5\ \nu/u_*'$, $u_*' = (gR'S)^{0.5}$ and $k_s \cong d_{65}$. The correction factor Y is shown on Fig. 7.4c. Further, $\beta_x = \log_{10}(10.6\ X/\Delta)$ and $\beta = \log_{10} 10.6$. (The partitioning of total drag into surface and form drag is discussed in Chapter 6.)

The total transport rate of the mixture is obtained by summing up the transport rates of individual components i_B/i_b within the mixture. For near uniform mixtures the total transport rate can be calculated using d_{35} as the effective diameter.

The formula incorporates a number of assumptions. For example, no evidence or justification is given in support of the assumption that $A_L = L/d \sim 100$ const. Available experimental evidence suggests that A_L increases with increasing $\rho u_*^2/\gamma_s{*}d$. On the "time of the grain", i.e. the duration of the jump $t_1 = A_3(d/w)$, Einstein (1950) writes: "No method exists today of determining experimentally the exchange time t. But experiments indicate that t is another characteristic constant of the particle, ..." How? The time frequency of the jumps must depend on turbulence characteristics and if anything time should be proportional to d/u_*. The "universal constants" A_*, B_* and η_o are not constants as emerges from the work by Bishop et al. (1965). Some queries also arise from the development of the probabilities and the

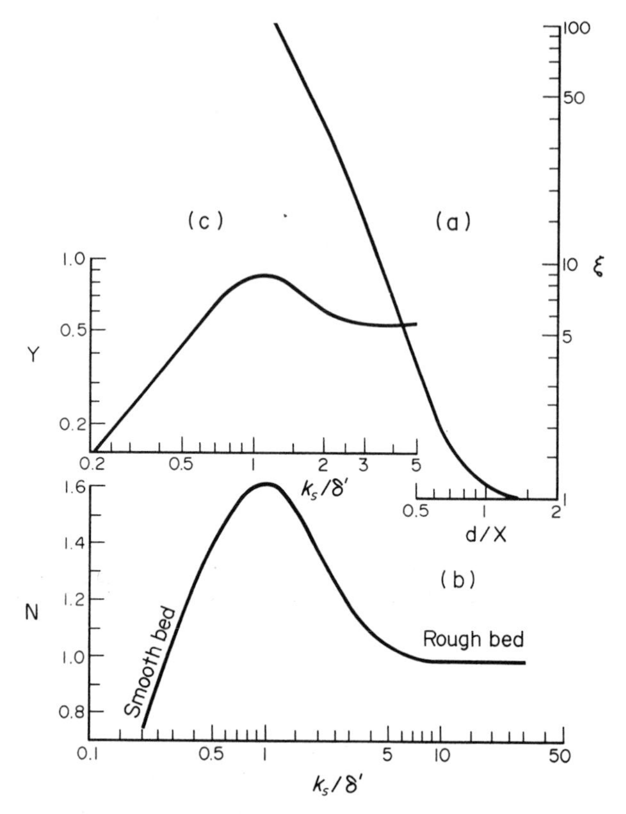

Fig. 7.4. Correction factors in the Einstein sediment transport formula.

assumption that the normal component of the fluctuating force always acts upwards. Yalin (1972) gives an extensive discussion of the Einstein formula.

Most of the concern has centred around the sheltering function ξ or hiding factor. In principle, it is intended to account for the difference in mobility of the various grain sizes in the mixture compared to their mobility in beds of respective uniform sized grains. The hiding factor as introduced by Einstein has been found to overpredict the transport rate of fine and under-predict that of coarse fractions of the mixture. Figure 7.5 shows a comparison by Misri et al. (1984) of the Einstein and Hayashi et al. (1980) functions, the function proposed by Pemberton (1972) from analysis of field data and laboratory data by the author, where d_a is the average grain size of mixture. Parker and Klingeman (1982) in their treatment of the hiding started from the transport of uniform gravel. Introduction of $W = \phi_B/\theta^{3/2}$ and W_r a low near threshold transport rate ($W = 0.002$) leads to $W/W_r = G(\chi)$, where $\chi = \theta/\theta_r$ and the function of $G(\chi)$ was fitted by $G(\chi) = \exp[14.2(\chi-1) - 9.28(\chi-1)^2]$ for $0.95 \leq \chi \leq 1.65$ (Andrews and Parker, 1985). They showed that for the surface layer $\theta_{ri}/$

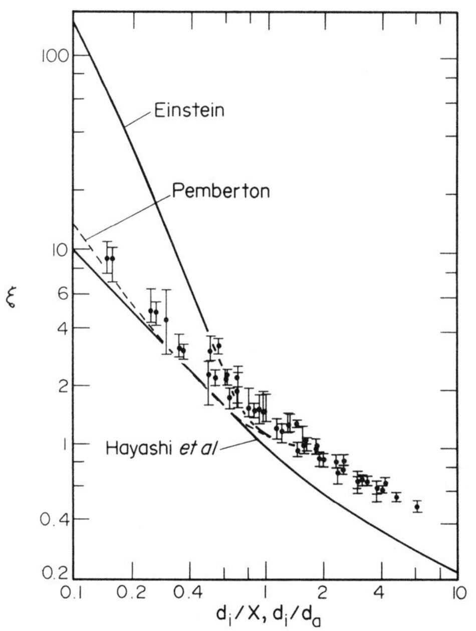

Fig. 7.5. The sheltering function ξ according to Misra et al. (1984). Vertical bars show the range of of scatter of data by Misra et al.

$\theta_{r50} = (d_i/d_{50})^{-0.982}$ which is an almost linear dependency. The bed shear stresses then become $\tau_{ri}/\tau_{r50} = (d_i/d_{50})^{0.018}$. In the absence of any hiding the ratio should be to the power of one. Diplas (1986) continuing from Parker and Klingeman and Andrews and Parker incorporated χ_{50} and d_i/d_{50} in the expression for hiding factor, which, based on Oak Creek data, is expressed as

$$\xi\left[\chi_{50}, d_i/d_{50}\right] = \left[\chi_{50}{}^{(d_i/d_{50})^{0.3214}-1}\right]\left[\left[\frac{d_i}{d_{50}}\right]^{-0.0606}\right]^{(d_i/d_{50})^{0.3214}} \qquad (7.26)$$

This is a family of curves through ($\xi = 1$, $d_i/d_{50} = 1$), the equal mobility point. The value of ξ remains approximately equal to one for $\chi_{50} \cong 1.2$. For $\chi_{50} > 1.2$ the particles larger than d_{50} are more mobile than those smaller, and vice versa when $\chi_{50} < 1.2$.

Considering the very complex interactions of grains in the surface layer, the formation of armour clusters and the flow structure over such a bed, the hiding factor can only be looked upon as a factor which could be evaluated from data. It also hides our inability to describe the principal process.

Brown (1950) found that the bulk of the data used by Einstein were satisfied by

$$\phi = 40(1/\psi)^3 \qquad (7.27)$$

Hence, the Einstein-Brown formula

$$\frac{q_B}{\sqrt{\{(S_s - 1)gd^3\}}F} = 40\left[\frac{\tau}{\gamma(S_s - 1)d}\right]^3 \qquad (7.27a)$$

where

$$F = \frac{2}{3} + \left[\frac{36\nu^2}{gd^3(S_s - 1)}\right]^{\frac{1}{2}} - \left[\frac{36\nu}{gd^3(S_s - 1)}\right]^{\frac{1}{2}}$$

Experimental data from uniform sediments are well described by eqn 7.27 for $0.01 \leq \phi \leq 0.1$. For $\phi > 0.1$ eqn 7.21 gives a better fit.

For low rates of transport, $\phi_B < 10$, the Einstein formula follows closely the simple relationship $\phi_B = 8(\theta - 0.047)^{3/2}$ by Meyer-Peter.

Engelund and Fredsøe (1976) also proposed a bed load transport relationship incorporating the probability p of particle movements in the surface layer which contains $1/d^2$ grains per unit area. Their result is

$$q_B = 9.3\,\frac{\pi}{6}\,dpu_* \,[1 - 0.7(\theta_c/\theta)^{1/2}]$$

or

$$\phi_B = 5\,p(\sqrt{\theta} - 0.7\,\sqrt{\theta_c}) \qquad (7.28)$$

where

$$p = \{1 + [(\pi/6)\beta/(\theta - \theta_c)]^4\}^{1/4}$$

and $\beta = \tan 27° = 0.51$ is the dynamic friction coefficient, assumed to be constant.

The probabilistic approach, of course, has been used by a number of researchers, for example, by Kalinske (1942,1947) and Frijlink (1952) who tried to determine the amount of shear taken by solid particles in motion. The formula by Frijlink was adapted by Bijker for calculation of coastal sediment transport rates and is discussed in the chapter on coastal processes.

Misri et al. (1984) modified the Einstein type transport relationships. They plotted from laboratory and field data ϕ_B versus θ' for uniform sediment and obtained a function with little scatter. This they fitted by

$$\phi_B = 4.6 \times 10^7(\theta')^8; \qquad \theta' \leq 0.065$$

$$\phi_B = \frac{8.5(\theta')^{1.8}}{\left[1 + \frac{5.95 \times 10^{-6}}{(\theta')^{4.7}}\right]^{1.43}}; \qquad \theta' \geq 0.065 \qquad (7.29)$$

where θ' is based on bed shear stress due to grain roughness ($\tau_o' = \rho g R'S$; $U' = (1/n)(R')^{2/3}S^{1/2}$; $n = d/24$). The data covered $10^{-6} < \phi_B < 20$ and $0.02 < \theta' < 1.1$.

For nonuniform sediment a new sheltering function was defined as

$$\xi_B = \frac{d_i}{d_a} = \frac{\tau_e}{\tau_o'} = f\left[\theta_i'; \ \frac{\tau_o'}{\tau_{oc}}; \ C_u\right] \tag{7.30}$$

where τ_e is the effective shear stress for transport of size fraction d_i as bed load, τ_{oc} is the critical shear stress based on d_a of the mixture and the Shields' criterion, τ_o' is based on size d_i and C_u is the uniformity coefficient (eqn 2.5d). This relationship accounts for sheltering of finer fractions and for the increased exposure of the coarse fractions. The data were fitted by

$$\xi_B = \frac{0.038\ K(\tau_o'/\tau_{oc})^{0.75}}{(\theta_i')^{1.2}[1 + 0.0031(\theta_i')^{-2.1}]^{0.33}} \tag{7.31}$$

where K and C_u are related by

C_u:	> 0.48	0.4	0.3	0.25
K :	1.0	1.1	1.3	1.6

The dimensionless bed load ϕ_B of fraction d_i is then calculated with $\tau_e = \xi_B\tau_o'$ from eqn 7.29 or from Fig. 7.6. This yields the transport rate $i_B g_B$ of each fraction from

$$\phi_B = \frac{i_B g_B}{\rho_s g d_i i_b}\left[\frac{\rho}{\rho_s - \rho}\ \frac{1}{g d_i^3}\right]^{1/2}$$

The total is given by summation.

Ranga Raju (1985) modified the method using the coefficients $K_B = f(\tau_o'/\tau_{oc})$; $L_B = f(C_u)$ and $K_B L_B \xi_B = f(\theta_i')$. The L_B - C_u relationship given is

C_u	< 0.25	0.3	0.35	> 0.4
L_B	0.625	0.769	0.909	1.0

The coefficient K_B is shown in Fig. 7.7. With known θ_i' the value of $K_B L_B \xi_B$ can be read from Fig. 7.8 yielding ξ_B. Thus with known $\xi_B\theta_i'$ the value of ϕ_{Bi} can be read from Fig. 7.6.

Van Rijn (1984) developed an analytical model for bed load transport in terms of the saltation height, particle velocity and bed load concentration. The saltation height and particle velocity were calculated using a computer model which was calibrated

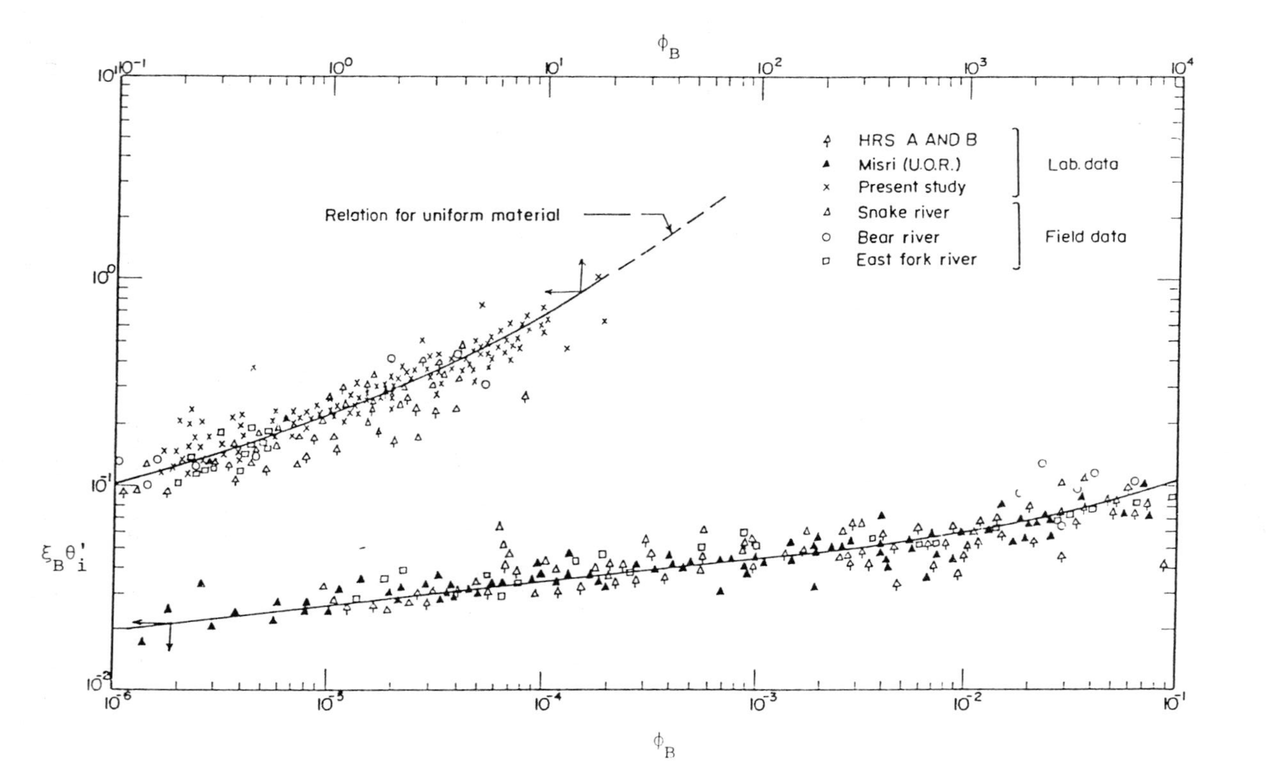

Fig. 7.6. Bed load transport ϕ_B according to Misra et al. (1984).

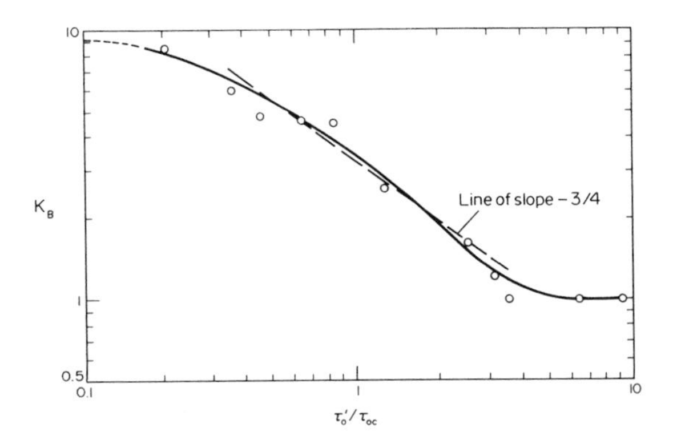

Fig. 7.7. Relation between K_B and τ_o'/τ_{oc}.

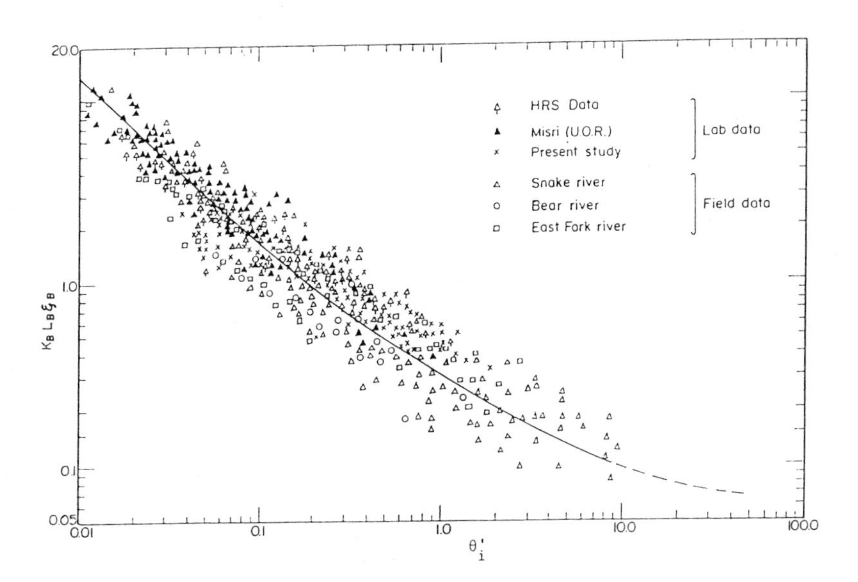

Fig. 7.8. Relation between θ_i' and $K_B L_B \xi_B$.

against laboratory data. The transport equation was expressed in terms of the dimensionless particle size D_* and his transport parameter T as

$$\frac{q_B}{[(S_s - 1)g]^{0.5}\, d_{50}^{1.5}} = 0.053\, \frac{T^{2.1}}{D_*^{0.3}} \tag{7.32}$$

for grain sizes from 0.2 to 2 mm, where q_B is in $m^2 s^{-1}$, $T = [(u_*')^2 - u_{*c}^2]/u_{*c}^2$, $D_* = [(S_s - 1)g/\nu^2]^{1/3} d_{50}$ and u_{*c} is the critical shear velocity according to Shields. The grain shear stress $\tau_o' = \rho g U^2/(C')^2$, or $u_*' = \sqrt{g}\, U/C'$ where $C' = 18 \log(12 R_b/3 d_{90})$. A verification analysis using published data shows that "about 77% of the predicted bed load transport rates are within 0.5 to 2 times the observed values".

A simplified formula when only the mean velocity, flow depth and particle size are known, was given as

$$\frac{q_B}{Uy_o} = 0.005 \left\{\frac{U - U_c}{[(S_s - 1)gd_{50}]^{\frac{1}{2}}}\right\}^{2.4} \left\{\frac{d_{50}}{y_o}\right\}^{1.2} \tag{7.33}$$

where

$$U_c = 0.19 d_{50}^{0.1} \log\left[\frac{12R_b}{3d_{90}}\right] ; \quad 0.1 \leq d_{50} \leq 0.5 \text{ mm}$$

$$U_c = 8.5 d_{50}^{0.6} \log\left[\frac{12R_b}{3d_{90}}\right] ; \quad 0.5 \leq d_{50} \leq 2.0 \text{ mm}$$

and R_b is the hydraulic mean radius for the bed (side wall correction applied).

7.1.3 Formulae based on work done

Bagnold (1956) proposed a formula based on the concept of work done. The mass of grains passing the cross section per unit width and time is

$$g_B = \rho_s \int_0^{y_o} cu\,dy \tag{7.34}$$

The available energy is taken to be proportional to $(\tau_o - \tau_c)u_*$ and the useful work done is

$$W = \Delta(\tau_o - \tau_c)\, u_* \eta \tag{7.35}$$

where η is the efficiency of the transport. Here, τ_c is the non-effective part of shear; that is, threshold and form drag. Strictly speaking, τ_c should be multiplied by velocity or shear velocity at threshold and not with the given value to have a clear physical meaning.

The final expression is

$$g_B = AB_b \rho_s d \left[\left[\frac{\rho_s - \rho}{\rho}\right] gd \cos\beta\right]^{\frac{1}{2}} \theta^{1/2}(\theta - \theta_c) \tag{7.36}$$

For usual values of river slopes this is

$$\phi = \frac{g_B}{g\rho_s}\left[\frac{\rho}{\rho_s - \rho}\frac{1}{gd^3}\right]^{\frac{1}{2}} = AB\theta^{1/2}(\theta - \theta_c) \tag{7.37}$$

The constant A is a function of Reynolds number and equal to 8.5 for $u_*k/\nu > 70$ and 5.5 for $u_*k/\nu < 5$, with a transition between these. The constant B is the ratio of the efficiency η to the friction coefficient $\tan\alpha$. Bagnold derived for bed load in water

$$B_b = \frac{[(2\tan\alpha)/3\,C_D]^{1/2}}{1 - \tan\beta/\tan\alpha} \rightarrow [(2\tan\alpha)/3\,C_D]^{1/2} \tag{7.38}$$

if $\tan\beta$, the bed slope, can be neglected, where C_D is the drag coefficient estimated from Fig. 2.8. For transport in air B is half the above value and θ_c becomes zero as soon as saltation sets in.

The work done was later (Bagnold 1973,1977) related to the stream power

$$P = \tau_o U = \rho g y_o S U = \rho u_*^2 U \tag{7.39}$$

where τ_o was assumed to act at the top of bed load layer at y = nd, where empirically $n = 1.4(u_*/u_{*c})^{0.6}$. Using u = U when y = $0.37y_o$ and a relative velocity between water and sediment of $u_r \cong w$, the fall velocity, lead to

$$\frac{g_{iB}}{P} = \frac{u_* - u_{*c}}{u_* \tan\alpha}\left[1 - \frac{5.75\,u_*\log(0.37\,y_o/nd) + w}{U}\right] \tag{7.40}$$

where g_{iB} is the immersed weight rate of transport. For τ_o = const. conditions near the bed remain constant. An increase in depth leads to an increase in U and the ratio U/u_n and a decrease in sediment transport efficiency where u_n is u at y = nd. This means that for the same stream power the transport rate decreases with increasing water depth or an increase in sediment load must lead to widening of the stream. Bagnold (1980) fitted a function to available bed load data according to which

$$\frac{g_B^*}{(g_B^*)_1} = \left[\frac{P - P_o}{(P - P_o)_1}\right]^{3/2}\left[\frac{y_o}{y_{o1}}\right]^{-2/3}\left[\frac{d}{d_1}\right]^{-1/2} \tag{7.41}$$

where values with subscript 1 refer to measured reference values used to make the equation dimensionless and g_B^* is immersed weight rate of bed load transport per unit width. The data cover three orders of magnitude in water depth and four in grain size.

Yalin (1963,1972) developed a bed load equation incorporating reasoning similar to Einstein but he interprets the transport rate as a rate of doing work, similar to Bagnold,

$$g_B = W_s \times u_s$$

where W_s is the weight of solids moving over unit area of the bed at velocity u_s. He then proceeds to develop expressions for u_s and W_s. His transport equation is

$$\frac{g_B}{\gamma_s{}^* du_*} = \text{const.} \left\{ s \left[1 - \frac{1}{as} \ln(1 + as) \right] \right\} \tag{7.42}$$

where

$$s = (\theta_o - \theta_c)\theta_c = u^2{}_* / u^2{}_{*c} - 1,$$

$$\theta_o = \rho u_*{}^2 / \gamma^*{}_s d,$$

$$a = 2.45\ (\rho u^2{}_{*c}/\gamma^*{}_s d)^{0\cdot 5} / (\rho_s/\rho)^{0\cdot 4} = 1.66\ \sqrt{\theta_c}$$

when $\rho_s/\rho = 2.65$, and contains one additional constant which has to be determined experimentally. By fitting his function to the Zürich and Gilbert data Yalin obtains a value of 0.635 for the constant.

His formula can be converted into the Einstein co-ordinates by multiplying through with $(\rho u^2{}_*/\gamma^*{}_s d)^{0\cdot 5} = 1/\sqrt{\psi}$, yielding

$$\phi = \frac{g_B \sqrt{\rho}}{(\gamma^*{}_s d)^{3/2}} = 0.635 \frac{s}{\sqrt{\psi}} \left[1 - \frac{1}{as} \ln(1 + as) \right] \tag{7.43}$$

or

$$\phi = f(\psi, \theta_c, \rho_s/\rho),$$

that is, Yalin's formula in the Einstein's co-ordinates implies a set of curves. This reduces to one family $\phi = f(\psi, \theta_c)$ for ρ_s/ρ = constant. Such a family with θ_c as parameter is shown on Fig. 7.9.

At the beginning of transport $\theta - \theta_c$ is small,

$$\frac{1}{as} \ln(1 + as) \cong 1 - \frac{1}{2} as,$$

and eqns (7.42) and (7.43) become

$$\frac{g_B}{\gamma^*{}_s du_*} \cong \frac{1}{2} \text{ const } as^2 \tag{7.42a}$$

and

$$\phi = \frac{1}{2} \text{ const. } \frac{as^2}{\sqrt{\psi}} = \frac{a \text{ const.}}{2\theta_c{}^2} \theta^{1/2} (\theta - \theta_c)^2 \tag{7.43a}$$

At large values of shear stress θ_o, as $as \to \infty$, and the term (1/as)-

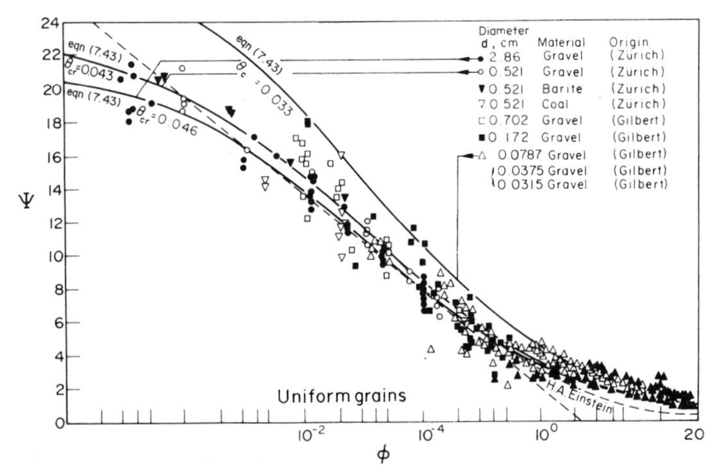

Fig. 7.9. Yalin's sediment transport formula.

ln (1 + as) approaches zero, eqns (7.42) and (7.43) yield

$$\frac{g_B}{\gamma^*_s d u_*} \cong \text{const. } s \tag{7.42b}$$

and

$$\phi = \text{const. } \frac{s}{\sqrt{\psi}} = \frac{\text{const.}}{\theta_c} \theta_o^{1/2}(\theta_o - \theta_c) \tag{7.43b}$$

The latter is the same as the Bagnold's eqn. (7.35).

The concept of stream power has been extensively applied by Yang to predict sediment transport rates. This will be discussed under total sediment load but the relationship for gravel transport (Yang, 1984) is quoted here for reference. Yang uses the "unit stream power", defined as the ratio of average velocity times energy slope to average fall velocity of sediment, US/w, and total sediment concentration, C, in parts per million by weight. The relationship for gravel is

$$\log C = 6.681 - 0.633 \log \frac{wd}{\nu} - 4.816 \log \frac{u_*}{w} + \left[2.784 - 0.305 \log \frac{wd}{\nu} - 0.282 \log \frac{u_*}{w}\right] \log\left[\frac{US}{w} - \frac{U_c S}{w}\right] \tag{7.44}$$

where U_c is the critical average velocity for incipient motion of sediment. The function was obtained by regression analysis using the data by Gilbert and Murphy (1914) and Casey (1935).

Paintal (1971) developed a stochastic model for bed load transport based entirely on probabilistic considerations. It is assumed that the particle exposures are uniformly distributed, the turbulence is normally distributed and the length of steps by the particles follows a negative exponential distribution. This leads to

$$q_{s*} = Af(B\theta_o) \tag{7.45}$$

where $q_{s*} = q_s\sqrt{\rho}/\{\gamma_s d\sqrt{[(\gamma_s - \gamma)d]}\}$, A and B are constants and θ_o is the mean dimensionless bed shear stress. Figure 7.10 shows the function fitted to experimental data.

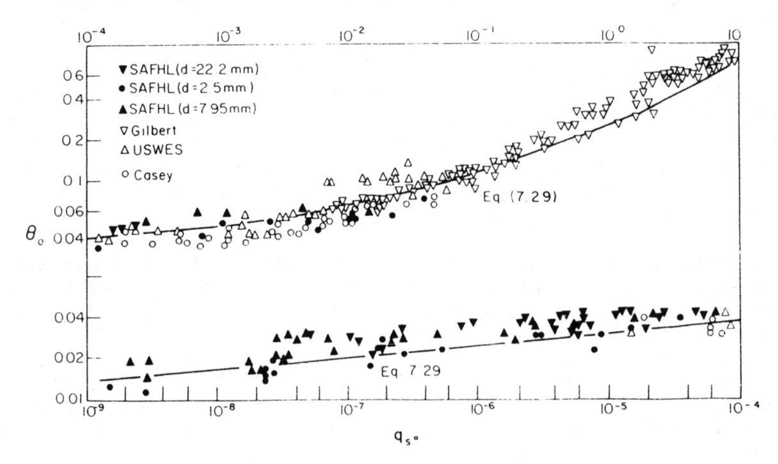

Fig. 7.10. Variation of q_{s*} with θ_o. (Paintal, 1971).

7.1.4 Comments

Most of the bed load transport formulae can be expressed as $\phi = f(\theta)$ or $q_B = f(\tau_o - \tau_c)$ or $q_B = f(q - q_c)$. For example, the Shields formula as

$$\frac{g_B}{\rho g u_* d} = 10\,\frac{U}{u_*}\,\theta(\theta - \theta_c) \tag{7.46}$$

or the Meyer-Peter and Müller formula, eqn 7.13 as

$$\phi = 8[\theta - 0.047]^{3/2} \tag{7.47a}$$

or

$$\left[\frac{\rho_s - \rho}{\rho_s}\right]\frac{g_B}{\rho_s u_*^3} = 8\left[1 - \frac{0.047}{\theta}\right]^{3/2} \tag{7.47b}$$

where $(k/k')^{3/2}$ is set equal to one.

The major problem is the determination of effective bed shear stress which is not equal to the grain roughness of the total bed shear stress. The form drag component of the total bed shear has an indirect effect. The form drag causes energy dissipation, i.e., increased turbulence near the bed which affects sediment entrainment. In general, bed load transport decreases with increasing roughness of the bed but there is as yet no unanimity on how to evaluate the effective bed shear stress.

The transport rates or shear stresses are expressed in terms of flow depths but since the transport relationships are non-linear the use of average depth is unsatisfactory. There are two alternate methods to calculate transport rates. One is to delineate approximately constant depth bands of flow and sum the transport rates, the other is to use an average depth. For example, if in the formula used the transport rate depends on depth cubed then the average depth $\overline{y}_o = (\overline{\Sigma y_o^3})^{1/3}$ could be used.

Bed load is usually the major component of transport in gravel rivers. Bed load formulae also imply steady state conditions but in gravel rivers this applies only if the bed shear stress is so high that no armouring can develop. At lower than the critical or limiting shear stress for the armour layer the transport rate decreases with time.

Further complications arise with transport at very small shear stress excess where the transport rate has been shown to be extremely sensitive to the shear stress. Taylor and Vanoni (1972) showed that the dimensionless transport rate $q_* = \theta^{17.5}$. A similar slope for the q_*-θ-function was found by Paintal, Fig. 7.10. Since then other researchers have shown that the exponent of θ is high at very low values of $(\theta - \theta_c)$ and decreases with increasing shear stress excess, e.g. Pazis and Graf (1977), $q_s \propto \theta^{16}$ for $g_{s*} < 0.0046$ and $q_s \propto \theta^{8.15}$ for $q_{s*} > 0.0046$

where $g_{s*} = \frac{g_s}{\rho_s g}\left[1/(S_s - 1)\frac{1}{g d_{50}^3}\right]^{1/2}$ and $q_* = q_s/q_s$ ref:

Bed load transport in the field is extremely difficult to measure even at great expenditure of money and effort and therefore the assessment of the "quality" of a formula is even more difficult. The U.S. Geological Survey built a concrete channel across the East Fork River bed, a trough 0.4 m wide and 0.6 m deep incorporating a conveyor belt which continuously extracted the sediment for analysis. It was returned to the river by another conveyor belt. The river is about 18 m wide and 1.2 m deep at bankfull discharge. Leopold and Emmett (1976) start the discussion of the results with "The most important general characteristic of bedload derived from this investigation is its variability. Bedload-transport rates are extremely variable in both space and time at a given river cross-section." The d_{50} of the bed load varied from 0.45 to 1.63 mm while the maximum size remained at about 16 mm. Sediment finer than 0.25 mm amounted only to a few percent. The data appear to confirm Bagnold (1966,1973) hypothesis that at high flows the transport rates are proportional to stream power. The efficiency of the transport increases with decreasing grain size, being only about 3% with 50 mm gravel and about 30% with 0.3 mm sand.

7.2 Suspended Sediment Transport

Only at relatively small values of shear stress excess, $\tau_o - \tau_c$, is the transport confined to bed load only. Increase in bed shear stress soon leads to suspension and transport as bed and suspended load. Since suspension is transported at approximately the velocity of flow the quantity of sediment transported as suspended load is usually very much greater than that of bed load which moves much more slowly.

The sediment is maintained in suspension, against the gravitational fall velocity, by the diffusion of turbulence from the bed. For this the rms values of the vertical turbulence components $(v'^2)^{\frac{1}{2}} = v^*$ need to be equal or greater than the fall velocity w. Boundary layer flow studies indicate that v^* is of the same order as the shear velocity u_*. Thus for initiation of suspension $u_*/w \geq 1$ or $\theta \geq w^2/(S_s - 1)gd$. Bagnold (1966), on the basis of experimental data, set $v^* \cong 0.8\, u_*$ and $\theta \geq 0.4\, w^2/(S_s - 1)gd$. Van Rijn (1984) proposed from his experimental data for the initiation of suspension

$$\frac{u_{*c}}{w} = \frac{4}{D_*} \text{ for } 1 < D_* \geq 10$$

$$\frac{u_{*c}}{w} = 0.4 \text{ for } D_* > 10 \tag{7.48}$$

where $D_* = [(S_s - 1)gd^3/\nu^2]^{1/3}$. Equation 7.48 sets a substantially lower limit than the criterion by Bagnold, by a factor of 5 to 6 for $D_* > 10$.

Observations show that the concentration decreases with the distance up from the bed. One of the aims of the theories of sediment suspension is to define this distribution of concentration. The literature on the subject of sediment suspension is voluminous but the physical and analytical description of suspension is still far from complete.

The initiation of suspension in a turbulent flow can be linked with the turbulence bursts and pressure fluctuations acting on the bed, as described in Section 3.2. These bursts and tongues also lead to formation of vortices which trap sediment and lift it like little tornadoes. However, the dilatation stresses are also likely to play a role in the initiation of suspension. Indeed, they are the only supporting mechanism in laminar flows of very high concentrations, e.g., mud flows.

The analytical models for sediment suspension may be subdivided into diffusion, energy, stochastic and numerical models, although the labelling is not clear cut. The diffusion models have found the greatest acceptance for practical applications and are usually also part of the numerical or mathematical models of suspended sediment transport.

7.2.1 The diffusion approach

Consider, for simplicity, the spreading of a "marked fluid" in another of the same density or of small, neutrally buoyant particles. The spreading may be caused by:

- (i) scattering due to random molecular motion of the host fluid,
- (ii) turbulent mixing, or
- (iii) it may result from systematic differences in velocity over the cross-section of the stream.

In the latter case elements of the marked fluid initially in the same cross-section travel at different velocities and to an observer who is moving at the mean velocity of the flow. Those near the boundary would appear to move upstream and those in the middle at the surface would move downstream, i.e. they disperse.

The spreading caused by random molecular action or by turbulence is referred to as *diffusion* or *dispersion* and that due to convective action of mean velocity as *convective dispersion*. The molecular diffusion, in other than Brownian motion, is of little direct interest here, but the solutions of the equations are important because of analogies between molecular and turbulent diffusion.

Analysis of molecular diffusion is based on the continuum hypothesis and Fick's law:

$$P = - D \frac{\partial c}{\partial y} \tag{7.49}$$

where

P = rate at which the quantity or property is transported across unit area normal to the y-direction;

D = coefficient of diffusion, or diffusivity;

c = concentration of some quantity transported by diffusion. However, it should be noted that, in terms of probabilistic theories, concentration also means probability of occurrence of a single particle in a given unit volume.

Equation (7.49), together with the requirement of conservation of matter, gives

$$\frac{\partial c}{\partial t} = - \frac{\partial P}{\partial y} = D \frac{\partial^2 c}{\partial y^2} \tag{7.50}$$

which has a solution

$$c(y,t) = \frac{B}{\sqrt{t}} \exp(- y^2/4Dt) \tag{7.51}$$

where B is a constant. Equation (7.51) gives the number of labelled molecules per unit volume at a distance y after time t.

The Fick's law can be generalized to

$$\frac{\partial u}{\partial t} + u\frac{\partial c}{\partial x} + v\frac{\partial c}{\partial y} + w\frac{\partial c}{\partial z} = D\frac{\partial^2 c}{\partial x^2} + D\frac{\partial^2 c}{\partial y^2} + D\frac{\partial^2 c}{\partial z^2} \tag{7.52}$$

or

$$\frac{\partial c}{\partial t} + u_i\frac{\partial c}{\partial x_i} = D\frac{\partial^2 c}{\partial x_i \partial x_i}$$

In turbulent flow $c = \bar{c} + c'$ and $u = \bar{u} + u'$, etc., wherein the mean values do not change with time in steady flow. Thus, using the analogy to molecular motion and substituting for c and u it is not difficult to show that eqn (7.52) could be written for turbulent flow as

$$\frac{\partial \bar{c}}{\partial t} + u_i\frac{\partial \bar{c}}{\partial x_i} = -\frac{\partial}{\partial x_i}\overline{(c'u_i)} + D\frac{\partial^2 \bar{c}}{\partial x_i \partial x_i} \tag{7.53}$$

where by the diffusion analogy

$$\overline{c'u'} = -D_{ij}\frac{\partial \bar{c}}{\partial x_j} \tag{7.54}$$

which on substitution and addition of the gravitational fall velocity term leads to

$$\frac{\partial \bar{c}}{\partial t} + \bar{u}_i\frac{\partial \bar{c}}{\partial x_i} = \frac{\partial}{\partial x_i}\left[D_{ij}\frac{\partial \bar{c}}{\partial x_j} + D_M\frac{\partial \bar{c}}{\partial x_i}\right] + \frac{\partial}{\partial y}(w_s c) \tag{7.55}$$

where D_M refers to molecular diffusion and D_{ij} is the turbulent eddy diffusivity tensor. It is important to note that the coefficient of turbulent diffusion D_{ij} is a second order tensor and that its values are a function of space, whereas the laminar diffusion coefficient is a scalar quantity. Generally D_{ij} is much larger than D_M so that D_M can be neglected. If the turbulence is *homogeneous*, D_{ij} reduces to D_{ii} (D_{xx}, D_{yy}, D_{zz}) and if the turbulence is *isotropic*, D_{ij} reduces to a scalar D_T. In analogy to the diffusion coefficient ε of fluid momentum (the eddy viscosity) the diffusion coefficient for sediment is denoted by ε_s.

The theoretical models for sediment suspension assume the concentration to be a function of elevation y only, and independent of longitudinal or lateral coordinates. Further, the continuity requirement for solids and fluid must be satisfied and usually a steady state is assumed, $\partial c/\partial t = 0$. This implies that at any elevation within the flow the downward convective movement at fall velocity is balanced by the upward conductive velocity due to diffusion of turbulence, consequently $\partial c/\partial y < 0$ everywhere. For steady flow, suspension of one particle size only, and *low concentration* $\partial w/\partial y = 0$, i.e., constant fall velocity w, eqn 7.55 reduces to

$$w \frac{\partial c}{\partial y} + \frac{\partial}{\partial y}\left[\varepsilon_s \frac{\partial c}{\partial y}\right] = 0 \qquad (7.56)$$

This is the usual starting point. Integrating once yields

$$wc + \varepsilon_s \frac{\partial c}{\partial y} + A = 0 \qquad (7.57)$$

The constant of integration A is set zero from the observation that up and down movements must balance. Setting A = 0 implies that the concentration in the water surface must go to zero since ε_s must be zero in the surface. The solutions then start from

$$wc + \varepsilon_s \frac{dc}{dy} = 0 \qquad (7.58)$$

or

$$\ln \frac{c}{ca} = - w \int_a^y \frac{dy}{\varepsilon_s} \qquad (7.59)$$

where y = a is a reference level at which $c = c_a$ is known, and differ mainly in the definition of ε_s under the integral. The form of c = f(y) depends on both the velocity distribution u = f(y) and the diffusion coefficient ε_s = f(y) used.

The simplest assumption is ε_s = *constant* as used in 1925 by Schmidt. This yields

$$\frac{c}{c_a} = \exp\left[-\frac{w}{\varepsilon_s}(y - a)\right] \qquad (7.60)$$

If $\varepsilon_s \cong \varepsilon$ (momentum diffusion in fluid) then, since ε is proportional to u_* and flow depth y_o, the exponent has the form $\left[\frac{w}{u_*}\right]\left[\frac{y}{y_o}\right]$. If $w/u_* << 1$ the concentration will become nearly uniform.

The momentum exchange coefficient ε in turbulent flow has a parabolic distribution over depth given by

$$\varepsilon = \kappa u_* y(1 - y/y_o) \qquad (7.61)$$

which follows from $du/dy = u_*/\kappa y$, $\tau = \rho\varepsilon du/dy$, $\tau = \tau_o(1 - y/y_o)$ and $\tau_o = \rho u_*^2$ for turbulent flow using the Prandtl mixing length concept. Equation 7.61 has a maximum $\varepsilon_{max} = 0.25\ \kappa u_* y_o$ at $y = y_o/2$ where κ is the Karman constant.

Setting $\eta = y/y_o$ and using the ε_s distribution given by eqn 7.61 yields

$$\ln \frac{c}{c_a} = - \frac{w}{\kappa u_*} \int_a^y \frac{d\eta}{\eta(1 - \eta)} \qquad (7.62)$$

where the integral can for practical purposes be replaced by a straight line which implies that ε_s may be assumed to be constant. (Strictly true for a velocity distribution described by a second order parabola.) The mean value of ε by eqn 7.61 for $\kappa = 0.40$ is

$$\varepsilon_{sm} = \frac{1}{y_o} \int_o^{y_o} \varepsilon \, dy = \frac{1}{15} u_* y_o$$

and inserting this in eqn 7.59 yields

$$\ln \frac{c}{c_a} = - \frac{15w}{u_* y_o} \int_a^y dy = - 15 \frac{w}{u_*} \frac{y - a}{y_o} \qquad (7.63)$$

A straight line at a slope $- 6.5\ w/u_*$ through c_a on a semilogarithmic paper (c on log scale) yields the concentration profile. Further, the shear velocity could be expressed through Manning or Chézy formula as $u_* = Un\sqrt{g}/y_o^{1/6}$ or $u_* = U\sqrt{g}/C$.

This is basically the solution by Makkaveer (1931) into which Karaushev later introduced

$$\varepsilon_s = \frac{gy_o}{MC} u_y \qquad (7.64)$$

where

$$u_y = u_{max} \left[1 - P \left[1 - \frac{y}{y_o} \right]^2 \right]^{\frac{1}{2}}$$

$$P = \frac{MU^2}{Cu_{max}^2} = 0.57 + 3.3/C$$

$$M = 0.7C + 6$$

$$U/u_{max} = 0.9(C - 1)/C$$

for $10 < C < 60$ and $M = 48$ = constant for $C > 60$, where C is the Chézy coefficient and u_{max} is surface velocity.

The most widely known solution is that published by Rouse (1937) although it has been reported that Ippen derived it independently earlier. Rouse set $\varepsilon_s = \varepsilon$ as given in eqn 7.61. The integrand dy/ε of eqn 7.59 then becomes

$$\frac{dy}{\varepsilon} = \frac{\rho\ du/dy}{\tau_o(1 - y/y_o)}$$

and integration yields

$$\frac{c}{c_a} = \left[\left[\frac{y_o - y}{y_o - a}\right]\frac{a}{y}\right]^z \tag{7.65}$$

$$z = \frac{w}{\kappa u_*} \tag{7.66}$$

Equation 7.65 is displayed in Fig. 7.11. For a given shear velocity z is proportional to w which means that fine-grained sediment has a small z-value and the particles are distributed fairly uniformly throughout the depth. Coarse grained sediments will show substantial variation in concentration over depth. Alternatively, for a given sediment with w = const. an increase in shear velocity leads to increased uniformity of concentration over depth.

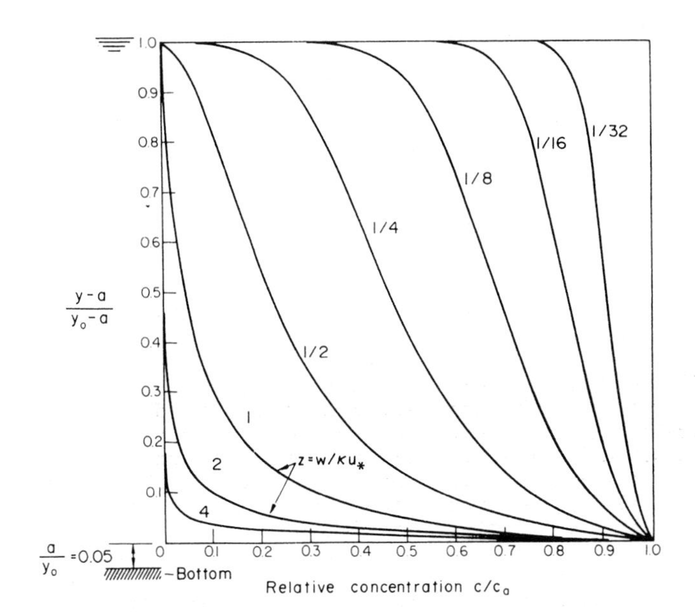

Fig. 7.11. Distribution of suspended load in a flow according to eqn 7.65.

Near the bed in the constant shear layer where $du/dy \cong$ constant, ε is approximately proportional to $u_* y$, i.e., $\varepsilon \cong \alpha u_* y$. In this layer $\varepsilon_s = \varepsilon$ assumption yields

$$\frac{c}{c_a} = \frac{y}{a}^{-\frac{w}{\alpha u_*}} \tag{7.67}$$

The concentration will tend to infinity for all positive values of w/u_* and

$$\int_a^y cdy = \frac{c_a}{(1 - \frac{w}{\alpha u_*})} y^{1-w/\alpha u_*}$$

shows that sediment will be mostly in the upper layers (above the constant shear layer) when $w/\alpha u_* < 1$. This implies that sediment put in suspension will tend to stay suspended. When $w/\alpha u_* > 1$ the sediment will be predominantly confined into the constant shear layer.

The Rouse solution of the diffusion equation is based on $\varepsilon = \varepsilon_s$ and the logarithmic velocity distribution, not withstanding that this velocity distribution is valid only for the lower 15 to 20% of flow depth (boundary layer thickness). Likewise, the Karman constant κ as for clear fluid flow was used. What value of κ should be used in sediment laden flow is still subject of debate. Earlier work by Vanoni and Nomicos (1959), and others, has indicated that κ decreases with concentration and could be as low as 0.21. Coleman (1985) found that κ "remained essentially constant over the range of Richardson numbers from zero (clear water) to about 100 (capacity of sediment suspension)" where the Richardson number is

$$Ri = \frac{-g\delta(\rho_{m\delta} - \rho_{mo})}{\bar{\rho}_m U_m^2} \tag{7.68}$$

where δ is the boundary layer thickness (flow depth y_o), $\rho_{m\delta}$ and ρ_{mo} are densities of sediment-water mixture at $y = \delta = y_o$ and $y = 0$, respectively, $\bar{\rho}_m$ is depth averaged density of mixture and U_m is velocity at $y = \delta = y_o$. For sediment laden flow he obtained $\kappa = 0.433$ with a standard deviation of 0.031. The clear-water values were $\kappa \cong 0.41$. Coleman also observed that the wake-strength coefficient Ω (eqn 6.8) was a function of sediment concentration. The wake-strength coefficient varied from clear-water asymptote of about 0.2 to about 0.9 at capacity of suspension, Fig. 7.12, but the wake function was not affected.

No definitive answers can be given on the behaviour of the sediment diffusion coefficient ε_s either. Results of measurements of the transfer coefficients are few, but for the momentum transfer they do indicate approximately the parabolic form as obtained from logarithmic velocity distribution. The distribution of ε appears to be a little asymmetric with the maximum being somewhat closer to the bed than predicted. The mass transfer coefficients,

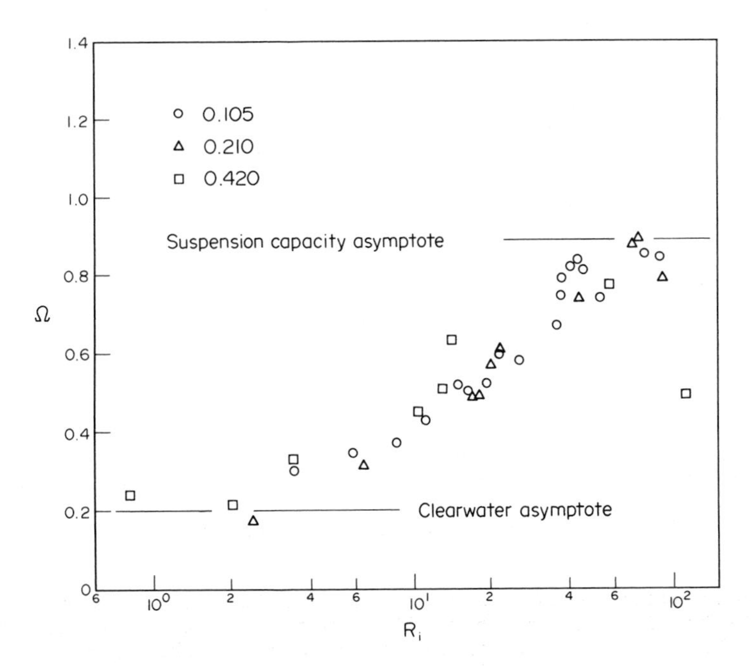

Fig. 7.12. The wake strength coefficient Ω plotted against the Richardson number.

ε_s, also indicate the above form, at least in the lower part of depth, but they generally do not go to zero at the surface and their values at times depart appreciably from those of momentum transfer. Here, the measurements by Coleman (1970) are typical. The laboratory results (Fig. 7.13) show that ε_s/u_*y_o varies with y/y_o over the lower 20-30% of depth, after which it tends to remain at approximately constant value up to water surface. A small increase in ε_s/u_*y_o with w/u_* is, however, apparent. These results indicate that ε_s has an inner and outer region. Figure 7.14 shows a similar result from field data, Table 7.1. A feature of these results is that coarser particles have larger ε_s-values than finer ones. A likely reason for this is that the larger grains acquire enough momentum to fly out of the eddy system of water turbulence and sample turbulence of neighbouring systems as well whereas the smaller grains remain locked within the eddy system of turbulence, i.e., a partial decoupling.

Notwithstanding all the shortcomings of the solution, eqn (7.65) has been shown to fit observed data quite well with suitable exponents z, as seen from Fig. 7.15 in which river and laboratory data are used, $0.044 \leq d \leq 0.295$ mm, $0.09 \leq y_o \leq 3.02$ m.

This led to numerous proposals for evaluation of the exponent z. The z-value would depend on the effects of sediment on turbulence, the effect of sediment and turbulence on fall velocity, on sediment grading, on turbulence due to bed features and secondary currents. An early approach was to express z as

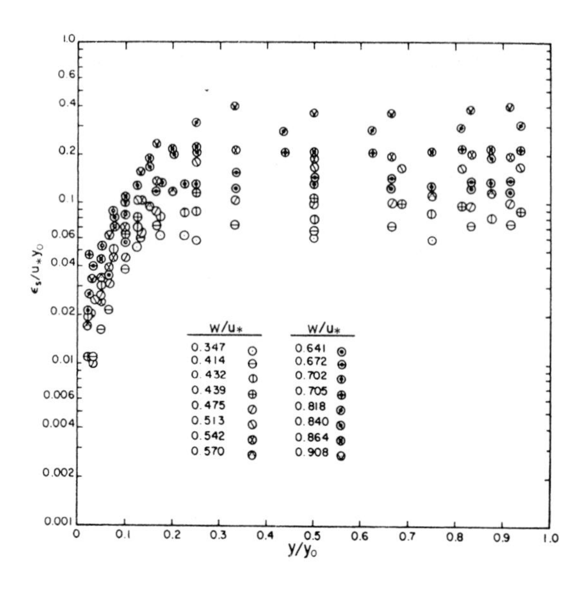

Fig. 7.13. Non-dimensional sediment transfer functions. Coleman, 1970.

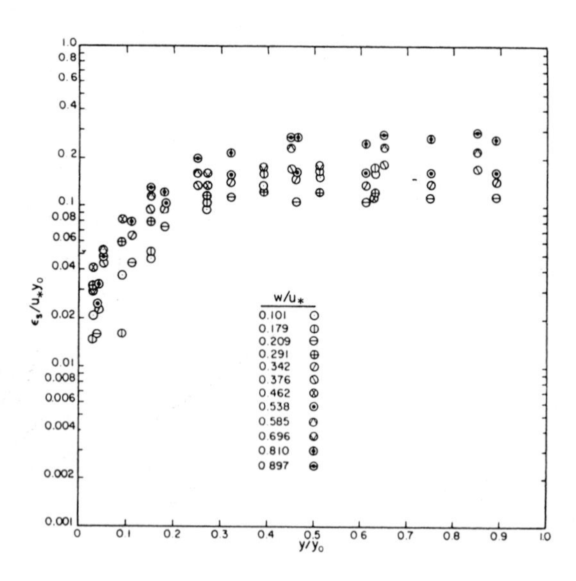

Fig. 7.14. Sediment transfer functions from Enoree River data, (Coleman, 1970).

TABLE 7.1 General conditions for Enoree River measurements (Anderson, 1942)

Diameter (mm)	w mm/s	y_o m	u_* mm/s
0.210	26.8	0.314	71.3
0.297	41.8	0.314	71.3
0.420	64.0	0.314	71.3
0.210	20.1	1.280	96.3
0.297	337.1	1.280	96.3
0.420	51.8	1.280	96.3
0.495	78.0	1.280	96.3
0.149	11.3	1.524	112.2
0.210	20.1	1.524	112.2
0.351	32.6	1.524	112.2
0.420	51.8	1.524	112.2
0.590	78.0	1.524	112.2

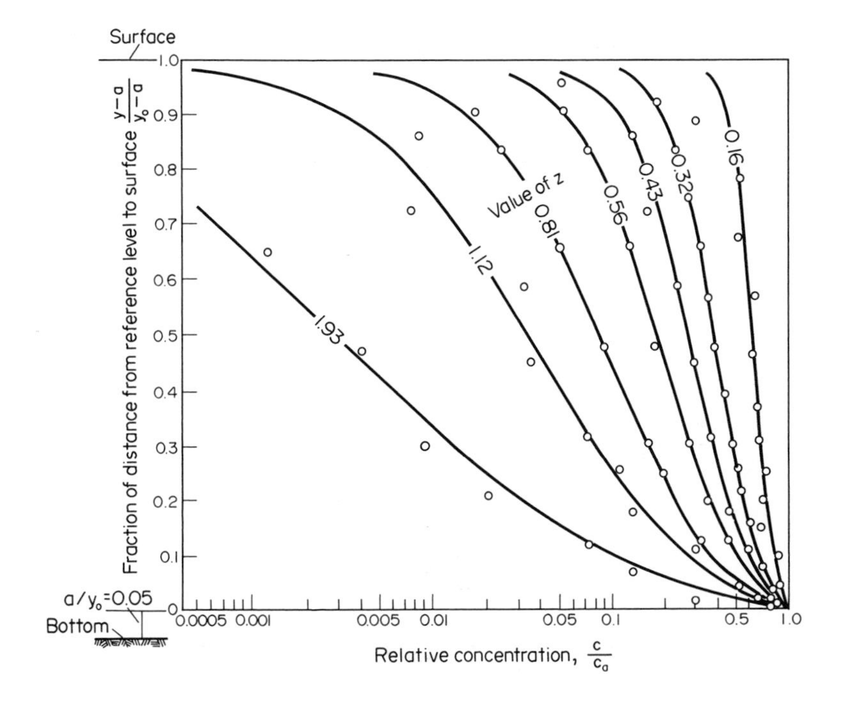

Fig. 7.15. Vertical distribution of relative concentration c/c, compared with eqn 7.65 for wide range of stream size and z values.

$$z_1 = \frac{w}{\beta \kappa u_*} \tag{7.69}$$

in which β and κ are still virtually unknown. Experimental data was interpreted to show that for fine sediment $\varepsilon \cong \varepsilon_s$ and $\beta \cong 1$, whereas for coarse sediment $\beta < 1$ and $\varepsilon_s < \varepsilon$. The latter deduction is not supported by the data by Coleman (1970). The data by Coleman were used by van Rijn (1984) to express β as

$$\beta = 1 + 2\left[\frac{w}{u_*}\right]^2; \quad 0.1 < \frac{w}{u_*} < 1 \tag{7.70}$$

according to which β is always greater than unity.

Paintal and Garde (1964) expresses z/z_1 from experimental data in terms of the Shields parameter θ. Their relationship could be described by

$$z/z_1 = 4.1\theta^{-1.67}; \quad 0.2 \leq \theta \leq 4 \tag{7.71}$$

and implies a variation in z/z_1 from 60 to 0.4 over the θ-range. The authors used $\kappa = 0.40$. The plotted data scatter appreciably.

Numerous refinements of the simple solution of the diffusion equation have been proposed. Hunt (1954) accounted for the space occupied by the sediment and obtained

$$\varepsilon_s \frac{\partial c}{\partial y} + c\frac{\partial c}{\partial y}(\varepsilon - \varepsilon_s) + (1 - c)cw = 0 \tag{7.72}$$

which for $\varepsilon = \varepsilon_s$ becomes

$$(1 - c)cw + \varepsilon_s \frac{\partial c}{\partial y} = 0 \tag{7.73}$$

Integration yields

$$\left[\frac{c}{1 - c}\right]\left[\frac{1 - c_a}{c_a}\right] = \left\{\left[\frac{1 - y/y_o}{1 - a/y_o}\right]^{1/z}\left[\frac{\beta_s - (1 - a/y_o)^{\frac{1}{2}}}{\beta_s - (1 - y/y_o)^{\frac{1}{2}}}\right]^z\right. \tag{7.74}$$

where $z = w/\kappa_s\beta_s u_*$. From data by Vanoni (1946), with κ_s ranging from 0.296 to 0.444, the β_s-value was found to be close to one. Equation (7.74) differs little from eqn (7.65) and the solution does not account for variation of fall velocity with concentration.

Antsyferov and Kos'yan (1980) report that both eqn (7.65) and (7.74) satisfy observations well in the upper 80-90% of flow depth, although the solution by Nikitin is stated to yield the best fit. Nikitin's solution is given as

$$\frac{c}{c_a} = \exp(-\phi) \tag{7.75}$$

where

$$\phi = \left[\left[\frac{2.8w}{u_*} + \frac{15.7\ \nu w}{y_o\ u_*}\right]\ln\left[\frac{y}{y_o - y}\ \frac{y_o - a}{a}\right] - \frac{15.7\ \nu w}{y_o\ u_*^2}\left[\frac{y_o}{y} - \frac{y_o}{a}\right]\right]$$

Numerous modifications were proposed by Einstein and Chien (1954), Chien (1956) and Tanaka and Sugimoto (1958) but the greatly increased complexities have not produced significant improvements.

Fukuoka and Kikkawa (1971) integrated eqn (7.57) and obtained

$$c = \left[\frac{y_o - y}{y}\ \frac{a}{y_o - a}\right]^z c_a + \frac{A}{w}\left[\left[\frac{y_o - y}{y}\ \frac{a}{y_o - a}\right]^z - 1\right] \tag{7.76}$$

$$\frac{A}{w} = \frac{c_b - \left[\dfrac{y_o - b}{b}\ \dfrac{a}{y_o - a}\right]^z c_a}{\left[\left[\dfrac{y_o - b}{b}\ \dfrac{a}{y_o - a}\right]^z - 1\right]}\ ; \quad z = \frac{w}{\kappa u_*}$$

where c_b is the sediment concentration at $y = b$. Thus, if $y = b = y_o$ and $c_b = 0$ eqn 7.76 reduces to eqn 7.65, i.e. $A = 0$.

Van Rijn (1984) solved eqn 7.73 by using a parabolic form of distribution of ε in the lower half of flow (eqn 7.61) and a constant ε in the upper half. The sediment diffusion was described by

$$\varepsilon_s = \beta\phi\varepsilon$$

where β is to account for differences in diffusion of solids and fluid and ϕ for damping of fluid turbulence by the solids. Integration of eqn 7.73 with $\phi = 1$ (no damping) yielded

$$\sum_{n=1}^{4}\left[\frac{1}{n(1 - c)^n}\right] - \sum_{n=1}^{4}\left[\frac{1}{n(1 - c_a)^n}\right] + \ln\left[\frac{c(1 - c_a)}{c_a(1 - c)}\right]$$

$$= \ln\left[\frac{a(y_o - y)}{y(y_o - a)}\right]^z \qquad \text{for } \frac{y}{y_o} < 0.5 \tag{7.77a}$$

and

$$= -z\left[\ln\left[\frac{a}{y_o - a}\right] + 4\left[\frac{y}{y_o} - 0.5\right]\right] \qquad \text{for } \frac{y}{y_o} \geq 0.5 \tag{7.77b}$$

where $z = w/\beta\kappa u_*$. For small concentrations, $c < c_a < 0.001$ these reduce to

$$\frac{c}{c_a} = \left[\frac{a(y_o - y)}{y(y_o - a)}\right]^z \qquad \text{for } \frac{y}{y_o} < 0.5 \tag{7.78a}$$

and

$$\frac{c}{c_a} = \left[\frac{a}{y_o - a}\right] \exp\left[-4z\left[\frac{y}{y_o} - 0.5\right]\right]\left[\text{for } \frac{y}{y_o} \geq 0.5\right] \qquad (7.78b)$$

The β-value is empirically expressed by eqn 7.70 and ϕ is replaced by an empirical adjustment of z as

$$z_1 = z + \psi \qquad (7.79)$$

where

$$\psi = 2.5\left[\frac{w}{u_*}\right]^{0.8}\left[\frac{c_a}{c_o}\right]^{0.4} ; \quad 0.01 \leq \frac{w}{u_*} < 1$$

and $c_o \cong 0.65$ is the maximum concentration in bed. The relationships were used to compute concentrations corresponding to published data and show very good agreement. For non-uniform sediment a representative size of the suspended material was empirically given as

$$\frac{d_s}{d_{50}} = 1 + 0.011\,(\sigma_g - 1)(T - 25) \qquad (7.80)$$

where σ_g is the geometric standard deviation of bed material and T is the transport parameter (eqn 5.54).

Observed concentration distributions frequently plot as if there were two zones, for example, Fig. 7.16 by Coleman (1969). From flume and field data Coleman expressed the thickness of the lower layer, a_*, as $a_* \cong 1.5\ \delta_D$ where δ_D is the displacement thickness of the boundary layer, or

$$\frac{a_*}{y_o} \cong 0.032(Uy_o/\nu)^{1/8} \qquad (7.81)$$

No correlation was found between a_*/y_o and u_*/U. Plotting the ratio ε_s/w from the diffusion equation, $\varepsilon_s/w = c/(dc/dy)$, Fig. 7.17, clearly shows that in the lower or inner region ε_s/w is proportional to y and constant in the outer region. Thus, for the outer region eqn (7.60) applies with y = a replaced by a_* and c_a by c_{a*}. No functional relationship for ε_s/w was established for the inner region and as a first approximation ε_s/w = y was used which yielded

$$\frac{c}{c_{a*}} = \frac{a_*}{y} \qquad (7.82)$$

and was shown to fit laboratory and field data, Fig. 7.18. The assumption of direct equality with y is questionable and $\varepsilon_s/w = \propto y$ appears more probable, which leads to eqn (7.60).

The two-layer system also emerges from the rather complex treatment by McTigue (1981) as well as by Antsyferov and Kos'yan (1980). The latter base their argument for the existence of the

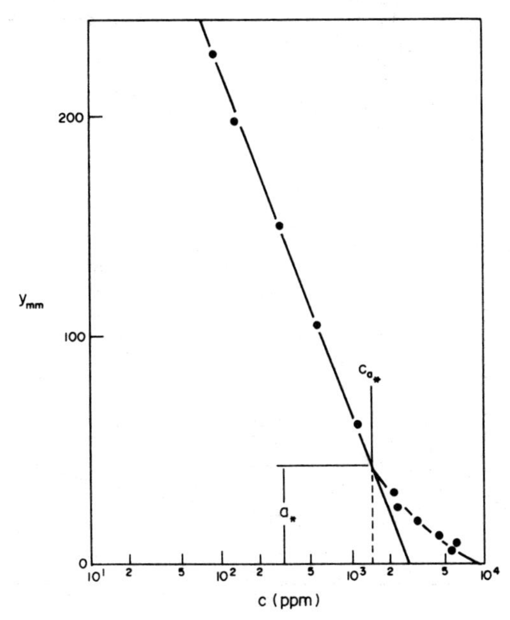

Fig. 7.16. A typical concentration profile, showing the method of defining a_* and c_*. Coleman, 1969.

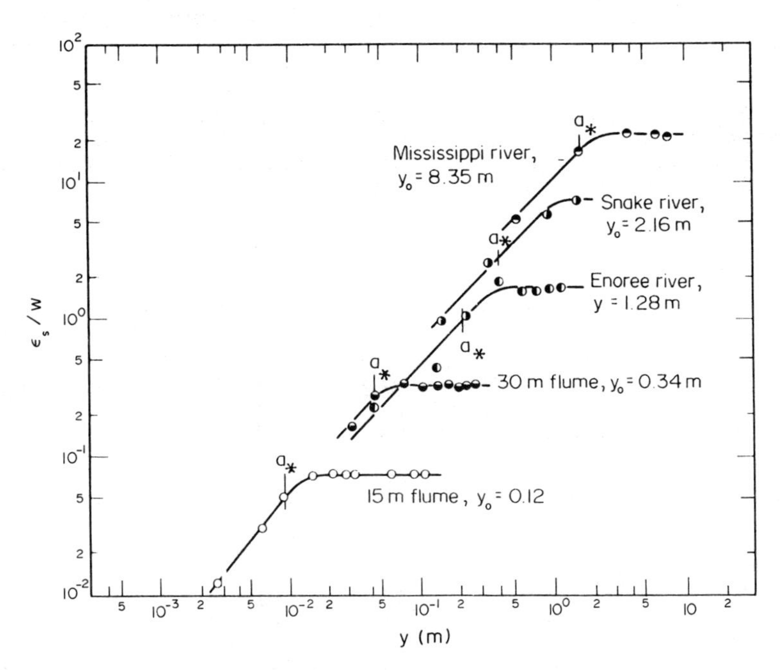

Fig. 7.17. Typical curves of the sediment transfer ratio ε_s/w. (Coleman, 1969).

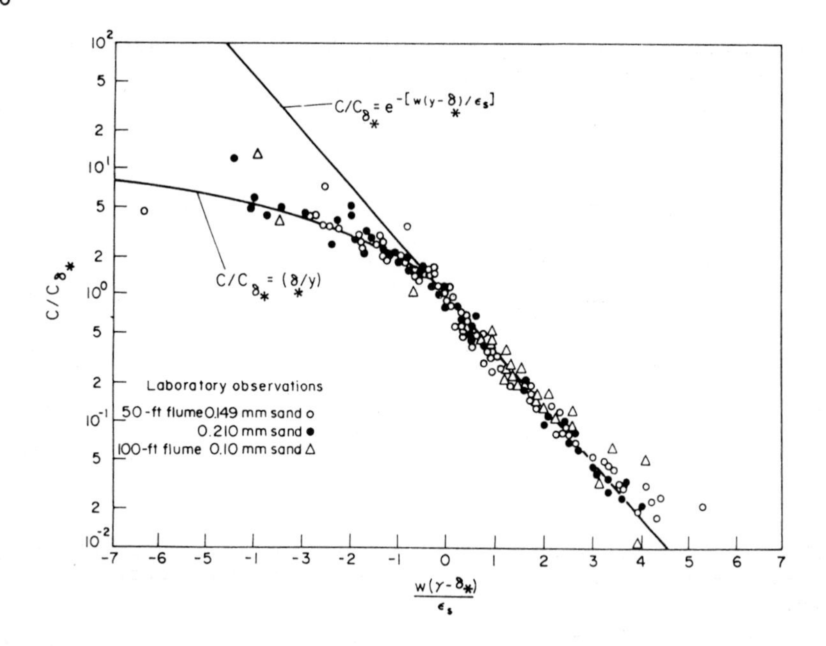

Fig. 7.18. Laboratory data plotted to verify eqns 7.82 and 7.60. (Coleman, 1969).

two zones on the observation that the rms values of vertical velocity fluctuations v* increase from zero at the bed to a maximum at about $y = (0.1 \text{ to } 0.2)y_0$ and then decrease to an almost constant value through the rest of depth of $v^* \cong 0.8\, u_*$. The turbulence diffusion model is valid for this upper part whereas in the lower layer the fall velocity of the particles plays a significant part. Saltation and dilatation stresses are bound to have an effect in this lower layer as well.

7.2.2 The energy approach

An energy approach to the suspension problem, known as the gravitational theory, was proposed by Velikanov (1954,1955,1958). It became the subject of bitter disputes, particularly in the Russian language publications. However, it has been demonstrated that the actual differences between the gravitational theory and the diffusion method are small and rest on the differences in assumptions made.

The features of the Velikanov's gravitational theory for a two-dimensional steady flow are summarised below. Continuity requires that the time-averaged vertical transport of water and sediment must be zero

$$\rho\overline{(1 - c)v} = 0 \tag{7.83}$$

$$\rho_s\, \overline{cV} = 0 \tag{7.84}$$

where the vertical velocity of solids $V = v - w$. Since both ρ and ρ_s are non-zero

$$\overline{(1 - c)v} = 0 \tag{7.85}$$

$$\overline{c(v - w)} = 0 \tag{7.86}$$

Since $\overline{cv} = \bar{c}\ \bar{v} + \overline{c't'}$

we obtain from eqns (7.85) and (7.86)

$$\bar{v} - \bar{c}\ \bar{v} - \overline{c'v'} = 0$$

$$\bar{c}\ \bar{v} + \bar{c}'v' - \bar{c}w = 0$$

Between these two

$$\bar{v} = \bar{c}w \tag{7.87}$$

$$\overline{c'v'} = \bar{c}w - \bar{c}\ \bar{v} = \bar{c}w(1 - \bar{c}) \tag{7.88}$$

In the latter expression $\overline{c'v'}$ is the average upward transport per unit area and time and this must be equal to the settling $- \bar{V}\ \bar{c}$. Hence,

$$\bar{V} = - w(1 - \bar{c}) \tag{7.89}$$

In the lifting process the flow must lose energy per unit volume at the rate of

$$E = \gamma_s^* \bar{c}\ \bar{V} = \gamma_s^* w\bar{c}(1 - \bar{c}) \tag{7.90}$$

where $\gamma_s^* = \rho g(S_s - 1)$. For clear water flow Velikanov writes

$$\rho g S u + \bar{u}\frac{d\bar{\tau}}{dy} = \frac{\bar{\tau}_o}{y_o}U + U\frac{d\bar{\tau}_o}{dy} = 0 \tag{7.91}$$

since the energy loss is dissipated by friction. For the mixture

$$\rho g S\bar{u}(1 - \bar{c}) + u\frac{d[\bar{\tau}(1 - \bar{c})]}{dy} = \gamma_s^* w\bar{c}(1 - \bar{c}) \tag{7.92}$$

and the shear stress is assumed to be affected by $\bar{c}$. Equation (7.92) can be written as

$$(1 - \bar{c})\left[\rho g\ S\bar{u} + u\frac{d\bar{\tau}}{dy}\right] - \bar{u}\ \bar{\tau}\frac{d\bar{c}}{dy} = \gamma_s^* w\bar{c}(1 - \bar{c}) \tag{7.93}$$

which for $c << 1$, observing eqn (7.91), yields

$$- \bar{u}\ \bar{\tau}\frac{d\bar{c}}{dy} = \gamma_s^* w\bar{c}$$

or
$$\overline{wc} + \frac{\bar{u}\ \bar{\tau}}{\gamma_s^*}\frac{d\bar{c}}{dy} = 0 \tag{7.94}$$

This equation is of the same form as obtained from the diffusion theory, except that the exchange coefficient has a different form. With the same assumptions of linear shear stress and logarithmic velocity distributions Velikanov obtains

$$\frac{c}{c_a} = \exp\left[-\frac{w}{u_*}\frac{\kappa\gamma_s^*}{\gamma S}\int_{\eta_a}^{\eta}\frac{d\eta}{(1-\eta)\ln \eta/\alpha}\right] \tag{7.95}$$

where $\eta = y/y_o$, η_a is a reference level just outside the limits of integration, and $\alpha = k_s/30y_o$. The agreement of eqn (7.95) with observed results is of the same order as of eqn (7.65) from the diffusion approach.

A detailed discussion of the gravitational theory in English is given by Bogardi (1974).

An energy based theory was also proposed by Bagnold (1956,1966). He reasoned that the excess weight of solids in motion must be supported by momentum transfer from solid or from fluid to solid, or both. The fluid must lift the solids at the rate that these are falling under gravity. Hence, the rate of work done by shear flow turbulence of the fluid is

$$\text{work rate of suspended load} = m_s^*\ gw = g_{is}\frac{w}{U_s}$$

where m_s^*g is the immersed weight g_{is} of sediment and U_s is the mean transport velocity of suspended soilds. Hence, w/U_s is analogous to the friction factor tan α. The fluid is pushing the grains up a slope w/U_s.

The available power supply per unit area is

$$P = \rho g y_o SU = \tau_o U \tag{7.96}$$

of which $\eta_b P$ is dissipated in bed-load transport, leaving $P(1 - \eta_b)$ for suspended load.

$$g_{is}\frac{w}{U_s}\ \eta_s P(1 - \eta_b)$$

or
$$g_{is} = \eta_s P\frac{U_s}{w}(1 - \eta_b) \tag{7.97}$$

This is combined with the bed-load transport. The result will be referred to under the heading Total Sediment Load. Note that the above approach tells us nothing about the distribution of sediment in suspension.

Bagnold estimates the maximum lifting power to be 0.266 ρv^*, where v^* is the root-mean-square value of vertical fluctuations. Thus, in terms of the stream power $\tau_0 U$, the efficiency of suspension is

$$\eta = 0.266\left[\frac{v^*}{u_*}\right]^2 \frac{v^*}{U} \tag{7.98}$$

where v^*/U is of the order of 1/20 and v^*/u_* is of the order of 0.8, or a little more as an average throughout the depth of flow, with a maximum of the order of unity. In suspending the sediment, work is done at a rate of

$$P = g_{is}\, w/\eta_s \tag{7.99}$$

which again could be related to the stream power.

A variant of the energy model of suspension developed by Bagnold (1962) and Vlugter (1962) has some interesting features. The mass per unit area of suspended sediment, m, is acted upon by gravitation and buoyancy, Fig. 7.19. The maintenance of suspension requires a power input F_B w cosα and the weight component provides the power $[(\rho_s - \rho)/\rho_s]mg \sin\alpha\, U_s$, where U_s is the velocity of the suspension. The difference has to come from flow, i.e. the net power demand is

$$P_N = \left[\frac{\rho_s - \rho}{\rho_s}\right] mg(w \cos\alpha - U_s \sin\alpha)$$

$$= \left[\frac{\rho_s - \rho}{\rho_s}\right] mg\, U_s \left[\frac{w}{U_s} \cos\alpha - \sin\alpha\right] \tag{7.100}$$

Here $P_N \to 0$ when $w = U_s \tan\alpha$, i.e. the power input from the weight component covers the power requirement for suspension. For $P_N \leq 0$ a state of auto-suspension exists.

The energy balance requires that the energy of the total mass available through the slope must equal the energy loss

$$(m_s\rho_s + m\rho)gSU_s = m_s(\rho_s - \rho)gw \cos\alpha + mgU_sS(U_s/U)^2 \tag{7.101}$$

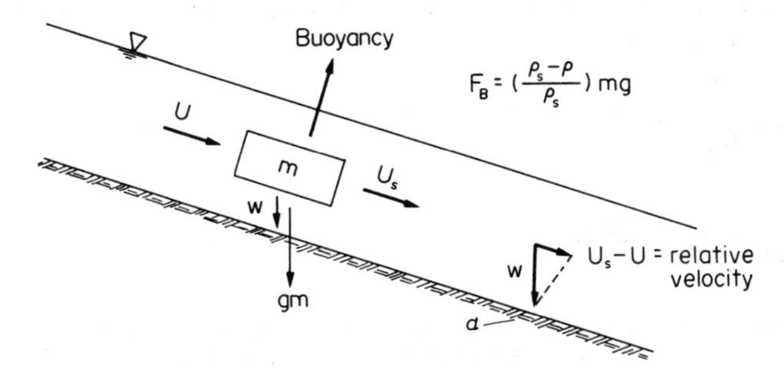

Fig. 7.19. The Bagnold-Vlugter model of suspension.

where (U_s^2/U^2) accounts for the ratio of energy loss in flow with sediment and without, including the energy transfer from water to sediment. Hence,

$$\rho_s m_s S U_s + \rho m S U_s = m_s(\rho_s - \rho) w \cos\alpha + m\rho U_s S\left[\frac{U_s^2}{U^2}\right]$$

$$m_s\left\{SU_s - \left[\frac{\rho_s - \rho}{\rho_s}\right] w \cos\alpha\right\} = m\,\frac{\rho}{\rho_s}\,U_s S\left\{\left[\frac{U_s^2}{U^2}\right] - 1\right\}$$

$$\frac{m_s}{m} = \frac{\dfrac{\rho}{\rho_s}\left\{\left[\dfrac{U_s}{U}\right]^2 - 1\right\}}{1 - \dfrac{\rho_s - \rho}{\rho}\,\dfrac{w}{SU_s}\cos\alpha} \qquad (7.102)$$

or for $\rho_s = 2.5$

$$\frac{m_s}{m} = \frac{0.4\{(U_s/U)^2 - 1\}}{1 - 0.6(w/SU_s)\cos\alpha}$$

or concentration

$$c_{vol} = \frac{(U_s/U)^2 - 1}{1 - 0.6(w/SU_s)\cos\alpha}$$

Equation (7.102) can also be solved for U_s. Attention is also drawn to the developments of these ideas by Schoemaker (1982).

A different energy approach was presented by Itakura and Kishi (1980). They start with the observation that the rate of energy input per unit volume, $\tau dy/dy$, for a flow of water-sediment mixture must equal the rate of power pressure fluctuations $g\overline{\rho'v'}$, the rate of energy dissipation through the motion and collisions of grains, including the effects of flow distortion, and the dissipation of energy by water flow:

$$\tau\frac{du}{dy} = g\overline{\rho'v'} + B\frac{\rho u_*^3}{\kappa L} + \rho\frac{u_*^3}{\kappa y} \qquad (7.103)$$

In pure water the first two terms on the right hand side are zero but in sediment laden flow these are assumed to be proportional to each other. The length L is defined as

$$\frac{1}{L} = \frac{\kappa g(S_s - 1)wc}{u_*^3} \qquad (7.104)$$

where for steady state $g\overline{\rho'v'} = \rho g(S_s - 1)wc$, and represents the ratio of buoyancy flux to dissipation. In meteorology L is known as the Monin-Obukhov length. After combination of the first two terms on the right hand side into $\alpha\rho u_*^3/\kappa L$, eqn (7.103) can be written as

$$\frac{du}{dy} = \frac{u_*}{\kappa y}\left[1 + \alpha \frac{y}{L}\right] \tag{7.105}$$

where the constant α was evaluated from measured velocity distributions using the mean concentration over the depth, $\bar{c}$, in eqn (7.104). A value of $\alpha \cong 7$ was used. Integration led to

$$\frac{u}{u_*} = \frac{1}{\kappa}\left[\ln \frac{u_* y}{\bar{\nu}} + \alpha \frac{y}{L}\right] + \left[B'\left[\frac{u_* k_s}{\bar{\nu}}, \frac{k_s}{L}\right] \cdot \frac{1}{\kappa} \ln \frac{u_* k_s}{\bar{\nu}}\right] \tag{7.106a}$$

and for clear water

$$\frac{u}{u_*} = \frac{1}{\kappa} \ln \frac{u_* y}{\nu} + \left[B\left[\frac{u_* k_s}{\nu}\right] - \frac{1}{\kappa} \ln \frac{u_* k_s}{\nu}\right] \tag{7.106b}$$

where $\bar{\nu} = \nu(1 + 2.5c)$ is the effective viscosity of mixture and k_s is roughness height. It was found from experiments that $B' \cong B$ and not a function of L. Thus,

$$\frac{u}{u_*} = \frac{1}{\kappa}\left[\ln \frac{u_* y}{\bar{\nu}} + \phi \frac{u_* y}{\bar{\nu}}\right] + \left[B\left[\frac{u_* y}{\bar{\nu}}\right] - \frac{1}{\kappa} \ln \frac{u_* k_s}{\bar{\nu}}\right] \tag{7.107}$$

where $\phi = \alpha\, \bar{\nu}/u_* L$. The second square bracket term for $u_* k_s/\bar{\nu} < 5$ is equal to 5.5 and for $u_* k_s/\bar{\nu} > 70$ eqn (7.107) becomes

$$\frac{u}{u_*} = \frac{1}{\kappa}\left[\ln \frac{y}{k_s} + \phi_1 \frac{y}{k_s}\right] + 8.5 \tag{7.108}$$

where $\phi_1 = \alpha\, k_s/L$.

The authors also derive a relationship for the Darcy-Weisbach friction factor f as

$$\left[\frac{8}{f_s}\right]^{\frac{1}{2}} = \left[\frac{8}{f}\right]^{\frac{1}{2}} + \frac{\alpha}{2\kappa} \frac{y_o}{L} \tag{7.109}$$

Then, following the same development as Rouse, eqn (7.105) leads to

$$\varepsilon = \kappa u_* y \left[1 + \frac{y}{y_o}\right]\left[1 + \alpha \frac{y_o}{L} \frac{y}{y_o}\right]^{-1} \tag{7.110}$$

and with $\varepsilon = \varepsilon_s$ eqn (7.59) yields

$$\frac{c}{c_a} = \left[\left[\frac{y_o - y}{y_o - a}\right]^{1+\phi_2} \left[\frac{a}{y}\right]\right]^z \tag{7.111}$$

where $\phi_2 = \alpha y_o/L$ and $z = w/\kappa u_*$. Application to laboratory data leads to good predictions but the method has not yet been adequately tested in the field.

7.2.3 Statistical models

Since suspension is maintained by turbulence which is random by nature, it is only natural that the distribution of suspended sediment should be subject to description by probabilistic methods. Even the equations of diffusion can be put in the form identical with those describing random functions. The studies of suspension which have led to the statistical models started with attempts to relate the particle motion to turbulence. One of the earliest and most quoted theoretical treatments is that by Tchen (1947). In Tchen's theory the following assumptions are made:

1. The turbulence of the fluid is homogeneous and steady.
2. The domain of turbulence is infinite in extent.
3. The particle is spherical and so small that its motion relative to the ambient fluid follows Stokes' law of resistance.
4. The particle is small compared with the smallest wavelength present in the turbulence.
5. During the motion of the particle the neighbourhood will be formed by the same fluid particles.
6. Any external force acting on the particle originates from a potential field, such as gravity field.

Here, the assumption 5 is the critical one and is not likely to be satisfied in a turbulent flow. The fluid elements in turbulent motion do not retain their identity for any length of time or distance and it is unlikely that a particle could be surrounded by the same fluid particles. It is implied that the solid particles do not overshoot and this could only be a reasonable assumption if the densities of particles and fluid are about the same. For a large density ratio the results are suspect. The equation of motion is put into the following form:

$$\underbrace{\frac{\pi}{6} d^3 \rho_p \frac{dv_p}{dt}}_{1.} = \underbrace{3\pi\mu d(v_f - v_p)}_{2.} + \underbrace{\frac{\pi}{6} d^3 \rho_f \frac{dv_f}{dt}}_{3.} + \underbrace{\frac{1}{2}\frac{\pi}{6} d^3 \rho_f \left[\frac{dv_f}{dt} - \frac{dv_p}{dt}\right]}_{4.}$$

$$\underbrace{+ \frac{3}{2} d^2 \sqrt{(\pi\rho_f\mu)} \int_{t_o}^{t} dt' \frac{\dfrac{dv_f}{dt'} + \dfrac{dv_p}{dt'}}{\sqrt{(t - t')}}}_{5.} + \underbrace{F_e}_{6.} \qquad (7.112)$$

where t_o is the starting time, suffixes f and p refer to fluid and particle repsectively, and the fluid velocity v_f is measured near the particle, but sufficiently far from it not to be disturbed by the relative motion of the particle. The meaning of the terms is as follows:

1. Force required to accelerate the particle.

2. Viscous resistance force according to Stokes' law.
3. Force due to pressure gradient in the fluid surrounding the particle, caused by the acceleration of the fluid.
4. Force required to accelerate the added mass of the particle relative to the ambient fluid.
5. The "Basset" term, which takes into account the effect of the deviation of flow pattern from steady state.
6. External potential force.

The terms 3, 4 and 5 are important only if the particles are about the same density as the fluid or lighter. The instantaneous coefficient of resistance of the particle may become many times that for steady motion if the particle is accelerated at a high rate by an external force.

Tchen's analysis showed that the ratio of particle diffusion coefficient to fluid diffusion coefficient $\varepsilon_s/\varepsilon$ is unity for infinitely long diffusion times and less than one for short diffusion times of heavy particles. However, for particles which are much lighter than fluid the intensity and amplitudes of the particle motion are greater than those of the fluid. These results are not supported by observations, e.g. Coleman (1970).

Since the pioneering work by Tchen numerous papers have been published on statistical methods for analysis of suspensions. (Hino, 1963; Liu, 1956; Bugliarello and Jackson, 1964; Chiu and Chen, 1969; Yalin, 1972; Ashida and Fujita, 1986) however, the statistical models have not yet established themselves in practical application of suspended sediment predictions.

7.2.4 Numerical models

A powerful tool for the analysis of suspended sediment problems is the numerical model. It enables the use of more complex descriptions of turbulent mixing. The basic equations are mass and momentum balance for fluid and sediment in turbulent flow. In addition a turbulence model is required which defines the fluid and sediment mixing coefficients. These are then solved numerically using a finite differences procedure and appropriate boundary conditions. Of these, the bottom boundary condition is still a basic unsolved problem, that is, the description of the exchange of sediment between the bed and suspension. The fluid mixing coefficient can be obtained from the Reynolds' equations by applying a K-Σ-turbulence model, where K represents the kinetic energy of turbulent motion per unit mass and Σ the rate of dissipation. (The more usual notation is the k-ε-model.) The aspects of such hydrodynamic models are discussed by Rodi (1978) and Celic and Rodi (1984a). The combination of Reynolds' equations with the K-Σ model has been shown to yield good descriptions for velocity fields. Hence, the mixing coefficients should also be realistic. Calculations of such coefficients in non-uniform flow were presented by Alfrink and van Rijn (1983). Celik and Rodi (1984b) combined the hydrodynamic model with a concentration distribution described by

$$u \frac{\partial c}{\partial x} + v \frac{\partial c}{\partial y} = \frac{\partial}{\partial y}\left[\frac{\varepsilon}{Sn} \frac{\partial c}{\partial y}\right] + \frac{\partial}{\partial y} (wc) \qquad (7.113)$$

where the diffusion coefficient is expressed as the ratio of momentum exchange coefficient to Schmidt number which was assumed to be equal to 0.5, i.e. $\varepsilon_s = 2\varepsilon$. The boundary condition at the surface is zero sediment flux across the surface; $\varepsilon_s \, \partial c/\partial y + wc = 0$. At the bed the net sediment flux is set equal to the change in the suspended load transport rate dq_s/dx as

$$\frac{dq_s}{dx} = \left[- \varepsilon_s \frac{\partial c}{\partial y} - wc \right] = E - D \tag{7.114}$$

where E and D represent entrainment and deposition rates, respectively. The rate of deposition was characterised by $c_a w$, which is equal to entrainment for equilibrium conditions. Celik and Rodi postulate that the part of energy consumed in keeping the particles in suspension, $\Delta y_o w\bar{c}$, is a constant fraction of the production of turbulence by shear $(\tau_o/\rho)\bar{U}$. Under saturation conditions

$$\bar{c}_{max} = \beta \frac{\tau_o}{(\rho_s - \rho) g y_o} \frac{\bar{U}}{w} \tag{7.115}$$

where β is an empirical constant. This was written as

$$\bar{c}_{max} = \beta \left[1 - \left[\frac{k_s}{y_o} \right]^n \right] \frac{\tau_o}{(\rho_s - \rho) g y_o} \frac{U}{w} \tag{7.116}$$

where the first bracket term is an empirical expression which accounts for bed roughness, with $\beta = 0.034$ and $n = 0.06$. The reference concentration, $c_{a\,max}$, was related to the transport capacity concentration, $\bar{c}_{max}$, with the aid of eqn (7.65), where $z = Sn\, w/\kappa u_*$, using the velocity distribution obtained from the hydrodynamic model. The $(\bar{c}/c_a)_{max}$ value was then used with eqn (7.116) to determine $c_{a\,max}$.

7.2.5. Reference concentration

A significant problem area with the evaluation of suspended sediment concentrations is the reference concentration c_a. The predictions are directly proportional to c_a and c_a is near the bed where the rate of change in concentration with elevation is high. The rate at which particles are entrained from the bed will be a function of turbulence and pressure fluctuations at the bed but the details of this process are still ill-defined.

Lane and Kalinske (1939) assumed that the concentration just above the bed is proportional to the rms value of the vertical velocity fluctuations v* and that v' is normally distributed. This led to

$$\frac{c_a}{c_B} = f \left[\frac{w}{u_*} \right] \tag{7.117}$$

where c_a is in ppm and c_B is the fraction of the same material in the bed in percentage by weight with fall velocity w. An approximate function through the scatter of data points could be drawn as follows:

w/u_*	:	1.05	0.80	0.35	0.06	0.03
c_a/c_B	:	0.1	1.0	10	100	1000

Generally, experimental data plotted as c_a versus w/u_* indicate at low values of w/u_* a dependance of c_a on w/u_* but for $w/u_* > \sim 0.3$ the trend disappears. A line could be drawn through the scatter of data points as follows:

w/u_*	:	0.60	0.55	0.50	0.35	0.30	0.20	0.06
c_a	:	10^{-4}	10^{-3}	$2x10^{-3}$	$6x10^{-3}$	10^{-2}	$2x10^{-2}$	0.1

This shows an about two orders of magnitude variation in c_a at only minimal variation of w/u_*. Thus, for at least for $w/u_* > \sim 0.2$ the definition of c_a is very imprecise.

The pick-up of fine sediments from a flat bed will also be influenced by the Reynolds number. Therefore, u_*d/ν has been included with w/u_* in the function supposed to define c_a/c_B. Such a relationship is shown in Fig. 7.20. In it both c_a and c_B are expressed as weight per unit volume.

Einstein (1950) simply assumed that bed load is confined to a layer of two grain diameters, i.e., $y = a = 2d$ and set

$$c_a = \frac{i_B q_B}{23.2\ u'_* d} \tag{7.118}$$

This result is somewhat illogical because the overall shear stress produces the turbulence and mixing at the bed, not only the grain roughness component of it.

The same restrictive assumption of $y = a = 2d$ was used by Engelund and Fredsøe (1976) who used equation (3.34) with $c_o = 0.65$, i.e., $c_o = 0.65/(1 + 1/\lambda)^3$ where λ is the linear concentration at the bed for which they calculate a maximum value of 3.74. For sand with $S_s = 2.65$ this leads to a maximum concentration $c_a = 0.32$, which was later modified to $c_a = 0.30$. This corresponds to the concentration at which grains will come to direct physical contact and the suspension tends to "freeze".

Even the flat bed assumption is very restrictive. The flat bed occurs only at the transition flat bed conditions which covers a very narrow range of flows. The bed is usually covered with bed features which not only complicate the problems with c_a but also make the definition of bed level difficult. The mixing process and turbulence in flow over a bed covered by bed features is much more intense than over a flat bed. The mixing is more intense still when the bed features are three-dimensional. Three-dimensional bed features encourage vortices with vertical axes which act like little tornadoes lifting sediment high up into the flow. Okeda and Asaeda (1981), for example, reported experiments with $d_{50} = 0.18$ mm sand at about $w/u_* \cong 0.6$. In one set of experiments a rigid flat bed was used and in other a loose bed. The β-factor in $w/\beta\kappa u_*$ for the rigid bed was calculated to be in the range of

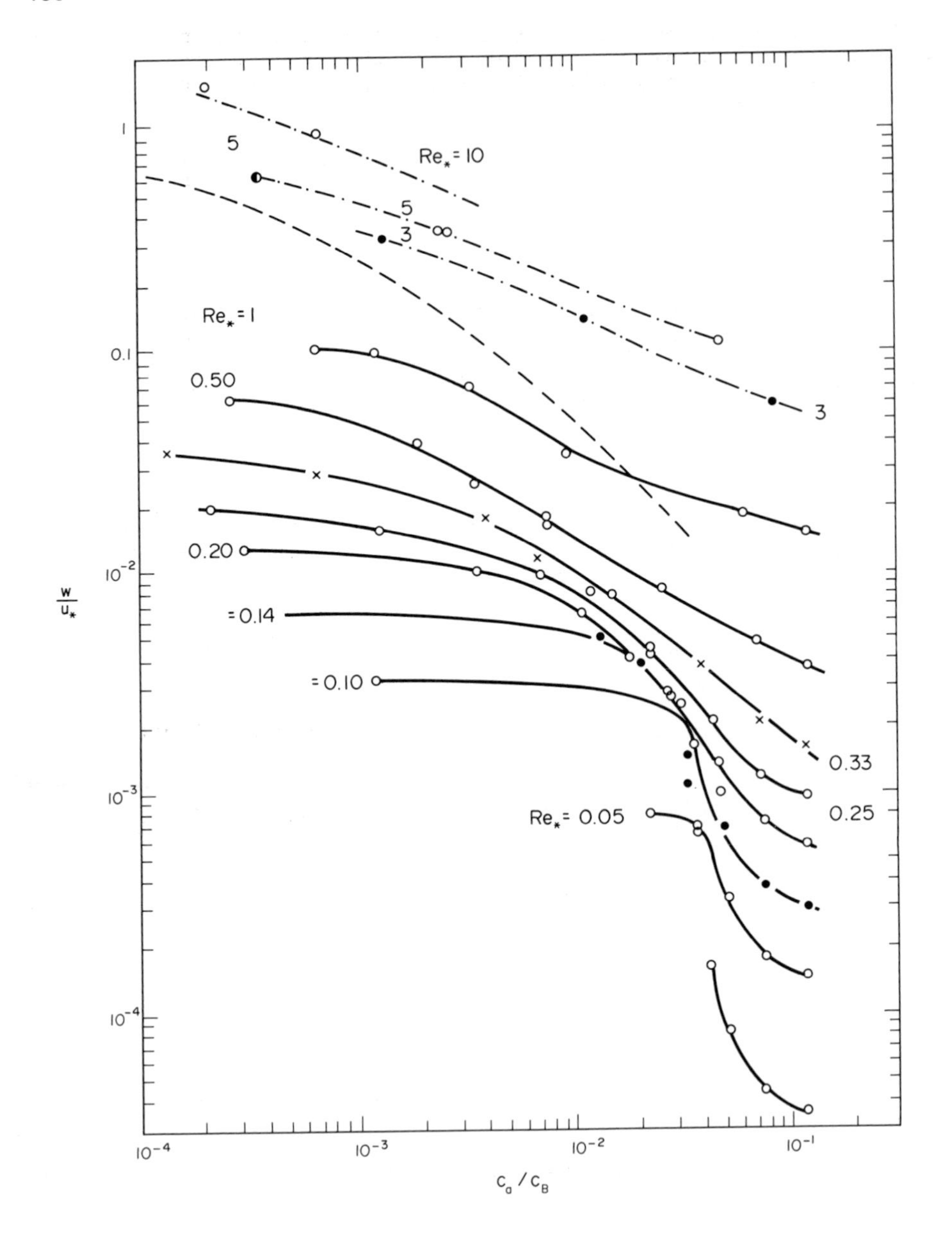

Fig. 7.20. Relationship between suspended and bed material, c_a/c_B, at equilibrium conditions as a function of w/u_* and Reynolds number $Re_* = u_* d/\nu$. Full lines data by Kalinske, chain dotted lines data by Pien (1941) and dashed line mean curve for field data according to Kalinske, after Kalinske and Hsia (1945).

1.3 - 1.8 and 2.4 over loose bed. The sediment concentration was observed to increase by a factor of 10 when the bed features became three-dimensional while the shear velocity remained approximately constant.

An expression which is simple to use and appears to yield good results is the semi-empirical relationship by van Rijn (1984)

$$c_a = 0.015 \frac{d_{50}}{a} \frac{T^{1.5}}{D_*^{0.3}} \tag{7.119}$$

where $y = a$ is assumed to be at the equivalent roughness height k_s or at half height of the bed features if known, i.e., $y = a = k_s$ or $0.5\ \eta$, with a minimum of $a = 0.01\ y_o$. Equation (7.119) was derived from experimental data using k_s as $y = a$. The parameters T and D_* are given by eqn (5.54) and (2.26), respectively.

7.2.6 Suspended sediment load

In principle, the suspended sediment transport is given by integration of the product of local velocity and concentration over the depth of flow. Thus, for unit width of flow

$$q_s = \int_{y=a}^{y_o} cudy \tag{7.120}$$

where c and u are time averaged concentration and velocity at elevation y.

Lane et al. (1941) suggested that for practical purposes eqn (7.63) was sufficiently accurate, which when introduced together with $(u-U)/u_* = 5.75 \log y/y_o + 2.5$ in the form of

$$\frac{u}{U} = 1 + \frac{u_*}{\kappa U} \left[\ln \frac{y}{y_o} + 1 \right] \tag{7.121}$$

in eqn (7.16) yields

$$q_s = qc_a \exp[15\ wa/u_* y_o]P \tag{7.122}$$

where $P = \int_0^1 (u/U)\exp[-\ 15(w/u_*)\eta]d\eta$, $\eta = y/y_o$, is shown in Fig. 7.21.

Einstein (1950) evaluated the suspended sediment load with c expressed by eqn (7.65) and the form of velocity distribution as given by Keulegan (1938)

$$\frac{u}{u_*} = 5.75 \log \left[\frac{30.2\ y}{\Delta} \right] \tag{7.123}$$

where Δ is the apparent roughness defined with eqn (7.24). Equations (7.65) and (7.123) in eqn (7.120) yield

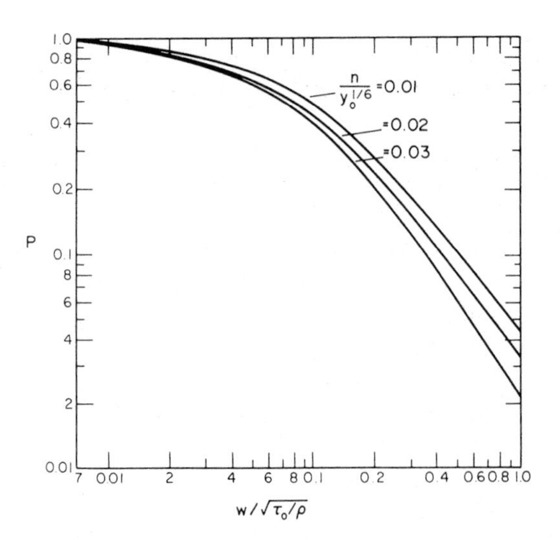

Fig. 7.21. Suspended sediment load function, according to Lane and Kalinske.

$$q_s = \int_a^{y_o} c_y u_y dy = \int_A^1 y_o c_y u_y dy$$

$$= y_o u_* c_a [A/(1-A)]^z \, 5.75 \int_A^1 [(1-y)/y]^z \log_{10}[30.2y/(\Delta/y_o)]dy$$

$$= 5.75 c_a y_o u_* [A/(1-A)]^z \, \{\log_{10}(30.2y_o/\Delta) \int_A^1 [(1-y)/y]^z \, dy$$

$$+ \int_A^1 [(1-y)/y]^z \, \log_{10} y dy\}$$

Changing the logarithm to base e leads to

$$q_s = 5.75 c_a y_o u_* [A/(1-A)]^z \, \{\log_{10}(30.2y_o/\Delta) \int_A^1 [(1-y)/y]^z \, dy$$

$$+ \, 0.434 \int_A^1 [(1-y)/y]^z \, \ln y dy\}$$

where $z = w/(\kappa u_*)$ and $\kappa \doteq 0.40$.

For practical application

$$q_s = 11.6 \; u_* c_a a [2.303 \; \log_{10}(30.2 \; y_o/\Delta) I_1 + I_2] \qquad (7.124)$$

where

$$I_1 = [0.216A^{z-1}/(1 - A)^z] \int_A^1 [(1 - y)/y]^z dy$$

$$I_2 = [0.2164^{z-1}/(1 - A)^z] \int_A^1 [(1 - y)/y] \ln y dy$$

and $11.6u_*$ is the velocity at the interface of the laminar sub-layer in the case of a hydraulically smooth bed, or the velocity at a distance of 3.68 roughness diameters from the wall in the case of rough boundaries. The values of I_1 and I_2 are given in graphical form. The integrals are given in Table 7.2.

Einstein assumed that suspension becomes impossible for $a = y < 2d$ and that the concentration of the bed load in this thin layer constitutes the limit of concentration c_a. The bed load of a given grain size fraction i_B is $i_B q_B$ and it moves at u_B. Its volume is $i_B q_B/u_B$ per unit area and concentration is $c_a \propto i_B q_B/u_B a$. Einstein wrote this as $c_a = i_B q_B/11.6u_* a$. Substitution in eqn (7.124) yields for the fraction i_s

$$i_s q_s = i_B q_B \, [2.303 \log(30.2y_o/\Delta)I_1 + I_2] \tag{7.125}$$

Brooks (1965) also used eqns (7.121) and (7.65) with $a = y_o/2$ which lead to

$$c = c_{md} \left[\frac{y_o - y}{y}\right]^z \tag{7.126}$$

where c_{md} is the concentration at mid-depth. Equation (7.126) becomes a straight line on logarithmic graph paper, Fig. 7.22 and eqn (7.120) becomes

$$q_s = c_{md} \int_a^{y_o} \left[\frac{y_o - y}{y}\right]^z \left[U + \frac{u_*}{\kappa}\left[1 + \ln \frac{y}{y_o}\right]\right] dy$$

$$= c_{md} U y_o \int_{\eta_a}^1 \left[\frac{1 - \eta}{\eta}\right]^z \left[1 + \frac{u_*}{\kappa U}(1 + \ln \eta)\right] d\eta$$

or

$$\frac{q_s}{q c_{md}} = \left[1 + \frac{u_*}{\kappa U}\right] \int_{\eta_o}^1 \left[\frac{1 - \eta}{\eta}\right]^z d\eta + \frac{u_*}{\kappa U} \int_{\eta_o}^1 \left[\frac{1 - \eta}{\eta}\right] \ln \eta d\eta \tag{7.127}$$

where the integrals are the same as in Einstein's treatment, $J_1(z,\eta_o)$ and $J_2(z,\eta_o)$ respectively, and are given in tabular form (Table 7.2). Thus,

$$\frac{q_s}{q c_{md}} = \frac{\bar{c}}{c_{md}} = J_1 + [J_1 - J_2] \frac{u_*}{\kappa U} = T\left[\frac{\kappa U}{u_*}, z, \eta_o\right] \tag{7.128}$$

Table 7.2.

Values of $J_1 = \int_A^1 \left[\frac{1-y}{y}\right]^2 dy$ and $J_2 = -\int_A^1 \left[\frac{1-y}{y}\right]^2 \ell n(y)\, dy$, (Einstein, 1950)

	$z = 0.2$		$z = 0.4$		$z = 0.6$		$z = 0.8$	
A	J_1	J_2	J_1	J_2	J_1	J_2	J_1	J_2
1.000	0	0	0	0	0	0	0	0
.90	.047736	.003029	.027451	.0017824	.015853	.0010544	.0091963	.0006271
.80	.11823	.014639	.077254	.010061	.051113	.0069698	.034211	.0048617
.70	.19844	.037872	.14166	.028791	.10287	.022085	.075852	.017071
.60	.28678	.076114	.21975	.062683	.17195	.052142	.13699	.043745
.50	.38286	.13380	.31211	.11826	.26079	.10571	.22249	.095397
.40	.48700	.21733	.42062	.20545	.37391	.19678	.34048	.19058
.30	.60027	.33688	.54901	.34122	.51953	.35108	.50574	.36606
.20	.72505	.51118	.70487	.55950	.71441	.62474	.74963	.70947
.16	.77925	.60429	.77833	.68576	.81399	.79601	.88466	.94185
.12	.83681	.71775	.86118	.84921	.93330	1.0316	1.0565	1.2815
.100	.86720	.78490	.90737	.95129	1.0035	1.1868	1.633	1.5175
.090	.88290	.82186	.93201	1.0093	1.0422	1.2779	1.2240	1.6606
.080	.89898	.86153	.95787	1.0731	1.0838	1.3806	1.2910	1.8258
.070	.91551	.90438	.98520	1.1440	1.1290	1.4977	1.3657	2.0177
.060	.93256	.95100	1.0143	1.2235	1.1786	1.6333	1.4502	2.2490
.050	.95023	1.0023	1.0455	1.3141	1.2338	1.7935	1.5477	2.5321
.040	.96866	1.0595	1.0795	1.4196	1.2964	1.9881	1.6633	2.8911
.030	.98809	1.1247	1.1172	1.5464	1.3698	2.2347	1.8060	3.3710
.020	1.00893	1.2018	1.1607	1.7073	1.4605	2.5707	1.9954	4.0729
.016	1.0178	1.2376	1.1805	1.7870	1.5047	2.7483	2.0937	4.4686
.012	1.0272	1.2777	1.2025	1.8810	1.5562	2.9687	2.2146	4.9858
.0100	1.0321	1.2999	1.2146	1.9356	1.5860	3.1031	2.2879	5.3165
.0090	1.0347	1.3117	1.2210	1.9655	1.6022	3.1788	2.3291	5.5084
.0080	1.0372	1.3240	1.2277	1.9975	1.6196	3.2618	2.3742	5.7233
.0070	1.0399	1.3370	1.2348	2.0320	1.6384	3.3536	2.4240	5.9674
.0060	1.0426	1.3508	1.2423	2.0697	1.6589	3.4567	2.4800	6.2494
.0050	1.0455	1.3655	1.2503	2.1114	1.6815	3.5746	2.5441	6.5830
.0040	1.0484	1.3814	1.2589	2.1583	1.7071	3.7129	2.6194	6.9907
.0030	1.0515	1.3990	1.2686	2.2127	1.7369	3.8816	2.7118	7.5141
.0020	1.0548	1.4189	1.2796	2.2788	1.7735	4.1014	2.8335	8.2451
.0016	1.0562	1.4279	1.2846	2.3105	1.7912	4.2137	2.8964	8.6428
.0012	1.0577	1.4377	1.2901	2.3470	1.8119	4.3497	2.9734	9.1499
.00100	1.0585	1.4430	1.2932	2.3678	1.8238	4.4310	3.0200	9.4675
.00090	1.0589	1.4458	1.2948	2.3791	1.8303	4.4763	3.0462	9.6497
.00080	1.0593	1.4487	1.2965	2.3911	1.8373	4.5255	3.0748	9.8520
.00070	1.0597	1.4517	1.2983	2.4038	1.8448	4.5794	3.1064	10.080
.00060	1.0602	1.4549	1.3002	2.4177	1.8530	4.6394	3.1419	10.340
.00050	1.0606	1.4583	1.3022	2.4328	1.8620	4.7073	3.1825	10.645
.00040	1.0611	1.4619	1.3044	2.4496	1.8722	4.7860	3.2303	11.013
.00030	1.0616	1.4658	1.3068	2.4689	1.8841	4.8807	3.2887	11.478
0.00020	1.0621	1.4702	1.3095	2.4919	1.8087	5.0019	3.3656	12.118
.00016	1.0624	1.4721	1.3108	2.5027	1.9057	5.0629	3.4053	12.460
.00012	1.0626	1.4742	1.3122	2.5151	1.9140	5.1361	3.4539	12.893
.00010	1.0627	1.4753	1.3130	2.5221	1.9187	5.1794	3.4833	13.161
.000090	1.0628	1.4759	1.3134	2.5259	1.9213	5.2034	3.4998	13.314
.000080	1.0628	1.4765	1.3138	2.5299	1.9241	5.2294	3.5179	13.484
.000070	1.0629	1.4772	1.3142	2.5341	1.9271	5.2578	3.5379	13.673
.000060	1.0630	1.4778	1.3147	2.5387	1.9303	5.2892	3.5603	13.889
.000050	1.0630	1.4785	1.3152	2.5436	1.9339	5.3245	3.5859	14.141
.000040	1.0631	1.4793	1.3158	2.5491	1.9380	5.3652	3.6160	14.442
.000030	1.0632	1.4801	1.3164	2.5554	1.9427	5.4138	3.6529	14.821
.000020	1.0633	1.4809	1.3171	2.5627	1.9485	5.4754	3.7014	15.336
.000016	1.0633	1.4813	1.3174	2.5662	1.9514	5.5062	3.7265	15.610
.000012	1.0633	1.4817	1.3177	2.5701	1.9546	5.5429	3.7572	15.954
.000010	1.0634	1.4820	1.3179	2.5723	1.9565	5.5645	3.7758	16.166

[1] Integrals calculated in closed form

Table 7.2. Continued

$z = 1.0$		$z = 1.2$		$z = 1.5$		$z = 2.0$[1]	
J_1	J_2	J_1	J_2	J_1	J_2	J_1	J_2
0	0	0	0	0	0	0	0
.005361	.000375	.0031405	.0002256	.0014223	.0001063	.000371	.000032
.023144	.003412	.015808	.0024074	.009065	.0014394	.00372	.000622
.056676	.013282	.042832	.010394	.028654	.0072628	.01523	.004076
.11083	.036968	.090824	.031440	.068744	.024911	.04501	.01727
.19315	.086806	.17014	.079551	.14382	.070593	.11370	.05927
.31630	.18633	.29873	.18367	.28118	.18159	.26742	.18462
.50399	.38603	.51204	.41108	.53992	.45829	.62537	.56917
.80955	.81741	.89522	.95351	1.0791	1.2246	1.5811	1.9350
.99269	1.1328	1.1437	1.3816	1.4719	1.9020	2.4248	3.3921
1.2404	1.6226	1.5008	2.0884	2.0905	3.1279	3.9728	6.4656
1.4027	1.9817	1.7476	2.6346	2.5535	4.1528	5.2948	9.3937
1.4981	2.2063	1.8974	2.9873	2.8481	4.8469	6.2052	11.539
1.6059	2.4722	2.0708	3.4152	3.2022	5.7205	7.3685	14.410
1.7294	2.7925	2.2751	3.9448	3.6366	6.8473	8.8972	18.376
1.8736	3.1869	2.5210	4.6178	4.1844	8.3469	10.980	24.080
2.0459	3.6875	2.8256	5.5028	4.9006	10.428	13.959	32.741
2.2590	4.3499	3.2188	6.7252	5.8863	13.494	18.522	46.942
2.5367	5.2838	3.7592	8.5433	7.3538	18.435	26.290	73.121
2.9323	6.7514	4.5860	11.614	9.8556	27.735	42.156	132.20
3.1514	7.6332	5.0743	13.580	11.480	34.276	54.214	180.77
3.4351	8.8472	5.7402	16.429	13.876	44.537	74.476	267.61
3.6155	9.6612	6.1840	18.433	15.590	52.276	90.780	341.25
3.7198	10.147	6.4484	19.664	16.657	57.244	101.68	302.04
3.8366	10.704	6.7511	21.108	17.919	63.267	115.34	457.18
3.9691	11.353	7.1033	22.832	19.446	70.741	132.93	543.32
4.1223	12.125	7.5224	24.944	21.343	80.301	156.43	661.79
4.3036	13.069	8.0356	27.617	23.787	93.032	189.40	833.56
4.5258	14.271	8.6905	31.160	27.103	110.98	238.95	1101.9
4.8125	15.895	9.5802	36.202	31.971	138.57	321.71	1571.3
5.2170	18.327	10.926	44.300	40.151	187.82	487.57	2570.7
5.4398	19.736	11.716	49.293	45.416	221.13	612.12	3359.1
5.7271	21.627	12.787	56.347	53.136	271.98	819.88	4728.0
5.9092	22.869	13.499	61.199	58.638	309.49	986.18	5862.0
6.0145	23.601	13.922	64.147	62.054	332.93	1097.1	6634.1
6.1322	24.434	14.406	67.570	66.093	361.50	1235.7	7614.8
6.2656	25.394	14.969	71.621	70.970	396.60	1414.0	8898.4
6.4196	26.525	15.638	76.530	77.021	441.03	1651.8	10645.0
6.6019	27.893	16.456	82.675	84.808	499.43	1984.8	13146.0
6.8249	29.614	17.499	90.721	95.359	580.83	2484.4	17001.0
7.1125	31.905	18.915	102.00	110.82	704.16	3317.1	23642.0
7.5180	35.276	21.054	119.80	136.78	920.55	4983.0	37516.0
7.7411	37.201	22.307	130.61	153.47	1064.6	6232.5	48303.0
8.0288	39.757	24.005	145.72	177.93	1282.0	8315.3	66821.0
8.2111	41.420	25.135	156.02	195.35	1440.9	9981.6	82024.0
8.3165	42.396	25.806	162.24	206.17	1541.1	11093.0	92312.0
8.4342	43.500	26.574	169.44	218.95	1661.0	12481.0	105332.0
8.5678	44.769	27.467	177.93	234.39	1807.7	14267.0	122296.0
8.7219	46.255	28.528	188.16	253.54	1992.4	16647.0	145261.0
8.9042	48.044	29.825	200.89	278.19	2234.2	19980.0	177974.0
9.1274	50.279	31.479	217.45	311.57	2568.6	24980.0	228060.0
9.4151	53.233	33.723	240.51	360.49	3071.3	33313.0	313700.0
9.8206	57.539	37.115	276.53	442.60	3943.2	49978.0	490870.0
10.054	59.979	39.101	298.24	495.39	4520.8	62478.0	627560.0
10.332	63.197	41.797	328.40	572.75	5386.5	83311.0	860760.0
10.514	65.279	43.588	348.86	627.85	6016.0	99977.0	1051160.0

A	$\int_A^1 \left(\frac{1-y}{y}\right)^z dy$				$\int_A^1 \log_e(y)\left(\frac{1-y}{y}\right) dy$			
	$z = 0$	3.0	4.0	5.0	0	3.0	4.0	5.0
1.0	0	0	0	0	0	0	0	0
.1	.90000	$.2851.10^{2}$	$.1758.10^{3}$	$.1237.10^{4}$	.66974	$.5560.10^{2}$	$.3632.10^{3}$	$.2602.10^{4}$
.01	.99000	$.4715.10^{4}$	$.3136.10^{6}$	$.2338.10^{8}$	.93495	$1.948\ .10^{4}$	$1.343\ .10^{6}$	$1.0198.10^{8}$
.001	.99900	$.4970.10^{6}$	$.3313.10^{9}$	$.2483.10^{12}$	.99209	$3.187\ .10^{6}$	$2.177\ .10^{9}$	$1.6535.10^{12}$
.0001	.99990	$.4997.10^{8}$	$.3331.10^{12}$	$.2498.10^{16}$	.99898	$4.353\ .10^{8}$	$2.955\ .10^{12}$	$2.239\ .10^{16}$
.00001	.99999	$.5000.10^{10}$	$.3333.10^{15}$	$.2500.10^{20}$	.99987	$5.508\ .10^{10}$	$3.723\ .10^{15}$	$2.816\ .10^{20}$

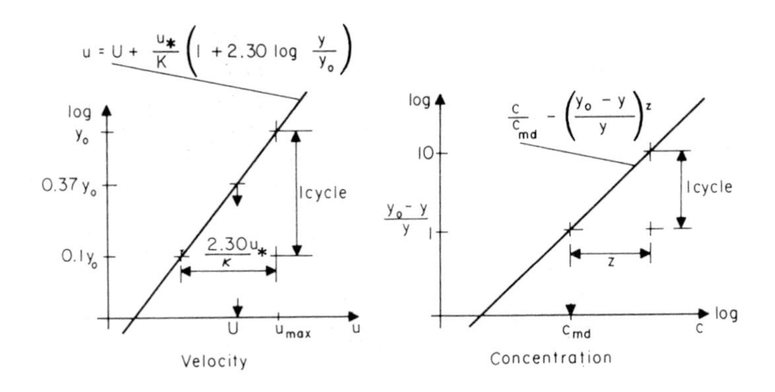

Fig. 7.22. Semi-logarithmic plots of velocity and concentration distributions.

The relationship to c_a is

$$\frac{\bar{c}}{c_{md}} = \frac{\bar{c}}{c_a}\left[\frac{y_o - a}{a}\right]^z \tag{7.129}$$

For evaluation of eqn (7.128), $2.30u_*/\kappa$ and U are read off the semilogarithmic plot of the velocity distribution and z from the concentration plot (Fig. 7.22). For the possible lower limit of integration Brooks lists

$$\eta_o = \frac{2d}{y_o}\ ; \quad \eta_o = \exp\ [-\ (\kappa U/u_* - 1)]\ \text{and}\ \eta_o = (c_{md}/c_b)^{\frac{1}{z}}$$

The first is the Einstein criterion, the second corresponds to y where u = 0 of the logarithmic distribution, and the third is the level at which the extrapolated concentration becomes equal to the concentration of the bed. Brooks favours the second which is in terms of the same parameters as eqn (7.128). Therefore, η_o can be eliminated and the transport can be expressed as a function of z and $\kappa U/u_*$ as shown on Fig. 7.23.

Van Rijn (1984) expressed the suspended load transport rate per unit width by using eqns (7.78(a) and (b)), (7.79) and the logarithmic velocity distribution $u/u_* = (1/\kappa)\ln(y/0.033k_s)$ is eqn (7.120). This yielded

$$q_s = \frac{u_* c_a}{\kappa}\left[\frac{a}{y_o - a}\right]^{z_1}\left[\int_a^{0.5y_o}\left[\frac{y_o - y}{y}\right]^{z_1}\ln\left[\frac{y}{y_1}\right] + \int_{0.5y_o}^{y_o} e^{-4z_1(y/y_o - 0.5}\ln\left[\frac{y}{y_1}\right]dy\right] = FUy_o c_a \tag{7.130}$$

where $y_1 = 0.033\ k_s$ and

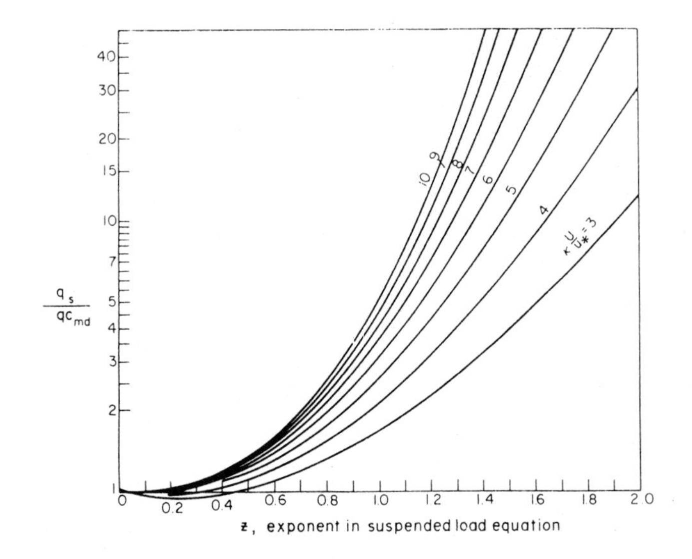

Fig. 7.23. Suspended sediment transport $q_s/qc_{md} = f(\kappa U/u_*, z)$ according to Brooks. Lower limit $\eta_o = \exp[-\kappa U/u_* - 1]$, $u(\eta_o) = 0$. (Brooks, 1965.)

$$F = \frac{\left[\frac{a}{y_o}\right]^{z_1} - \left[\frac{a}{y_o}\right]^{1.2}}{1 - \frac{a}{y_o}^{z_1} (1.2 - z_1)} \tag{7.131}$$

which is said to be within about 25% accuracy by $0.3 \leq z_1 \leq 3$ and $0.01 \leq a/y_o \leq 0.1$.

A simplified expression was proposed for the case where only mean flow velocity, depth and particle size are known:

$$\frac{q_s}{Uy_o} = 0.012 \left\{ \frac{U - U_c}{[(S_s - 1)gd_{50}]^{\frac{1}{2}}} \right\}^{2.4} \left[\frac{d_{50}}{y_o}\right] D_*^{-0.6} \tag{7.132}$$

where U_c is given with eqn (7.33). The correlation was obtained with data covering y_o = 1-20 m, U = 0.5 - 2.5 m/s, and d_{50} = 0.1 - 2.0 mm, σ = 2 and T = 15°C.

The integration methods of suspended sediment load are basically design tools for prediction purposes. Another set of problems is the estimation of the sediment discharge, including wash load, for a given water discharge from measured (usually depth-integrated) suspended sediment samples. This involves the estimation of the *unmeasured suspended load* below the lowest measuring elevation y_1. Colby and Hembree (1955) adapted the Einstein method for this purpose using empirical and partly arbitrary assumptions. The results have become known as the modified Einstein method.

Since the integration methods of calculation of suspended load are rather involved, engineers have been looking for simpler relationships between suspended sediment load and water discharge. Rubey (1933) showed, for example, that the average sediment concentration, $\bar{c}$, is proportional to $R^{1/2}S^{2/3}$. Combining with the Chézy formula shows that the suspended sediment load; $q_s = q\bar{c}$, is proportional to $R^2S^{7/6}$. This led to expressions of the form

$$q_s = aq^b \tag{7.133}$$

where usually $1.9 \leq b \leq 2.2$, with $b = 2.0$ as an approximate value when depth changes are moderate and shear velocity relatively high. However, values of b up to 3.0 have been reported. At low values of shear velocity or high values of w/u_*, the sediment discharge is also a function of w/u_*.

Ranga Raju (1985) presented an empirical method of calculating suspended load. The method is similar to the bed load estimation by Misri et al. (1984). Accordingly, for uniform sediment the dimensionless suspended sediment transport is

$$\phi_s = 28\theta \tag{7.134}$$

and non-uniformity is accounted for by

$$\xi_s = \tau_{eff}/\tau_o = f(\theta, \tau_o/\tau_c, C_u) \tag{7.135}$$

where τ_o is average bed shear stress, τ_c is critical shear stress according to Shields criterion based on average grain size and C_u is the uniformity coefficient (eqn 2.5d). The effect of τ_o/τ_c is accounted for by a coefficient K_s presented as a graph with a few points in the range of τ_o/τ_c of 9 to 17 and extrapolated to smaller values:

τ_o/τ_c:	< 2	4	5	7	9	10	11	14	17	> 20
K_s:	2.2	2.0	1.85	1.65	1.5	1.35	1.25	1.1	1.01	1.0

The effect of C_u is given as:

C_u:	< 0.25	0.30	0.40	> 0.50
L_s:	0.80	0.90	0.96	1.00

The product $K_sL\xi_s$ is given as a function of $\theta_i = \tau_o/\rho g(S_s - 1)d_i$ for the size fraction d_i. The function is based on laboratory data with mean diameters of 0.25, 0.42, 0.55 and 0.57 mm. The line drawn through the band of data could be represented by

θ_i :	0.2	0.3	0.4	0.5	0.6	0.7	0.8	0.9	1.0	1.5	2	3	4	6
$K_sL_s\xi_s$:	2.3	1.65	1.3	1.1	0.95	0.84	0.77	0.71	0.66	0.51	0.42	0.34	0.28	0.23

since K_s and L_s are known ξ_s can be calculated. For $\xi_s\theta_i \geq 0.035$ the relationship between $\xi_s\theta_i$ and ϕ_s is given for $\phi_s \geq 0.1$ by

$$\log \xi_s\theta_i = -1.2412 + 0.1733 \log \phi_s \qquad (7.136a)$$

and for $0.01 \leq \phi_s \leq 0.1$ by

$$\log \xi_s\theta_i = -1.2441 + 0.1705 \log \phi_s \qquad (7.136b)$$

The transport rate for d_i is then from

$$\phi_s = \frac{i_s g_s}{i_b \rho_s g d_i}\left[\frac{1}{(S_s - 1)}\,\frac{1}{g d_i^3}\right]^{\frac{1}{2}} \qquad (7.137)$$

and weight rate of transport

$$g_s = \Sigma\, i_s g_s$$

7.2.7 Wash load

A definition widely used is in the wording of the American Geophysical Union

> Wash load is that part of the sediment load of a stream which is composed of particle sizes smaller than those found in appreciable quantities in the shifting portions of the stream bed.

Wash load effectively arises from the dirt brought in by overland flow and consists of fine material. The amount depends on catchment conditions. Some wash load may also be produced by wear of gravel in transport, particularly of graywacke origin. Although a suspended load, wash load is not in a functional relationship with flow rate like the suspended load, except that rain washes the material into the stream and rain also leads to an increase in stream flow. If a suspended load sample is analysed, it will be found that fractions greater than a given size can be related to flow rate according to suspended load concepts but not the finer fractions. The dividing size is not fixed but depends on flow intensity. A rule of thumb is that the grain size separating wash and suspended load is about the 10% finer size given by sieve analysis of bed material. Some wash load can occasionally settle on the stream bed but the amount is only a small fraction of the total wash load in transport.

There is no real upper limit for the wash load transport capacity. The increasing concentration of the colloidal fractions in the flow gradually leads to change in flow properties. The fluid-wash load mixture does no longer approximate to a flow of a newtonian fluid but a Bingham fluid (a term used in rheology), pseudo plastic. In the Bingham fluid there is a yield stress before the fluid starts to deform. Generally, in the pseudo plastic fluids the shear stress-velocity gradient relationship is not linear. At high concentrations of colloidal material the flow is referred to as *hyperconcentrated* flow. In the Yellow River system concentrations as high as 1500 kg/m³ have been reported. In the July to September season average concentrations of 223 kg/m³ were reported by Wan Zhaohui and Xu Yian (1984). For the behaviour of such flows reference is made to Engelund and Wan (1984), Wan (1985) and Beverage and Culbertson (1964).

Heavy concentrations of wash load affect the density of flow and the density gradient near the boundary. This can lead to behavioural changes of the sediment laden flow which are not accounted for in the usual flow models, in particular changes in resistance to flow. Simons and Richardson (1960) reported up to 40% reduction of resistance in subcritical flows at concentrations of 40 000 ppm and an increase in resistance in supercritical flow. Heavy concentrations of wash load also affect the bed features. The onset of development is delayed, the bed features are flatter and they disappear sooner than in a flow of clear water, i.e. the span between threshold and transition flat bed is shortened and the height is reduced.

7.2.8 Comments

A fundamental difficulty with estimation of suspended load arises from the effects of macro-turbulence and secondary comments, discussed in Chapter 5. The concentrated eddies and boils lift quantities of sediment into the flow where it becomes dispersed. This is not a diffusion process in the sense discussed earlier. Concentration profiles predicted by the diffusion relationships have been shown to fit the observed data well for fine sediment over a flat rigid bed in laboratory flumes. However, over a loose boundary the observed distributions are more uniform over depth. In the presence of three-dimensional bed features the amount of suspended sediment increases markedly while the mean shear stress remains essentially constant. Sediment is entrained into the outer flow from the side slopes of the three-dimensional bed features. The measured concentration profiles even differ with the location of measurement.

The Reynolds number dependence of suspension is expressed in terms of the dimensionless distance from the boundary $y^+ = u_* y/\nu$. Observations show a maximum turbulence intensity $(\overline{u'^2})^{\frac{1}{2}} u_*$ at $y+ = 10$-15. Blinco and Partheniades (1971) showed that the relationship between y^+ and turbulence intensity is fairly insensitive to boundary roughness. In terms of the grain size, they found that the peak intensity occurred at about $Re_* = u_* d/\nu = 35$-55. An increase in Re_* lead to an increase in relative turbulence intensity at Re_* less than the peak value and to a decrease when Re_* was greater (Taylor and Vanoni, 1972). The consequences of this appear with temperature changes. A decreasing temperature increases kinematic viscosity and reduces Re_*. This could lead to either an increase or decrease in turbulence intensity, depending on Re_*. An increase in turbulence intensity increases suspended sediment load.

The discussion of suspension has been confined to a steady state condition. Problems where the sediment inflow and outflow do not balance, that is, erosion or deposition problems, are best treated with the aid of numerical models, e.g., Celik and Rodi (1985), Kerssens and van Rijn (1977), Kerssens et al. (1979), van Rijn (1985).

7.3 Total Sediment Transport Rate

More often than not, we are interested in the total sediment transport rate, not in how much is transported in which mode. The total sediment discharge is an integral part of any problem involving alluvial channels because these channels do not just carry water but water and sediment. The sediment discharge is part of stability or instability of the channel, silting of reservoirs, etc, i.e., alluvial channels have to be designed to carry certain water *and* sediment rates. These two transports cannot be separated. They are interrelated and equally important, whatever the project.

The total sediment load in nature is the sum of bed load, suspended load *and* wash load. In laboratory studies the wash load is almost invariably absent and frequently total load amounts to bed load only. Predictions of bed load and suspended load lead by addition to what is frequently called total load. However, wash load has to be added to this and the latter may not be insignificant.

Data from field measurements give the suspended and wash load above a lower limiting elevation $y = y_1$, depending on the type of the sampler. The unmeasured load below this elevation has to be estimated. The bed load too has to be estimated or measured with bed load measuring devices. Bed load transport in the field is extremely difficult to measure, even with the most elaborate methods.

Typical of the additive methods of bed and suspended load is that by Einstein (1950). From eqn (7.129) the total transport rate for any size fraction is

$$i_T q_T = i_B q_B [2.303 \log(30.2 y_o/\Delta) I_1 + I_2 + 1] \qquad (7.138)$$

and the total sediment discharge is given by summation of the $i_T q_T$ terms. However, any of the various bed load and suspended load transport formulae could be used in the same manner.

The Einstein method was adapted by Colby and Hembree (1955) to account for the unmeasured sediment load when using field data, the modified Einstein method. The calculation is based on measured mean velocity and depth instead of hydraulic mean radius. The exponent z is determined from measured data for dominant grain size, and for other grain sizes is assumed to vary as the 0.7 power of the fall velocity w. Einstein (1964) wrote the equation given by Colby et al. in a simpler form as

$$\frac{i_{Ts} g_{Ts}}{i_{sm} g_{sm}} = \left[\frac{\eta_1}{\eta_0}\right]^{z-1} \left[\frac{1-\eta_0}{1-\eta_1}\right]^{z} \frac{(1 + P_E I_1 + I_2)\eta_0}{(P_E I_1 + I_2)\eta_1} \qquad (7.139)$$

where g_{Ts} is the total sediment discharge per unit width and g_{sm} the measured suspended sediment discharge, $\eta_o = a/y_o = A$; $\eta_1 = y_1/y_o$, where y_1 is the lower limit of sampling, z is the modified exponent and $P_E = 2.30 \log(30.2 y_o/\Delta)$ has a modified value. Colby et al. discuss the procedure in detail and give a numerical example. The U.S. Bureau of Reclamation too issued a user manual, *Step Method for Computing Total Sediment Load by the Modified Einstein Procedure*, July 1955, and an accompanying note, "Computation of z-s for use in the modified Einstein procedure", June 1966.

The unmeasured sediment load in sandy rivers, g_{su}, increases with the mean velocity of flow to a power slightly more than three (Colby 1957), i.e., $g_{su} \cong 355U^{3.1}$, where g_{su} is in kNm^{-1} per day. This is based on data from rivers in the U.S.A.

Bishop et al. (1965) simplified the Einstein method for calculation of the total load and presented a graph of A_* and B_* values as a function of grain size. White et al. (1973) modified the method further and expressed A_* and B_* as a function of the dimensionless particle size D_* as

$$\begin{aligned} A_* &= 7.2 \log D_*^{2.88} ; & D_* < 40 \\ A_* &= -101.96(\log D_* - 2)^2 + 43.5 ; & D_* > 40 \end{aligned} \qquad (7.140)$$

$$\begin{aligned} B_* &= 0.045 \log\left[\frac{u_*' d}{\nu} - 2\right]^2 + 0.143 ; & \frac{u_*' d}{\nu} < 10 \\ B_* &= -0.07 \log\left[\frac{u_*' d}{\nu} - 2\right]^3 + 0.12 ; & \frac{u_*' d}{\nu} > 10 \end{aligned} \qquad (7.141)$$

The associated hiding factor was

$$\begin{aligned} \xi &= 10^{3.27(\log d_{35}/d)^2} & \text{if } d_{max} > u_*^2/0.1(S_s - 1)g \\ \xi &= 1 & \text{if } d_{max} < u_*^2/0.1(S_s - 1)g \end{aligned} \qquad (7.142)$$

where d is the median and d_{max} the largest particle size in the bed sediment. The total load relationship was expressed as eqn (7.23) except that the lower limit was set at $-\infty$, and $\psi_* = \xi\psi'$ where ψ' is in terms of d_{35}. The relationship is shown in Fig. 7.24, in terms of ϕ_{50}.

Laursen (1958) proposed from his laboratory data a total sediment transport formula. His formula is for mean sediment concentration in terms of weight

$$\bar{c} = \Sigma i\left[\frac{d_i}{y_o}\right]^{7/6}\left[\frac{\tau_o'}{\tau_{ci}} - 1\right] f(u_*/w_i) \qquad (7.143)$$

where the function $f(u_*/w_i)$ is given in graphical form (Fig. 7.25); w_i is the fall velocity in water of particles of mean grain size d_i; τ_o' is bed shear stress due to grain resistance and is expressed with the aid of the Manning-Strickler formula, $U/u_* = 7.66(y_o/d)^{1/6}$. Since $\tau_o = \rho u_*^2$,

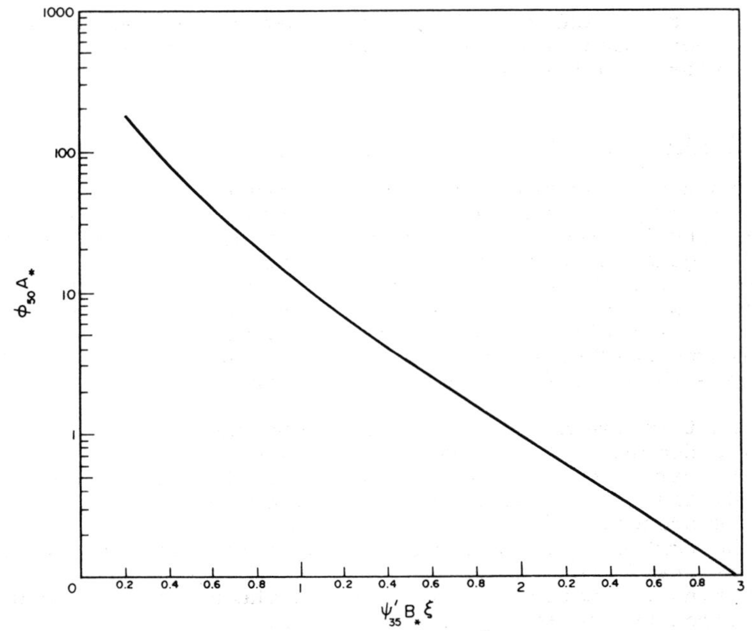

Fig. 7.24. Relationship $A_*\phi_{50} = f(\psi'_{35}B_*\xi)$ according to White et al. (1973).

$$\tau'_o = \frac{\rho U^2}{7.66^2}\left[\frac{d}{y_o}\right]^{1/3} \cong \frac{\rho U^2}{58}\left[\frac{d_i}{y_o}\right]^{1/3}$$

The critical shear stress for the grains of size d_i is

$$\tau_{ci} = \theta_c(\gamma_s - \gamma)d_i$$

and the weight rate of transport is

$$g_{Ts} = \bar{c}S_s q$$

where for most sediments $S_s = 2.65$. The Σi signifies that the contributions of all the size fractions are added to give total transport, that is when the grading is strongly non-uniform and has to be subdivided. Since the relationship is entirely based on laboratory data wash load is not included.

A similar relationship was developed by Bogardi (1965)

$$\frac{\bar{c}}{\left[\dfrac{d}{R}\right]^{7/6}\left[\dfrac{\tau'_o}{\tau_{cr}} - 1\right]} = f\ \frac{gd}{u_*^2}, d \qquad (7.144)$$

where R is the hydraulic mean radius. The function is shown on

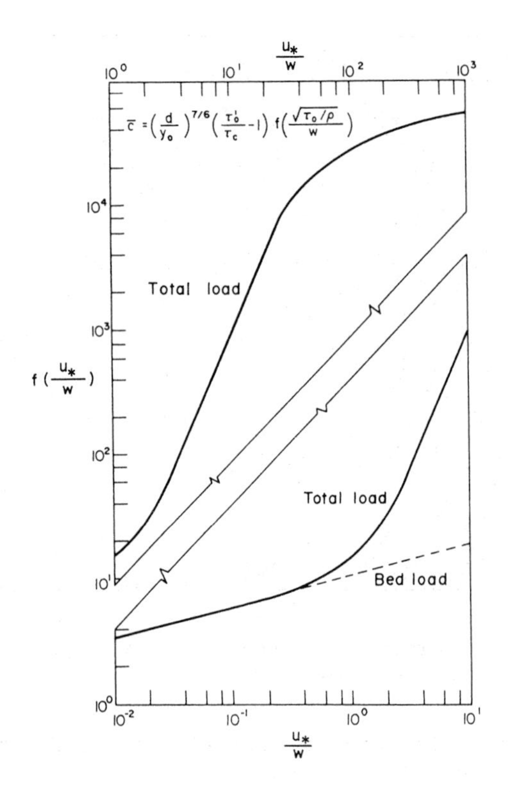

Fig. 7.25. Laursen sediment transport function. (Laursen, 1958).

Fig. 7.26. The areas on Fig. 7.26, between the parallel lines sloping upwards, Bogardi labels: smooth bed, ripples, dunes, transition and antidunes.

Bagnold (1966) work done concept leads to

$$g_{iTs} = g_{iB} + g_{is} = P\left[\frac{\eta_b}{\tan\alpha} + \eta_s \frac{U_s}{w} (1 - \eta_b)\right] \tag{7.145}$$

where U_s is the mean velocity of the suspended grains approximately equal to U, P is the stream power unit width and the η-s are the respective efficiencies. For $d < 0.5$ mm, $\eta_s(1 - \eta_b)$ is estimated to be 0.01 and $\eta_b/\tan\alpha \cong 0.17$ when $\gamma_s{}^* = 1.65$. For d > 0.5 mm and $\theta > 1$ Bagnold gives special graphs for estimation of these terms.

An energy approach was also used by Engelund and Hansen (1967) for total load over a dune bed. The moving sediment particle is lifted the height of the dune η. Thus, energy is required. The relationship obtained is

$$f\phi = 0.1\theta^{5/2} \tag{7.146}$$

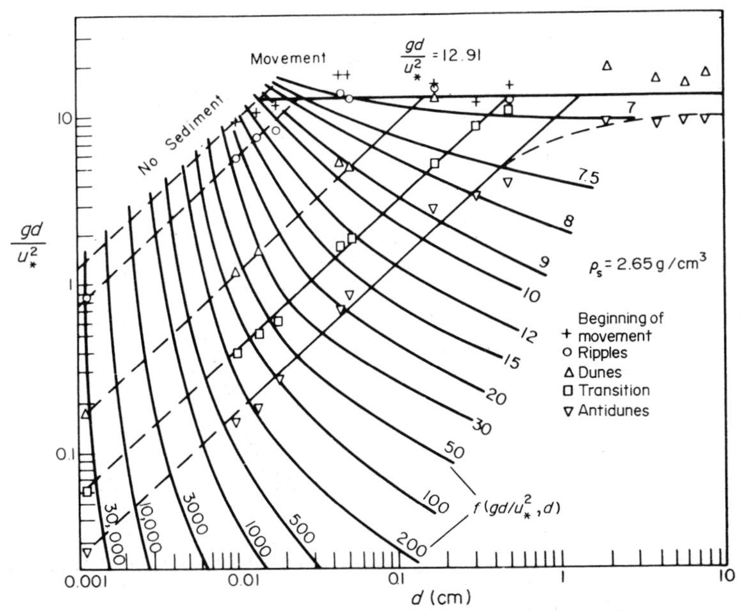

Fig. 7.26. Correlation between gd/u_*^2, and $f(gd/u_*^2, d)$. (After Bogardi, 1965).

where the transport function ϕ (eqn (7.18)) is for total transport and the friction factor

$$f = \frac{2gy_oS}{U^2} \text{ or } f = \frac{2g}{C^2}$$

where C is the Chézy coefficient. (Notice that $f \neq 8g/C^2$ as defined earlier.) The formula is based on experiments with uniform sands and $u_*/d_{50}\nu > 12$. It has been found to give good results when C is based on actual observations and the suspended load is significant, although the concepts used in the derivation of the formula do not relate to suspension or to transition flat bed. The total sediment discharge can also be expressed as

$$q_{Ts} = 0.05U^2 \left[\frac{d_{50}}{g(S_s - 1)}\right]^{\frac{1}{2}} \left[\frac{\tau_o}{(\rho_s - \rho)gd_{50}}\right]^{3/2} \qquad (7.147)$$

using f as defined above.

The energy concept in relation to stream channel geometry and sediment transport has been studied extensively by Yang (1972, 1973, 1976, 1977, 1979, 1982, 1984) and Song and Yang (1982). He correlated experimental data with the rate of expenditure of potential energy which was expressed as unit stream power, US. Multiple regression yielded

$$\log c_T = 5.435 - 0.286 \log wd/\nu - 0.457 \log u_*/w$$

$$+ (1.799 - 0.409 \log wd/\nu - 0.314 \log u_*/w) \log(US/w - U_c S/w) \tag{7.148}$$

where c_T is the total sediment concentration in parts per million (ppm) by weight, w is average fall velocity, d median diameter of sediment, U the average stream velocity, S energy slope, and U_c the critical velocity for initiation of motion eqn (3.15).

At higher concentrations the incipient motion criteria can be neglected, yielding

$$\log c_T = 5.165 - 0.153 \log wd/\nu - 0.297 \log u_*/w$$

$$+ (1.780 - 0.360 \log wd/\nu - 0.480 \log u_*/w) \log US/w \tag{7.149}$$

Yang demonstrates that the relationships work well both for field and laboratory data.

Total sediment transport in terms of concentration was also correlated by Brownlie (1981). From regression analysis of laboratory and field data c_T in ppm by weight was expressed as

$$\bar{c}_T = 7115 C_F (F_g - F_{gc})^{1\cdot978} S^{0\cdot6601} \left[\frac{R}{d_{50}}\right]^{-0\cdot3301} \tag{7.150}$$

where

$$F_g = V/[S_s - 1) g d_{50}]^{\frac{1}{2}}$$

$$F_{gc} = V_c/[S_s - 1) g d_{50}]^{\frac{1}{2}} = 4.596 \theta_c^{0\cdot5295} S^{-0\cdot1405} \sigma_g^{-0\cdot1606}$$

$$\theta_c = 0.22Y + 0.06(10)^{-7\cdot7Y}$$

$$Y = [(\Delta g d_{50}^{3})^{\frac{1}{2}}/\nu]^{-0\cdot6}$$

$$\Delta = (\rho_s - \rho)/\rho$$

C_F = 1 for laboratory data
= 1.28 for field data

σ_g = geometric stand. dev. of particle size distribution

In a simplified form the equation reads

$$\bar{c}_T = 7100 C_F (S^{1/3} F_g - S^{1/3} F_{gc})^2 (R/d_{50})^{-1/3} \tag{7.151}$$

The difference in C_F-values is ascribed to irregular cross-sections in the field. The argument is supported by a somewhat tenuous theoretical proof. The difference could equally well arise from wash load which has not been separated from field data.

Regime type of total sediment transport formulae have been given by Blench (1966) and Maddock (1976).

The Hydraulic Research Station, Wallingford, carried out extensive studies of the available sediment transport formulae. The work led to the Ackers and White (1973) formula. The formula introduces through the so-called mobility parameter F_{gr} the different response of different grain sizes to total applied shear stress. The mobility number

$$F_{gr} = \frac{u_*^{\,n}}{(g\Delta d)^{\frac{1}{2}}}\left[\frac{U}{\sqrt{32}\,\log \alpha y_o/d}\right]^{1-n} \qquad (7.152)$$

which is a modified Shields' parameter, for example for n = 1, $F_{gr} = \sqrt{\theta}$. The coefficient $\alpha \cong 10$ was obtained from turbulent rough boundary flow data, and $\sqrt{32}\,\log 10y_o/d \cong 7.66(y_o/d)^{1/6}$ for relatively shallow flows $y_o/d = 400$, i.e., eqn (6.27). The dimensionless grain size D_* was expressed as

$$D_{gr} = d_{35}(g\Delta/\nu^2)^{1/3} \qquad (7.153)$$

and a sediment transport parameter

$$G_{gr} = \left[\frac{u_*}{U}\right]^n \frac{Xy_o}{S_s d} = \left[\frac{u_*}{U}\right]^n \frac{q_T}{Ud} \qquad (7.154)$$

where $X = \rho_s q_T/\rho q$. These dimensionless terms were linked by

$$G_{gr} = C[(F_{gr}/A) - 1]^m \qquad (7.155)$$

where the coefficients, A, C, m and n were determined from experimental data, using a numerical optimisation technique, and are as follows:

(i) Coarse sediments: $D_{gr} < 60$ or $d > 2.5$ mm

$n = 0;\quad A = 0.170;\quad m = 1.500;\quad C = 0.025.$

(ii) Transition range; $60 > D_{gr} > 1$

$n = 1.00 - 0.56 \log D_{gr}$

$m = (9.66/\sqrt{D_{gr}}) + 1.34$

$A = (0.23/\sqrt{D_{gr}}) + 0.14$

$C = 10^{[2.86 \log D_{gr} - (\log D_{gr})^2 - 3.53]}$

or

$\log C = 2.86 \log D_{gr} - (\log D_{gr})^2 - 3.53$

(iii) Fine sediments: $D_{gr} < 1$ or $d < 0.04$ mm

$n = 1.$

The constant A represents a critical value of the mobility number F_{gr} at threshold conditions. For $D_{gr} < 1$, $F_{gr} = \sqrt{\theta_c} = 0.37$. Between $D_{gr} = 1$ and 60 the Shields' curve is approximated by the transition function and the coarse sediments θ_c is set at $0.17^2 = 0.029$ which is lower than the usual values used.

The method was tested by White et al. (1975) with the aid of data from over 1000 flow experiments and 260 field measurements. It was shown that 68% of data points fell within $0.5 < X_{calc.}/X_{obs} < 2$ range, compared to 63% by eqn (7.146), 46% by the Einstein method, etc.

Building on the above relationships White and Day (1982) propose that the transport rate of a size fraction d_i of mixture is proportional to its percentage in mixture, i.e., $(G_{gr})_i = G_{gr}\ 100/\Delta p_i$, and that the threshold of the fraction is given by $A_i/A_r = 0.4(d_i/d_r)^{-0.5} + 0.6$, where d_r is a representative particle size which as a uniform sediment would start to move under the same conditions. They concluded from laboratory data that $d_r < d_{50}$ for a widely graded sediment and $d_r > d_{50}$ for a narrowly graded sediment. Over the range of $\sigma_g = (d_{84}/d_{16})^{1/2}$ of 1.45 to 4.28 they suggested $d_r/d_{50} = 1.6\sigma_g^{-0.56}$. However, an extrapolation of this result for $\sigma_g \to 1$ leads to an illogical result.

7.3.1 Example

An example for total load calculation by the Einstein method is shown below:

A stream with slope $S = 9 \times 10^{-4}$ has the following grading

$\bar{d} = 0.50$ mm, 30%, $w_i = 54.0$ mm/s

$\bar{d} = 0.35$ mm, 40%, $w_i = 37.2$ mm/s

$\bar{d} = 0.20$ mm, 30%, $w_i = 20.4$ mm/s

$d_{35} = 0.30$, $d_{50} = 0.35$, $d_{65} = 0.41$ mm and $S_s = 2.65$. The bar refers to average value of class i and w_i is its fall velocity.

Calculate the localised scour depth in the contraction $B_1/B_2 = 1.25$ for an upstream flow rate of $q_1 = 1.0\ m^3/s$ per metre.

Hydraulic calculations to find the stage/discharge curve for the upstream reach are shown in the table. The Einstein-Barbarossa curve No. 1 of Fig. 6.6 has been used.

The bedload transport rates are calculated according to the Einstein method and are shown in the table below. The parameters involved, such as X, Y, ξ, ϕ_*, Ψ_*, I_1, - I_2, are given in text books and in Einstein (1950).

$$U = 5.75\ u_*'\ \log(12.27\ m'/\Delta) \qquad m' \equiv y'$$

$$\psi' = \frac{\rho_s - \rho}{\rho} \frac{d_{35}}{y'S} = \frac{0.55}{y'} \qquad \psi = \frac{\rho_s - \rho}{\rho} \frac{d_i}{y'S} = 1.83 \frac{d_i \text{ (mm)}}{y'}$$

$$X = 0.77\Delta \text{ for } \Delta/\delta > 1.8 \text{ and} \qquad \psi_* = \xi \, Y \left(\frac{\beta}{\beta_x}\right)^2 \psi$$

$$X = 1.39\delta \text{ for } \Delta/\delta < 1.8$$

$$\Delta = k_s/\delta', \; k_s = d_{65}, \; \delta' = 11.6 \, \nu/u'_*$$

$$\beta_x = \log(10.6 \, X/\Delta), \; \beta = \log 10.6$$

$$P_E = 2.303 \log(30.2 \, y_o/\Delta)$$

$$i_B q_B = i_b \, \phi_* \, \rho_s \, g^{3/2} \, d^{3/2} \sqrt{S_s - 1}$$

$$= 0.10459 \, i_b \, \phi_* \sqrt{d^3} \text{ when d is in 0.1 mm}$$

$$i_{T_s} g_{T_s} = i_B \, g_B \, [P_E \, I_1 + I_2 + 1]$$

$$\nu = 1.145 \times 10^{-6} \text{ m}^2/\text{s}$$

Hydraulic calculations

'	$u'_* = (gy'S)^{1/2}$	$\delta' = \frac{11.6\nu}{u'_*}$	$\frac{d_{65}}{\delta'}$	x	$\Delta = k_s/x$	U	ψ'	$\frac{U}{u''_*}$	u''_*	$y'' = \frac{u''^2_*}{g S}$	Y_o	u_*	$q = U Y_o$	X	Y	$\left(\frac{\beta}{\beta_x}\right)^2$	P_E
m	m/s	m				(m/s)			m/s	m	(m)	m/s	$m^2 s^{-1}$				
		$\times 10^{-4}$			$\times 10^{-3}$									$\times 10^{-4}$			
0.2	0.042	3.16	1.297	1.58	0.259	0.960	2.75	16.8	0.057	0.37	0.57	0.071	0.547	4.40	0.82	0.667	11.106
0.4	0.059	2.24	1.834	1.42	0.289	1.445	1.37	29.6	0.049	0.27	0.67	0.077	0.968	3.11	0.64	0.941	11.158
0.6	0.073	1.82	2.247	1.32	0.311	1.831	0.92	42.5	0.043	0.21	0.81	0.085	1.483	2.54	0.58	1.196	11.275

Sediment transport rate per metre width

d	i_b	y'	ψ	$\frac{d}{X}$	ξ	ψ_*	ϕ_*	$i_B g_B$	$\Sigma i_B g_B$	$A = \frac{2d}{Y_o}$	$z = \frac{w_i}{0.4 u'_*}$	I_1	$-I_2$	$P_E I_1 + I_2 + 1$	$i_{T_s} g_{T_s}$	$\Sigma i_{T_s} g_{T_s}$
m	%	m						N/s	N/s						N/s	N/s
$\times 10^{-4}$										$\times 10^{-4}$						
5	30	0.2	4.58	1.136	1.08	2.67	2.2	0.77		17.54	3.21	0.10	0.80	1.311	1.01	
		0.4	2.29	1.608	1.00	1.38	5.2	1.82		14.93	2.27	0.17	1.39	1.507	2.75	
		0.6	1.53	1.968	1.00	1.06	7.0	2.46		12.35	1.85	0.25	1.98	1.839	4.52	
3.5	40	0.2	3.21	0.796	1.40	2.46	2.45	0.67	1.44	12.28	2.21	0.18	1.04	1.959	1.32	2.33
		0.4	1.60	1.125	1.09	1.05	7.05	1.93	3.75	10.45	1.57	0.36	1.85	3.167	6.12	8.87
		0.6	1.07	1.378	1.02	0.76	10.0	2.74	5.20	8.64	1.28	0.60	2.95	4.815	13.19	17.71
2	30	0.2	1.83	0.455	4.80	4.80	0.70	0.06	1.50	7.02	1.21	0.73	3.50	5.607	0.35	2.68
		0.4	0.92	0.643	1.75	0.97	7.8	0.69	4.44	5.97	0.86	2.35	8.30	18.921	13.09	21.96
		0.6	0.61	0.787	1.40	0.59	13.2	1.17	6.37	4.94	0.70	5.15	16.10	42.966	50.36	68.07

For q_1 = 1.0 m³/s per metre width the depth y_1 = 0.68 m, U_1 = 147 m/s and the specific energy $y_1 + U^2{}_1/2g$ = 0.79 m. The total sediment transport rate q_{s1} = 24.8 N/s per metre width and the rate q_{s2} = 1.25 q_{s1} = 31 N/s-m. For this, Fig. 7.27 shows a depth of y_2 = 0.705 m; U_2 = 1.775 m/s and $y_2 + U^2{}_2/2g$ = 0.865 m. Hence, the bed is lowered to 0.075 m.

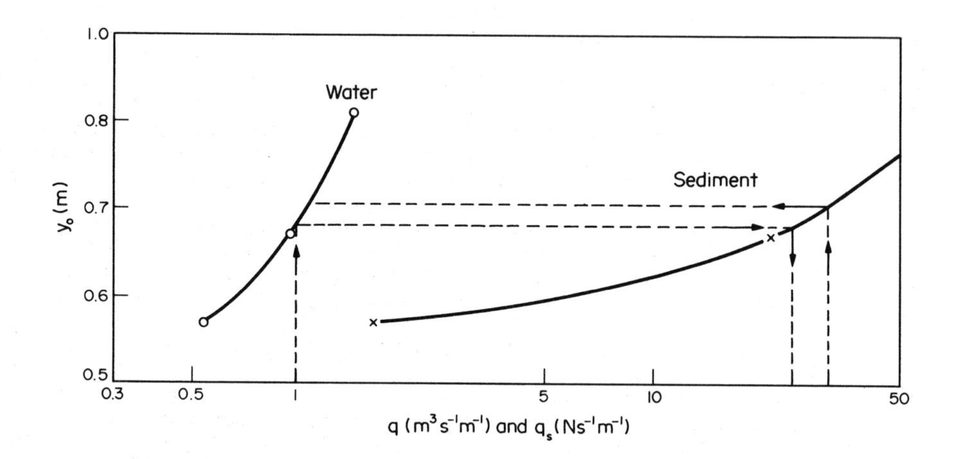

Fig. 7.27. Depth versus water and sediment discharge (Example 7.3.1).

From Fig. 6.11 the depth of flow at q_2 = 1.25 m²/s is y_2 = 0.74 as compared to y_2 = 0.705 m above. This reflects the reduction of the f value at the transport rate determined by availability of sediment rather than transport capacity.

7.3.2 Comments

The question "which formula do I use?" cannot be answered directly All formulae are to some extent empirical and none can claim to rest on sound theoretical foundations. They are also all based on experimental data. The measurements of sediment transport rates in laboratory can be quite accurate but the flow conditions are "unnatural", i.e., straight rectangular flumes. The measurement of sediment transport in field is extremely difficult and the accuracy is questionable in all but the most elaborate measuring programmes. There is also the vexed question of calibration of the various measuring devices. For description of measuring devices in general, reference is made to Vanoni (1975). Laboratory data show that in flumes carrying sand the transport rate can actually decrease as the flow increases, for example, as shown by Willis et al. (1972). The total shear stress τ_o, or θ in flow over sand beds can change appreciably, as shown in Fig. 6.3, as the bed features change. The fraction of the total shear stress effective in sediment transport has not yet been adequately defined. The problem area is the form drag which contributes only indirectly to transport through increased turbulence in the vicinity of the bed.

The sediment transport in the field is not uniformly distributed in the cross-section. Suspended load is sensitive to the bed features which vary over the cross-section and all transport formulae show that the transport rate is non-linearly related to flow depth and bed shear stress. The proportions of bed and suspended load vary with cross-sectional shape.

The transport rate in gravel rivers is also a function of time for constant discharge whenever armouring can develop.

These are but a few on the list of problems but should suffice to show that the assessment of the quality of a formula is a subjective task, dependent on selection of data and the method of comparison.

White et al. (1975) compared eight formulae using 1000 flume and 260 field measurements. The discrepancy ratio X_{calc}/X_{obs} was plotted against the dimensionless grain size ($X = \rho_s q_s/\rho q$) and the percentages within the 0.5 to 2 range were as follows:

Formula	% in $0.5 \leq X_{calc}/X_{obs} \leq 2$ range
Ackers and White	68
Engelund and Hansen	63
Rottner (1959)	56
Einstein (total load)	46
Bishop et al. (1965)	39
Toffaletti (1968)	37
Bagnold (total load)	22
Meyer-Peter and Müller	10

The laboratory data include particle sizes from 0.04 to 4.94 mm and field data from 0.095 to 68 mm.

The comparison of formulae by Yang and Molinas (1982) also used laboratory and river data encompassing mean grain sizes from 0.15 to 1.71 mm, channel widths 0.134 to 532 m, flow depths 0.01 to 15.2 m, temperature 0° to 34.3°C, average velocity 0.23 to 1.97 m/s and slopes from 4.3×10^{-5} to 2.79×10^{-2}. The range of data is the same as given by Yang (1973) for the data from which the formula was derived. The discrepancy ratio, defined as the ratio between computed and measured values, is given as follows

Formula	Data		
	Lab.	River	All data
Colby (1964)	0.31	0.61	0.34
Yang (eqn 7.148)	1.01	1.31	1.03
Yang (eqn 7.149)	1.02	1.12	1.03
Shen and Hung (1971)	0.91	1.18	0.95
Engelund and Hansen	0.88	1.51	0.96
Ackers and White	1.28	1.50	1.31
Maddock (1976)	0.99	0.49	0.92

A different picture is painted by the comparative study carried out by van Rijn (1984), also using field and laboratory data.

The field data were from U.S. Corps of Engineers, Middle Loop river, Niobrara River and canals in India and Pakistan; a total of 486. Particle sizes ranged from 0.08 to 0.40 mm, flow depths 0.3 to 3.6 m and velocities 0.4 to 2.4 m/s. In addition 783 flume data were used ranging for particle size from 0.1 to 0.48 mm, depths 0.1 to 0.4 m and velocities from 0.4 to 1.3 m/s. The discrepancy ratio, r, defined as the ratio of predicted to observed transport rates in percentage were as follows:

	$0.75 \le r \le 1.5$				$0.5 \le r \le 2$				$0.33 \le r \le 3$			
Data	1	2	3	4	1	2	3	4	1	2	3	4
US rivers Corps Engrs	53	39	32	6	79	67	61	24	94	87	78	44
Middle Loop River	39	13	37	63	78	37	74	94	96	80	98	100
Indian Canals	30	15	27	3	60	45	48	6	90	73	70	24
Pakistan Canals	23	37	34	13	56	71	71	29	91	94	91	48
Niobrara River	55	13	29	86	95	67	58	98	98	95	98	98
	45	32	32	22	76	64	63	39	94	88	84	55
Flumes												
Guy et al.	40	67	56	68	70	89	85	90	91	98	99	98
Oxford	37	20	31	45	84	38	59	89	96	70	81	96
Stein	54	73	81	56	70	95	97	97	97	97	100	100
Southampton A	64	49	46	49	85	73	79	82	97	91	94	94
Southampton B	18	12	82	91	81	82	96	97	94	97	100	100
Barton-Lin	35	60	30	40	65	100	50	65	100	100	100	100
	41	46	52	59	77	74	77	89	95	89	94	98
	43	37	40	36	76	68	68	58	94	88	88	71

In the above table columns 1 are values by the method of van Rijn (eqns 7.32 and 7.129); 2 by Engelund-Hansen formula; 3 by Ackers -White formula and 4 by Yang formula. The results show poor results by the Yang formula for canals in India and Pakistan which have the deepest flows of the above data. Since the other formulae produce reasonable results van Rijn concludes that "the method of Yang must have serious systematic errors at large flow depth. On the average the predicted values are much too small".

Both of the above comparisons are for sand sizes. Only White et al. have a few data points for gravels both from laboratory and field. In general, data for comparison of transport in gravel rivers is scarce and the data itself is highly variable. The formulae do not adapt well to the variation of particle size grading of really coarse sediments. Any predictions for grain sizes other than used to derive the formula should be viewed with caution.

The quality of predictions generally improves with increasing suspended load transport.

An essential precaution in selection of sediment transport formulae for use on a particular stream is that the formula should have been developed from data similar to the situation at hand, i.e., gravel, sand, etc. Extrapolation of a formula beyond the data range on which it is based should be avoided. This applies to grain sizes as well as flow parameters. Predictions should be made with the aid of several formulae. Finally, attention is also drawn to comments under Section 7.1.1.

Chapter 8

Stable Channel Design

The problems of design and maintenance of stable channels in alluvial soils are central to all irrigation schemes. In countries like India and Pakistan, for example, there are thousands of km of unlined canals through the landscape, some of these carrying 350 to 400 m^3/s of water. It is likely that in the future even larger canals will be built to improve the spatial distribution of water on earth. However, the same problems occur in design of navigation channels and improvement or stabilisation of rivers. The requirement in all cases is that the channel must be stable and it must be able to transport both water *and* sediment, frequently at widely varying rates.

The methods of design of stable channels can be divided into two categories; the empirical and the tractive force method. The latter uses the boundary shear stress and hydraulic concepts to define a stable channel whereas the former is based on rules derived from observation. Lane (1953) defined a stable channel as follows: "A stable channel is an unlined channel for carrying water, the banks and bed of which are not scoured by the moving water, and in which objectionable deposits of sediment do not occur." Such a channel may scour to some extent or form deposits but over a yearly cycle the net effect must be zero. An alluvial channel is said to be *in regime* when its average values of width, depth and mean bed level do not show a definite trend with time. Blench (1957) suggested for the averaging interval "a score or two of years".

8.1 The Empirical Stable Channel Design

The empirical, so-called regime theory is mainly a product of the Anglo-Indian school of hydraulic engineering. Throughout Middle Asia, India and Egypt canals in fine-grained soils, of less than 1 mm particle size, are wide-spread and stability against erosion and sedimentation of such canals is of great practical importance. It may be said that the *regime method* started with the Kennedy (1895) formula

$$V_o = aD^b = 0.548\ D^{0.64}\ (ms^{-1}) \tag{8.1}$$

or

$$V_o = 0.84\ D^{0.64}\ (ft/s)$$

where D is the mean depth. This formula with the given constants applies only to the cross-sections used in the Bari Doab canal system in the Punjab.

Note that in the regime method nearly all of the numerical coefficients are *not dimensionless* - as a check of dimensional homogeneity will reveal - and therefore their magnitude depends on the units of measurement. In here the original formula in ft/lb units are recorded for reference but are always put in brackets. Lindley (1919) concluded that "the dimensions, width, depth, and gradient of a channel to carry a given supply loaded with a given silt-charge, were all fixed by nature".

Among the subsequent contributions the best known is that by Lacey (1929). Lacey did not produce any new data but rearranged the then existing field data in so clear and able a manner that the rules that emerged have often been taken to be basic laws. The data came from three canal systems. The Upper Bari Doab canals studied by Kennedy, the Lower Chenab canals reported on by Lindley and the Madras canals. From the Madras and Bari Doab canals velocity and hydraulic mean radius were known, from Chenab hydraulic radius and slope. One bed material sample is reported from Bari Doab. No data existed on sediment transport rates. Lacey wrote the Kennedy type formula in terms of the hydraulic mean radius as

$$V_o = a_1 R^{b_1} \tag{8.2}$$

and found that Kennedy's data were satisfied by $a_1 = 0.646\ (1.17)$ and $b_1 = 0.5$, i.e.

$$V_o = 0.646\ R^{0.5}$$

$$\{V_o = 1.17\ R^{0.5}\} \tag{8.3}$$

The relationship for regime for any type of sediment, other than "Kennedy's standard silt", was obtained by introducing a factor f, known as Lacey's silt factor.

The silt factor expresses the relationship to the "standard silts" for which $a_1 = 0.646$ as

$$f = \left[\frac{a}{0.646}\right]^2; \quad \left\{f = \left[\frac{a}{1.17}\right]^2\right\} \tag{8.4}$$

where a is the value of a_1 for the particular material.

Thus, the general formula is

$$V_o = 0.646 \sqrt{(fR)}; \quad \{V_o = 1.17 \sqrt{(fR)}\} \tag{8.5}$$

where V_o is the stable or non-silting velocity, f is the silt factor for which Lacey gave the following values:

	f
Massive boulders (d = 600 mm)	39.60
Large stones	38.60
Large boulders, shingle and heavy sand	20.90
Medium boulders, shingle and heavy sand	9.75
Small boulders, shingle and heavy sand	6.12
Large pebbles and coarse gravel	4.68
Heavy sand	2.00
Coarse sand	1.56-1.44
Medium sand	1.31
Standard Kennedy silt (Upper Bari Doab)	1.00
Lower Mississippi silt	0.357

The relationship between the silt factor and sediment size were based only on four pieces of information:

	f	n	V_o (m/s)	R(m)	Bed
Bari Doab Canals	1.0	0.0225	-	-	0.4 mm sand
Thrupp	4.68	0.033	1.9	1.86	Coarse gravel
Griffith	9.75	0.040	4.0	3.96	Boulders and sand
Song River	39.6	0.055	7.8	3.66	635 mm boulders

where n is a roughness coefficient in a formula for velocity, similar to Manning formula. The n-value for Bari Doab was assumed since no slope measurements were made.

It follows from eqn (8.5) that the product (fR) is constant in silt-stable canals with the same mean velocity and it was assumed that this also applies to the product (fP), where P is the length of wetted perimeter. Consequently, the product f^2A is a function of velocity alone. Available data plotted as V_o versus Af^2 yielded.

$$Af^2 = 134.2\, V_o^5; \quad \{Af^2 = 3.8\, V_o^5\} \tag{8.6}$$

$$Qf^2 = 134.2\, V_o^6; \quad \{Qf^2 = 3.8\, V_o^6\} \tag{8.7}$$

It should be noted that the range of values plotted was

$$0.46 < Af^2 < 278.7; \quad \{5 < Af^2 < 3000\}$$

$$0.30 < V_o < 1.22; \quad \{1 < V_o < 4\}$$

This is a relatively limited range, since for rivers the mean velocity may substantially exceed 3 m/s, and Af^2 may be 10^4 or more.

The mean velocity V_o may be calculated from Manning's formula

$$V_o = (1/n)R^{2/3} S^{1/2} \tag{8.8}$$

With n replaced from

$$n = 0.022 f^{0.2} \tag{8.9}$$

a relationship suggested by Lacey, R and V_o may be eliminated by means of eqns (8.5) and (8.7) to give

$$S = f^{1.51}/(3844 Q^{1/9}); \{S = f^{1.51}/(2587 Q^{1/9})\} \tag{8.10}$$

It is interesting that with known Q and f the designer can calculate the equilibrium slope without dimensioning the channel, and thus see whether or not an equilibrium slope is possible. However, in river work, Q, f and S are all given, and they may not satisfy eqn (8.10).

Basically for a known discharge, and assumed shape of cross-section, the dimensions of the channel can be calculated after assuming a suitable value of the silt factor f. With f and Q known eqn (8.7) yields V_o, and Q/V_o yields the area of cross-section A. Equation (8.5) yields the hydraulic mean radius

$$R = 2.4 V_o^2/f; \quad \{R = 0.7305 V_o^2/f\} \tag{8.11}$$

Dividing eqn (8.6) by eqn (8.11) and rearranging yields

$$Pf = 56 V_o^3 \quad \{Pf = 5.2 V_o^3\} \tag{8.12}$$

and

$$P/R = 23.3 V_o; \quad \{P/R = 7.1 V_o\} \tag{8.13}$$

Equations (8.7) and (8.12) yield

$$\underline{P = 4.8326 \sqrt{Q};} \quad \{P = 2.668 Q^{1/2}\} \tag{8.14}$$

The result given by eqn (8.14) is remarkable since it shows that for a given discharge the wetted perimeter of a stable channel is constant and independent of the fineness of the silt. Equation (8.13) shows that the shape of the channel cross-section depends on the velocity alone. However, it is well known that the silt size influences the shape of the cross-section. This tendency is evident in nature and it is diagrammatically illustrated in Fig. 8.1.

Equation (8.14) is useful for determining the required waterway. Lacey suggested that a stable cross-section of a watercourse is semi-elliptical. If the periphery is not composed of the same material the cross-sectional shape will not tend to be semi-elliptical. For example, where the banks are of stiffer material the bed is nearly horizontal.

Lacey also proposed a relationship between the particle size and the silt factor as

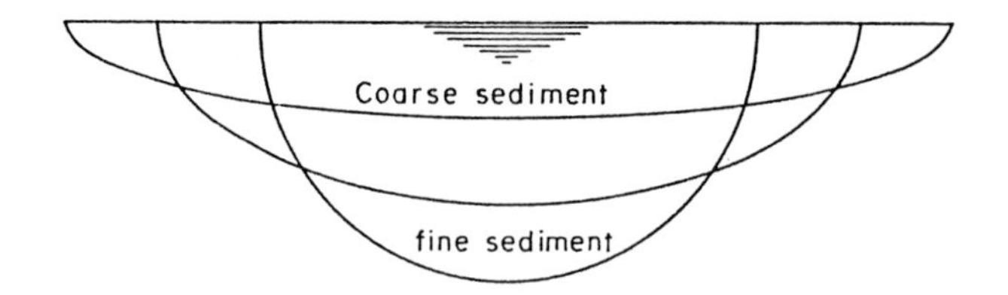

Fig. 8.1. Diagrammatic representation of the variation of cross-sectional shape with the size of sediment for constant discharge and length of wetted perimeter.

$$\underline{d_{mm} = f^2/2.52} \ ; \quad \{d'' = f^2/64\} \tag{8.15}$$

With eqn (8.9) this yields

$$n = d^{1/10}/41.44; \ \{n = d^{1/10}/30\}$$

which could be compared with the Stickler formula eqn (6.25). The relationship for n was based on very limited data.

In his reply to the discussions and in a later paper (1953) Lacey altered several of his equations and added others. The more significant formulae in the latter form are:

$$P = 4.832\ Q^{1/2}; \quad \{P = (8/3)Q^{1/2}\} \tag{8.16}$$

$$A = 2.282\ Q^{5/6}/f^{1/3}; \quad \{A = 1.26\ Q^{5/6}/f^{1/3}\} \tag{8.17}$$

$$R = 0.4725\ Q^{1/3}/f^{1/3} \tag{8.18}$$

$$V_o = 0.4382\ Q^{1/6}f^{1/3}; \quad \{V_o = 0.794\ Q^{1/6}f^{1/3}\} \tag{8.19}$$

$$S = f^{5/3}/(3169.8\ Q^{1/6}); \quad \{S = f^{5/3}/(1750\ Q^{1/6})\} \tag{8.20}$$

$$P/R = 23.355\ V_o; \quad \{P/R = 6.9896\ V_o\} \tag{8.21}$$

$$\underline{V_o = 10.846\ R^{2/3}S^{1/3} \cong 10.8\ R^{2/3}S^{1/3}}; \tag{8.22}$$

$$\{V_o = 16.116\ R^{2/3}S^{1/3} \cong 16\ R^{2/3}/S^{1/3}\}$$

$$n = 0.0225\ f^{1/4} \tag{8.23}$$

$$f = 1.76\ \sqrt{d_{mm}} \tag{8.24}$$

$$f = 281.6\ R^{1/3}S^{2/3} \tag{8.25}$$

$$\{f = 190\ R^{1/3}S^{2/3}\}$$

The last expression for S is based on $V \propto R^{3/4}S^{1/2}$, whereas eqn (8.10) was based on Manning's formula. Lacey (1946,1958) proposed a "general theory of flow in alluvium" which was an attempt to derive the regime equations with the aid of dimensional considerations.

It must also be noted that the development of eqns (8.5) and (8.22) was based on measured values of the hydraulic mean radius R and slope S, and not on measured velocity. Observations indicate that the Lacey's first equation $V \propto \sqrt{R}$ is not well founded, the exponent of R varies widely. However, the relationship

$$m^{1/2}S = \text{const.} \tag{8.26}$$

is well supported by observations. A general examination of the equation $V = CR^xS^y$ was carried out by Liu and Hwang (1959) and Henderson (1961). See also eqn (6.20) and its discussion. Relationships of the form $R^xS^y = \text{const.}$ can also be derived from the bed-load equations for constant d and q_s/q.

Criticism of Lacey's formulae centred on two points: that they took no account of the magnitude of sediment discharge, and that they were based on observations covering a narrow range of silt sizes only.

It may have been noticed that the sediment discharge q_s does not appear in any of these equations but it is easy to demonstrate with eqn (8.5), the Manning's equation and a bed-load equation that the Lacey silt factor f must incorporate the sediment load and specific gravity effects. Therefore, the value of f may differ appreciably from canal to canal. The total sediment concentration in the irrigation canals of India appear to vary little and ranges between 1000-2000 ppm. In principle, eqn 8.5 must also represent sediment transport. This implies that all regime canals in a system linked to a primary canal must have the same sediment concentration and sediment size. The silt factor f is thus also a sediment concentration parameter. These aspects have been discussed in detail, e.g., by Mahmood and Shen (1971).

Observation has shown that the sediment load has appreciable effect on the wetted parameter. With low sediment load and fine sediment P has been found to be $3.58\ Q^{1/2}$ or less and with high discharge of sand $P = 6.07\ Q^{1/2}$.

Inglis (1941-2) reported on the range of departure from Lacey's mean values found by the Punjab Irrigation Research Institute. They made a statistical analysis of a mass of data obtained from the lower Chenab canal and found that

P values varied from 0.82 to 1.45 $\bar{P}$; stand. dev. 0.178,
V values varied from 0.89 to 1.21 $\bar{V}$; stand. dev. 0.095,
S values varied from 0.69 to 1.31 $\bar{S}$; stand. dev. 0.177,

where the bar refers to mean values.

They further reported that "there was much evidence to show that the divergences were mainly caused by variations in the sand charge entering different channels". Inglis also pointed out that silt factors computed from different combinations of the Lacey equations lead to different values:

$$f_{VR} = 2.52\ V^2/R \tag{8.27}$$

$$f_{RS} = 291\ (RS^2)^{1/3} \tag{8.28}$$

and

$$f_{RSV} = 3127\ RS/V \tag{8.29}$$

Chien Ning (1957) showed that

$$f_{VR} = 0.061\ c^{0.715} \tag{8.30}$$

$$f_{RS} = 2.2\ d^{0.45}\ c^{0.05} \tag{8.31}$$

where c is the concentration of bed material load (q_{Ts}/q) in ppm where the data relate to low concentrations, generally less than 500 ppm. His study showed that f_{VR} increases rapidly with concentration whereas f_{RS} is fairly insensitive to it. From equations 8.27 and 8.28

$$f_{VR}\ f_{RS}{}^{3} = 733.3\ (VS)^{2} \tag{8.32}$$

Inglis (1946/7) proposed regime formulae which incorporated sediment discharge. In the discussion of this paper White proposed a family of dimensionless regime equations. However, the coefficients have not been evaluated for either set of equations.

Blench (1957,1966) introduced the bed and side factors into the design of stable canals by the regime method:

$$\text{Bed factor}\ \frac{V^2}{D} = F_b \tag{8.33}$$

which is basically the same as eqn (8.5), and

$$\text{Side factor}\ \frac{V^3}{B} = F_s \tag{8.34}$$

where B is defined as the breadth that multiplied by average depth, D, gives the area of the mean section. When multiplied by $\rho^2\nu$ the term $(\rho^2\nu V^3/b)$ is dimensionally $[F^2L^4]$ and it is suggested that F_s is a factor in the expression for the square of the mean tractive force intensity on the hydraulically smooth sides. A plot of data obtained from irrigation canals as gDS/V^2 versus VB/ν on double logarithmic paper was found to have the slope of minus 1/4. Thus, it was concluded that the regime boundary is a smooth one. The Blasius equation for smooth circular pipes was modified to

$$\frac{V^2}{gDS} = 3.63 \left[\frac{VD}{\nu}\right]^{0.25}$$

and the equivalent for open-channel flow was written as

$$\frac{V^2}{gDS} = 3.63\ (1 + ac) \left[\frac{VB}{\nu}\right]^{0.25} \tag{8.35}$$

becoming the third equation in Blench's system. The term (1 + ac) is an empirical addition, by means of which it was found that the above equation could be made to satisfy the classic laboratory flume data of Gilbert for dunes on the bed.

Here c is a bed-load charge in parts per million by weight (10 ppm is inserted as 10) and a is approximately 1/4000 for Gilbert's sands. A value of 1/2330 is recommended for non-uniform sands.

The three independent regime equations for design purposes were rearranged as

$$B = (F_b Q/F_s)^{1/2} \tag{8.36}$$

$$D = (F_s Q/F_b^2)^{1/3} \tag{8.37}$$

$$S = \frac{F_b^{5/6} F_s^{1/12} \nu^{1/4}}{3.63(1 + ac) g Q^{1/6}} \tag{8.38}$$

where $(F_b F_s)^{\frac{1}{2}}$ has the dimensions $(T/L)^{\frac{1}{2}}$, $(F_s F_b^2)^{1/3}$ the dimension $(T)^{1/3}$, d is the median bed material size by weight in mm, and in subcritical flow

$$F_b = F_{bo}(1 + 0.012\ c) \tag{8.39}$$

where the values given for F_{bo} are

$$F_{bo} = 1.9\ \sqrt{d}_{mm} \tag{8.39a}$$

for d < 2 mm and

$$F_{bo} = 1.75\ d_{mm}^{1/4}\ (\nu_{20}/\nu)^{1/6} \tag{8.39b}$$

where ν_{20} is kinematic viscosity at 20°C. Blench (1969) also gives

$$F_{bo} = 0.58\ w_{cm/s}^{11/24}\ (\nu_{20}/\nu)^{11/72} \tag{8.39c}$$

which generally yields larger values than the expressions above. The values for F_s are 0.1, 0.2 and 0.3 for bank material of very light, medium and high cohesiveness, respectively. When using the above values of F_b and F_s the three equations in SI units become

$$B = 1.811\ (F_b/F_s)^{\frac{1}{2}}\ \sqrt{Q} \tag{8.36a}$$

$$D = (F_s/F_b^2)^{1/3}\ Q^{1/3} \tag{8.37a}$$

$$S = \frac{F_b^{5/6} F_s^{1/12} \nu^{1/4}}{1.11(1 + 4.3 \times 10^{-4} c)} \tag{8.38a}$$

Simons and Albertons (1963) carried out an extensive study of the regime equations using Indian and U.S.A. data. The range of data used is indicated in Table 8.1.

Table 8.1

	No. of reaches studied	Discharge (m^3/s) (min.)	Discharge (m^3/s) (max.)	Slope x 10^3 (min.)	Slope x 10^3 (max.)	Average sed. conc. (ppm)	d (mm)
San Luis Valley, Color.	15	0.48	42.5	0.79	9.7		20-80
Punjab, India	42	0.14	254.9	0.12	0.34	238	0.43
Sind, India	28	8.81	256.5	0.059	0.100	156-3590	0.0346-0.1642
Imperial Valley, Calif.	4	-	-	-	-	2500-8000	-
Irrig. Can., Wy, Col., Neb.	24	1.22	29.4	0.058	0.387	-	0.028-7.6

These canals are divided into five types:

1. Sand bed and banks.
2. Sand bed and cohesive banks.
3. Cohesive bed and banks.
4. Coarse non-cohesive material.
5. Sand bed, cohesive banks and heavy sediment load 2000-8000 ppm (mainly wash load).

The data used related to sediment load less than 500 ppm.

Each type can be fitted by the following equations:

$$P = K_1 Q^{1/2} \tag{8.40}$$

$$B = 0.9 P = 0.92 B_s - 0.61 \tag{8.41}$$

$$R = K_2 Q^{0.36} \tag{8.42}$$

$$D = 1.21 R; \quad R < 2.1 \text{ m} \tag{8.43}$$

$$D = 0.61 + 0.93 R; \quad R > 2.1 \text{ m} \tag{8.44}$$

$$V = K_3 (R^2 S)^n \tag{8.45}$$

$$\frac{c^2}{g} = \frac{V^2}{gDS} = K_4 \left[\frac{VB}{\nu}\right]^{0.37} \tag{8.46}$$

$$A = K_5 Q^{0.87} \tag{8.47}$$

Table 8.2

	Channel Type				
Coefficient	1	2	3	4	5
K_1	6.34	4.71	4-4.7	3.2-3.5	3.08
K_2	0.4-0.6	0.48	0.41-0.56	0.25	0.37
K_3	9.33	10.8		4.8	9.7
K_4	0.33	0.53	0.88		
K_5	2.6	2.25	2.25	0.94	
n	0.33	0.33		0.29	0.29

Above P is wetted perimeter, B = A/D is the mean channel width, A is cross-sectional area of flow, D is mean flow depth, B_s is water surface width, S is slope, R is hydraulic mean radius, C is Chézy coefficient and Q is flow rate. The agreement of data with eqn 8.46 was found to be satisfactory when $VB/\nu < 2 \times 10^7$. The study did not include sediment size as a variable. One of their conclusions was that in regime channels the Froude number $F_r < 0.3$.

Chitale (1966) analysed data from Indian subcontinent and fitted

$$P = 4.30\ Q^{0.523} \qquad (8.48)$$

$$R = 0.499\ Q^{0.341} \qquad (8.49)$$

$$S = 0.00028\ Q^{-0.165} \qquad (8.50)$$

$$U = 7.34\ R^{1/2}\ (R^{1/2}S)^{0.293} \qquad (8.51)$$

In analysing data from India, Pakistan, America and Egypt Chitale (1976) found that the error of estimate in channel width reduced from about 20% to 10% by correlation with additional parameters and fitted

$$P = 6.592^{0.209}S^{-0.097}Q^{0.414}d^{0.115} \qquad (8.52)$$

It should be noted that the regime method evolved for design of canals for a given flow rate.

Maddock (1969) gave an empirical formula for the "mid-velocity range in which most natural streams and canals flow" (1-4 ft/s) as follows:

$$U = 1.13\ k_1 q^{1/16} q_s^{\ 1/4}\left[\frac{\gamma_s^* d}{\rho w^2}\right]^{1/16} \frac{(wg)^{1/16}}{\rho^{1/4}} \qquad (8.53)$$

$$k_1 = \exp\left[\frac{1.55}{\pi}\right] \cot^{-1}\,[-5.55 + 3.8\,\log(\gamma_s^* d/\rho w^2)]$$

in ft-lb-sec system of units, with d measured in feet. In SI-system of units the factor for U is 5.80.

Parallel with the development of the regime method in India, similar work was going on in Egypt. Chaleb (1929-30), for example, published a formula for non-silting conditions; $\{V_0 = 0.39D^{0.73}\}$. In *Irrigation Practice in Egypt* by Molesworth and Yenidunia of 1922 one finds equations of the Kennedy type. The authors recommended depth relationships which also included the channel slope and the bed width B. An account of these formulae is given by Leliavsky (1955). Large discrepancies were observed between calculated values and field data as well as between the results from various formulae when applied under different conditions. This was demonstrated by Lane (1935) as seen from Fig. 8.2, reproduced from his paper. Lane also dealt with the shape of the cross-section of silt-stable canals and introduced the term "form factor" which is the ratio of the flow cross-sectional area to that of the rectangle enclosing the channel cross-section below water level. This ratio would be 0.50 for a triangle, 0.67 for a parabola, $\pi/4$ for a semi-ellipse and unity for a rectangle by definition. Lane's study yielded ratios from 0.56 to 0.92, the value depending on the size and grading of the boundary material. It indicates that a parabola would be a better shape than an ellipse. However, the main point is that the assumption of a single type of cross-sectional shape, such as Lacey's semi-ellipse, is not substantiated.

One can find local regime theories developed in every area where irrigation is practised. The fact that these regime rules have not "transplanted" is an indication that not all of the physical parameters defining the problem are correlated by the regime method of design.

8.2 Tractive Force Method of Stable Channel Design

The tractive force method of design is based on use of boundary shear stress and sediment transport relationships. The method was substantially advanced by the work of US Bureau of Reclamation under the direction of Lane (1955). In any channel where the shear stress at any point on its periphery exceeds the critical stress for grain movement sediment is set in motion. Such a channel is stable only if the sediment transport capacity of the channel equals the rate of sediment supply.

The earlier attempts made use of simple bed load functions like the Du Boys formula. In terms of slope and sediment discharge per unit width and time it may be written as

$$q_s = C_s \tau(\tau - \tau_c) \tag{8.54}$$

where $\tau = \gamma y_o S$ and $\tau_c = \gamma y_o S_c$.

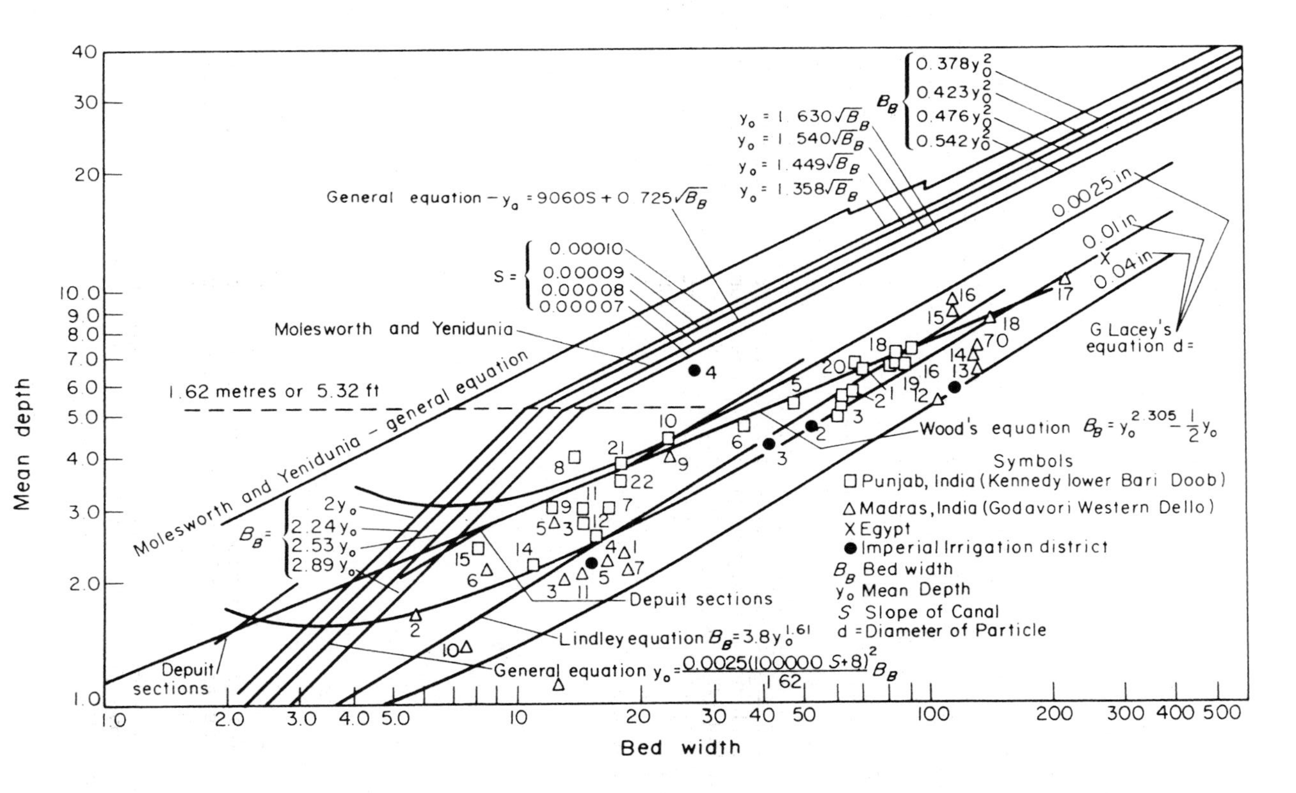

Fig. 8.2. Bed width-depth relation for a non-silting non-scouring canal. (By permission of the American Society of Civil Engineers, from *Proc. A.S.C.E.* 61, 1314.)

Hence,

$$q_s = C_s \gamma^2 y_o^2 S^2 \left[1 - \frac{S_c}{S}\right] \tag{8.55}$$

Empirical values for the sediment discharge coefficient C_s are given in Chapter 7. The water discharge is given by the Chézy equation. The ratio of these two relationships defines the relative intensity of sediment movement

$$\frac{q_s}{q} = \frac{C_s \gamma^2}{C} y_o^{1/2} S^{3/2} \left[1 - \frac{S_c}{S}\right] \tag{8.56}$$

The critical slope was expressed by Straub as

$$S_c = 0.00025(\bar{d}_{mm} + 0.8)/y_o \tag{8.57}$$

Expressing the slope in terms of Manning's formula we can solve for the equilibrium velocity.

Any of the bed-load functions could be treated in similar fashion by using empirical coefficients. For example, the Shields bed-load equation

$$\frac{g_B}{\gamma q} \frac{S_s - 1}{S} = 10 \frac{\tau - \tau_c}{\gamma(S_s - 1)d}$$

could be written as

$$\frac{g_B}{\gamma q} = \frac{10}{(S_s - 1)^2 d} y_o S^2 \left[1 - \frac{S_c}{S}\right]$$

or, since in the turbulent region

$$\frac{\tau_c}{\gamma(S_s - 1)d} = 0.056 \tag{8.58}$$

$$\frac{g_B}{\gamma q} = \frac{10S}{(S_s - 1)} \left[\frac{y_o S}{(S_s - 1)d} - 0.056\right] \tag{8.59}$$

For example, with $S_s = 2.6$ this becomes

$$\frac{g_B}{\gamma q} = S \left[3.91 \frac{y_o S}{d} - 0.35\right]$$

With the same value of 2.6 for specific gravity eqn (8.58) gives for the critical conditions the particle size as

$$d = \frac{\tau_c}{0.056(S_s - 1)\gamma} = \frac{\tau_c}{0.09\gamma} \cong 11 \; y_o S \tag{8.60}$$

This is a very simple criterion for determination of the particle size which will be on the point of movement in the stream. This relationship is derived from theoretical reasoning and the constants involved are well supported by experimental results.

At threshold conditions the grain roughness could be characterised by the particle size d or 2d. Then with $y' = k/30.2$ in the logarithmic velocity distribution and $u \equiv V$ at $y = 0.4\, y_o$ the Chézy coefficient becomes

$$C = 18 \log \frac{y_o}{d} + 18.8 = 18 \log \frac{11.1\, y_o}{d}$$

with

$$V = C\sqrt{(y_o S)} \text{ and } Q = BCy_o\sqrt{(y_o S)}$$

width, slope, depth, and sediment size are related to the discharge.

In the region where $y_o/d \cong 30$ Manning's n is related to the median particle size by the Stickler formula, eqn (6.26). Slightly different values of n are quoted by Keulegan (1938), Lane and Carlson (1953) and Irmay (1949). Note that here n refers to the surface roughness of the flat threshold bed of the channel. Combining eqn (8.60) and $n = 0.04168\, d^{1/6}$ (eqn (6.26)) where d is measured in metres, yields

$$n = 0.062(y_o S)^{1/6} \tag{8.61}$$

Inserting this in the Manning's formula gives

$$V = 16.1\, y_o^{1/2} S^{1/3} \tag{8.62}$$

and

$$Q = 16.1\, B y_o^{3/2} S^{1/3} = 0.441\, \frac{Bd^{3/2}}{S^{7/6}} \tag{8.63}$$

Equation (8.62) could be compared to the Lacey equation $V_o = 10.8\, R^{2/3} S^{1/3}$ by plotting $V/S^{1/3}$ versus y_o.

The above are values for a two-dimensional channel at threshold conditions. At live bed conditions the resistance to flow has to be expressed as a function of shear stress. For this, for example, Engelund and Hansen (1967) combined their resistance equation for loose bed with their transport relationship, i.e.,

$$f\phi = 0.4\theta^{5/2} \text{ and } \theta' = 0.06 + 0.4\theta^2$$

using eqn 6.67 for velocity. However, any of the sediment transport and resistance relationships could be used, for example, van Rijn (1984). Ackers (1980) describes the use of the Ackers and White formula together with the resistance relationships by White et al., eqns 6.60, 6.61 and 6.63.

The major limitation of these two-dimensional approaches is that these do not allow for the

(i) variation of drag over the periphery,
(ii) variation of resistance to displacement of a particle with position on the periphery, and
(iii) soil type and the associated capacity to withstand fluid drag.

Additional problems arise with channels subject to scour, deposition or both at various times.

8.2.1 Drag distribution and resistance to motion

Generally the boundary shear varies from a maximum at the deepest point to zero just above the water's edge (Fig. 8.3). Drawing orthogonals to the isovels of a cross-section generally reveals the varying nature of the velocity gradient in the boundary region. From such an isovel pattern the boundary shear-stress distribution could be calculated. It is generally assumed that the orthogonals to the isovels are surfaces of zero shear. This is true only on the average. At any instant, however, there is a turbulence shear stress on these orthogonals caused by the turbulent momentum transfer. If the orthogonals are assumed to be surfaces of zero shear, then the component of weight of the water contained between the orthogonals and the boundary $\delta W = \gamma \delta AS$ is balanced by the shear stress, so that

$$\tau_o = \frac{\gamma \delta AS}{ab}$$

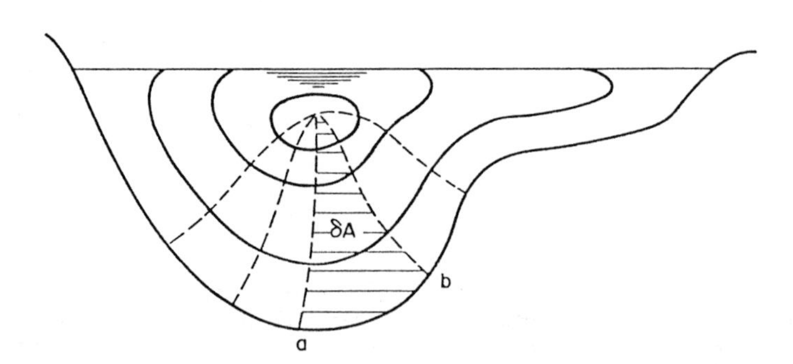

Fig. 8.3. Diagrammatic velocity distribution.

When the velocity distribution is not known, as in design, determination of the boundary shear stress distribution has to be obtained by analytical methods, mostly based on the logarithmic velocity distribution. These results are both unwieldy and often in poor agreement with observation, the latter due to the secondary currents and varying roughness over the periphery.

Not only the shear stress, but also the resistance to displacement of a particle depends upon its location on the periphery of the cross-section. The following analytical treatment was marked out in principle by Forchheimer (1924) and was introduced into the tractive force designs by Lane (1953). With ϕ being the angle of repose, the criterion for equilibrium in still water would be $W \sin\alpha = W \cos\alpha \tan\phi$

$$\therefore \tan\alpha = \tan\phi$$

In flowing water, Fig. 8.4, the weight component of the particle down the slope, T, combines with the applied fluid shear stress τ_o. The resultant R of these forces is resisted by a force which cannot exceed the value of normal weight component times the friction factor, that is $W \cos\alpha \tan\phi$.

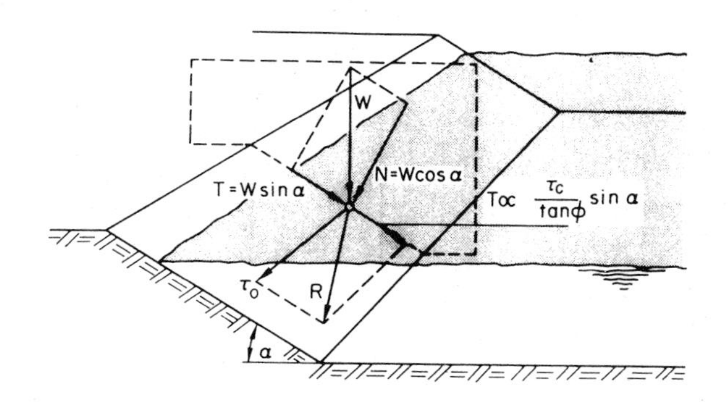

Fig. 8.4. Forces on a particle on the slope.

On the horizontal, the critical shear stress is

$$\tau_c \propto W \tan\phi$$

or

$$W \propto \tau_c/\tan\phi$$

where W is the weight of particles per unit area. On the slope the weight term is replaced by the normal load $N = W \cos\alpha$ and the limiting resisting force per unit area will be equal to the resultant force R. Hence,

$$R \propto \tau_c \cos\alpha$$

and

$$T = W \sin\alpha \propto \frac{\tau_c}{\tan\phi} \sin\alpha$$

The constant of proportionality is assumed to remain constant throughout.

Hence,

$$\tau_c^2 \cos^2\alpha = \tau_o^2 + \tau_c^2 \frac{\sin^2\alpha}{\tan^2\phi}$$

$$\therefore \frac{\tau_o}{\tau_c} = \cos\alpha \left[1 - \frac{\tan^2\alpha}{\tan^2\phi}\right]^{\frac{1}{2}} \qquad (8.64a)$$

This is the form given by Lane.

With

$$\cos^2\alpha - \frac{\sin^2\alpha}{\tan^2\phi} = 1 - \sin^2\alpha \left[1 + \frac{1}{\tan^2\phi}\right]$$

it becomes

$$\frac{\tau_o}{\tau_c} = \left[1 - \frac{\sin^2\alpha}{\sin^2\phi}\right]^{\frac{1}{2}} \qquad (8.64b)$$

Both of these expressions give the ratio of shear stress required to start motion on the slope to that required on the level surface of the same material.

8.2.2 Design values for boundary shear

Efforts have been made to relate the capacity of soils to withstand drag force to the type of soil, the aim being to provide tables or charts which would specify the permissible boundary shear for any kind of material.

These design values for non-cohesive material are classified according to particle size. The available information is not conclusive. There is no agreement about the particle size to be used when dealing with non-uniform materials. The U.S. Bureau of Reclamation recommends d_{75} for particles larger than about 5 mm diameter and the median size for finer non-cohesive materials (see Fig. 8.5).

A reduction from 10% to 40% of these design values for shear stress is recommended with increasing horizontal curvature of the channel. Design values for cohesive soils are discussed in Chapter 10.

8.2.3 The minimum stable cross-section

The minimum cross-sectional shape is largely dependent upon the soil type. For channels in non-cohesive material the weight component down the slope has to be combined with the applied fluid shear, whereas in cohesive materials the rolling-down effect of the particles is negligible.

One of the most widely used cross-sections is the trapezoidal one.

In designing such a channel by the tractive force method the design value of the boundary shear occurs only over parts of the wetted perimeter, Fig. 8.6. The slope correction (eqn (8.64)) has to be applied on the side slopes.

The minimum stable hydraulic cross-sections would have the stage of impending motion at all points of the cross-section at the same time. For a given soil and discharge this cross-section has the least excavation, least width, and maximum mean velocity. The shape of such a cross-section in non-cohesive material was derived by Glover and Florey (1951). The derivation assumes that the shear stress on an element of the boundary is due to the weight of water vertically above; that is, the weight component down the slope of the channel. The lateral shear forces between the adjacent currents of different velocity are neglected by this assumption, but alternative studies by the Bureau of Reclamation have shown that this is not serious. The method does not allow for secondary currents which are present in straight as well as meandering channels.

The force equilibrium yields, Fig. 8.7,

$$\gamma y \Delta x S = \tau_o \frac{\Delta x}{\cos\alpha}$$

$$\therefore \tau_o = \gamma y S \cos\alpha$$

If $\tau_{o(max)}$ is the maximum shear stress at $y = y_o$, then

$$\frac{\tau_o}{\tau_{o(max)}} = \frac{y}{y_o} \cos\alpha$$

At the threshold of particle movement $\tau_{o(max)} = \tau_c$ and by substitution for τ_o/τ_c from eqn (8.64)

$$\frac{y}{y_o} = \left[1 - \frac{\tan^2\alpha}{\tan^2\phi}\right]^{\frac{1}{2}}$$

From Fig. 8.7

$$\tan\alpha = \frac{dy}{dx}$$

Substituting and rearranging yields

$$\left[\frac{dy}{dx}\right]^2 + \left[\frac{y}{y_o}\right]^2 \tan^2\phi = \tan^2\phi \qquad (8.65)$$

or

$$\frac{dy}{\sqrt{[1 - (y/y_o)^2]}} = \tan\phi dx$$

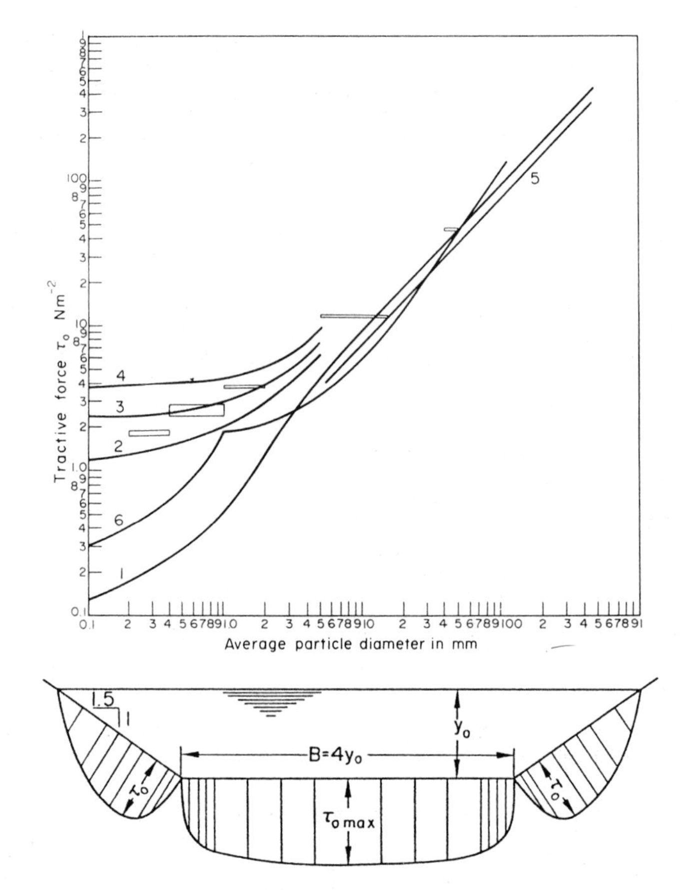

Fig. 8.5. Recommended permissible tractive force values for non-cohesive material. 1, Shields curve; 2-5 U.S. Bureau of Reclamation values, 2 for canals with clear water, 3 for canals with low concentration of fine sediment and 4 for canals with high concentration of fine sediment in the water, 5 for canals in coarse material with 25% of material larger than the indicated value; 6 U.S.S.R. canals with clear water; the horizontal bars are values recommended by the Nuernberg Kulturamt.

Fig. 8.6. Shear stress distribution over the periphery of a trapezoidal channel, Lane (1952). $\tau_{o(max)}$ equals 0.89; 0.97 and 0.99 times $\gamma y_o S$ for B equals 2; 4 and 8 times y_o, respectively. The maximum value on the sides, τ_o, equals 0.735; 0.750 and 0.760 times $\gamma y_o S$, respectively, and occurs at 0.1 to 0.2 of the depth and varies slightly with the slope of the sides.

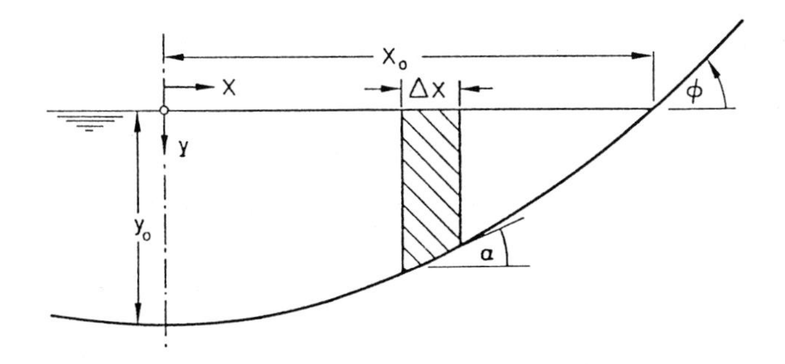

Fig. 8.7. Definition sketch.

This yields

$$y_o \cos^{-1} \frac{y}{y_o} = x \tan\phi \qquad (8.66)$$

or

$$\frac{y}{y_o} = \cos\left[\frac{x \tan\phi}{y_o}\right] \qquad (8.67)$$

Thus, the above assumptions lead to a cosine curve for the stable cross-sectional shape.

The half-width is

$$x_o = \frac{\pi y_o}{2 \tan\phi} \qquad (8.68)$$

The area of the cross-section is

$$A = 2y_o \int_0^{\pi y_o/2 \tan\phi} \cos\left[\frac{x \tan\phi}{y_o}\right] dx = \frac{2y_o^2}{\tan\phi}\left[\sin\left[\frac{x \tan\phi}{y_o}\right]\right]^{\pi y_o/2 \tan\phi}$$

$$= 2y_o^2/\tan\phi \qquad (8.69)$$

The perimeter P is

$$P = 2\int_0^{x_o}\left[1 + \left[\frac{dy}{dx}\right]^2\right]^{\frac{1}{2}} dx = 2\int_0^{x_c}\left\{1 + \tan^2\phi\left[1 - \cos^2\left[\frac{x \tan\phi}{y_o}\right]\right]\right\}^{\frac{1}{2}} dx$$

Substituting $\theta = x \tan\phi/y_o$ and transcribing the integrand as

$$\sqrt{(\sec^2\phi - \tan^2\phi \cos^2\theta)} = \sec\phi\sqrt{(1 - \sin^2\phi \cos^2\theta)}$$

yields

$$P = \frac{2y_o}{\sin\phi}\int_o^{\pi/2} \sqrt{(1 - \sin^2\phi \cos^2\theta)}d\theta$$

With limits from 0 to $\pi/2$, $\cos^2\theta$ may be replaced by $\sin^2\theta$ without changing the result. Hence,

$$P = \frac{2y_o}{\sin\phi}\int_o^{\pi/2} \sqrt{(1 - \sin^2\phi \sin^2\theta)}d\theta$$

This is an elliptic integral of the second kind and is complete with the limits $\pi/2$ and 0. The value is readily obtained from tables for any given value of ϕ. Designating the integral by the symbol E yields

$$P = \frac{2y_o}{\sin\phi} E \tag{8.70}$$

From known area and wetted perimeter

$$R = \frac{A}{P} = \frac{2y_o^2}{\tan\phi}\frac{\sin\phi}{2y_oE} = \frac{y_o \cos\phi}{E} \tag{8.71}$$

Leliavsky (1955) gives a slightly different approach to the same problem.

Bretting (1958) assumed that the shear stress is proportional to the distance between the bottom and the water surface, measured perpendicular to the bed, whence

$$\frac{\tau_o}{\tau_{o(max)}} = \frac{y}{y_o \cos\alpha}$$

Combined with logarithmic velocity distribution formulae, differential equations were set up and the boundary shape and isovel patterns calculated.

8.2.4 Design by tractive force method

Watercourses which are very wide relative to their depth can be treated by the two-dimensional approach.

Narrow watercourses have to be treated as stable in cross-section, and the simple expression of $d = 11y_oS$ obtained in the two-dimensional approach must now be written in terms of the hydraulic mean radius.

The general discussion of this problem will be greatly simplified if the angle of repose is given a fixed value, which can be selected according to particular conditions. For example, the stable channel parameters for $\phi = 30°$ and $\phi = 35°$ become:

		$\phi = 30°$	$\phi = 35°$	
Surface width	$B = 2x_o$	$= 5.45\ y_o$	$B = 4.49\ y_o$	(8.68 a,b)
Area	A	$= 3.45\ y_o^2$	$A = 2.86\ y_o^2$	(8.69 a,b)
Perimeter	P	$= 5.87\ y_o$	$P = 4.99\ y_o$	(8.70 a,b)
Hydraulic mean radius	R	$= 0.59\ y_o$	$R = 0.57\ y_o$	(8.71 a,b)
	$\frac{P}{R} = 10$		$\frac{P}{R} = 8.75$	(8.72 a,b)

Thus eqn (8.60) is replaced by

$$d = \frac{11RS}{0.59} \doteq 19RS \tag{8.73}$$

and eqn (8.62) modified by a factor $\left[\frac{11}{19}\right]^{1/6}$, yields

$$V \cong 14.7R^{1/2}S^{1/3} \tag{8.74}$$

Eliminating S between eqns (8.73) and (8.74) yields

$$V \cong 5.5\ d^{1/3}R^{1/6} \tag{8.75}$$

and would be an expression corresponding to Lacey's first equation

$$V \cong 0.646\ \sqrt{(fR)}$$

These relationships define the limiting minimum size of the channel for given discharge, slope and particle size. By the use of eqn (8.74) it has been implied that the channel is not rippled or covered by dunes, etc; that is, one is designing for the threshold of particle movement. For nonuniform sediment the dominant size has to be used, about the d_{80} to d_{90} size. However, due to the one-sixth power dependence of resistance on particle size the effect of errors in estimating the particle size is small.

The coefficient of the mean velocity relationship could be raised by about 10-15%.

Other necessary relationships for the design can be derived from those above.

Thus VA = Q and PR = A, yield

$$Q \cong 14.7\ PR^{3/2}S^{1/3} \tag{8.76}$$

and substituting from eqn (8.73) for **R** and eqn (8.72a) for P = 10R we obtain

$$S \doteq 0.335d^{1 \cdot 15}Q^{-0 \cdot 46} \tag{8.77}$$

This could be compared with the Lacey expression, eqn (8.20). The channel width for

$$\phi = 30° \quad B = 5.45y_o = \frac{5.45d}{11S} \cong 0.5\frac{d}{S} \tag{8.78a}$$

$$\phi = 35° \quad B = 4.49y_o = 0.41\frac{d}{S} \tag{8.78b}$$

or

$$\phi = 30° \quad B = 1.48d^{-0.15}Q^{0.46} \tag{8.79a}$$

$$\phi = 35° \quad B = 1.22d^{-0.15}Q^{0.46} \tag{8.79b}$$

and

$$\phi = 30° \quad P = 1.59d^{-0.15}Q^{0.46} \tag{8.80a}$$

$$\phi = 35° \quad P = 1.35d^{-0.15}Q^{0.46} \tag{8.80b}$$

This could be compared with the Lacey equation for P, eqn (8.14).

These formulae have been worked for assumed values of specific gravity and angle of repose and are valid for sediment larger than 6 mm; that is, where the plot of Shields entrainment function versus particle Reynolds number levels out to a value of about 0.056. However, the same treatment can be followed for any given conditions by introducing appropriate values and the coefficients are changed accordingly. General expressions can be written, obviously more cumbersome or correction factors could be worked out (Henderson, 1961).

If the slope is less than the value computed by eqn (8.77) the channel will aggrade. Then equilibrium can be established only by increasing the slope.

In canal design this may mean re-siting, in river control work it would suggest a cut-off. However, it must be kept in mind that these relationships are based on a straight channel. Therefore, in river control work allowance must be made for bend losses.

It must be stated that only limited field and laboratory data is available to test the equation for minimum slope, eqn (8.77). The analysis also shows that the width to depth ratio depends only on the angle of repose and is about five. In natural rivers in coarse alluvium it is much greater and increases with discharge. But eqn (8.77) does provide the engineer with some guidance, when considering an aggrading river, as to whether it is because of lack of slope or because the river has been allowed to spread into too wide a channel.

In river work an added complication arises in defining the design discharge. It should be the discharge, which at constant steady flow would have the same overall effect upon the river channel, as the natural fluctuating discharge. Such a discharge has been termed the *dominant or bank-full discharge*, but this discharge is still rather ill-defined.

The results of the analysis can be displayed for comparison, as well as for design purposes, for a given discharge on a B versus S plot with the particle size d as a parameter (Fig. 8.8).

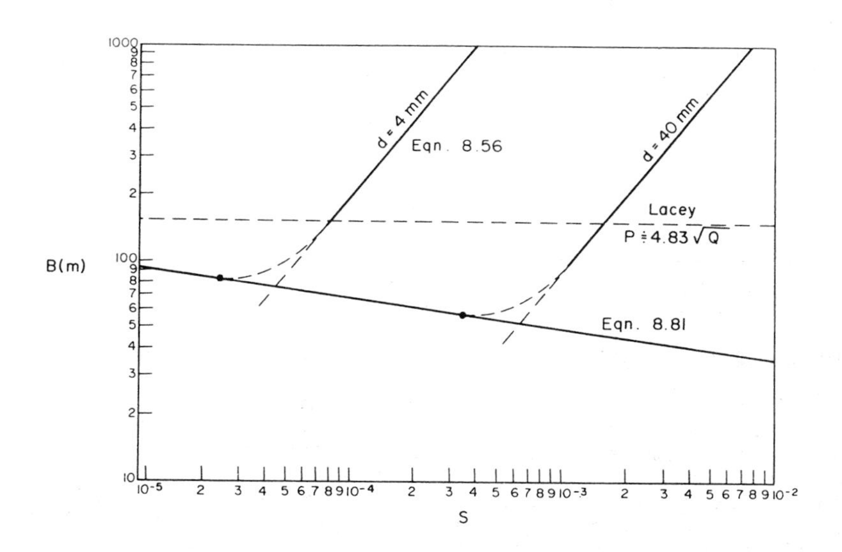

Fig. 8.8. Diagrammatic presentation of the width B versus slope S by tractive force design method for a constant discharge $Q = 1000\ m^3s^{-1}$.

For a given Q, eqns (8.77) and (8.79) define a line on the B versus S-plane. This line defines the limiting conditions for the stable channel, that is the limiting narrow section.

Below this line the B-S relationship cannot be satisfied. From eqn (8.79) for a given Q

$$d = (N_1/B)^{13/2}$$

and from eqn (8.77)

$$d = (S/N_2)^{13/15}$$

Whence,

$$S = N_2 \left[\frac{N_1}{B}\right]^{15/2} \tag{8.81}$$

where N_1 and N_2 are numerical constants for the particular conditions.

For the wide section, the two-dimensional design, Q,B,d and S, were related by eqn (8.63), or in terms of B

$$B = P = \frac{2.27QS^{7/6}}{d} \tag{8.82}$$

For a given Q this equation defines a family of straight lines on the log-log plane of B versus S. For a given discharge and particle size, the narrowest channel with the minimum cross-sectional area is characterized by a point on the line defined by eqn (8.81). Expressing it in other words, eqns (8.77) and (8.79) define the point where the relationship between B and S for a given d and Q lies on the line defined by eqn (8.81). Between the eqn (8.63), which defines the wide channel, and eqn (8.81), for the limiting narrow one, there will be a transition. The foregoing treatment does not determine the transition region.

It is important to realize that the y-α relationship

$$\frac{y}{y_o} = \left[1 - \frac{\tan^2\alpha}{\tan^2\phi}\right] \text{ or } \frac{y}{y_o} = \cos\left[\frac{x \tan\phi}{y_o}\right]$$

will be satisfied if a section of constant depth is inserted into the centre of the cosine curve.

It is likely that, if an exact relationship is used instead of the approximation $\tau \propto y \cos\alpha$, a continuous flat curve would result instead of the straight centre part. Yet, this is a small discrepancy and could be overshadowed by the effects of the secondary currents. The above cross-section could be considered to be the transition between the minimum one and the infinitely wide (two-dimensional) cross-section.

This also shows that there is an infinite number of cross-sections which satisfy the tractive force design criteria, and consequently, with d and Q fixed, an unlimited number of the ratios of P/R. It should be recalled that the Lacey approach postulates a unique relationship between P/R and the mean velocity (eqn (8.21)). Limited laboratory and field data appear to indicate that channels designed according to this method are stable. When flow is increased, scouring commences on the banks.

In principle, the threshold stable channel design concept "fails" when sediment transport starts. Then the applied stress on the bank exceeds the resistance to movement and the sediment starts to move in the direction of the resultant stress which is directed downstream and inwards towards the channel bed (Fig. 8.4). Consequently the banks will continue to erode as long as there is sediment transport. Parker (1978) used the concept of lateral transfer of downstream momentum by turbulent diffusion (Lundgren and Jonsson, 1964) together with singular perturbation techniques to explain "the coexistance of stable banks and mobile beds in straight reaches of coarse gravel rivers". The downstream velocity (momentum) is greater at the centre than near the banks and turbulence produces a net lateral flow of this momentum towards the banks. Superimposed is the lateral transport of longitudinal momentum by the secondary currents which move momentum to the water's edge.

The extension of the threshold method of stable channel design to live-bed conditions of active sediment transport has so far not made a great deal of progress. However, alternative methods have evolved. Thus, Chang (1980) building on the work by Yang (1974) hypothesised that width, depth and slope of stable channels adjust to make the stream power ρgQS a minimum. Using a trapezoidal channel, Engelund's resistance expression and DuBoys transport equation in a numerical model he obtained realistic results, see Fig. 5.5.

A related approach by Ramette (1979,1980) postulates that the channel carries a maximum amount of sediment of known size for known discharge and that the dimensions adjust to make the ratio of kinetic energy recovery to potential energy spent a maximum when the channel slope is equal or less than valley slope. The analysis led to equations for cross-sectional areas of stable channels in keeping with eqn 8.6 and width proportional to $Q^{0.375}S^{-0.19}$.

8.3 Discussion

The regime method of design is for a channel with a live bed that may scour or deposit at times, but over a climatic cycle the net result of scour and deposition is zero. It is empirical and based on data from canals carrying low sediment loads. The empirical nature requires that when used for the design the conditions must be within the range of data from which the coefficients in the design formulae were determined. The regime methods were also intended for a constant water discharge. The regime method is characterised by

$$RS^2 = \text{constant}$$

The tractive force methods of design are related to shear stress which is proportional to RS, i.e., characterised by

$$RS = \text{constant}$$

for given conditions.

The three Lacey equations

$$V = 0.646\ \sqrt{(fR)} \quad (8.5)$$

$$V = 10.8R^{2/3}S^{1/3} \quad (8.22)$$

$$P = 4.83\ \sqrt{Q} \quad (8.16)$$

involve five variables, i.e. Q,P,R,S and f and therefore only two quantities may be fixed initially. But in river work, for example, Q,S and f are usually given and cannot be altered to suit the design. In such a case the three Lacey equations can be satisfied simultaneously only if Q,S and f satisfy

$$S = \frac{f^{5/3}}{3169.8Q^{1/6}} \tag{8.20}$$

where the silt factor incorporates the effects of particle size and sediment transport rate or concentration at given water discharge. In the light of the discussion relating to Fig. 8.8 it may be concluded that in river work eqn 8.20 gives the minimum slope requirement. If the slope is greater the channel could be widened and thus, its sediment transporting power reduced. If the slope is less a stable channel cannot be formed. The equation yields the verification that the actual values of P and S are greater than minimum values.

There are a great many similarities between the velocity- depth or hydraulic radius relationships of the regime equations and the bed load equations, as was illustrated by Laursen (1958) and Gill (1968).

The tractive force method of stable channel design can, in principle, be extended to live bed by using a resistance relationship applicable to live bed, for example, the equivalent roughness height definition by van Rijn, eqn 6.50 or formulae proposed by White et al., eqns 6.64, 6.65 or 6.67 together with an appropriate sediment transport relationship. Such an approach was outlined by White et al. (1982) using the condition of *maximum sediment transporting capacity* to define and equilibrium channel. They show that this is equivalent to the minimum stream power definition used by Yang and Chang.

Chapter 9

Erosion and Deposition

The term erosion can refer to a multitude of physical processes, such as soil erosion, beach erosion or erosion of a river bank. For localised phenomena the term scour is used. It may result from flow of water (or air) or from wave action. Scour is most pronounced in alluvial materials but even deeply weathered rock could be vulnerable at some circumstances. Scour may also occur over a reach of stream channel but usually a local scour is implied, such as scour at a bridge pier or abutment, scour at a head of a groyne or spur dike, scour in a constriction of waterway, scour at stilling basins or culvert outlets, etc.

Deposition of sediment may occur in a reach of a river but usually a deposition in reservoirs and of river deltas is meant. Beach is said to accrete not to deposit.

Scouring, in the sense of soil erosion, means loss of valuable asset. Scouring at a structure may lead to its failure. The physics of scour at structures is particularly complex because, here, in addition to all the other complexities of sediment transport, the flow pattern is usually three-dimensional with zones of acceleration and deceleration, local vortices etc.

The scour at a structure may be

(i) *general scour* of the stream which would occur irrespective whether the structure was there or not;

(ii) *constriction scour* arising from constriction of waterway, for example, when flows over flood plains have been intercepted by bridge causeways and channelled through the bridge waterway, or

(iii) *local scour* resulting from the effect of the structure on local flow pattern. The local scour may be superimposed on constriction and/or general scour.

In addition, the scour at the structure may be

(i) clear-water scour or

(ii) live-bed scour.

Clear water scour refers to conditions when the bed material upstream of the scour area is at rest. The bed shear stresses in areas beyond the direct influence of the structure are at or below the critical or threshold shear stress for initiation of bed material movement. Wash load, however, could be present.

Live-bed scour occurs under conditions of general sediment transport and there is a sediment input into the scour zone from upstream.

Due to the complex physics of the scouring process methods of scour prediction are essentially empirical. Many specific projects require model studies. For this small scale hydraulic models are a powerful tool. The modelling aspect is not covered here but reference is made to Yalin (1971), Kobus (1980) and Kolkman (1982).

Aspects of general scour and river geometry were discussed in Chapter 5 and those of beach erosion in Chapter 11.

9.1 Soil Erosion

Soil erosion is strongly linked to local climatic conditions and to vegetal cover. At any given location the most dominant single factor is land management. From the same type of land in the same area, the erosion rates can easily differ a thousand times, depending on land use management. The rates of soil erosion are important for long term management of land and rivers and in particular for estimation of the rate of siltation of reservoirs.

The processes of soil erosion in global scale are so varied due to the climatic and physiographic conditions, land use and management practices, soil types, etc., that it will be futile to attempt to arrive at a universal method for evaluation of the erosion rate. In many parts of the world, soil loss data has been accumulated and is continuously being added to form field measurements and from experimental plots, particularly for arable land. Regional formulae have been developed with the aid of local data, for example, the Musgrave (1947) equation or the "universal soil loss equation" developed by Wischmeier and Smith (1965)

$$E = R\ K\ L\ S\ C\ P \tag{9.1}$$

where E is the average annual soil loss, R is a factor which relates the erosion potential to the local average rainfall, K is a soil erodibility factor, L and S are factors which allow for the effects of the length of the area and its slope, C is a cropping management factor and P is a conservation practice factor. For regions for which these empirical factors have been established, reported correlations between predicted and measured erosion rates have been good. The formula will not help if such local data is not available, particularly since the errors through multiplication may accumulate. A detailed discussion of the various methods of soil loss estimation is given by Kirkby and Morgan (1980).

Concentrated erosion occurs in gulleys and again, the rates of erosion are strongly dependent on soil type. Temporary erosion peaks arise from construction sites, strip mining, logging operations etc. but the magnitudes are very sensitive to measures taken to minimise erosional losses.

Any calculations of sediment volumes should be compared to estimates using other methods. The estimates also have to incorporate assumptions on the future use and management of the catchment. For some regions, empirical relationships for the delivery ratio have been established. These frequently have the form

$$\log D = a - b \log cA^d - e \log L/E - f \log B \qquad (9.2)$$

where A is area, L is catchment length along the main stream, E is elevation difference over the distance L and B is a mean bifurcation ratio which is the ratio of the number of streams of a given order to the next higher order. The symbols a to f represent constants. For Piedmont region, USA, Roehl (1962) obtained $a = 45$, $b = 0.23$, $c = 10$, $d = 1$, $e = 0.51$, and $f = 2.79$. For Kansas, Western Texas Maner (1958) obtained

$$\log D = 2.943 - 0.824 \log L/E \qquad (9.3)$$

The average trends of sediment yield can also be represented by a relationship (Fleming 1968) of the form

$$Y = aA^b \qquad (9.4)$$

where the exponent b is a small negative number. Fleming had $b = -0.04$, whereas data for smaller areas in the Upper Mississippi Basin show slopes of the order of $b = 0.12$. For moderate sized areas, both sediment yield and sediment delivery ratios vary approximately with -1/8 to -1/10 power of the area involved. Data from experimental plots can be considered as point source with 100% delivery. Assuming that 10% of the total material eroded is transported to the sea then the delivery ratio becomes

$$D = 0.1\, aA^b \qquad (9.5)$$

and a can be evaluated by using the point source data from known area and 100% delivery.

Langbein and Schumm (1965) correlated data on mean annual precipitation, temperature and general type of vegetation with sediment concentration. The results show that in dry regions where the vegetation is sparse and the sediment is readily available, the infrequent runoffs carry high concentrations of sediment. At about 300 mm of mean annual precipitation and 10°C annual mean temperature the sediment yield is shown to reach a maximum, Fig. 9.1. The sediment yield decreases from thereon with increasing mean annual precipitation due to the increasing protective effects of vegetation, particularly grass and forest cover. However, the peak sediment yield - mean annual precipitation relationship is a function of the annual mean temperature as shown schematically in Fig. 9.1. The full lines refer to data from large area sediment stations and the dotted line to small areas as determined from reservoir siltation data. The plotted functions are indicative of average values only and tenfold departures in individual cases are not uncommon. The picture is further complicated for regions with highly seasonal rainfall. There the peak could move to mean annual rainfall values greater than 1000 mm.

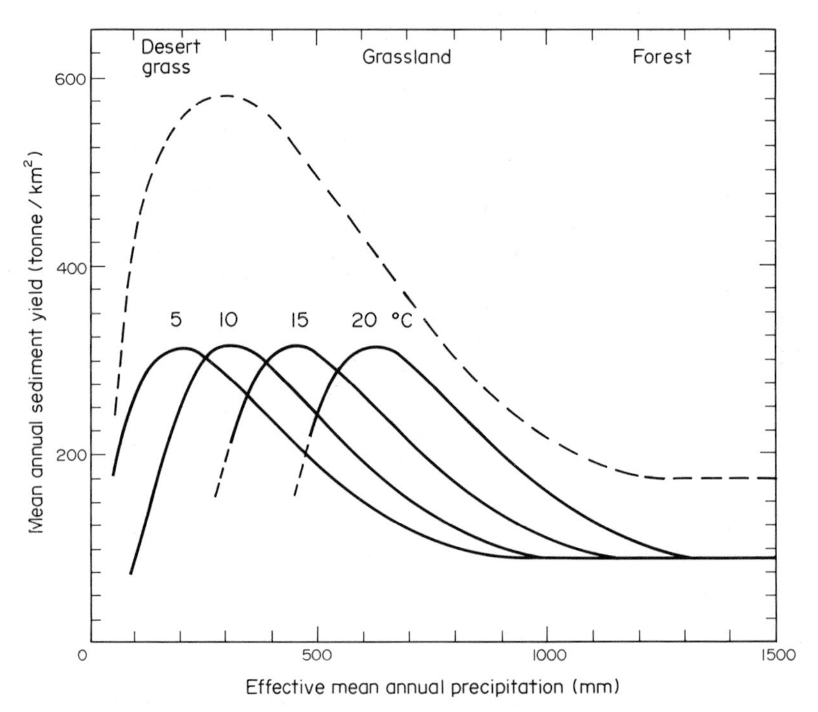

Fig. 9.1. Typical relationship between sediment yield, annual mean precipitation and temperature, after Schumm (1965). Full lines refer to data from large area sediment stations, dotted line to reservoir accumulations from small areas.

Figure 9.2 shows average values of sediment concentration in runoff, which in turn could be related to mean annual precipitation and temperature, Fig. 9.3, after the relationship by Langbein et al. (1949).

The above relationships are only indicative of trends. For any particular project local data has to be collected and the sediment yield has to be estimated for the given conditions. The actual estimation of sediment yield for a particular water resources project is frequently made more difficult by the project itself.

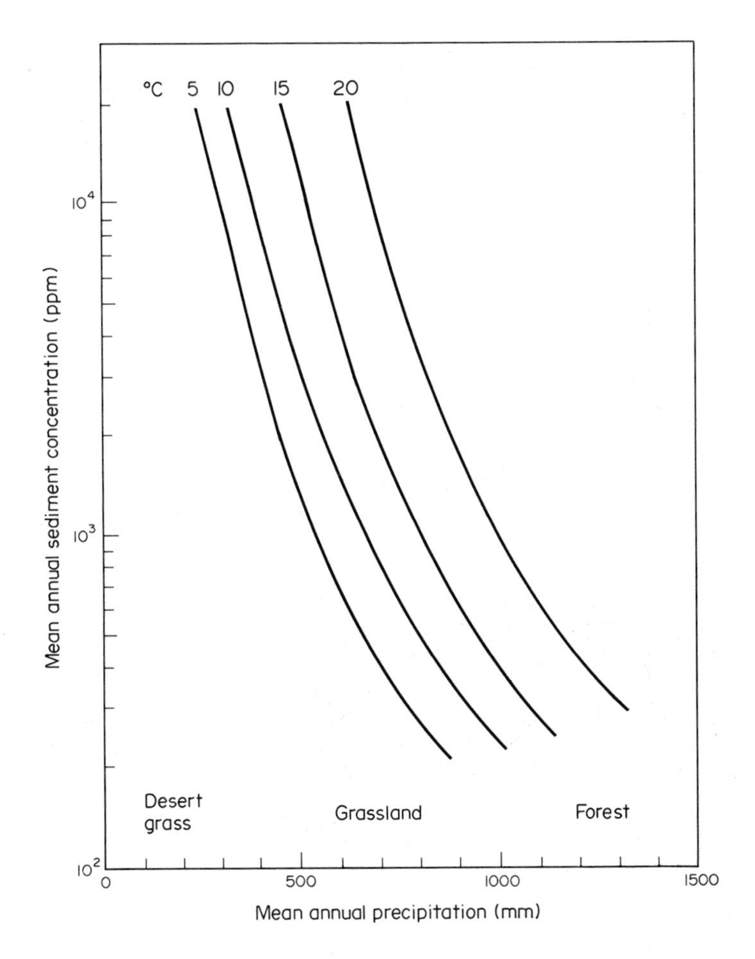

Fig. 9.2. Relationship between mean sediment concentration in runoff, annual mean precipitation and mean annual temperature, according to Schumm (1965).

Often the construction of a dam and reservoir, for whatever purpose, leads to changes in land usage of the catchment. The standard of execution and management of these changes may be very difficult to predict, particularly in some of the new developing countries. If the management of the conversion processes is lax very high sediment yields may occur. For small catchments increases more than a thousandfold during the transition have been reported.

Such very high rates of erosion from relatively small areas can lead to severe environmental problems. Therefore, the reduction and control of erosion from areas of land under preparation for urban or industrial or mining developments is important. It is generally more effective and less costly to reduce erosion on

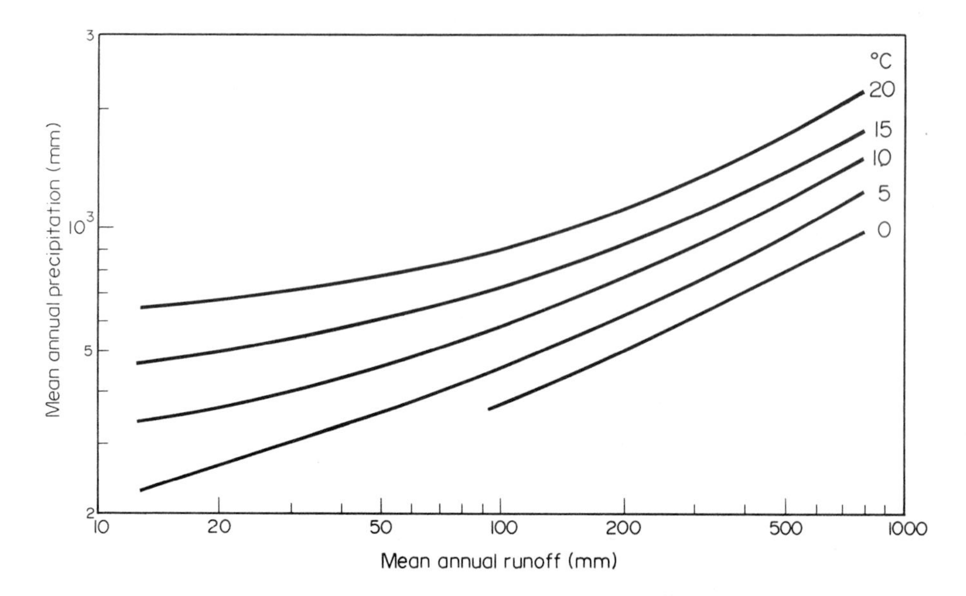

Fig. 9.3. Mean annual runoff as a function of mean annual precipitation and temperature, after Langbein et al. (1949).

site and transport of erosion products away from the site than to remedy the effects downstream. The sediment yield, Q_S, from an area A is the sum of sediment yields from areas under construction, q_{SC} and A_C, and undisturbed areas, q_{SU}, i.e., $Q_S = D[Q_{SC}A_C/A + q_{SU}(1 - A_C/A)]A$. where D is the delivery ratio, for example, eqn (9.2). Brandt et al. (1972) obtained for data from Maryland, USA, slightly higher delivery ratios, $D \cong 0.5A^{-0.15}$ where A is in km^2. This could be used as an indicative value in moderate zones. The major difficulties arise with estimation of erosion rates q_{SC}. Usually some local data is available for erosion rates from undisturbed areas. For pasture and woodland mixtures the rates are usually in the range of about 200 to 400 tonne per km^2 per year, the rates increasing with decreasing mean annual rainfall and with increasing amount of arable land but seldom more than 750 tonne per km^2 per year. Data presented by Chen (1974) from construction areas from Maryland and Virginia range 12 500 to 50 000 tonne per km^2 per year. Chen also presents an empirical method for estimation of cost-effectiveness of control measures which could be adapted for other regions.

An important part of any water resources project is the design of effective sediment control measures aimed at reduction of sediment yield from the catchment. The topic of soil conservation commands a literature on its own right. Here only the need and importance of works to reduce soil erosion can be emphasized. The most important of all the measures to reduce soil erosion is the land use management and cropping practices which range from arable farming to forestry. Depending on topography and intensity of land use additional engineering works may be necessary. Terracing, contour farming, etc. are some of the age-old methods still in use, to reduce soil loss. In the upper parts of many catchments debris dams are effective in trapping and retaining sediment and reducing sediment runoff. Lower down the whole spectrum of river improvement measures may enter the picture. However, the most important aspect is the prevention, that is reduction, of soil erosion at the source rather than trapping or transporting the eroded material. Soil conservation engineering relates to works and structures of various kinds aimed at reduction of soil loss and regulating runoff from the land.

9.2 Constriction Scour

A constriction of a watercourse will lead to increased velocity of flow through the reduced cross-section and scouring of the bed. It may also lead to backwater effects until the deepening of the constricted cross-section alleviates the damming up of water.

The well-known treatment by Laursen (1958) assumed that the channel was reduced to a narrower width, the long contraction, Fig. 9.4. It was assumed that the sediment transport over the berms by Q_o is zero. The conditions to be satisfied are

$$Q_t = Q_c + Q_o = \frac{1}{n_2} B_2 y_2^{7/6} \sqrt{(y_2 S_2)} \tag{9.6}$$

$$Q_{s1} = \dot{Q}_{s2}$$

Both Q_c and the total flow Q_t are given by the Manning formula

$$Q_s \propto \bar{c}_1 Q_c = \bar{c}_2 Q_1 \tag{9.7}$$

The mean concentration of sediment transport $\bar{c}$ (in per cent by weight) is given by the Laursen formula, eqn (7.143), in which $(\tau_o'/\tau_c - 1)$ is approximated for flood conditions by $\tau_o'/\tau_c = V_o^2/36d^{2/3}y_1^{1/3}$ and the $f(u_*/w)$ is replaced by $K(u_*/w)^a$. The exponent a is said to be $a = 1/4$ for $u_*/w < 1/2$, $a = 1$ for $u_*/w = 1$, and

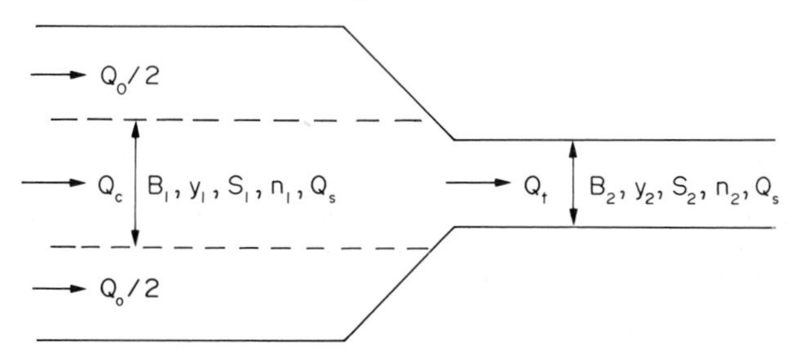

Fig. 9.4. Definition sketch of a long contraction with overbank flow.

$a = 9/4$ for $u_*/w > 2$. In here $u_* = \sqrt{(gyS)} = Q\sqrt{g}.n/By^{7/6}$. When eqn (9.7) is written out in full, the coefficients and sediment characteristics cancel out and an algebraic rearrangement yields

$$\frac{y_2}{y_1} = \left[\frac{Q_t}{Q_c}\right]^{6/7} \left[\frac{B_1}{B_2}\right]^{\frac{6}{7}\frac{2+a}{3+a}} \left[\frac{n_2}{n_1}\right]^{\frac{6}{7}\frac{a}{3+a}} \tag{9.8}$$

The last factor can be neglected, because the ratio of the Manning n is close to 1 and the power is not more than 0.37. If the contraction is for overbank flow only the main channel flow conditions will not be affected and the ratio of widths is unity. Thus, eqn (9.8) reduces to

$$\frac{y_2}{y_1} = \left[\frac{Q_t}{Q_c}\right]^{6/7} \tag{9.9}$$

for all conditions of transport. Alternatively, if the contraction narrows the main channel, the discharge ratio is one since $Q_c = Q_t$, and eqn (9.8) reduces to

$$\frac{y_2}{y_1} = \left[\frac{B_1}{B_2}\right]^m \tag{9.10}$$

where $m = 0.59$, 0.64, and 0.69 for $u_*/w < \frac{1}{2}$, 1 and > 2, respectively, by using the Laursen sediment transport relationship. The shear velocity u_* is for upstream conditions. This form of long constriction is illustrated in Fig. 9.5.

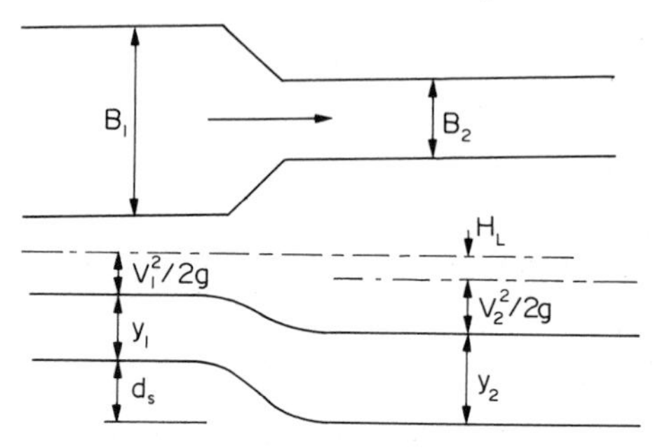

Fig. 9.5. Definition sketch for long contraction.

Straub (1935) solved this problem by using the Du Boys sediment transport formula and the Manning equation for wide channels. The result is

$$\frac{y_{o2}}{y_{o1}} = \left[\frac{B_1}{B_2}\right]^{3/7} \left\{ \frac{-\frac{\tau_c}{\tau_1} + \left[\left[\frac{\tau_c}{\tau_1}\right]^2 + 4\left[1 - \frac{\tau_c}{\tau_1}\right]\frac{B_1}{B_2}\right]^{\frac{1}{2}}}{2(1 - \tau_c/\tau_1)} \right\}^{/7} \tag{9.11}$$

where τ_c is the critical shear stress of the bed material. For $\tau_1 >> \tau_c$ eqn (9.11) reduces to eqn (9.10) with m = 9/14 = 0.642 and when $\tau_1 = \tau_c$, m = 6/7 = 0.857. Intermediate values for τ_c/τ_1 = 0.2, 0.4, 0.6 and 0.8 are m = 0.668, 0.700, 0.733, and 0.782, respectively. For $\tau_1 < \tau_c$

$$\frac{y_{o1}}{y_{o1}} = \left[\frac{B_1}{B_2}\right]^{6/7} \left[\frac{\tau_1}{\tau_c}\right]^{3/7} \tag{9.12}$$

Laursen (1963) obtained the same relationship for threshold conditions in the contraction, except that τ_1 is replaced by the grain roughness value τ_1' using the Manning-Strickler formula. Laursen expressed from continuity, $V_1y_1B_1 = V_2y_2B_2$, and assuming equal slopes, $S_1 = S_2$, the scour depth d_s as

$$d_s = (y_2 - y_1) + (1 + K)\left[\frac{V_2^2}{2g} - \frac{V_1^2}{2g}\right] \tag{9.13}$$

where K is a headloss factor. This, via the Manning-Strickler formula, lead to

$$\frac{d_s}{y_1} = \left[\frac{\tau_o'}{\tau_c}\right]^{3/7}\left[\frac{B_1}{B_2}\right]^{6/7} - 1 + 1.83(1 + K)\left[\frac{d}{y_1}\right]^{2/3}\left[\frac{(B_1/B_2)^{2/7}}{(\tau_o'/\tau_o)^{6/7}} - 1\right]\left[\frac{\tau_o'}{\tau_c}\right] \tag{9.14}$$

and if the energy loss and difference in velocity heads are neglected

$$\frac{d_s}{y_1} = \left[\frac{\tau_o'}{\tau_c}\right]^{3/7}\left[\frac{B_1}{B_2}\right]^{6/7} - 1 \tag{9.15}$$

Laboratory data (Webby, 1984) show that eqn (9.12) consistently underpredicts the depth y_{o2} by about 20%. This depth occurs in the constrictions about 1.5 B_1 downstream of the entrance. At the entrance, a local scour, y_2', develops around the convex "corners" of the transition. The depth y_2', for equilibrium conditions depends on the shape of the contraction. For a streamlined contraction $y_2' \cong 1.5\ y_2$ and about $y_2' = 2.6\ y_2$ for a blunt one, where y_2 is the actual equilibrium depth in the long contraction. In rivers where the width/depth ratio is usually large the lowering of the bed will not be uniform over the width (Neill, 1975) and allowance for extra scour depth should be made.

9.3 Scour at Bridge Piers

Failure of bridges due to scour at their supports is a common occurrence. It may be due to scour at a pier or at an abutment. At times even scour of the river bank, leading to the stream changing its course altogether and outflanking the bridge. Major scouring usually occurs during floods, that is, when the flow is unsteady and may even be angled to low flow direction. Additional problems are caused by floating debris or ice packs.

Local scour at a pier commences when the shear velocity u_* or velocity U exceeds about half the critical or threshold value for movement of sediment.

Chabert and Engeldinger (1956) appear to have been the first to describe the bahavioural pattern of scour at a cylindrical pier (Fig. 9.6) in terms of development with time and flow velocity. They showed that the clear-water scour approaches equilibrium asymptotically, over a period of days, whereas the live-bed scour develops rapidly and its depth fluctuates in response to the passage of bed features. Shen et al. (1969) suggested that the mean value of the live-bed scour depth was about 10% less than the maximum clear-water scour depth. Raudkivi (1982) suggested that a second peak exists, as shown by the dashed line in Fig. 9.6(b).

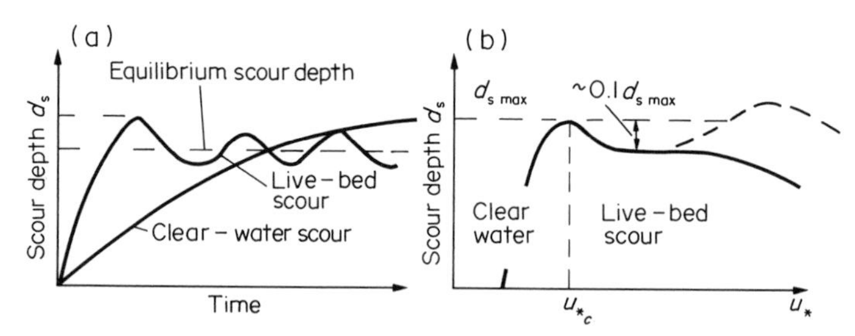

Fig. 9.6. Scour depth for a given pier and sediment size, (a) as a function of time and (b) as a function of shear velocity or approach velocity.

The treatments of the scour problem usually start with the statement that the depth of scour depends on variables which characterise the fluid, bed material in the stream and at bridge crossings (grading, layering, particle size and shape, alluvial or cohesive), the flow and the geometry of the bridge pier and stream and end with an empirical formula. A literature survey of the various formulas for estimating scour depth was published by Breusers et al. (1977) and Raudkivi and Sutherland (1981). The formulas give widely differing estimates. These disparities have been discussed by Melville (1974), Anderson (1974) and Hopkins et al. (1983).

9.3.1 Flow pattern at a cylindrical pier

The *flow pattern past a cylinder*, protruding vertically from a horizontal plane boundary, in uniform open channel flow is complex in detail, and the complexity increases with the development of a scour hole at the base of the cylinder. Studies of flow patterns for such a case have been reported by Hjorth (1972,1975), Melville (1975), and Melville and Raudkivi (1977). The component features of the flow pattern are the downflow in front of the cylinder, cast-off vortices and wake, boundary layer, horseshoe vortex and bow wave, as shown diagrammatically in Fig. 9.7.

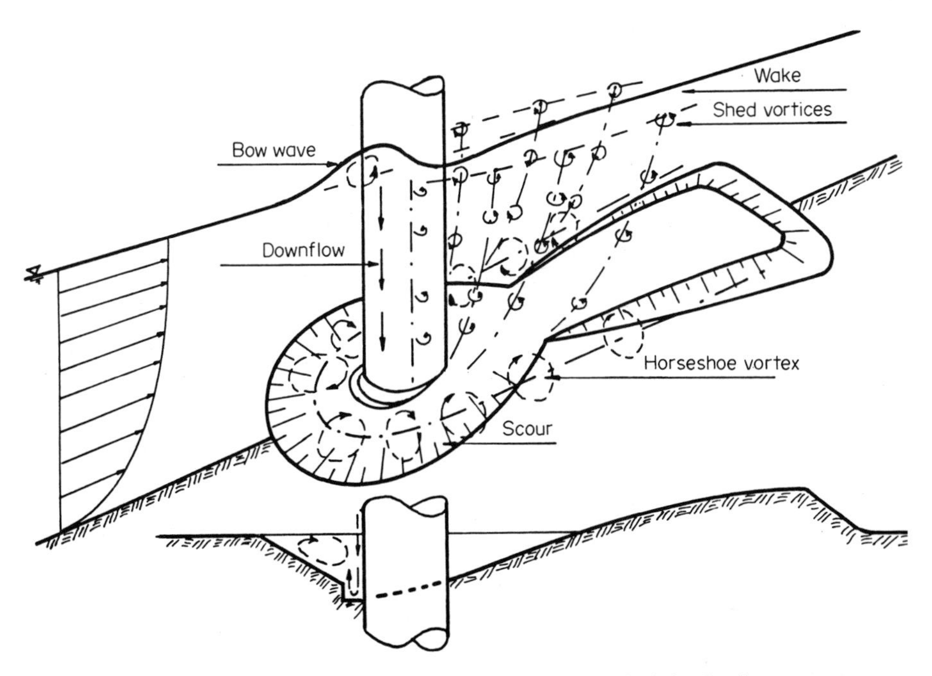

Fig. 9.7. Diagrammatic flow pattern at cylindrical pier.

The approach flow goes to zero at the upstream face of the cylinder, in the vertical plane of symmetry, and since the approach flow velocity decreases from the free surface downward to zero at the bed, the stagnation pressure, $\rho u^2/2$, also decreases. This downward pressure gradient drives the downflow. The downflow, in the vertical plane of symmetry, has at any elevation a velocity distribution, with zero in contact with the cylinder and again some distance upstream of it. The v_{max} plotted in Fig. 9.8 based on experimental data by Ettema (1980), is the maximum distribution at any elevation and occurs at 0.05 to 0.02 cylinder diameters upstream of it, being closer to the cylinder lower down. The maximum downflow velocity at a cylinder protruding from a flat bed ($d_s/b = 0$) is about 40% of the approach flow mean velocity U. The maximum of v_{max}/U occurs when the depth of scour is about 2.3 times the pier diameter or more. This maximum is located at about one pier diameter below the approach flow bed and is about 80%. The dashed line in Fig. 9.8 shows the relationship for the maximum downflow distribution as proposed by Shen et al. (1969) for the same constant approach flow velocity U.

The so-called horseshoe vortex develops as the result of separation of flow at the upstream rim of the scour hole; it is a lee eddy similar, for example, to the eddy or ground roller downstream of a dune crest. The horseshoe vortex is a consequence of scour, not the cause of it. The horseshoe vortex extends downstream, past the sides of the pier, for a few pier diameters before losing its identity and becoming part of general turbulence. The horseshoe vortex also pushes the maximum downflow

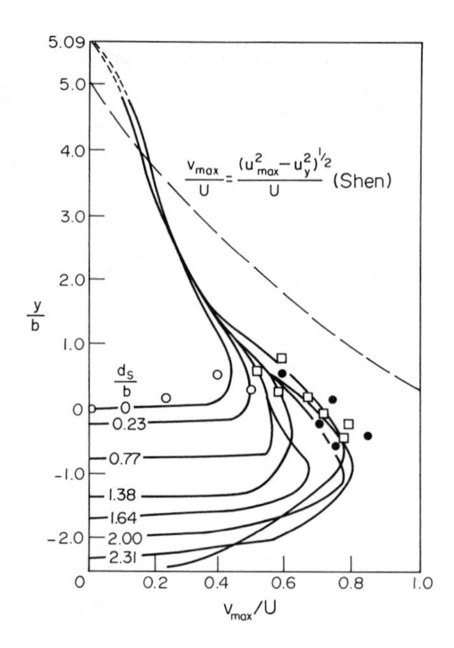

Fig. 9.8. Downflow velocity v_{max} in front of pier according to Ettema (1980) with points from measurement by Melville (1975) and relationship (dashed line) by Shen et al. (1963) for constant approach flow velocity U.

velocity within the scour hole closer to the pier.

The stagnation pressure causes not only downflow but also sidewards acceleration of flow past the cylinder. The separation of flow at the sides of the cylinder creates the wake with the cast-off vortices at the interfaces to the main stream. The cast-off vortices are translated downstream with the flow and interact with the horseshoe vortex at the bed causing it to oscillate laterally and vertically. The cast-off vortices also act as little tornadoes lifting sediment from the bed.

A bow wave develops at the surface with rotation in the opposite sense to that in the horseshoe vortex. The bow wave becomes important in relatively shallow flows where it interferes with the approach flow and causes a reduction in the strength of the downflow.

9.3.2 Scour in uniform sediment

When the velocity in a laboratory experiment over a flat granular bed with a cylindrical pier is gradually increased, scouring starts at the flanks of the pier when the approach velocity U or its shear velocity u_* reaches about half the threshold value for the beginning of transport of the given sediment, U_c or u_{*c}. The scour propagates upstream around the perimeter and soon meets in front of the pier. The downflow is the primary scouring agent and acts as a vertical jet. A groove forms around the front perimeter of the pier in which the jet is turned 180°. The upward flow combines with the horseshoe vortex, the lee eddy in the scour hole, and the spiraling motion carries the eroded sediment past the pier. During the active erosion stages, the upstream face of the groove is almost vertical with a sharp rim to the slope of the scour hole. Parts of this rim collapse in a random, irregular fashion, sending local avalanches of sediment into the groove from where it is removed by the downflow.

Under clear-water scour conditions, $u_* < u_{*c}$, the equilibrium depth of scour is reached very slowly. In laboratory experiments, reported by Raudkivi and Ettema (1977), typically about 50 hrs of continuous running produced, at near threshold conditions, a scour depth which did not change measurably thereafter. As reported in the above reference, the maximum equilibrium scour depth depended on whether the sediment was ripple or non-ripple forming, uniform or non-uniform.

The general features of local scour at a cylindrical pier in uniform sediment and subcritical flow conditions are illustrated in Fig. 9.9 for conditions which are not affected by flow depth or sediment size relative to the pier size. These effects will be discussed later. The use of velocity ratio U/U_c or u_*/u_{*c} instead of a Froude number collapses data from experiments with different sediment sizes. The effect of flow depth relative to pier size can be described separately. At near threshold conditions the equilibrium scour depth in non-ripple forming sediments reaches a maximum of about 2.3b, where b is the diameter of the cylindrical pier or thickness of a pier. With ripple-forming sediments ($d < \sim 0.7$ mm) it is not possible to maintain a flat sand bed; ripples develop on the bed of approach flow and a small amount of sediment transport takes place, i.e., live-bed conditions develop before the condition $u_* = u_{*c}$ is exceeded.

After the onset of live-bed conditions the equilibrium scour depth initially decreases and then increases again, reaching a second peak at transition flat bed conditions, and its magnitude fluctuates with time in response to the passage of bed features past the pier. *The equilibrium depth now must be defined in terms of a time-average scour depth*, not a maximum. The range of the fluctuations in scour depth is a function of the size of bed features, which is a function of flow depth, not pier size. The range from most laboratory data is 0.75 of dune height, h, or approximately ±0.5h. The fluctuations of scour depth do not vanish at transition flat bed conditions because of the avalanching of sediment into the scour hole. The minimum scour depth occurs approximately when the dunes have a maximum steepness, and in that region also the functions of scour in ripple and non-ripple forming sediments merge. The initial drop in the local scour depth reflects the very rapid increase in sediment transport rate with bed shear stress when the latter is only slightly in excess of τ_c, whereas the stagnation pressure and local scouring power increase only gradually. The subsequent increase in scour depth with bed shear stress indicates that the scouring power increases faster than the sediment transport rate. However, this increase is at least partly due to the longer

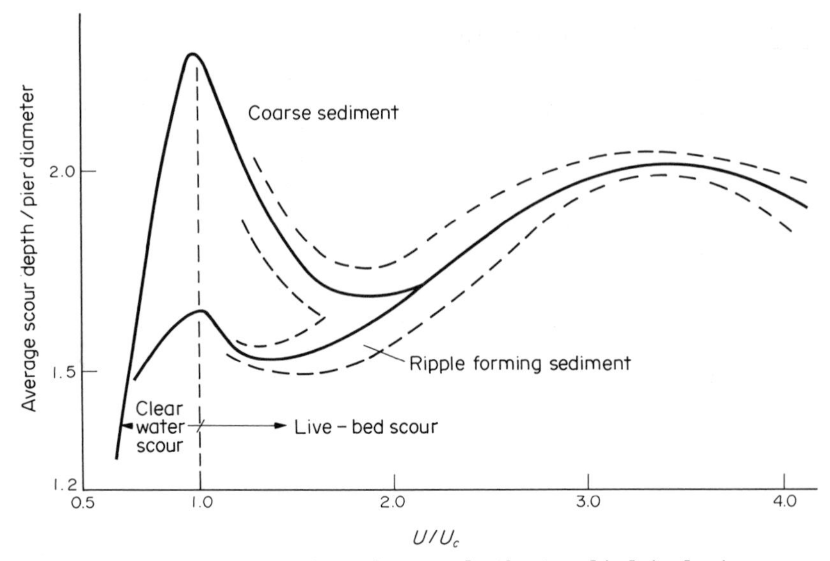

Fig. 9.9. Variation of scour depth at cylindrical pier in uniform sediment and in deep flow relative to pier width.

troughs of the bed features at higher shear stress excesses, which give more time for the scour hole to deepen before the next crest arrives. Only a small number of data points are available for scour depth beyond the transition flat bed condition, and these show a decrease from the transition flat bed peak value.

Laboratory data relating to the above have been presented by Raudkivi (1986), Chiew (1984) and Melville (1984). The scour depths shown are relative to local bed level which may additionally be subject to general and/or constriction scour. Further local variations in bed level arise from three-dimensionality of flow and associated bed features. Local variations in bed level may also arise from general erosion of river beds during a flood, for example, in relative narrow valleys where water depth variations are large. Gravel rivers usually have a network of moving channels and gravel banks. These banks lead to partial bifurcation of flow at their upstream ends and confluence at the downstream ends. Particularly deep local channels can develop at such confluences due to spiralling flows, like at the outer banks of river bends. The lowering of the channel bed locally could be substantial. Field measurements at Ohau River in New Zealand, Thompson and Davoren (1983), showed a local channel depth of about 3 m in a broad gravel river with an average depth of the order of 0.75 m and a mean velocity of 2.38 ms^{-1} (bed material d_{50} = 20 mm, d_{65} = 35 mm d_{84}/d_{50} = 3.5. This definition, instead of $(d_{84}/d_{16})^{\frac{1}{2}}$, was used because the coarse fractions of sediment are dominant). The local scour depth at a 1.5 m diam cylindrical pier was an extra 1.28 m.

9.3.3 Effect of sediment grading

Raudkivi and Ettema (1977) showed that sediment grading has a strong influence on the equilibrium depth of clear-water scour. Figure 9.10 shows the relationship of the maximum clear-water scour depth to a geometric standard deviation of sediment grading, $\sigma_g = (d_{84}/d_{16})^{0.5}$. The ordinate, K_σ, is the ratio of equilibrium scour depth in graded sediment to that in uniform sediment. It is seen that the clear-water scour in typical river gravels with $\sigma_g \sim 3.5$ is only about 20% of the depth in uniform sediment.

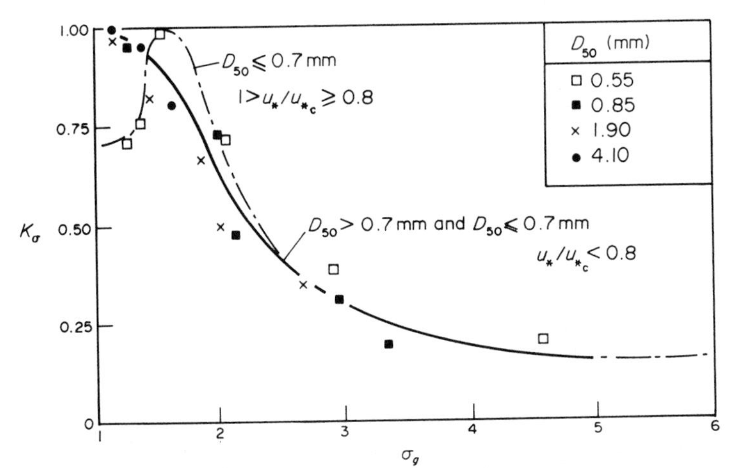

Fig. 9.10. Ratio, K_σ, of local clear-water scour depth in graded sediment to that in uniform sediment as function of geometric standard deviation of sediment size distribution.

The effect of sediment grading on the depth of live-bed scour is considerably more complex. Figure 9.11 shows the mean scour depth function for non-uniform sediment ($\sigma_g = 3.5$) with that for coarse uniform sediment, as shown in Fig. 9.9. The function for the graded sediment was fitted to data obtained by Chiew (1984) with a 45 mm diam pier, sediment with $d_{50} = 0.80$ mm, and a flow depth of 170 mm. Experimental points have been omitted. However, the spread of data points was generally less than 0.2 d_s/b, and less in individual series. The depth of scour when $U/U_c \cong 1$ is given by the clear-water relationship (Fig. 9.10), e.g., 0.23 x 2.3b = 0.53b, where U_c is based on d_{50}. However, the d_{50}-size of a non-uniform sediment is only an order of magnitude indicator; selective sediment transport commences at lower than threshold values of velocity for the d_{50}-size and a process of armouring of the bed surface commences. The armouring gradually increases the effective critical velocity or shear velocity of the bed surface and leads to a deeper equilibrium scour depth, up to the limiting critical shear velocity of the armoured surface, u_{*a}. At shear velocities greater than u_{*a}, the armour layer fails and the bed surface does not armour anymore.

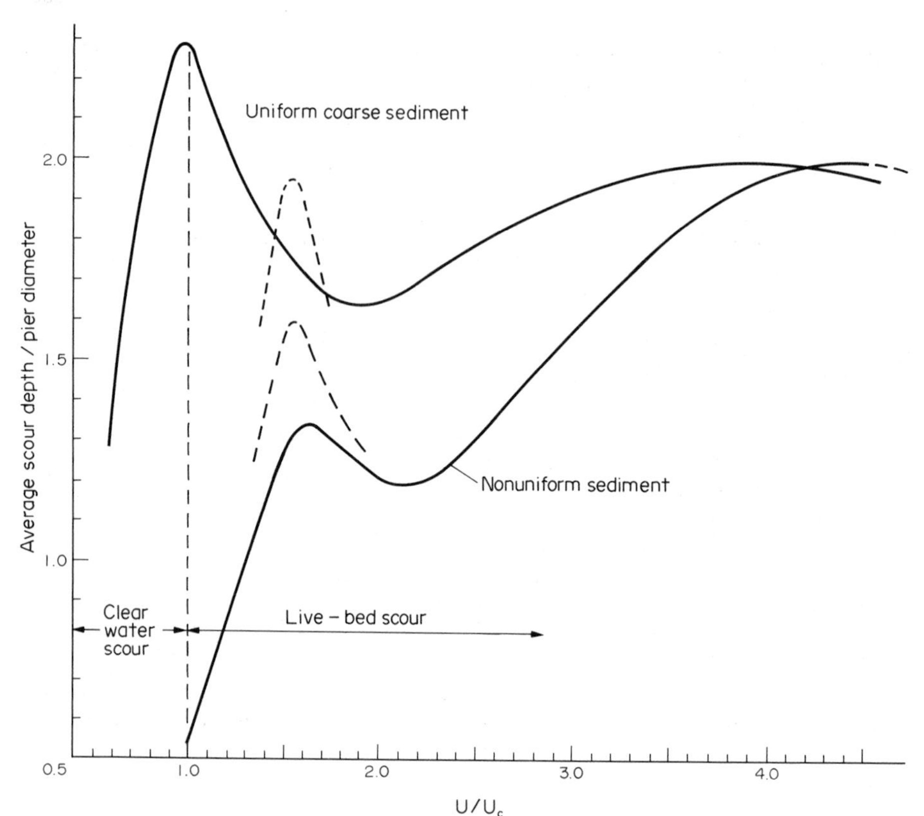

Fig. 9.11. Average local equilibrium scour depth at 45 mm diameter pier in uniform (d_{50} > 0.7 mm and σ_g < 1.3) and non-uniform ($\sigma_g \cong 3.5$) sediment with d_{50} = 0.8 mm in 170 mm deep flow in sediment re-circulating flume (dashed peak refers to scour when bed is armouring and sediment transport goes to zero; dotted line shows adjustment of dashed line for relative grain size effect).

At shear velocities $u_{*c} < u_* < u_{*a}$ the sediment transport could be unsteady, approaching zero as the armour layer stabilizes, or it could be in an equilibrium state. The latter exists when there is a continuous supply of sediment from upstream which moves over the armoured bed. The critical shear velocity of an armour layer is ill-defined but indicative values are given in Section 5.4.

If the flow conditions persist long enough at the critical shear velocity of the armoured surface and the sediment transport rate goes to zero, the local scour depth approaches that for maximum clear-water scour conditions. Since clear-water scour conditions prevail, U_c goes to U_{ca} of the abscissa as U/U_{ca}. The local scour depth under an equilibrium state of sediment transport depends on the amount of finer sediment in transport and on the grading of the parent sediment, and is less than the clear-water peak value. The full line in Figure 9.11 for non-uniform sediment shows the first or armour peak for the equilibrium state. The peak indicated by the dashed line is for the case when sediment transport goes to zero. The peak shown in the dotted line indicates an adjustment for the relative grain-size effect described later.

On passing the critical shear stress, τ_{ca}, the surface layer fails and the sediment transport rate increases rapidly, leading to a reduction of local scour depth, analogous to transition from clear-water scour to live-bed scour.

After the initial decrease, the scour depth starts to increase again with increasing applied shear stress, up to the transition flat bed condition. First, the coarser fractions of sediment are still effective in armouring the bed of the local scour hole but their effectiveness decreases with increasing bed shear stress and vanishes at the transition flat bed conditions altogether, that is, if the largest stones in the sediment are not large relative to the width of the groove excavated by the downflow (not larger than about one-tenth of the pier diameter).

Limited laboratory data indicate that the effect of sediment grading is too small to be separated from the usual scatter present when $\sigma_g < 2$ and the data follow the uniform sediment function. On the other hand, field data indicate that fluid shear stresses in gravel rivers during floods seldom exceed about double that of the critical armour layer shear stress. Thus, if prolonged periods of flow at just below τ_{ca} are not likely, the actual scour depths observed may be appreciably less than the peak values, according to Fig. 9.11.

The critical velocity over a limiting armour layer, U_{ca}, can be estimated with the aid of the d_{50}-size of this armour layer, d_{50a}. According to the studies by Chin (1985) the maximum value of d_{50a} is

$$d_{50a} = d_{max}/1.8 \tag{9.16}$$

where d_{max} is the characteristic maximum size of the bed material and has to be estimated, as discussed in Section 5.4. The limiting armour layer d_{50a}-size yields from the Shields criterion the u_{*ca}-value and the corresponding critical mean velocity could be estimated from

$$U_{ca} = u_{*ca}[5.75\log(y_o/2d_{50}a) + 6] \tag{9.17}$$

The critical velocity at which the armour peak occurs when there is some sediment transport over the armour layer was found by Baker (1986) to be about 0.8 U_{ca}. The peak then occurs at $[U - (0.8\ U_{ca} - U_c)]/U_c = 1$, where U_c is the critical velocity for the bed material with d_{50}.

9.3.4 Effect of pier to sediment size ratio

It was shown that the equilibrium depth of clear-water scour is affected by the relative size of the pier to sediment, b/d_{50}. Figure 9.12 combines the above data for clear-water scour, involving six pier sizes and sediment from 0.24 mm to 7.80 mm, with data obtained by Chiew (1984) and Chee (1982) for live-bed scour. The abscissa in Fig. 9.12 gives the ratio of scour depth affected by the relative size b/d_{50} to that independent of the relative size. The results show that the local scour depth becomes independent of sediment size for $b/d_{50} > 50$. For smaller ratios the scour depth decreases because the sediment becomes too large relative to the width of the groove excavated by the downflow. The coarser grains, which accumulate on the bed of the groove, make it also more porous and allow more of the downflow to penetrate and dissipate its energy in the bed. The grain-size effect would be rare in nature but is quite frequent with laboratory data.

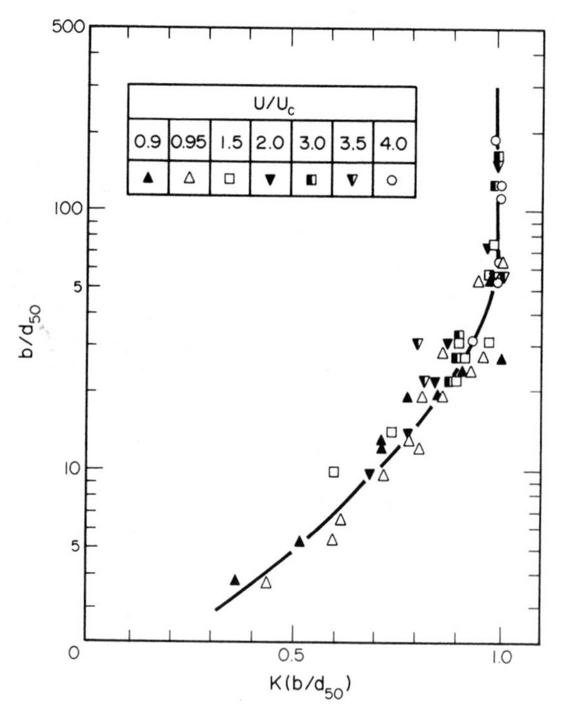

Fig. 9.12. Ratio, $K(b/d_{50})$, of local equilibrium scour depth affected by relative size b/D_{50} to that independent of relative size as function of b/d_{50}.

The pier size itself affects the time taken for the equilibrium depth to be reached since the volume of the scour hole around the upstream half of the perimeter increases proportional to the pier diameter cubed.

9.3.5 Effect of flow depth

There are several references in the literature stating that the scour depth is affected by the flow depth relative to the pier size; for example, the scour depth becomes independent of flow depth greater than three pier diameters. A decreasing depth of flow leads to increasing prominence of the surface roller (bow wave), which interferes with the downflow and reduces its strength. No exact dependence has been obtained for the relative flow depth effect but data collected, both at clear-water scour conditions and live-bed scour at transition flow bed conditions, are shown in Fig. 9.13 as relative flow depth y_o/b versus K_y, where K_y is the factor by which the scour depth at the given pier in relatively deep water (unaffected by flow depth) is to be multiplied to obtain the scour depth in shallow flow. For $b/d_{50} > 75$ and $y_o/b < 3$ the K_y factor is approximately given by

$$K_y = 0.77\ (y/b)^{0.21} \tag{9.18}$$

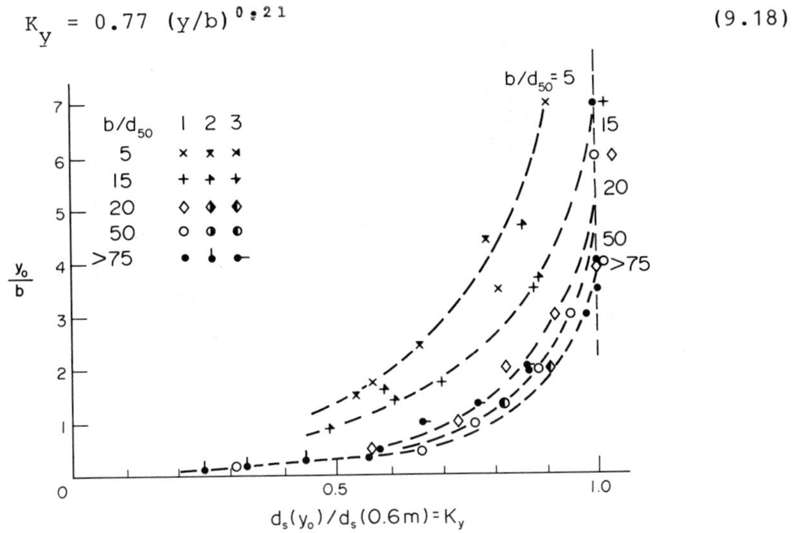

Fig. 9.13. Ratio, K_y, of local equilibrium scour depth in shallow water to that in deep water relative to pier size (independent of depth) as function of y_o/b with relative sediment size as parameter (series 1 are clear-water scour data).

9.3.6 Effect of alignment and shape

For pier shapes other than cylindrical, the depth of local scour depends on the alignment of the pier with flow. The local scour depth is related to the projected width of the pier, and this width increases rapidly with the angle of attack of flow. With an increasing angle of attack, the location of maximum scour depth moves along the exposed side of the pier from the front to the rear end of the pier. The downflow on the exposed side combines with velocity into a strong spiralling current, which is a very powerful scouring agent, and excavates a deep hole near the end of the pier. In general, angles of attack greater than 5-10°

should be avoided. Where this is not possible the use of a row of cylindrical columns would produce a shallower scour; for example, with five-diameter spacing the local scour can be limited to about 1.2 times the local scour at a single cylinder. The graph of multiplying factors for the depth of local scour as a function of angle of attack and the length-to-thickness ratio of the pier by Laursen and Toch (1956) is still commonly in use and indicates right orders of change, Fig. 9.14.

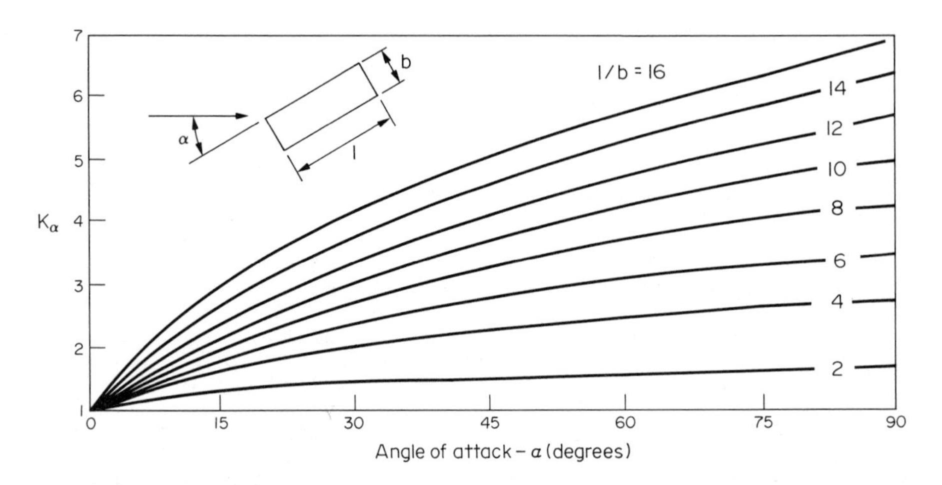

Fig. 9.14. Alignment factor K_α for piers aligned with flow.

There are also multiplying factors for pier shape, ranging from about 0.7 to 1.2, but the shape effect will be overshadowed by small changes of alignment or even a little debris caught on the pier.

Apart from the cross-sectional shape of the pier, the slope of the leading edge affects the scour depth. For example, laboratory results show that a 22½° slope from vertical in the upstream direction increases the scour depth by about 20% whereas the same slope in the downstream direction decreases the scour depth by about 25% because the flow "spills" sideways.

The problems of local scour are frequently complicated by the presence of footing, a caison, or capping blocks on piles. The data on scour depths is predominantly based on well defined shapes, such as cylinders, rectangular piers, etc without the three-dimensional effects that for example, a capping block may cause.

A footing or caison with the top below the general bed level, and a diameter 1.6 times or more of that of the pier, Fig. 9.15, is effective in reducing the local scour depth by interception of the downflow. However, if the top of the caison comes to bed level, or even above, the scour depth is increased, i.e. the local scour depth is now essentially governed by the diameter or

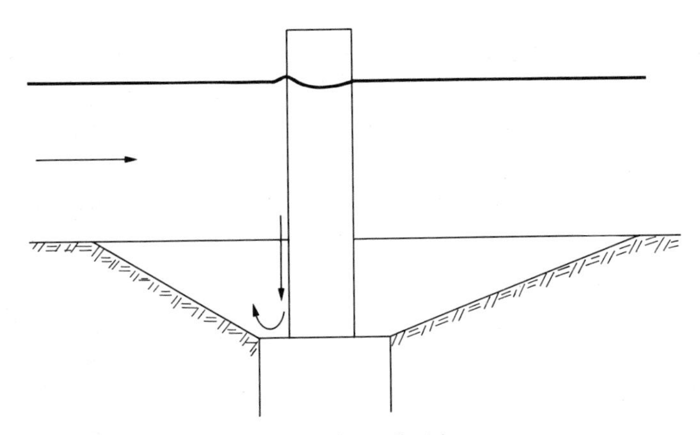

Fig. 9.15. Diagrammatic caison footing.

projected width of the caison rather than of the pier. The bed level here means the level of the troughs of the largest bed features. The footing effectively limits the scour depth. The practical difficulty is the determination of the "lowest bed level" which is a highly variable quantity. The bed features vary in size and shape and three-dimensionality can cause localised low spots, or the river in a flood may erode (cut) its bed in the given stretch of river, or the location of the pier may coincide with a moving stream channel in the cross-section. Thus, reliance on such a method of scour reduction is dangerous. Tapering of the transition to pier reduces scour depth.

9.3.7 Local scour in layered sediments

Many sedimentary deposits are non-homogeneous and frequently, distinct layers are present. In some instances, coarse sediments cover deposits of fine grained sediment. It is of great practical interest to be able to predict what happens when the local scour at the bridge pier penetrates the layer of coarse sediment but very little in field or laboratory data is available on this aspect.

Deeper than usual scour holes have to be expected when the local scour penetrates the top layer of coarse sediment and exposes a layer of very fine sediment, particularly if this is associated with oblong piers misaligned with flow. Under those conditions the highly turbulent flow in the wake of the pier can suspend and carry away large quantities of fine sediment from the scour hole. Such a case occurred at the Bulls Bridge on the Rangitikei River in New Zealand in 1973. The scour depth exceeded 12m. The top gravel layer was about 1m thick. The pier was about 2m thick and 10m long and at the time of failure angled to approach flow by about 45°. Note that the failure depth was only about twice the effective pier width.

9.3.8 Local scour under unsteady flow

Unsteady flows occur during the passage of flood waves in rivers, demand surges downstream of hydro-power stations and under wave action. These are all waves but of very different wave lengths. The passage of waves increases the number of variables affecting the local scour. The waves may be translating, like a flood, or oscillatory like the wind waves. The scour may be due to the orbital velocity or it may be superimposed on a current.

Flood waves frequently have sufficiently low rates of change so that the flow could be approximated to a constant value over several hours. With very rapidly changing flood flows the duration of the flood may become significant, i.e. when the time is too short for completion of the scouring to a steady state equilibrium depth. The rates of scouring observed in laboratory were discussed by Raudkivi and Ettema (1977).

The oscillatory wave on its own leads to relatively small net scour depth at a cylindrical pile, although the temporary scour with fairly long waves may be significant (Raudkivi (1976)). Zanke (1981) introduced the parameter A/b, where $A = H/\sinh kh$ is the double amplitude of the orbital motion at the bed and b is the diameter or width of structure. This parameter combines the depth of flow, wave height, wave length and size of structure in a descriptive manner. For values of $A/b > 100$, the conditions of steady flow are approached, provided the bed shear stress generated is substantially greater than the critical value for sediment entrainment. The net local scour depth, d_s/b, decreases rapidly with decreasing A/b, and the maximum depth moves to the flanks. The instantaneous scour depth decreases with increasing b since the half period of the orbital motion is too short to transport large volumes of sediment. With waves superimposed on a current, the net local scour depth approaches the values due to the current only, except that net scour develops at bed shear velocities down to about a third of the critical value for entrainment, i.e. as long as the mean shear stress due to orbital velocities is higher than the critical value.

9.3.9 Scour protection

The idea of bed protection and prevention of scour at the pier has attracted a good deal of attention. Reduction of scour depth would mean shallower foundations and reduced cost.

The most common method of scour "prevention" is the dumping of stones on the river bed around the pier, a rip-rap protection, or dumping of stones into the scour hole around the pier. A rule of thumb is that the width of the protection measured from the pier should be at least twice the width (projected) of the pier. Bonasoundas (1973) recommended a rip-rap protection for a cylindrical pier in the shape of longitudinal cross-section of an egg, with the blunt end facing the flow. The overall recommended width is 6b and length 7b of which 2.5b is upstream of the upstream face of the pier. The recommended thickness of rip-rap is $b/3$ and minimum stone size is $d(\mathrm{cm}) = 6 - 3.3U + 4U^2$ where U is in m/s.

In general, scour starts at the pier at about half the threshold velocity of the sediment on the upstream bed. Hence, the critical velocity for the rip-rap is $U_c \cong 2U$ where U is the mean approach flow velocity at design discharge. The combination of Strickler and Manning formulae yields $U/u_* = 7.66(R/d)^{1/6}$ and the entrainment conditions according to the Shields criterion are given by $\theta \cong 0.04$. Thus, substituting for the hydraulic radius R, the flow depth y_o leads to

$$U = 4.8(S_s - 1)^{1/2}d^{1/3}y_o^{1/6} \cong 6d^{1/3}y_o^{1/6} \tag{9.19}$$

and for stones with specific gravity $S_s \cong 2.6$, where d is measured in m. These stone sizes are smaller than recommended in codes of practice which incorporate safety factors. A simple working relationship is

$$U \cong 4.9\sqrt{d} \tag{9.20}$$

from empirical data.

Rip-rap is successful where the river bed level does not vary much, that is where the trough elevations of largest bed features can be estimated with confidence, and the rip-rap can be placed at the trough elevation. Such rock protection is best placed in a pre-excavated scour hole. Where the bed is subject to substantial lowering (cut) during a flood, the rip-rap protection could be destroyed by undermining from the sides as the river bed around the protection is lowered.

The rip-rap should further be placed on a good inverted filter or a suitable filter matt to prevent the winnowing out of the bed material and disintegration of the protection.

If the rip-rap is piled high around the pier, it may become part of a pier of larger diameter and actually aggravate conditions. Neill (1964) reported a case where "the deep-water piers had been surrounded by large heaps of stone extending up to nearly low-water stage, about 10 m above the bed. One pier collapsed in a flood, apparently as a result of undermining from downstream". Such a footing presents a much wider "pier" to the flow and the severity of erosion will inevitably increase.

Rip-rap protections become useless in rivers with substantial internal stream channels which move about as, for example, in braiding rivers.

Chabert and Engeldinger (1956) investigated the effectiveness of an array of piles in front of a pier for scour reduction. The effectiveness is a function of the number of piles, their protrusion and spacing from each other and from the pier, and the angle made by the two lines of piles. The authors reported up to 50% reduction in scour depth but gave no design criteria other than model tests. Again, such a protection is of value when the general bed level during a flood can be predicted with confidence, and where the bed level movements are relatively small.

9.3.10 Scour at pile group piers

A bridge engineer has direct control over the geometry of his bridge foundations. It is, therefore, important that he has an appreciation of the influence exerted by different possible configurations, e.g. circular or rectangular piers or pile groups. Scouring at piers has been more thoroughly investigated than that at pile groups. Useful insight to this aspect of local scour is provided by an investigation by Hannah (1978) who studied local scour at groups of cylindrical piles with steady uniform flow and clear-water conditions. A series of tests was first performed on single piles and with a pier (length:width = 6:1) to provide a basis against which pile group scour could be evaluated. Pile groups of various spacings and with different angles of attack were then investigated for one flow condition. The sediment used in all tests had a d_{50} = 0.75 mm and σ_g = 1.32.

Tests showed that scour depths were 80% of equilibrium scour depths after seven hours. Further, after seven hours, only minor changes occurred in scour and deposition patterns. Consequently, all tests reported herein, were run for seven hours. This contrasts with the work of Dietz (1972,1973) in which the tests were stopped after two hours.

Single Pile

For a cylindrical pile 33 mm in diameter, the scour depth after seven hours was 62 mm with a flow of 0.285 m/s (92% of threshold) at a depth of 140 mm. All pile group measurements were made with this same flow velocity and depth so that the results could be compared directly with those of a single pile.

Pile Groups

Four of the mechanisms which affect pile at scour groups are not present in scouring at a single pile:

(i) Reinforcing: Causes increased scour depths at the front pile and is a consequence of the dynamic equilibrium which exists in the base of a scour hole when stable conditions have been reached. Bed material is continuously lifted from the base of the hole by the flow which is not, however, capable of removing this material from the scour hole. Should a downstream pile be so placed that the scour holes overlap, then the bed level is lowered at the rear of the upstream scour hole. It is thus, easier for the flow to remove material from this hole and it deepens. As pile separation increases, the reinforcing effect reduces and disappears when the maximum bed level between the piles returns to the undisturbed bed level.

(ii) Sheltering: The presence of an upstream pile can cause a reduction in the effective approach velocity for downstream piles. This reduction decreases the effect of the 'horseshoe vortex' and thereby reduces scour at downstream piles. A second form of sheltering occurs if material scoured from the upstream pile is deposited on the bed in front of the downstream pile. Flow is then deflected up from the bed and around the downstream pile. The strength and thus the effectiveness of the 'horseshoe vortex' at this pile is thereby reduced. As pile separation increases, the velocity deficit in the wake of the upstream pile disappears and the sheltering effects will become negligible.

(iii) Shed Vortices: Vortices shed from an upstream pile are convected downstream following paths as described for piers. Should a second pile be placed close to one of these paths, the vortices, by virtue of their velocity and pressure distributions, assist in lifting material from the scour hole. The scouring potential of the shed vortex is a function of the intensity of its convection speed and of the distance between the path and the effected pile. This effect will, therefore, reduce more rapidly for piles in line with the flow than for those at angles of attack which place downstream piles on the paths traced by vortices shed by upstream piles.

(iv) Compressed 'Horseshoe Vortex': When piles are placed transverse to the flow, each will have, except at very close spacings, its own 'horseshoe vortex'. As pile spacing is decreased, the inner arms of the 'horseshoe vortices' will be compressed. This causes velocities within the arms to increase with a consequent increase in scour depths. This effect will also occur for other (non-zero) angles of attack with its importance for any particular angle being strongly dependent on the pier spacing.

Two Piles

In Line (Angle of Attack 0°). Figure 9.16 shows how the scour depths at each of the piles and the bed level between the piles depend upon the relative spacing a/b, where a is the distance between the centre lines of the piles and b is the common pile diameter.

For two piles touching (a/b = 1), the scour depth at the front pile is the same as for a single pile but with increasing separation, the front pile experiences the reinforcing effect which reaches a maximum at a/b = 2.5 and is evident until a/b = 11. For larger spacings, the scour depth is the same as for a single pile.

The bed elevation between the piles is a minimum (d_m is a maximum), see Fig. 9.16, at a/b = 1 when it is the same as the base of the rear scour hole. With larger separation (up to a/b = 10), the bed level increases as some of the material from the front hole is deposited between the piles. At a spacing of a/b ≐ 10, the scouring in front of the second pile is balanced by deposition from the front scour hole and a section of the bed between the piles remains at its original level ($d_m = 0$). For larger spacings, $d_m < 0$, implying that deposition from the front hole has become

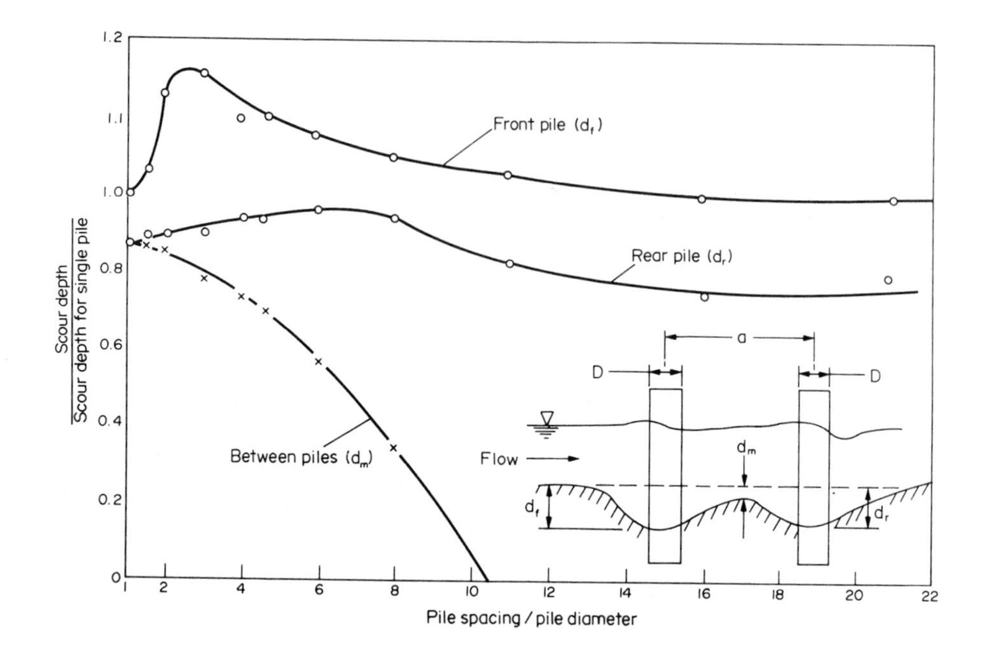

Fig. 9.16. Scour depths for two piles in line with flow as a function of pile spacing.

dominant. For these spacings, (a/b > 10), the reinforcing effect at the front pile is weak and must disappear completely as the bed profile assumes that arising from scour at a single pile.

Reduced scour depths at the rear pile result primarily from sheltering by the front pile reducing the effective approach velocity. At a/b = 1, the sheltering is complete with the two piles acting as an entity. The maximum scour depth in the resultant scour hole occurs at the front pile with the scour depth at the rear pile being only 87% of the maximum. As separation increases, the sheltering afforded by the front pile reduces and at a/b = 2, a 'horseshoe vortex' forms around the rear pile and increases in strength with separation. At these small separations, vortices shed from the front pile pass close enough to the rear pile to aid scouring. Scour depths at the rear pile thus increase with separation.

At a/b = 6, the scour depth at the rear pile reaches a maximum. At larger separations, the shed vortices pass further from the pile and are therefore less effective in removing material from the scour hole. In addition, the bed level between the scour holes is building up and interfering more with 'horseshoe vortex' formation at the rear pile. Scour depths reduce as a result of these two effects reaching an apparent minimum at a/b = 17. With further separation, only the wake sheltering effect remains and this progressively weakens. In these tests, it was still present at a/b = 25. Eventually, the pile will not be influenced by the upstream conditions and the scour depth will be that of a single pile.

With three piles in line at equal spacings up to a/b = 6, the scour at the middle pile was deeper and at the third pile, shallower than at the downstream pile of the two-pile group.

Transverse (Angle of Attack 90°). Figure 9.17 shows results of tests in which two piles were set at right angles to the flow. Both piles scoured to the same depth (± 2 mm) and the results have been averaged to define a single curve. The mid-distance scour is also shown.

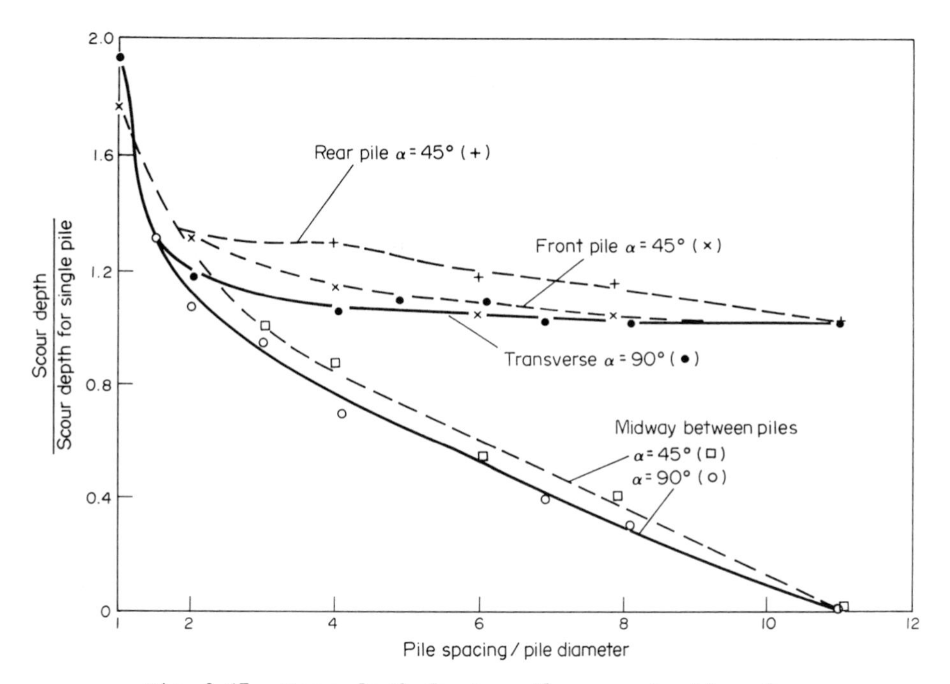

Fig. 9.17. Scour depth for two piles as a function of pile spacing. Transverse piles - solid lines. Angle of attack 45° - dashed lines.

For the twin pile, a/b = 1, the scour depth is 1.93 d_s. This is in accordance with the concept of scour depth being proportional to the frontal width of the pile. The scour depth reduces rapidly with separation to 1.3 d_s at a/b = 1.5 and with a/b greater than 8, the scour depth is essentially that of a single pile. The scour holes become separate entities at a/b = 11 when the bed midway between them regains its original level.

At close spacings, ($a/b < 2$), the increased scouring results from the increase in effective pile diameter. At $a/b > 2$, separate 'horseshoe vortices' form. Between the piles, these are compressed thus, creating higher velocities and greater scouring potential which reduces as a/b increases, reaching zero at $a/b = 8$.

Non-aligned (Angle of Attack 45°). Figure 9.17 shows results of tests with two piles having their line of centres at 45° to the approach flow. The twin pile scour depth was 1.77 d_s. This is slightly more than would be suggested by a linear dependence of scour depth on frontal width, namely 1.71 times.

Scour depth at the rear pile exceeded that at a front pile for all a/b between 1 and 11. At greater separations, the piles will act independently having scour depths equal to that of a single pile and the bed-level midway between the piles will be unaffected. Increased depths at the rear pile are caused by a combination of the action of the shed vortices from the front pile and compression of the 'horseshoe vortices' between the two piles. Evidently, these two processes overcome any sheltering effects and have their maximum influence at $a/b = 4$ where the difference between front and rear scour depths was a maximum.

Effect of Angle of Attack, α. Tests with $a/b = 5$ were done for $0° < \alpha < 90°$ at 15° intervals with the results shown in Fig. 9.18. Scour at the front pile was not very sensitive to angle of attack varying by less than 5% of its value at $\alpha = 0°$. Scour depths at the rear pile are much more sensitive to changes in angle of attack as the various scouring mechanisms, described above, come into effect.

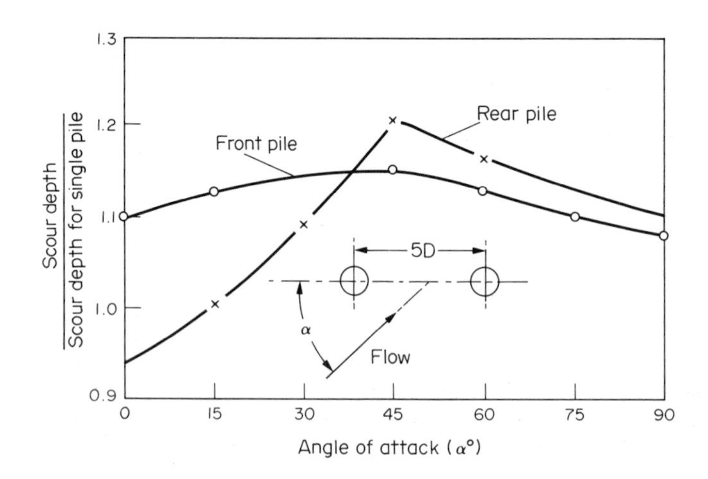

Fig. 9.18. The effect of angle of attack, α, on scour depths at two piles spaced five pile diameters apart.

At small angles, ($\alpha < 15°$), the dominant effect at the rear pile is sheltering by the front pile. As the angle increases, sheltering is reduced and the pile is affected by shed vortices. Consequently, scour depths increase reaching a maximum (for $a/b = 5$) at approximately 45°. Scour depths reduce when the pile moves clear of the shed vortices and approach that of a single pile as angle of attack approaches 90°.

9.3.11 Summary of pier scour estimation

As a general guide the expression

$$\bar{d}_s = 2.3\ K_\alpha b \tag{9.21}$$

gives the first order estimate of maximum average local scour depth in a flow which is deep relative to the pier width or diameter. Only when confident predictions can be made of flow velocities, depths and directions during a design flood are refinements warranted. The effect of pier shape is overshadowed by small angularities of the approach flow, i.e., K_α, and small debris rafts but flow depth effect could be significant.

This maximum scour depth will not occur at the clear water scour conditions in fine sand beds but can occur at transition flat bed conditions where the live-bed scour is only marginally less.

In gravel rivers the first peak (armour peak) of scour depth is a function of σ_g of sediment. Laboratory data indicate average scour depth $\bar{d}_s/b$ of 2 at $\sigma_g = 2$, reducing to about 1.6 at $\sigma_g = 5$. The transition flat bed conditions are rare in gravel rivers.

Note that the actual scour depth fluctuates due to passage of bed features and allowance has to be made for this as well as for any general or constriction scour.

9.4 Scour at Abutments and Spur Dikes

Groynes or dikes, and frequently the abutments of bridges, protrude into flow and lead to the formation of a substantial local scour. Since the scouring mechanism at the head of the groyne and at the abutment is the same discussion of this type of scour is done jointly.

Scour at short abutments has been likened to that at a pier, divided at the centreline in the direction of flow. There are some significant differences. The imaginary plane dividing the pier scour hole is a real boundary for the abutment with its associated boundary layer flow. This leads to a region of semi-stagnant water, a "dead-water region", upstream of the abutment, Fig. 9.19. The size of the region depends on the length and alignment of the abutment or spur. The curvature of the flow along the "interface" between the stagnation region and flow causes a secondary current which together with the flow leads to a spiral motion or vortex motion like in flow through river bends. This vortex in flows past an abutment is more localised and it

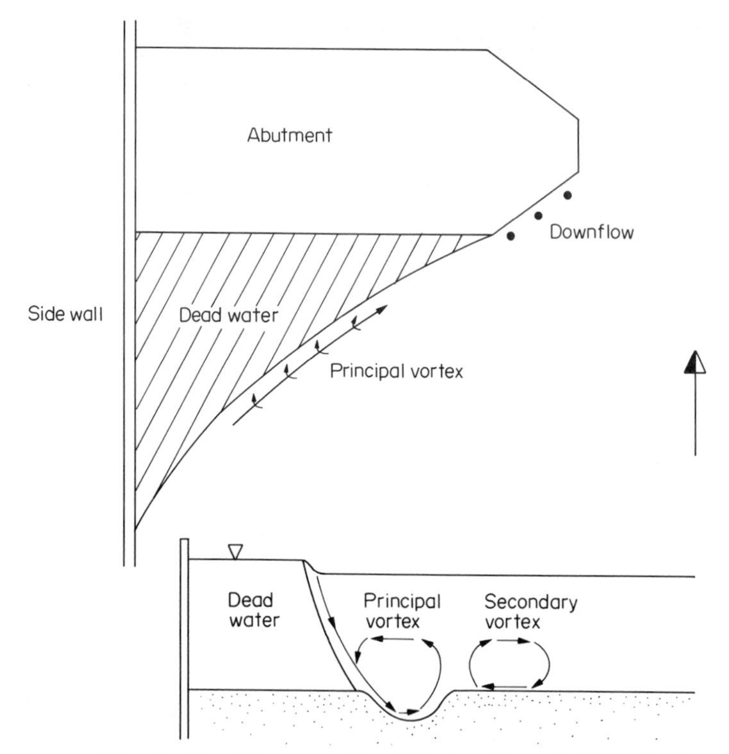

Fig. 9.19. Diagrammatic illustration of flow features past an abutment or spur dike.

has a strong scouring action. The vortex erodes a groove along its path, Fig. 9.20, and it also induces a complex system of secondary vortices. At abutments with wing walls the flow impinging on the wall also creates a downflow, like at bridge piers, which excavates a locally deepened scour hole at the wall. The effect of downflow is reduced with the spill-through type abutments. (Wing wall type has vertical walls with central piece parallel to flow and wings angled back; spill-through type has a sloping protected end rounded back.)

Prediction of the scour at abutments is routinely done with the help of the methods developed by Laursen (1958,1960,1963), Liu et al. (1961), Field (1971) or the regime depth formulae.

The Laursen method is an adaptation of the long contraction. The local scour depth at the inlet to the contraction, d_s, is assumed to be related to the contraction scour and is called the abutment scour, i.e. the contraction scour depth is set equal to d_s/r, which is equal to the d_s shown in Fig. 9.5. Only one side of the contraction is considered, the other boundary is assumed to be straight forming a $B = 2.75\ d_s$ wide long contraction. The channel flow in eqn (9.9) is then $Q_B/B = Q_C/2.75\ d_s$ and substituted in eqn (9.9) leads to

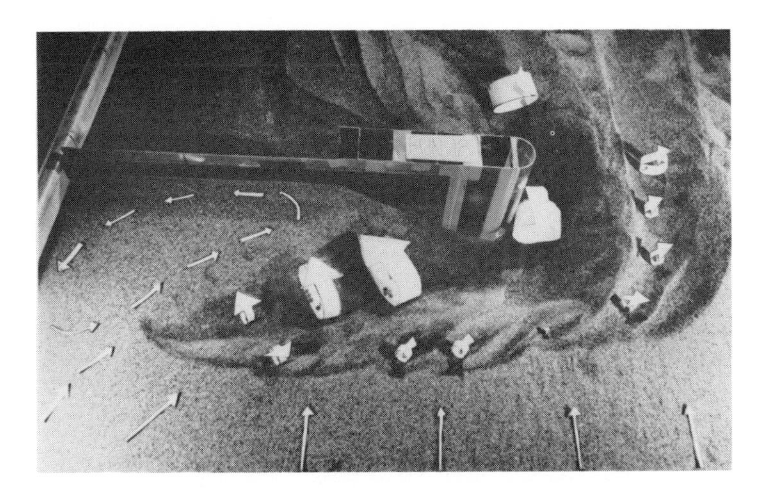

Fig. 9.20. Scour pattern at an abutment head. Arrows indicate flow path and vortex motion, Kwan (1984).

$$\frac{Q_o B}{Q_B y_o} = 2.75 \frac{d_s}{y_o}\left[\left[\frac{d_s}{r y_o} + 1\right]^{7/6} - 1\right] \tag{9.22}$$

Where Q_o is the flow over the flood plane on the side of the abutment, and Q_B that over the width B of the channel. The width B is estimated and adjusted to satisfy $B = 2.75 d_s$. Equation (9.22) is shown in Fig. 9.21. The curve defining the lower limit of the shaded area (r = 4.1) is recommended when overbank flow Q_o is small and there is no cross flow from the channel to the flood plain immediately upstream of the contraction.

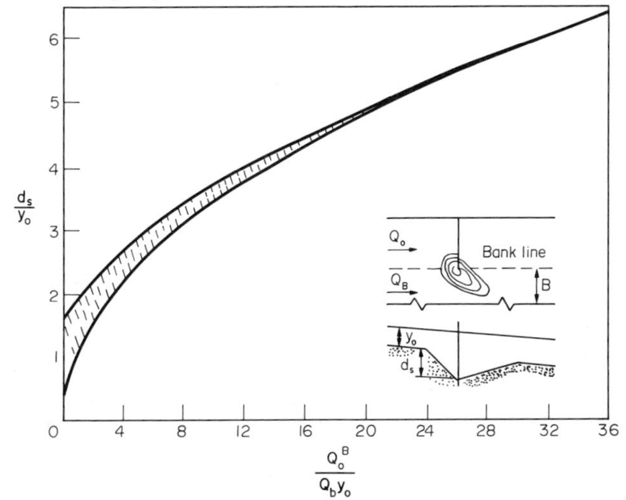

Fig. 9.21. Basic design curve for an overbank bridge constriction, Laursen (1958).

For abutments which encroach into the main channel the treatment is based on eqn (9.10). The effective length of the abutment is designated with L and it obstructs a flow Q_L. Using $Q_L/Ly_o = Q_B/By_o$; $B_1 = L + 2.75d_s$; $B_2 = 2.75d_s$; m = 0.59 and adjusting width B to $B_2 = 2.75d_s$ leads to

$$\frac{L}{y_o} = 2.75\frac{d_s}{y_o}\left[\left[\frac{d_s}{ry_o} + 1\right]^{1.7} - 1\right] \tag{9.23}$$

Equation (9.23) is shown in Fig. 9.22 for r = 11.5. For m other than 0.59 the d_s value is adjusted by a multiplication factor K_τ which accounts for changes in sediment transport as the ratio of u_*/w changes. Values of $u_*/w < 0.5$ imply sediment transport in contraction as bed load, $u_*/w = 1$ as saltation and $u_*/w > 2$ as suspended load. The Fig. 9.22 also shows a multiplication factor, K_θ, for alignment of the abutment relative to flow. However, the studies by Kwan (1984) show that the maximum equilibrium scour depths at the same protrusion distance from the bank occur at abutments perpendicular to the flow.

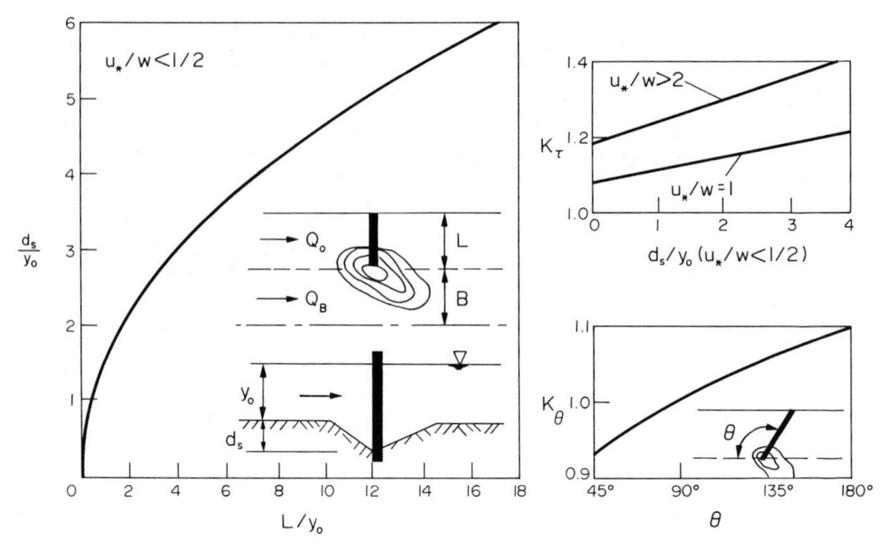

Fig. 9.22. Live-bed scour depths at abutments according to Laursen (1963), also showing correction factor K_τ for the type of sediment transport and K_θ for angle of attack.

Substitution of the width of the pier for 2L in eqn (9.23) leads to one of the Laursen formulae for pier scour.

An equation equivalent to eqn (9.23) for clear-water scour can be obtained from eqn (9.15) for the case when no general sediment transport takes place. Substituting $B_1 = L + 2.75d_s$ and $B_2 = 2.75d_s$ leads to

$$\frac{L}{y_o} = 2.75 \frac{d_s}{y_o}\left[\frac{(d_s/ry_o + 1)^{7/6}}{(\tau_o'/\tau_c)^{1/2}} - 1\right] \tag{9.24}$$

and is shown in Fig. 9.23. Substitution of b = 2L yields the Laursen formula for clear-water scour at piers.

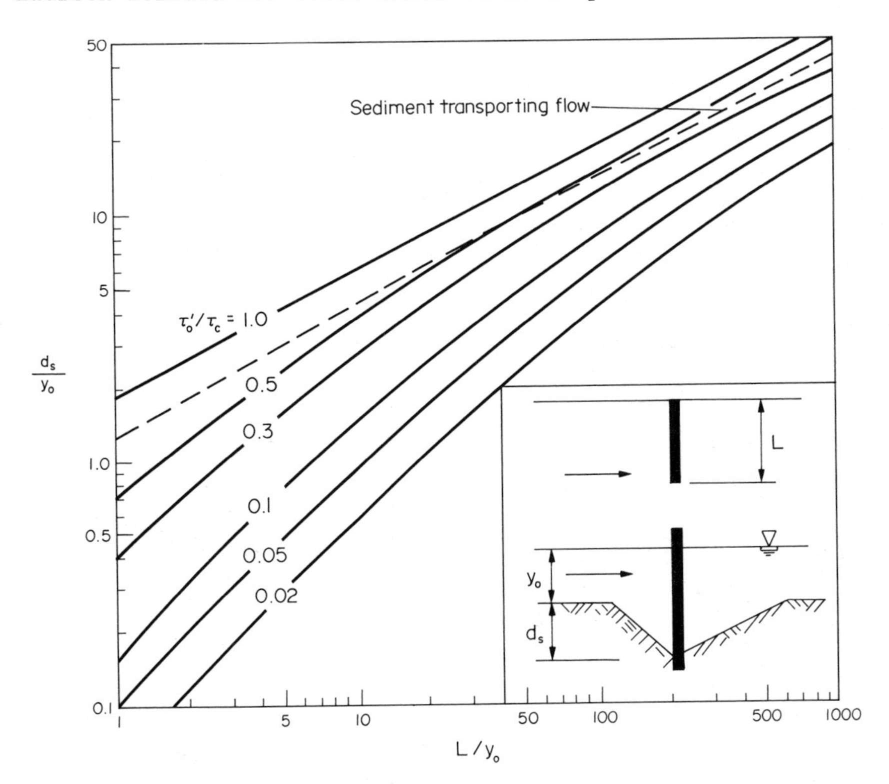

Fig. 9.23. Clear-water scour depths at an abutment according to Laursen (1963) formula.

Liu et al. (1961) proposed on the basis of dimensional analysis and laboratory tests

$$\frac{d_s}{y_o} = 2.15 \left[\frac{L}{y_o}\right]^{0.4} Fr^{0.33} \tag{9.25}$$

where L is the length of the abutment at bed level normal to flow. They recommend a 30% increase of d_s/y_o for maximum scour depth. From comparisons with field data, obtained at rock dikes on the Mississippi River, Richardson et al. (1975) limit the maximum equilibrium scour depth to

$$\frac{d_s}{y_o} = 4\ Fr^{0.33} \tag{9.26}$$

where for both formulae Fr is based on upstream flow depth and velocity.

Field (1971) prepared a design chart (Fig. 9.24) based on Liu's data. Field's parameters are

$$\frac{D_s}{y_o} M, \quad \frac{L}{y_o} M \text{ and } FrM^{1/2}$$

where D_s is scour depth measured from water surface, M is the opening ratio defined as the width of the bridge opening at half the normal depth of flow to the average top width of the approach channel (B'/B). Multiplying factors for abutment and location shape (Table 9.1), and for angle of attack are the same as those by Liu et al. The same design curves are said to be valid for clear-water using y_c and V_c in the formula. These are the depth and velocity in the approach channel at threshold conditions.

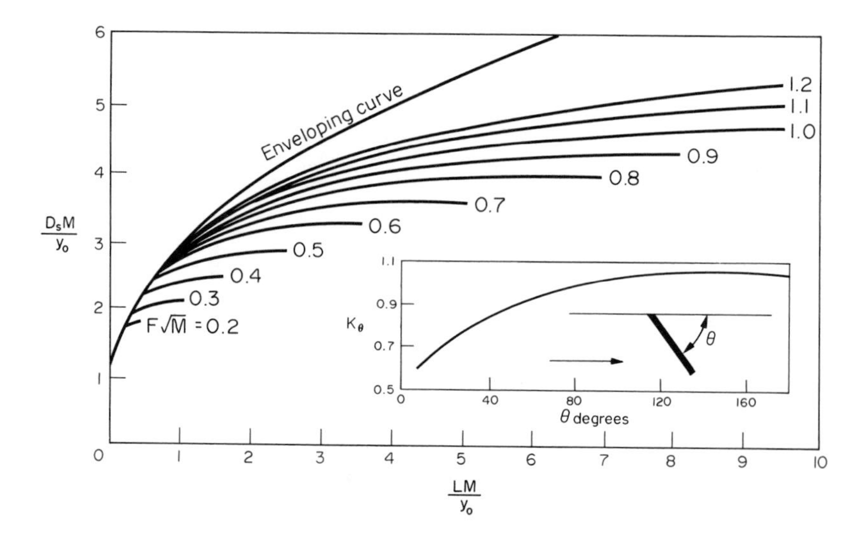

Fig. 9.24. Field's (1971) design chart for equilibrium scour depth, measured from the water surface. Inset: Multiplying factor K_θ for angle of attack.

Tey (1984) compared measured and predicted scour depths using Fig. 9.24 and found that nearly all measured depths were greater up to about 30%.

The regime methods of scour depth prediction at the head of a dike or an abutment rest on correlations with field observations. Inglis (1949) related the total depth $D_s = y_o + d_s$ to the three-dimensional regime depth of Lacey

$$y_{3r} = 0.47(Q/f)^{1/3} \tag{9.27}$$

Table 9.1. Multiplying factors for abutment shape and position

Shape	K_s	Position	K_b
Vertical board	1.00	Straight channel or entry to bend	1.00
Narrow vertical wall	1.00	Concave side of bend	1.10
45° Wing wall	0.85	Convex side of bend	0.80
1½:1 Full spill through	0.65	Below bend, concave side	
1½:1 Spill through with vertical wall below normal bed level	0.80	(i) Sharp bend	1.40
		(ii) Moderate bend	1.10

where the silt factor $f = 1.76 \sqrt{d_{mm}}$. The ratios of D_s/y_{3r} recommended for various conditions are:

(i) straight spurs facing upstream with sloping heads 1.5 to 1 horizontal, K = 3.8

(ii) as in (i) but heads sloping 1V:20H, K = 2.25

(iii) scour at noses of large radius guide banks, K = 2.75

(iv) scour at spurs along river banks, K = 1.7 - 3.8, depending on severity of attack.

Similar formulae but in terms of the Lacey two-dimensional regime depth

$$y_{2r} = 1.34(q_1^2/f)^{1/3} \qquad (9.28)$$

were proposed by Ahmad (1953), where q_1 is discharge per unit width in the contraction. The associated coefficient $K = D_s/q_1^{2/3}$ values in terms of angle of spur measured from downstream flow direction were given as

α:	30	45	60	90	120	150
K:	1.8	2.0	2.15	2.25	2.4	2.45

For spurs below a bend on concave side with strong swirl behind the bend K should be increased to 3 - 3.35 and 2.2 - 2.6 when there is no swirl.

From data available a reasonable estimate of scour depths at spur dikes is given by

$$D_s = Kq_1^{2/3} \qquad (9.29)$$

where K is

(i)	in straight channels and $\alpha = 90°$,	$K = 1.8 - 2.4$
(ii)	in moderate bend, no swirl	$K = 2.2 - 2.4$
(iii)	below a bend on concave side and swirl behind the bend	$K = 3.0 - 3.3$

9.4.1 Discussion

From dimensional reasoning one can deduce that the depth of equilibrium scour at an abutment, a spur dike or a groyne is primarily a function of

$$\frac{d_s}{y_o} = f\left[\frac{u_*}{u_{*c}}, \frac{L}{y_o}, \frac{d_{50}}{y_o}, \sigma, K_s, \beta\right]$$

where flow depth y_o and length L could be interchanged, σ is either the stand. deviation of sediment grading or a sorting parameter, K_s is a shape factor and β is inclination of the abutment to flow.

The shear velocity dependence of abutment scour depth for uniform sediment is shown in Fig. 9.25. The similarity to pier scour in uniform sediment is apparent although the clear-water scour peak is not quite as prominent.

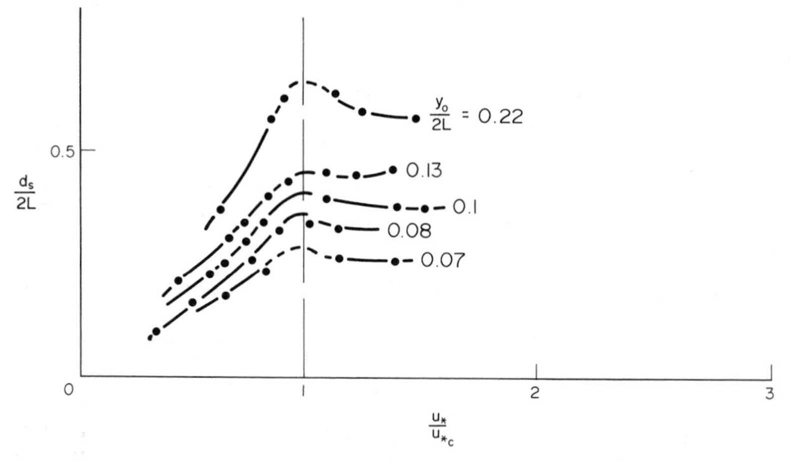

Fig. 9.25. Scour depth in uniform sediment versus shear velocity at abutments. Abutments $2L/d_{50} > 50$.

The ratios y_o/L or L/y_o define the relative size of the abutment. When the abutment is short, relative to flow depth, the abutment approximates to a half pier and L or 2L is the appropriate parameter for normalising. Figure 9.26 by Tey (1984) shows such a plot of laboratory data, where Line 1 is fitted to data from wing-wall (WW), spill-through (ST) and vertical wall pointing downstream (TS1) type abutments. Line 2 is for semi-circular (SCE) and vertical plate (TS2) perpendicular to flow type abutments,

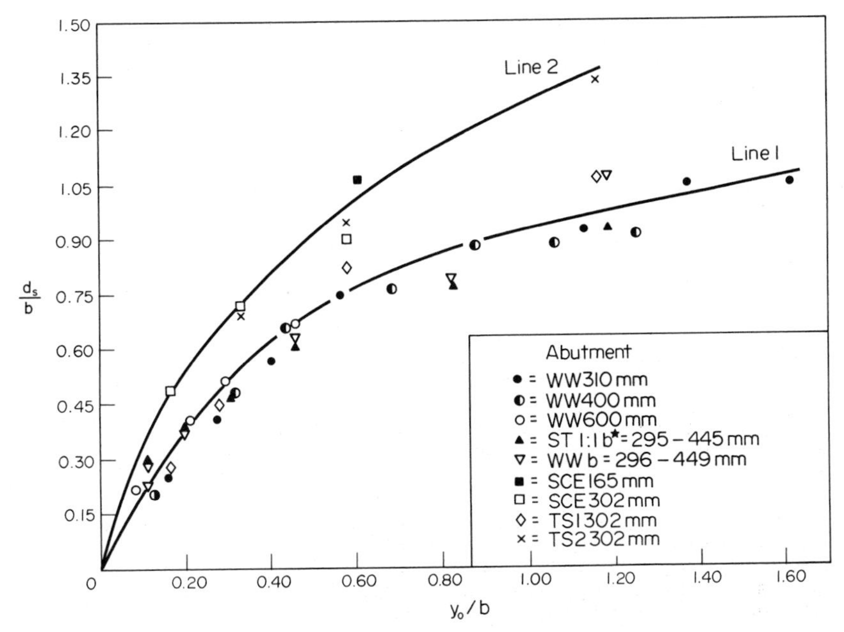

Fig. 9.26. Dimensionless scour depth versus flow depth, Tey (1984).

the lengths of which are given in the legend. Generally, abutments are at low values of y_o/L and piers at larger values. The semi-circular abutment, pier and plate data show deeper scour because of the stronger downflow on vertical faces. A similar plot in terms of d_s/y_o and $2L/y_o$ is shown in Fig. 9.27. It is seen that the scour depth with increasing abutment length becomes insensitive to length.

If the maximum clear-water scour depths of piers in deep water are plotted as d_s versus b the data will follow the $y_s \cong 2.3b$ line. The abutment data, in terms of y_s versus L or 2L, and of piers in relatively shallow water will curve away from this deep water asymptote. Abutment scour is generally at conditions approximating pier scour in relatively shallow flow. However, Fig. 9.27 indicates that for very long abutments the maximum clear-water scour is quite substantial, i.e., about seven times the flow depth. In principle, the scour depth at the abutment is defined by a surface function $d_s = f(y_o, L)$.

Laboratory data show that the proportion of flow noticeably affected by the presence of the abutment extends only a small distance out into the flow beyond the abutment, Fig. 9.28. According to Kwan (1984) the affected extra width, B_o, is about 10-15% greater than L. The contraction ratio affects the scour depth for clear-water scour conditions because then in the $B-B_o$ part of the cross-section the shear velocity increases with increasing contraction and local scour depth increases up to the state when the contracted section starts to scour. At live-bed conditions

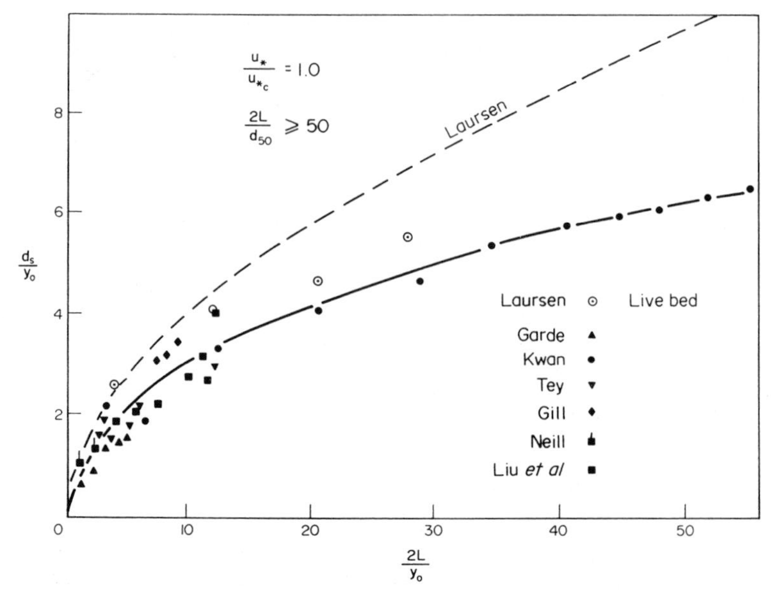

Fig. 9.27. Laboratory data for maximum clear water scour depth versus abutment length at vertical wall abutments. Points by Laursen are for live-bed scour.

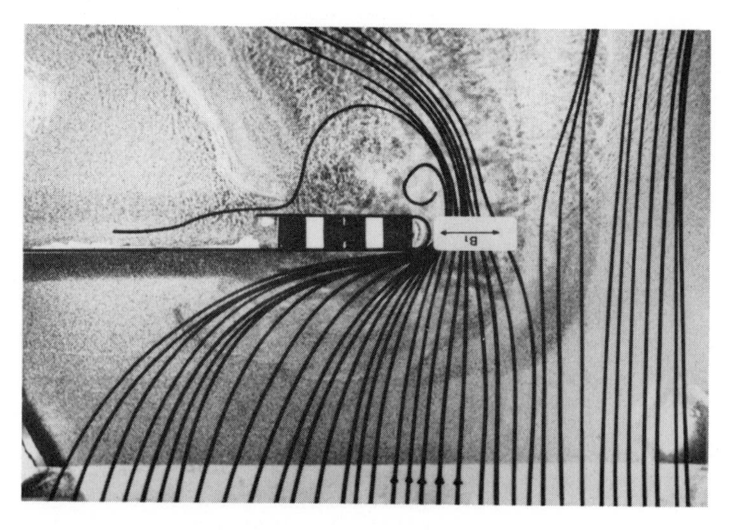

Fig. 9.28. Surface pathline pattern for flow past an abutment, Kwan (1984).

the contraction affects the constriction scour but not necessarily the local scour at the abutment. In analogy with live-bed clear-water scour at piers the live-bed scour at abutments is less than the maximum clear-water scour depth. This maximum is affected by the sediment grading in the same manner as at piers, Fig. 9.7, and the live-bed scour depth will fluctuate with the passage of bed features.

A frequent mode of abutment failure is outflanking, i.e., erosion of the causeway while the abutment structure remains intact. This results from the causeway material sliding into the scour hole and thus creating a cove in the causeway behind the abutment. Progressively an eddy develops in that recess which draws the interface to the stagnation region back from the head of the abutment, more of the flow impinges, downflow strength increases and the scour depth increases. The end result can be a break-through behind the abutment head.

A method of protection against the abutment scour is to pre-excavate the area on which the vortex is located and where the downflow impinges and to fill it with rip-rap on a sutiable filter. The rip-rap needs to be designed for the locally higher velocities. Laboratory and field data indicate that this velocity is about 1.25V, where V is the approach velocity upstream, and the protection needs to extend about 1.5 times the depth out from the abutment.

9.5 Scour at Culvert Outlets

Culverts have many different forms and occur in vast numbers in every country. At unprotected culvert outlets scour holes can develop which may through undermining endanger the culvert and the embankment. The scour which develops when a circular pipe discharges on an unprotected bed is characterised by its depth, y_s, length, L_s, and width, B_s, all of which denote the maximum dimensions. The material scoured will initially form a mound downstream which in time will be carried away by the flow. The size of the scour hole depends on the type of soil, on the exit velocities and tailwater depth. When the ratio of tailwater depth at the outlet to water depth in the culvert at the outlet is less than 0.3 the ratio L_s/y_s is about 5, but it increases to about 11 to 12 for the depths ratios about 0.9 or greater.

Bohan (1970) studied the development of scour in a sand bed downstream of a culvert flowing full in most tests and correlated the results with $F_r = U/(gD)^{\frac{1}{2}}$, where D is the culvert diameter. His results covered $1.06 \leq F_R \leq 6$, and can be described by

$$L_s/D = 15 \quad Fr^{2/3} \tag{9.31}$$

$$B_s/D = 7.5 \quad Fr^{2/3} \tag{9.32}$$

$$y_s/D = 1.52 \; Fr^{0.63} \tag{9.33}$$

The development of the scour hole was observed to be proportional to $(\text{time})^{0.1}$. The minimum stone size, d_r, for protection was given as

$$d_r/D = 0.25\ Fr \qquad (9.34)$$

for low tailwater, i.e. $y_{ot} < 0.5D$, and

$$d_r/D = 0.25\ Fr - 0.15 \qquad (9.35)$$

for tailwater higher than half the diameter, i.e. $y_{ot} > 0.5D$.

Breusers expressed the data by Bohan, Ruff et al. (1982) and Abt et al. (1984) in terms of U/u_{*c} as

$$L_s/D = 0.12(U/u_{*c})^{2/3} \qquad (9.36)$$

$$B_s/D = 0.06(U/u_{*c})^{2/3} \qquad (9.37)$$

$$y_s/D = 0.65(U/u_{*c})^{2/3} \qquad (9.38)$$

where u_{*c} is given by the Shields criterion.

However, scour dimensions cannot be related uniquely to U/u_{*c} or the exit Froude number Fr due to the free surface and complex eddying effects. Additional complications arise from sediment grading, particularly when armouring is possible. Observations indicate that the scour in cohesive soils will in time develop to about the same size as in noncohesive materials for the same conditions.

The effect of culvert slope, according to Abt et al. (1985) is to increase the scour dimensions. Slopes up to 10% increase depth up to 15% and scour length up to 25%.

Various forms of bed protection at culvert outlets were studied by Fletcher and Grace (1974) and Simons and Stevens (1972). The latter gave detailed recommendations for the design of non-scouring bed protection and bed protection which is allowed to scour, i.e., the thickness of the initial protection is greater than the equilibrium scour depth.

Culverts are usually finished at the downstream end with a vertical headwall. The bed at this headwall at culvert outlet gradually erodes down to about the same depth as the maximum scour depth for the given conditions (Mendozo et al. 1983). Therefore, the foundations of the headwall must extend below the maximum scour depth.

Scour basins are undesirable downstream of culverts on steep slopes because these can lead to the formation of a local hydraulic jump, i.e., on a steep slope downstream scour hole should be avoided with the aid of appropriate protection.

In summary, initial estimates of the scour dimensions can be obtained from eqns (9.32), (9.33) and (9.38). For design of basins and rock protection at culvert outlets reference is made to procedures outlined by Simons and Stevens (1972).

9.6 Scour due to Plunging Jets

The energy of excess water discharged from reservoirs over dams and spillways is reduced in special stilling basins or in the river downstream. The latter is usually the case when the excess water is spilt as a jet which lands in the river, well downstream of the dam. These high energy jets develop a scour hole and this may endanger the dam itself or the stability of the valley slopes. The mound of eroded material downstream of the scour hole can also significantly increase the tailwater level at the dam. The effect of impinging jets on the river bed depends mainly on the nature of the bed material which may be fractured rock, weathered rock, gravel, etc. The spillways of the jet type may have an overflow jet, a jet discharging under pressure or have a so-called ski-jump apron at the end of spillway. The scouring by jets from high head dams is usually associated with rocky river beds and valleys. The scouring of these is a specialised and complex problem and is discussed by Häusler (1983).

The jets impinge on the river at an angle which depends on the type of the spilling. The treatments of the resulting scour problem are based on studies of scour by submerged jets, i.e., the entry of jet is taken as the starting point. The characteristics of submerged jets are described in detail by Rajaratnam (1976), List (1982). References to scour studies associated with jets are Rouse (1940), Clarke (1962), Johnson (1967), Westrich and Kobus (1973), Westrich (1974), Kobus et al. (1979), Rajaratnam (1981) and Rajaratnam (1982). The influence of the angle of entry was studied by Rajaratnam (1981). The results showed that the angle had little effect on the scour depth but the scour profile became increasingly more asymmetric as the angle departed from vertical.

The various prediction formulae are based on small scale laboratory studies and non-cohesive sediments. The scaling of these to predict scour at high head structures has to be viewed with caution. The use of the jet dimensions and velocities at the impact creates no problems in laboratory since there the air entrainment is negligible. However, in the field air entrainment can lead to substantial changes to jet dimensions and properties. In practice air entrainment is usually neglected when a solid core is still present in the jet. Air entrainment would decrease the scour depth. The velocity of round jets decreases with distance s travelled in the water as s^{-1}, and that of plane jets as $s^{-\frac{1}{2}}$. The depth of penetration according to Häusler (1983) could be taken for practical purposes as $20D_o$ for round jets and $20(2B_o)$ for plane jets, where D_o is the diameter of round jet at impact with water surface and $2B_o$ the thickness of plane jet.

The equilibrium scour due to a plunging jet in coarse loose material in a laboratory experiment develops very quickly. About 2/3rd of the equilibrium depth is reached in the first half a minute and changes in depth after 2 to 3 hours amount to only a few percent.

Breusers described the data by Clarke, Westrich and Rajaratnam for maximum scour depth due to a vertical submerged jets by

$$\frac{y_s}{D_o} = 0.75 \frac{U_o}{u_{*c}} \quad ; \quad \frac{U_o}{u_{*c}} < 100 \tag{9.39a}$$

$$\frac{y_s}{D_o} = 0.035 \left[\frac{U_o}{u_{*c}}\right]^{2/3} \quad ; \quad \frac{U_o}{u_{*c}} > 100 \tag{9.39b}$$

The average width to depth ratio of the scour holes was five.

One of the frequently quoted model studies of scour by plunging jets is that by Veronese (1937) whose prediction formula is

$$y_s + y_o = 3.68H^{0.225}q^{0.54}d_m^{-0.42} \tag{9.40}$$

which Jaeger (1939) modified to

$$y_s + y_o = 6H^{0.25}q^{0.5}(y_o/d_{90})^{0.33} \tag{9.41}$$

In these H is the difference in head between upstream and downstream water levels, q discharge per unit width, d_m the mean grain size where d_m and d_{90} are in mm, y_o the downstream water depth and y_s depth of scour measured from bed level. The experiments cover $q = 0.01 - 0.07$ m³/s and d = 9, 14, 21 and 36 mm. These experiments were repeated by Hartung (1957) who proposed

$$y_s + y_o = 12.4H^{0.36}q^{0.64}d_{85}^{-0.32} \tag{9.42}$$

For ski jumps spillways Damle et al. (1966) obtained from model and field studies the expression

$$y_s + y_o = 0.54(qH)^{0.5} \tag{9.43}$$

which could be compared with the expression

$$y_s + y_o = 1.18q^{0.51}H^{0.235} \tag{9.44}$$

obtained from model and field data in Taiwan by Chian Min Wu (1973). Martins (1975) proposed a similar empirical relation from prototype observations

$$y_s + y_o = 1.5q^{0.6}z_2^{0.1} \tag{9.45}$$

where z_2 is the difference between upstream water level and the elevation of the lip of the flip bucket.

A detailed study of the model and prototype data was carried out by Mason (1984) and Mason and Arumugam (1985). They expressed the model data by

$$y_s + y_o = 3.27q^{0.6}H^{0.05}y_o^{0.15}g^{-0.3}d_m^{-0.1} \tag{9.46}$$

and prototype as well as model data by

$$y_s + y_o = (6.42 - 3.1H^{0\cdot 1})q^{(0\cdot 6 - 0\cdot 0033H)}H^{(0\cdot 05 + 0\cdot 005H)} y_o^{0\cdot 15}g^{-0\cdot 3}d_m^{-0\cdot 1} \quad (9.47)$$

Equation (9.46) is dimensionally correct and satisfies Froude number scaling. The data include free overflow, flip buckets, low level outlets and tunnel outlets.

The formulae quoted predict scour depths of the right order although substantial differences still exist. An extensive literature survey on scour prediction for energy dissipators of high head structures was published by Whittaker and Schleiss (1984).

9.7 Scour at Stilling Basins

Prediction of scour downstream of river barrages and stilling basins for spillway discharges is an integral part of design of such structures. Most of the scour studies of this type have been associated with particular projects. In a more general sense the studies could be subdivided into studies of *equilibrium scour* in coarse alluvial material and of the *rate of development* of scour, particularly of fine sediments. The scour at stilling basins is associated with horizontal jets, so called wall jets. The outer regions are similar to free jets but along the bed a boundary layer develops (Rajaratnam, 1976). Laursen (1952) studied the scour patterns caused by two-dimensional horizontal jets and showed that the scour patterns were similar. Their length increased with U_o/w. The length of such scour is given by Whittaker and Schleiss (1984) as

$$\frac{L_{st}}{2_{Bo}} = 35(U_o/w)^{0\cdot 57}U_o^{0\cdot 86} \quad (9.48)$$

where L_{st} is measured to the crest of the bar downstream of the scour hole. The observation that the shape of the scour is almost independent of flow velocity and bed material for given geometric arrangement has been verified by several researchers since Laursen, that is, when the velocity is greater than the critical value for movement of bed material.

The deepest scour develops in uniform noncohesive sediments. The scour in beds of sediment with a broad range of grain sizes is less. The flow suspends and removes the finer fractions leaving the bed covered with layers of coarser grains. This has led to the use of d_{85} to d_{95} of the bed material in the formula for scour depth. The scour depth has been observed to decrease with increasing percentage of the fines fractions, $d < 60$ μm, and with decreasing voids ratio. No methods exist for reliable prediction of these effects.

A schematic illustration of scour at weirs and at stilling basins is shown in Fig. 9.29. At the discontinuity between the bed and the apron the bed erodes usually more gradually than in the basic

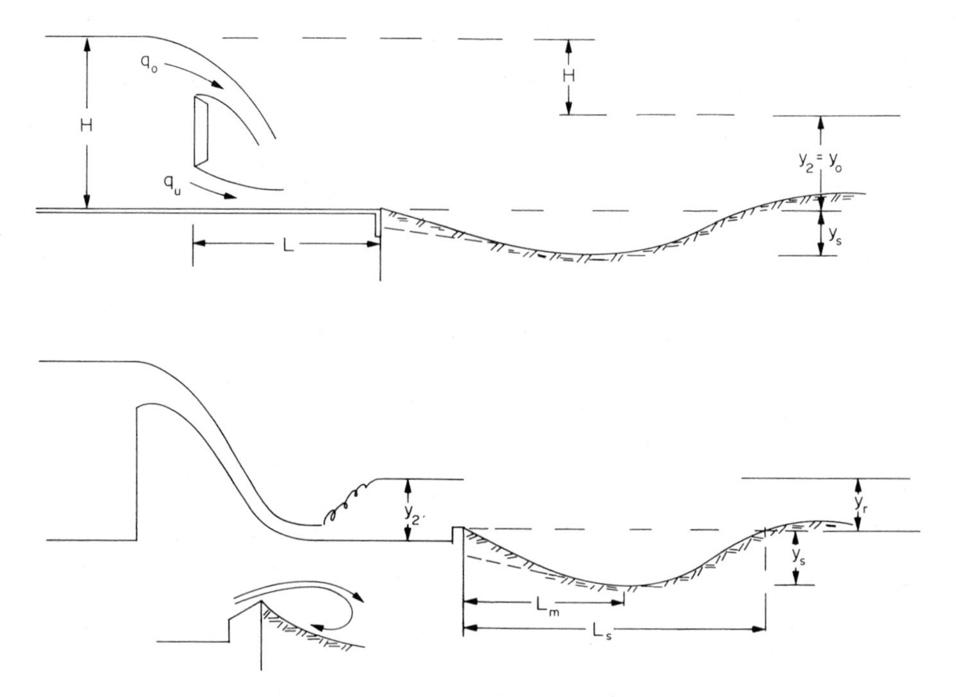

Fig. 9.29. Illustration of scour at weirs and stilling basins.

scour hole changing its shape as shown by the dotted line. As a rule of thumb this lowering is about ½ y_s. Aprons which have upward sloping end sills create a ground roller downstream and prevent scour at the end of the apron or basin. The scour shapes vary with the type of end sill. A vertical end sill (step) at the end of the basin shifts the location of the maximum scour depth downstream and also leads to a deeper scour than is observed with upward sloping end sills (1:3 to 1:6). The end sills downstream of the hydraulic jumps are in addition very effective in maintaining the jump on the apron (in the stilling basin) when flow conditions vary.

Generally, the published formulae are characterised by a low level of predictive quality. Comparisons lead to differences by a factor four or more and usually model testing is necessary.

The empirical regime method (Blench, 1957) uses the two-dimensional regime depth y_{2r}, eqn (9.28), to estimate the scour depth downstream of a stilling basin with hydraulic jump:

$$y_s = (0.75 \text{ to } 1.25)y_{2r} \tag{9.49}$$

The formulae by Schoklitsch (1932,1935) for scour downstream of a weir were derived from model tests. These are for scour caused by overflow

$$y_s + y_o = 4.75H^{0.2}q_o^{0.57}d_{90}^{-0.32} \tag{9.50}$$

when $q_u = 0$ and apron length $L = 0$; and for scour by underflow

$$y_s = 0.378H_1^{0.5}q_u^{0.35} + 2.15a \tag{9.51}$$

when $L = 1.5H$. The term a accounts for difference between apron and downstream bed levels, i.e., apron is higher than bed by a. The scour due to overflow can be compared with a scour by plunging jets, eqns (9.40), (9.41) and (9.42).

The influence of a horizontal apron of length L on the scour depth caused by an underflow was studied by Shalash (1959) who proposed the formula

$$y_s + y_o m = 9.65H^{0.5}q_u^{0.6}d_{90}^{-0.4}(L_{min}/L)^{0.6} \tag{9.52}$$

where $L_{min} = 1.5H$. However, it is not clear where the hydraulic jump was located or what form it had.

The scour downstream of a hydraulically rough apron due to underflow was studied by Dietz (1969) who expressed the maximum scour depth as

$$(y_s)_{max}/y_o = (V_{max} - V_c)/V_c \tag{9.53}$$

where $V_{max} = \alpha V$, V_c is the critical mean velocity for the bed material according to the Shields criterion ($V = (8/f)^{\frac{1}{2}}u_{*c}$ or $V_c = (1/n\sqrt{g})y_o^{1/6}u_{*c}$), V is the mean velocity and α depends on turbulence intensity, $\alpha = 1 + 3(\overline{u'^2})^{\frac{1}{2}}/\bar{u}_m$ and is affected by upstream flow geometry. Over a rough apron $\alpha \cong 1.5$. The data by Dietz and Bell (1980) indicate that the scour depth reaches a maximum for given conditions, i.e., downstream a horizontal apron and no sediment input from upstream. Using the parameter $\delta = [(V - V_c)/w][(S_s - 1)gd^3/\nu^2]^{1/3} = [(V - V_c)/w]D_*$ of Dietz, Phillips (1984) fitted

$$\left[\frac{L_m}{y_s}\right]_{max} = \begin{cases} 65.5\ \delta^{-0.509} & ;\ \delta \leq 150 \\ 5.1 & ;\ \delta \geq 150 \end{cases} \tag{9.54}$$

where L_m is the distance to maximum scour depth y_s.

The laboratory studies refer to clear-water scour. There is usually no sediment input from upstream into the scour hole. A live-bed situation would lead to a somewhat reduced scour depth.

Data which is specifically applicable to scour downstream of stilling basins is limited and the scour depth is significantly affected by the hydraulic performance of the stilling basin. Novak (1956,1961) expressed the scour depth downstream of stilling basins as

$$y_s + y_2 = K[6H^{0.25}q^{0.5}(y_2/d_{90})^{1/3} - y_2] \tag{9.55}$$

where H is the head difference between upstream and downstream water levels, y_2 is the downstream water depth and d_{90} is in mm. The coefficient K = 0.45 for y_2/y_c = 1.6 and K = 0.65 for y_2/y_c = 1 where y_c is the critical depth of flow.

The effect of the depth of tailwater, y_r, relative to conjugate depth, y_2, on the depth and length of scour downstream of a horizontal apron at the toe of a spillway was given from laboratory data by Farhoudi and Smith (1982, 1985) as

	$(L_m/y_s)_{max}$	$(L_s/y_s)_{max}$
y_r/y_2 = 1.25	3.04	6.82
= 1.0	3.2	7.2
= 0.78	4.8	9.8

Dietz as well as Breusers (1966,1967) studied the slope of the scour β. The experimental arrangement of Breusers is shown diagrammatically in Fig. 9.30. The angle is steepest when the apron is smooth because the velocities near the bed are then highest. Data from Delft Hydraulics Laboratory (1972) show the following slopes:

y_D/y_o	L/y_o	Apron smooth	Apron rough
		cos β	
0	-	4	6
0.3	1	4.5	7
0.3	5	3	3.7
0.3	15	3.7	4.5
0.6	3	3.2	4.5
0.6	7	3.5	3
0.6	12	2.5	3.5

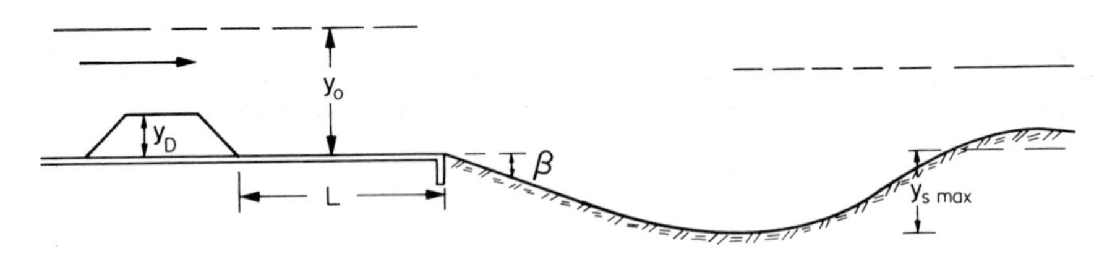

Fig. 9.30. Definition sketch.

Both Breusers and Dietz found that the maximum scour depth could be expressed as

$$\frac{y_{s\ max}}{y_o} = \frac{t}{t_1}^{0.38} \tag{9.56}$$

where t_1 is time in hours when $y_{s\ max} = y_o$ for which Breusers gave

$$t_1 = 330\ \Delta^{1.7} y_o^{2} (\alpha V - V_c)^{-4.3} \qquad (9.57)$$

No equilibrium was indicated by the data obtained by Breusers. At the end of the aprons below spillways Farhoudi and Smith (1982) obtained a rate of development proportional to $(t/t_1)^{0.19}$. Working with fine uniform gravel $d_{50} = 2.11$, $\sigma_g = 1.25$, Bell (1980) obtained for the development of scour after initial about 5 minutes $y_{s\ max}$ proportional to $t^{0.31}$. Three-dimensional features tend to increase the scour depth and the exponent in eqn (9.56) becomes greater than 0.38 at early stages of scour. The function slowly drifts away from the early slope and is better fitted by a logarithmic expression.

Scour protection downstream of weirs and barrages has been extensively used. The empirical Bligh's formula for the length of apron has stood the test of time in many locations. It simply states that the apron should be n-times the height difference between head and tailwater, where n = 12-15 in Indian applications and 18-24 on the Nile. A similar empirical formula for the apron length is

$$L = 5(H.\Delta H)^{\frac{1}{2}}$$

where H is the upstream head over the crest and ΔH is the head difference between upstream and downstream water levels.

The bed protection downstream of aprons or stilling basins can be of rock blocks or of rip-rap, both placed on inverted filter.

9.8 Rip-rap Scour Protection

Layers of rock, the so-called rip-rap, are extensively used to protect river banks, coastline, etc. against erosion. The criteria for the selection of rock size have evolved from experience and the recommended sizes differ appreciably from one controlling authority to another.

In principle, a rip-rap rock in unidirectional flow should be heavier than given by the threshold of movement criterion. The combination of the Manning and Strickler formulae $[U/u_* = (24/\sqrt{g})(y_o/d)^{1/6}]$ with the Shields threshold value θ_c yields

$$U_c \cong 24(S_s - 1)^{1/2} \sqrt{\theta_c}\ d^{1/3} y_o^{1/6} \qquad (9.58a)$$

or

$$d = \frac{U^3}{24^3 (S_s - 1)^{3/2} \theta^{3/2} y_o^{1/2}} \qquad (9.58b)$$

Using $\theta_c = 0.04$ leads to

$$U_c = 4.8(S_s - 1)^{1/2} d^{1/3} y_o^{1/6} \qquad (9.59a)$$

or

$$d = \frac{U^3}{110.6(S_s - 1)^{3/2} y_o^{1/2}} \qquad (9.59b)$$

or

$$d_{(m)} = 4.5 \times 10^{-3} U^3/\sqrt{y_o} \qquad (9.59c)$$

for stones with $S_s = 2.6$. Equation (9.59a) could be compared with eqn (3.14) by Neill and has a factor of 2.35 instead of 2.0. The latter implies a $\theta_c = 0.034$.

The stone sizes given above are generally substantially smaller than by the various codes of practices which incorporate safety factors.

The threshold approach to rip-rap stability is not only affected by the choice of the critical threshold value of θ, which varies substantially with exposure of stones, Fig. 3.12, but also by the shape of the stones. The drag coefficient for the drag force exerted by fluid varies with the shape and roughness of the stones. On rounded rock the C_D- value drops substantially on exceeding a certain value of the Reynolds number. For example, for spheres the C_D-value approximately halves when $Re = Ud/\nu$ exceeds about 2×10^5. For cylinders this change is even larger, C_D drops to about a quarter of its earlier value. However, quarry rocks with sharp edges are less likely to experience this type of change because separation is fixed at the edges. A decrease in C_D-value implies a larger θ_c-value. The asymmetric shapes of rocks also introduce an unknown lift force effect.

The simplest of the working relationships is

$$V = 4.92 \sqrt{d} \qquad (9.60)$$

where V is velocity near the bed in m/s, d is the diameter of the equivalent sphere in m of specific gravity $S_s = 2.65$. This applies for horizontal bed. Most empirical relationships give d proportional to V^2 rather than V^3 which follows the observation by Brahms in 1754 that V_c is proportional to $(\text{mass})^{1/6}$, i.e. $V_c^2 \propto d$. The reduction in stability on a slope has to be allowed for additionally and may include the reduction due to seepage forces.

Following a procedure similar to that which led to eqn (8.64), using the symbols shown in Fig. 9.31, yields in the plane of the slope that $F_D a_3 \sin\delta = W_s a_1 \sin\alpha \sin\beta$ or

$$\frac{\sin\beta}{\sin\alpha} = \frac{a_3}{a_1} \frac{F_D}{W_s \sin\alpha}$$

In the vertical plane through the resultant R the balance is given by $a_2 W_s \cos\alpha = a_4 F_L + a_3 F_D \cos\delta + a_1 W_s \sin\alpha \cos\beta$ and the ratio of the two sides, with $a_2 W_s \cos\alpha$ as numerator, represents a safety factor. In a no flow situation F_L and F_D are zero and assuming

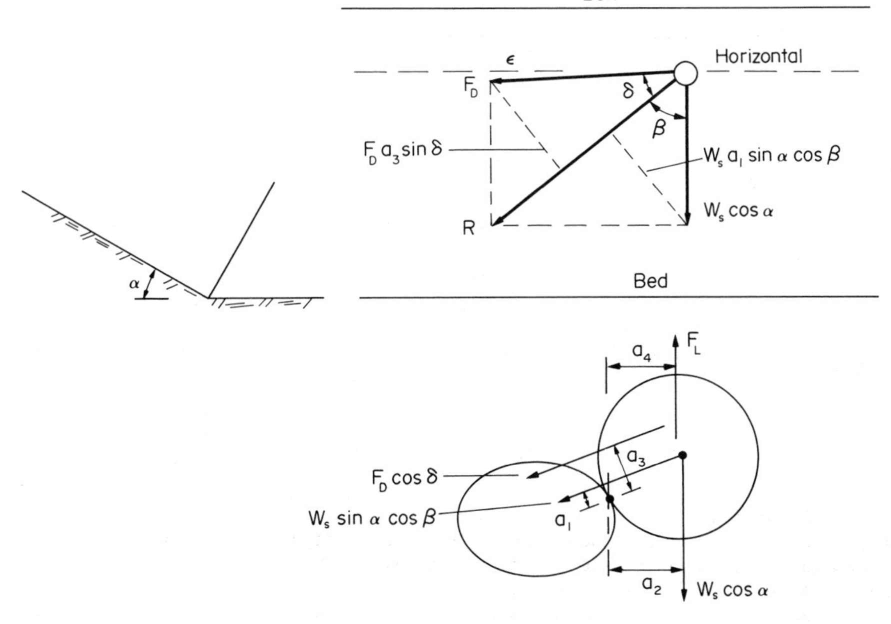

Detail in the plane of R

Fig. 9.31. Definition sketch.

that on the average $a_2/a_1 = \tan\phi$; the angle of repose of the material, the ratio is seen to reduce to $\tan\phi/\tan\alpha$ as required. For the general case the safety factor F_s can be written as

$$F_s = \frac{\cos\alpha}{\dfrac{a_4}{a_2}\dfrac{F_L}{W_s} + \dfrac{a_3}{a_2}\dfrac{F_D}{W_s}\cos\delta + \dfrac{a_1}{a_2}\sin\alpha\cos\beta}$$

$$= \frac{\cos\alpha \tan\phi}{K_s \tan\phi + \sin\alpha\cos\beta} \tag{9.61}$$

where K_s stands for the first two terms in the denominator. When $\varepsilon = 90°$ both β and δ are zero. This yields F_s for a plane bed sloping at angle α, as over an apron, with $K_s \rightarrow K$ where K = [illegible] $(a_3 F_D + a_4 F_L)/a_2 W$; W is the weight force of the element and a_2 the moment arm of W cos α to the pivotal point, a_3 and a_4 are the moment arms of drag force, F_D, and lift force, F_L, respectively; tan ϕ is the angle of repose, which for quarry rocks by an empirical expression is

$$\tan\phi = 1/(0.3 + 0.59\, d/k) \tag{9.62}$$

where d is the rip-rap "size" and k is the size (roughness) of the layer on which the rip-rap rests. Note, that rip-rap stability is strongly dependent on the roughness of the surface it rests. Hence, the minimum of two layers requirement. The threshold condition is given by $F_S = 1$. From eqn (3.14)

$$F_s = 1 = \frac{0.5\ V_c^{\ 2}}{(S_s - 1)gd_{50}} \frac{d_{50}}{y_o} \tag{9.63}$$

if d_{50} is assumed to be representative.

The relationship for rip-rap site given by the Shore Protection Manual (1973) is

$$\frac{V}{\sqrt{2g}} = Y(S_s - 1)^{\frac{1}{2}}(\cos\alpha - \sin\alpha)^{\frac{1}{2}}d^{\frac{1}{2}} \tag{9.64}$$

where α is the slope in flow direction, Y is a constant equal to 1.20 for embedded rock and 0.86 for nonembedded rock (Isbach constants),

$$V = 5.75u_*\log(12.27y_o/d)$$

and

$$d_{(m)} = 1.24(\text{Mass of rock/density})^{1/3}$$

The rip-rap should be composed of a well graded mixture of rocks so that the voids between large stones are filled by smaller ones and the stones support each other. The thickness should be two layers or more. The rocks are characterised by a representative size

$$d_r = \left[\frac{1}{10} \sum_{1=1}^{10} d_i^{\ 3}\right]^{1/3} \tag{9.65}$$

where $d_1 = \frac{1}{2}(d_o + d_{10})$; $d_2 = \frac{1}{2}(d_{10} + d_{20})$ etc., and a sorting coefficient. Frequently used in the form

$$C_s = d_{60}/d_{10} \tag{9.66}$$

The d_r-size is approximately equal to d_{67} by weight. An illustration of a well-graded rip-rap is given by the following values:

d/d_{50}	0.2	0.33	0.4	0.7	1.0	1.15	1.5	2
% by weight:	0	10	16	30	50	60	84	100

The rip-rap protects the filter layer which protects the bed or bank material from being winnowed away by the currents.

Rock or concrete block protection is at times used in the form of a pavement. The design of such pavements must consider the stability of blocks when subjected to stagnation pressure from underneath. Whenever one block protrudes the stagnation pressure thus created at its upstream face propagates underneath and creates uplift. This is further augmented by the lift force created by the protrusion.

9.9 Degradation and Aggradation

9.9.1 Degradation

In contrast to the term scour, which is a localised lowering of river bed, *degradation* implies a lowering of the river bed over long distances. Degradation may progress downstream or upstream or both. The most common example of downstream degradation is that downstream of dams which trap the sediment in transport and release clear water. For example, the sand and gravel bed below the Hoover dam in Colorado degraded 7.1 m in 30 years and the degradation extended over 52 km. This is the highest reported degradation but those of a few metres are quite common. Degradation progressing upstream occurs when the downstream water level is lowered, e.g. lowering of a lake level or removal of a control and is more rapid than downstream progressing degradation.

In general, a downstream progressing degradation leads to a flattening of slope, i.e. the end result could be looked at as a rotation of the bed profile about a downstream control point, e.g. a lake or sea. It can be caused by a *reduction of sediment in transport*, *a reduction in size of the bed material* or *an increase in water discharge*. The rate of degradation decreases with time and the equilibrium conditions are approached very slowly. Profile adjustments occur only when the threshold conditions of the armour layer on the bed are exceeded. An imposed steepening of the slope by lowering of water level, by cutoffs etc leads to upstream propagating degradation and continues until equilibrium is established by coarsening of bed material or by change in river geometry. An excellent discussion of a large number of case histories is given by Galay (1983).

Degradation of river bed can lead to slumping of the valley slopes and to undermining of structures along the river, such as bridge foundations.

The analysis of such non-equilibrium problems rests basically on two equations, the continuity equation for water and sediment transport and the equation of motion for water and sediment. The continuity equation for sediment is

$$\frac{\partial g_{ST}}{\partial t} + \frac{\partial g_{ST}}{\partial x} = 0 \qquad (9.67)$$

which, in terms of the elevation z of river-bed above a datum can be written as

$$\frac{\partial z}{\partial t} + \frac{1}{(1 - n)} \frac{\partial q_{ST}}{\partial x} = 0 \tag{9.68}$$

where n is the porosity of bed sediment, g_{ST} and q_{ST} are the total sediment transport rate in terms of mass and volume per unit width, respectively.

The transport equation for sediment can be expressed as

$$q_{TS} = f(U) \text{ or } q_{TS} = f(\tau_s) \tag{9.69}$$

Most analytical treatments have favoured dependence on velocity rather than bed shear stress.

The continuity equation for water is

$$\frac{\partial y_o}{\partial t} + U \frac{\partial y_o}{\partial x} + y_o \frac{\partial U}{\partial v} = 0 \tag{9.70}$$

which for a two-dimensional steady flow is simply

$$q = Uy_o \tag{9.71}$$

The equation of motion for water may be written as

$$\frac{\partial U}{\partial t} + U \frac{\partial U}{\partial x} = - \frac{1}{g} \frac{\partial y_o}{\partial x} - gS_f + gS_o \tag{9.72}$$

where S_o is the bottom slope ($- \partial z/\partial x$) and S_f is the friction slope. These are for a simple two-dimensional flow case, without lateral velocities and in or out flows. The friction slope can be evaluated by Chézy or the Strickler-Manning formula $S_f = U^2/C_{yo} = n^2U^2/y_o^{4/3}$. This sytem of four equations was reduced by de Vries (1973) to

$$- c^3 + 2Uc^2 + (gy_o - U^2 + gdq_s/dU)c - Ugdq_s/dU = 0 \tag{9.73}$$

where $c = dx/dt$ and q_s stands for $q_{ST}/(1 - n)$, or in dimensionless form

$$\phi^3 - 2\phi^2 + (1 - Fr^{-2} - \psi Fr^{-2})\phi + \psi Fr^{-2} = 0 \tag{9.74}$$

where

$\phi = c/U$ = relative celerity
$Fr = U/(gy_o)^{\frac{1}{2}}$
$\psi = (1/y_o)dq_s/dU$ = a transport parameter.

The transport parameter ψ becomes proportional to concentration q_s/q when it is assumed that $q_s = aU^b$, i.e.,

$$\psi = bq_s/q \tag{9.75}$$

The three roots of eqn (9.74) are shown in Fig. 9.32 in terms of

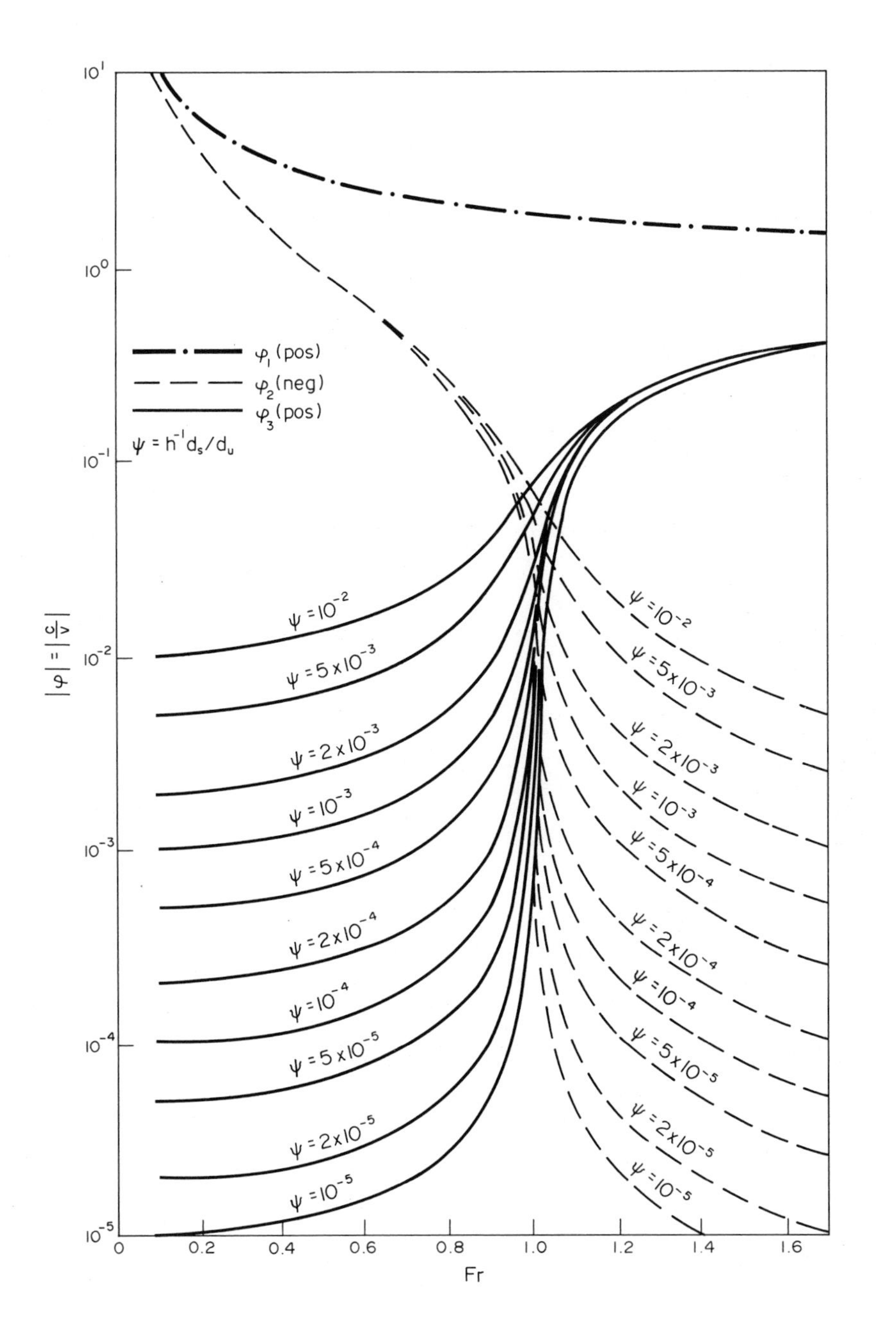

Fig. 9.32. Relative celerities according to eqn 9.74 (de Vries, 1973).

Fr and ψ. The first dimensionless velocity $\phi_1 = 1 + Fr^{-1}$ or $c = U + (gy_o)^{\frac{1}{2}}$ is of small water surface waves moving in the direction of flow and is practically independent of the mobility of bed. The influence of sediment transport can also be neglected on $\phi_2 = 1 - Fr^{-1}$, surface waves moving against the flow, when $Fr < 0.7$. The third velocity is that of bed features, given by

$$\phi_3 = \frac{\psi}{1 - Fr^2} \text{ or } c_3 = \frac{dq_s/dU}{1 - Fr^2} \cdot \frac{U}{y_o} \cong \frac{bq/q_s}{1 - Fr^2} \tag{9.76}$$

when the Froude number is not close to unity. When $Fr < 1$, $\phi_3 > 0$ and for $Fr > 1$, $\phi_3 < 0$. At critical flow conditions

$$\phi_{2,3} \cong \neq \pm \sqrt{\tfrac{1}{2}\psi} \tag{9.77}$$

For $Fr < 1$ it is seen from Fig. 9.32 that ϕ_3 is very much smaller than $|\phi_{1,2}|$, which means that bed waves can be analysed by assuming $|\phi_{1,2}| \to \infty$, an essentially steady water flow in which $\partial U/\partial t$ and $\partial y_o/\partial t$ can be neglected. In the same sense, when analysing water waves, like flood routing, $\phi_3 \to 0$ and the bed may be assumed to be stationary.

These assumptions for $Fr \neq 1$ lead to the following simplification

$$\frac{\partial z}{\partial t} + \frac{\partial q_s}{\partial x} = 0 \tag{9.78}$$

$$qs = f(U) \tag{9.79}$$

$$U \frac{\partial U}{\partial x} + g \frac{\partial y_o}{\partial x} + g \frac{\partial z}{\partial x} = - g \frac{U|U|}{C^2 y_o} \tag{9.80}$$

$$Uy_o = q(t) \tag{9.81}$$

which yield one differential equation

$$\frac{\partial z}{\partial t} + \frac{gdq_s/dU}{gq/U^2 - U} \frac{\partial z}{\partial x} = - \frac{g|U|U}{c^2 y_o} \cdot \frac{dq_s/dU}{gq^2/U^2 - U} \tag{9.82}$$

or with eqn (9.76)

$$\frac{\partial z}{\partial t} + c_3 \frac{\partial z}{\partial x} = f(U) \tag{9.83}$$

where f(U) is a measure of damping of the disturbance and is zero for frictionless flow. For a steady uniform motion of water eqn (9.80) reduces to the last two terms. Replacing y_o from eqn (9.71), differentiating and combining with eqns (9.78) and (9.79) leads to

$$\frac{\partial z}{\partial t} - K \frac{\partial^2 z}{\partial x^2} = 0 \tag{9.84}$$

where

$$K = \frac{1}{3}\frac{C^2 q\,dq_s/dU}{U^2} = \frac{1}{3}\frac{U\,dq_s/dU}{S_o}\left[\frac{U_o}{U}\right]^3$$

in which U_o and S_o refer to original uniform conditions. For $U \cong U_o$

$$K \cong \frac{1}{3}\frac{U_o\,dq_s/dU}{S_o} = \frac{1}{3}\,b\,\frac{q_s}{S_o} \tag{9.85}$$

when $q_s = aU^b$. Equation (9.84) is the basis of the so-called parabolic models of bed level variation and is applicable when $x > 3y_o/S_o$.

Without the uniform flow assumption the result is a hyperbolic equation (Vreugdenhil and de Vries, 1973):

$$\frac{\partial z}{\partial t} - K\frac{\partial^2 z}{\partial x^2} - \frac{K}{c}\frac{\partial^2 z}{\partial x \partial t} = 0 \tag{9.86}$$

where after use of $\partial z/\partial x \cong S_o$; $U \cong U_o$ and $q_s = aU^b$ the constant K reduces to eqn (9.85).

There is no unique sediment transport equation available and it is a matter of choice which one to use. A simplifying assumption is that the sediment is mainly transported as a bed load. Combining the two continuity and two equations of motion and the varied flow equation yields an equation for $\partial z/\partial t$ which can be used to prediction of change of the bed elevation over short time increments. Higher order derivatives of z may be important when the time increments are large.

Mostafa (1957) used the Einstein formula and obtained for the slope

$$S = \frac{0.06\Delta}{R}\,\frac{k}{Y} \tag{9.87}$$

where $\Delta = (\rho_s - \rho)/\rho$, R is hydraulic mean radius, k is roughness height (Nikuradse) and $Y = 0.06/\theta$, where $\theta = \tau_c/g(\rho_s - \rho)k$. The 0.06 arises from Shields' entrainment function $\tau_c/g(\rho_s - \rho)d_{50} = 0.06$. The slope expression is related to the mean velocity

$$U = \frac{Q}{A} = 5.75(0.06g\Delta k/Y)^{\frac{1}{2}}\,\log[12.27(R/k)N] \tag{9.88}$$

where N is the roughness correction factor used by Einstein and is equal to one for $k/\delta' \sim > 8$ where $\delta' = 11.6\,\nu/u_*$. Mostafa used $k = d_{98}$. The analysis can be started by putting N = Y = 1 and computing the related values of R and Q, then check for N and Y.

Komura and Simons (1967) used the Kalinske and Brown sediment transport formula from which, with a number of assumptions, they derive from the final equilibrium profile

$$z_f = \sum_{i=1}^{n} \left[C_s \left[\frac{d_{sf}}{yf} \right]_m \Delta x_i + \frac{2}{7} \left[1 - \left[\frac{y_c}{y_f} \right]^3 \right] \left[\frac{y_f}{d_{sf}} \right]_m \Delta d_{sfi} + \right.$$

$$\left. \frac{1}{7} \left[\frac{y_f}{B} \right]_m \left[6 + \left[\frac{y_c}{y_f} \right]^3 \right] \Delta B_i \right] \tag{9.89}$$

where x = 0 and z = 0 at the downstream control section and x is measured upstream. The formula gives the elevation difference between the ends of each of the n increments Δx of the total stream length summed to station n. The subscript m refers to mean value in the reach Δx_i, Δd_{sfi} is the difference between the values of the final mean particle size at the two ends of each Δx_i, ΔB_i is the difference in width of the stream at the ends of the reach Δx_i, y_c is the critical depth

$$y_c^3 = \frac{\alpha Q^2}{gB^2{}_m} \tag{9.90}$$

where α is the kinetic energy coefficient ($\alpha \cong 1.3$), B_m is the mean width of the reach, y_f is the depth

$$y_f = \left[\frac{n^2 Q^2}{B^2{}_m C_s d_{sf}} \right]^{3/7} \tag{9.91}$$

C_s, the coefficient of armouring, is written as (S_s - 1) times the Egiazaroff mobility number

$$C_s = \frac{0.1(S_s - 1)}{[10g(19d_{50}/d_{av})]^2} \tag{9.92}$$

where $d_{av} = \frac{1}{2}[d_{85}(\text{moving}) + d_{85}(\text{total})]$ as recommended by Egiazaroff and the moving load is given by the fraction for which $\tau_o/\tau_c > 1$.

For constant width, the final equilibrium profile is

$$z_f = z_o + \left[\frac{7}{9} \left[\frac{C_s}{a} \right] \left[\frac{d_{sfo}}{y_{fo}} \right] [e^{(9/7)ax} - 1] + y_{fo}[1 - e^{-(2/7)ax}] \right.$$

$$\left. - \frac{1}{2} y_{fo} \left[\frac{y_c}{y_{fo}} \right]^3 [e^{(417)ax} - 1] \right] \tag{9.93}$$

where y_{fo} is the final depth at x = 0, d_{sfo} is the final particle size at x = 0, which, when x = 0 is an end control, would remain at its original size. The exponent a can be estimated from river-bed samples, for example, as $d_o\exp(ax_1) = 11$ and $d_o\exp(ax_2) = 24$,

$x_2 = x_1 = 24.4$ km gives $a = 0.032$ km^{-1} and $d_o = 0.0047$ m, i.e. $d_{sf} = 0.0047 \exp(0.032\ x)$, i.e. 11 and 24 mm are the given d_{50}-sizes at the two stations Δx apart.

Apart from the problem that the use of different transport formulae gives different answers, there is the problem of effective particle size and armouring for which there is no satisfactory solution. The size distribution of the bed surface material varies as a function of time and as a function of distance along the stream. The rate of coarsening of bed material is rapid at the start of degradation but it slows down very soon. The transport of bed material becomes intermittent and occurs only when the bed shear stress exceeds the threshold value of the armour layer. Only when the shear stress exceeds the limiting value for armour formation does an indiscriminate transport occur. As the flow subsides the armour layer establishes itself again and the transport rate diminishes. Since the degradation processes are very slow it may be better to solve the $\partial z/\partial t$ relationship as a function of time, particularly as in the latter stages, the threshold conditions for the armour layer are less and less frequently exceeded.

Degradation can be counteracted by release of sediment from reservoirs. It can also be controlled to some extent in any reach of a river, for example, by reduction of slope by ground sills which convert the river into a cascade. A sequence of reservoirs, although trapping sediment and causing degradation, also has a reducing effect in that the flood peaks are reduced. Since most of the transport and degradation occurs during the short periods of floods this reduction of transport capacity can be significant.

9.9.2 Aggradation

Aggradation occurs when the stream bed elevation is building up. However, the accumulation of sediment in a reservoir, which is also a form of aggradation, is called siltation. Aggradation of stream beds is usually associated with excessive sediment supply, for example, from a badly eroding catchment, or reduction of water flow as by water diversion structures which exclude sediment. Although aggradation is basically the same problem as degradation, it is frequently made more difficult through spatial variation of the sediment supply, i.e. sediment may be supplied from both banks overland and by tributaries over a considerable distance of the stream, not just an input boundary condition at the upstream end. A tributary, for example, may also bring in an excess amount of sediment at a section x_1 of a river. The flow in the river will then continue to change until a new sediment transport equilibrium is established over that section of the river, no deposition or scour. The upstream reach will slowly aggrade to the new slope imposed by the equilibrium at x_1 and its hydraulic conditions change to carry the incoming sediment at the new slope. The bed elevation changes logarithmically with time. From the input station downstream, the aggradation proceeds as a propagating wedge over which equilibrium conditions prevail. Downstream of the front of this wedge of sediment, flow remains essentially unaffected, i.e. at the initial conditions, (Soni et al. 1980).

The parabolic and hyperbolic methods can both be applied to aggradation, e.g. Jain (1981), Ribberink and van der Sande (1985). The predictions by these models, however, are in poor agreement with observations. The reasons for this are likely to be found in the use of oversimplified resistance and transport relationships. The results would also be affected to a lesser extent by the linearisation of the equations, and use of uniform flow relationships.

The aggradation in the form of delta formation in reservoirs is one of the most common problems. Most of the papers and reports on this topic, however, are qualitative in nature. Examples of analytical approaches to this problem are Yücel and Graf (1973) and Chen et al. (1978). Chang (1982) and Mertens (1986) describe experimental data. The major effects of the delta formation are the reduction of reservoir capacity and the rise in water level in the stream, as illustrated in Fig. 9.33. The loss of reservoir capacity can be rapid, particularly in semi-arid regions where seasonal heavy rainfalls on more or less bare soils lead to very high sediment loads in streams.

The delta form shown in Fig. 9.33 develops in narrow valleys. If the stream flows into a body of water, substantially wider than the jet entering, the delta grows both in length and in width.

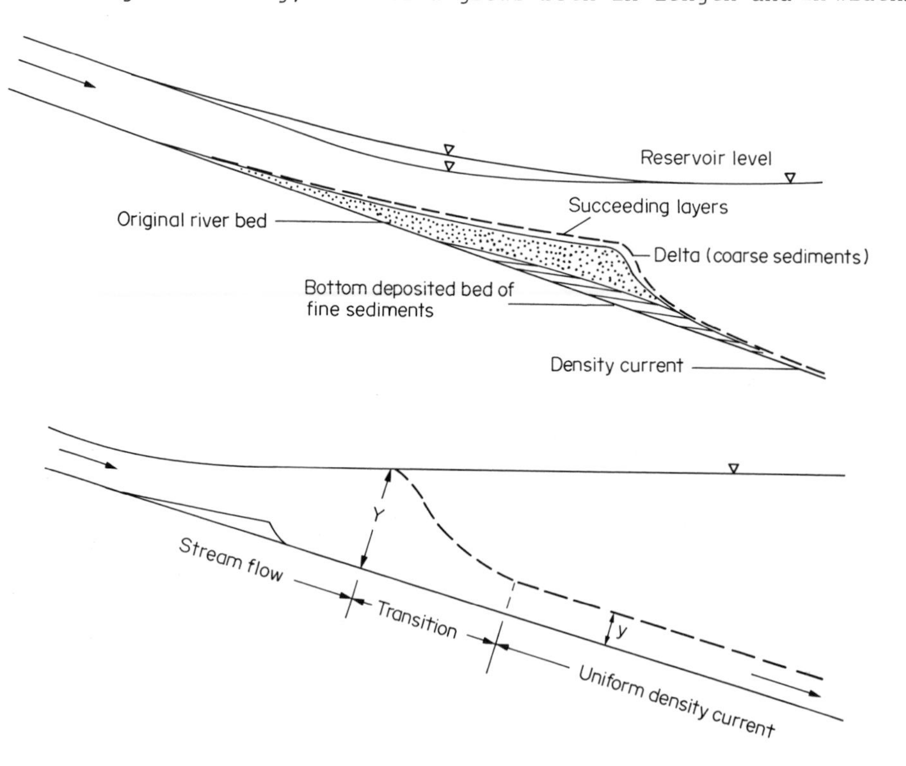

Fig. 9.33. Diagrammatic illustration of a reservoir delta, plunge point and density current.

The initial development is a strongly elongated delta in flow direction, over which the flow depth shallows. This encourages sidewards flow. The main stream may swing to one side or bifurcate and carry sediment sidewards until the friction losses again grow to an extent which causes a break-out at some other location. The end result is a more or less circular delta, as illustrated in Fig. 9.34.

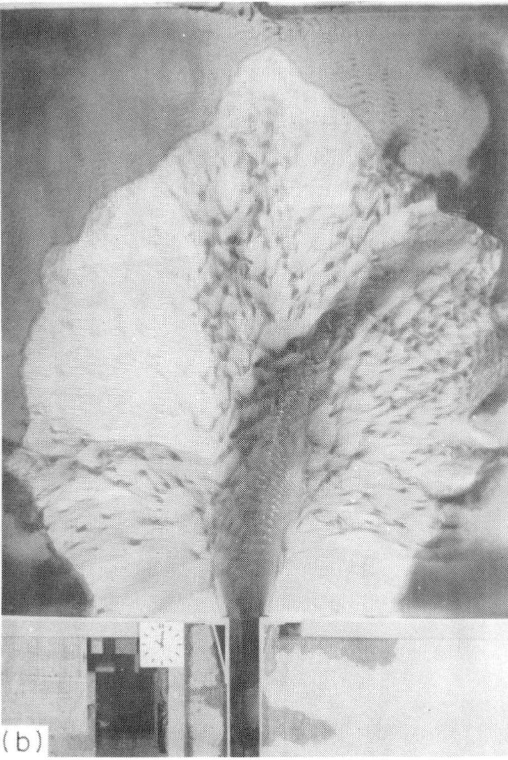

Fig. 9.34. Formation of a sand delta in laboratory experiment, (a) after 2 hours, (b) after 46 hours, Mertens (1986).

The distribution of sediment within the reservoir and the retention of sediment depend on sediment size and grading, size and shape of reservoir, the inflow and outflow rates and the type and location of outflow. Discrete non-flocculating particles settle according to their fall velocity and coarser fractions arrive as bed load. The settling of the finer fractions of the suspended load is strongly affected by the flow pattern in the reservoir. An inflow heavily laden with suspended sediment can also continue as a density current through the reservoir. A density current is the flow of a fluid of slightly different density under, through or over another fluid without loss of identity through mixing at the interfaces, although the fluid of the density current is miscible with the other fluid. The density differences may arise from sediment concentration, temperature or salinity. The sediment laden stream entering a nearly static water of the reservoir has higher density and momentum. It penetrates a distance into the reservoir and then plunges under the lighter water of the reservoir. It may continue to flow along the bed. Such a flow exhibits most of the features of open channel flow. The major differences are the much greater surface drag at the interface than between water and air and the mixing across the interface. The flow continues until it runs out of momentum and has deposited its load or until stopped by the dam.

The reservoir is basically a settling basin, but due to geometry and widely varying flow rates a very complex settling basin. The simple settling basins are used in many forms to separate solids from liquids. The basic concept is to provide a basin long enough and large enough in cross-section to enable solids to settle before the flow leaves the basin, and of course to provide adequate space for temporary storage of solids. The performance of the basin depends also on the design of the inlet and outlet transitions. These must keep the turbulence production to a minimum, must avoid flow separation and formation of large eddies, and must achieve an evenly distributed flow through the basin. The velocity through the basin must be less than the threshold value for movement of the sediment size to be removed from flow. An important field of application of such simple basins is the trapping of products from soil erosion on construction sites or mining operations. More complex settling basins occur at water diversion structures.

In concept, the rate at which water is freed from sediment is $Q = Aw$ where $A = BL$ is the surface area of the basin and w the fall velocity of sediment. The time for the particle to fall through the depth D is $t = D/w$ and in that time the particle moves forward a distance $L = tV$, which is the minimum length of the basin. In practice, the sediment has a size distribution, i.e. a distribution of fall velocities which depend further on the concentration. The design particle size and its hindered fall velocity, w_d, has to be selected according to requirements. All particles with $w > w_d$ will be totally removed. Hazen (1904) put forward the hypothesis that the removal ratio for particles with $w < w_d$ is proportional to their fall velocities, i.e. the removal ratio or trap efficiency $E = w/w_d = BLw/Q$, where B is the width of the basin. This is illustrated in Fig. 9.35. The area under the curve will give the removal ratio for the design w_d and the basin dimensions.

A very simple design relationship was put forward by the U.S. Bureau of Reclamation (1949):

$$W/W_o = \exp[- wL/q] \qquad (9.94)$$

where W/W_o is the ratio of sediment leaving to sediment entering the basin in terms of weight or mass, L is basin length and q is discharge per unit width. An extension of this concept was published by Cecen et al. (1969).

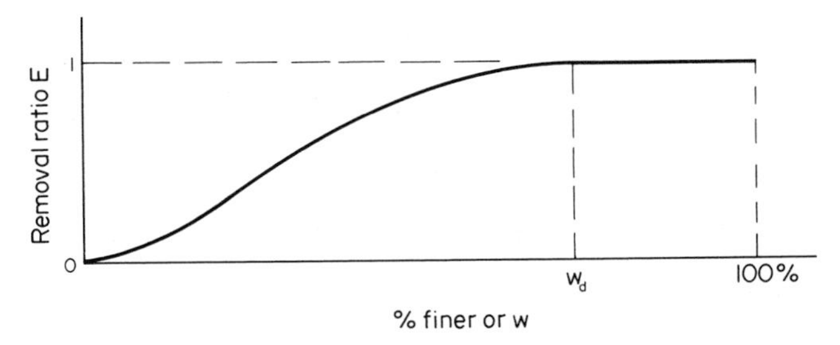

Fig. 9.35. Diagrammatic illustration of the removal ratio.

Hazen assumed a rectangular basin with uniform velocity and that the concentration of suspended particles of each size is the same at all points in the vertical at the inlet section. Camp (1946) and Dobbins (1944) extended the idea by introducing mixing. Camp assumed a parabolic velocity distribution with constant mixing $\varepsilon = 0.075\ Du_*$ throughout the depth, D, and expressed u_* with the aid of Manning formula and E as a function of wA/Q and $wD/2\varepsilon$. Brown (1950) converted the latter into a parameter $wD^{1/6}/Vn\sqrt{g}$. Chen (1975) related E for highly turbulent flow conditions (low values of $wD/2\varepsilon$) to wA/Q:

$$E = 1 - \exp(-\ wA/Q) \qquad (9.95)$$

and prepared a series of curves for a range of particle sizes, Fig. 9.36, which emphasises the effect of sediment size on the A/Q-ratio required.

In large reservoirs the amount of sediment deposited is expressed in terms of the *trap efficiency*. It is the amount of sediment deposited in percentage of the total sediment inflow over a selected period of time. The trap efficiency depends on the ratio of storage capacity to inflow, the shape and age of reservoir, the type and location of outlets, sediment composition and reservoir operation.

Brune (1953) related the trap efficiency empirically to capacity/inflow ratio, the so-called Brune's curves, Fig. 9.37, where capacity is measured at the mean operating level and inflow is the mean annual inflow from the catchment. These apply to storage reservoirs, not to settling basins or flood control reservoirs where large flow rates are allowed to pass through the reservoir. For additional methods reference is made to U.S. Bureau of Reclamation (1977), Borland (1971).

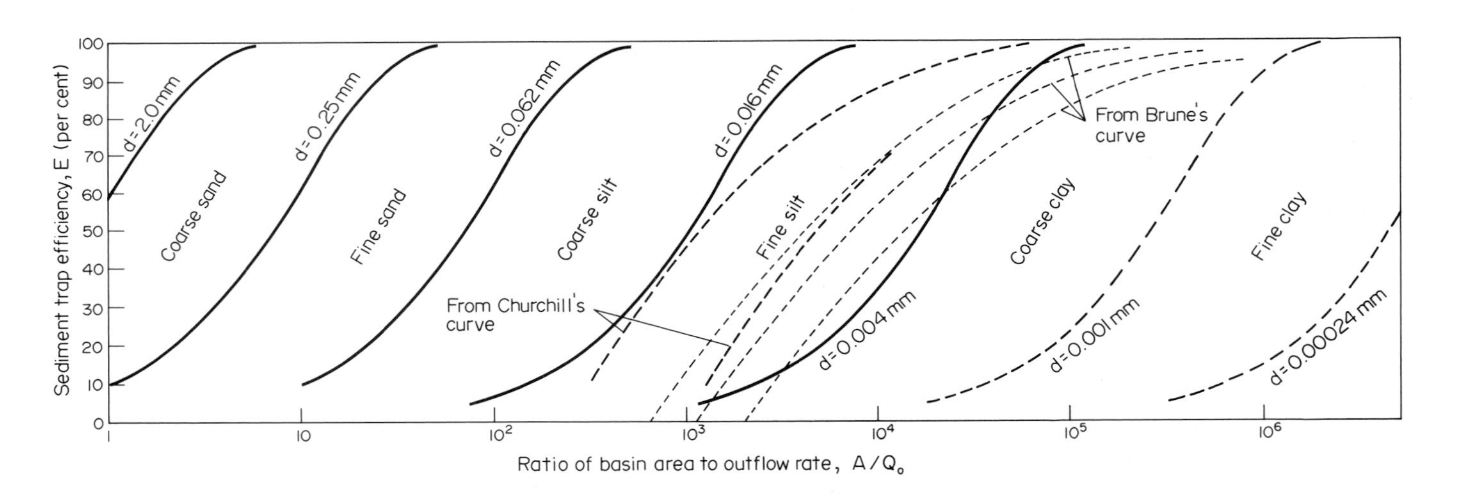

Fig. 9.36. Trap efficiency versus ratio of basin area to outflow rate with particle size as parameter, after Chen (1975).

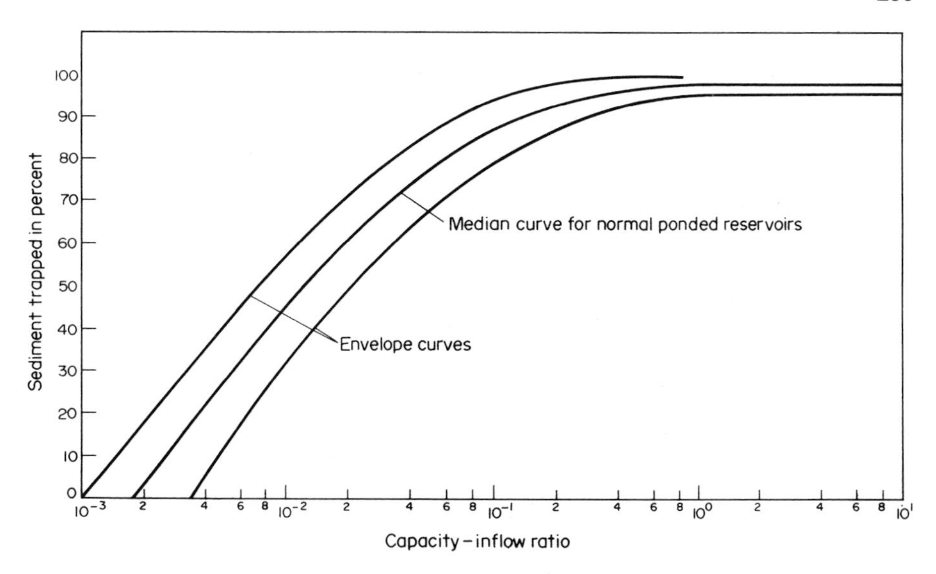

Fig. 9.37. Reservoir trap efficiency curves, after Brune (1953).

Chapter 10

Cohesive Sediments

A characteristic feature of many deposits is that they form a coherent mass and not just a collection of individual particles in contact with each other. Such coherent deposits are said to be cohesive, a term like climate which is used to describe the combined effect of complex interactions of many factors but cannot be defined. Cohesive sediments contain significant amounts of clay minerals which assume control of the properties of the soil. Only about 10% of clay will suffice to assume control of the soil properties. The cohesive properties arise from electrochemical forces in the clay-water medium. These forces usually dominate and may be orders of magnitude larger than the weight force of individual particles. The study of forces in a system of clay minerals and water based electrolytes forms part of the domain of physical chemistry. The advances are greatest in the study of suspensions, the so-called colloidal systems, but consolidated clays are still to a large extent "terra incognita". The mechanics of sediment transport relates almost exclusively to noncohesive sediments.

10.1 Clay Minerals

Most of the clay minerals have a layered structure and consist of two types of so-called sheets; the silicon-oxygen sheet and the Al, Fe or Mg-O-H sheet.

The silicon-oxygen sheet has a tetrahedral structure in which the silicon atom is equidistant from four oxygens or hydroxyls. Three of the four oxygen atoms of each tetrahedron are shared by three neighbouring tetrahedra, forming the silica sheet $Si_4O_6(OH)_4$, illustrated in Fig. 10.1. The oxygen to oxygen distance is 0.255 nm and the thickness of the sheet is 0.493 nm (nm = 10^{-9}m).

The other sheet consists of two layers of closely packed oxygens or hydroxyls in which aluminium, magnesium or iron are embedded. The metal atoms are surrounded by six oxygen atoms or OH groups. These have their centres on the six corners of a rectangular

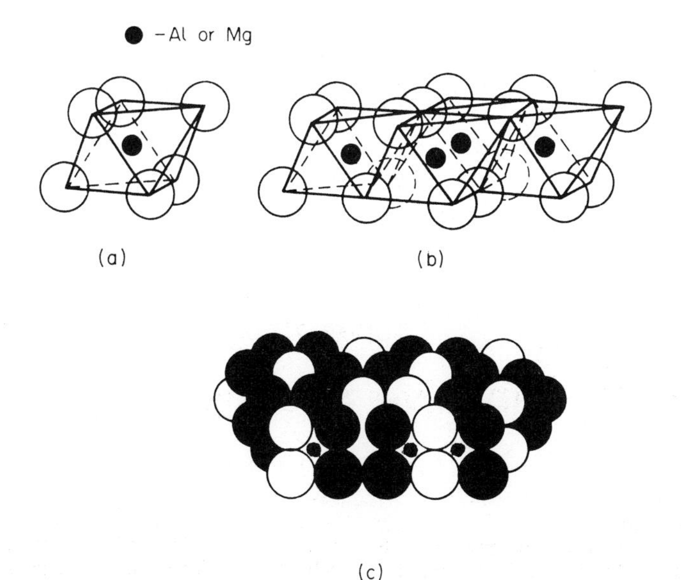

Fig. 10.1. Diagrammatic structure of the tetrahedral sheet. (a) Tetrahedral arrangement of Si and O. (b) Sheet structure of tetrahedral unit. (c) Sheet structure projected on the base plane of the tetrahedron.

octahedron. The shearing of oxygen atoms by neighbouring octahedrons leads to the octahedral sheet, illustrated in Fig. 10.2. If the metal is aluminium, the sheet is referred to as the alumina sheet. In it only two-thirds of the possible positions are filled by Al atoms for balance conditions. The alumina sheet is also called the gibbsite layer $Al_2(OH)_6$. Magnesium as the metal fills all the spaces and the sheet is the magnesium or brucite sheet $Mg_3(OH)_6$. Average oxygen distances are 0.26 nm and the thickness of the sheet is 0.505 nm.

The almost identical dimensions and patterns allow sharing of oxygen atoms between the two types of sheets. This may be between one silica and one alumina sheet, one Al or Mg sheet with a silica sheet on each side, etc. Most clay minerals are built up as stacks of such sheets.

The major types of clay minerals are the

(i) two-layer (1:1) clays like kaolinite,
(ii) three-layer (2:1) clays like montmorillonite, illite, a.o., and the
(iii) four-layer (2:1:1) clays like chlorites

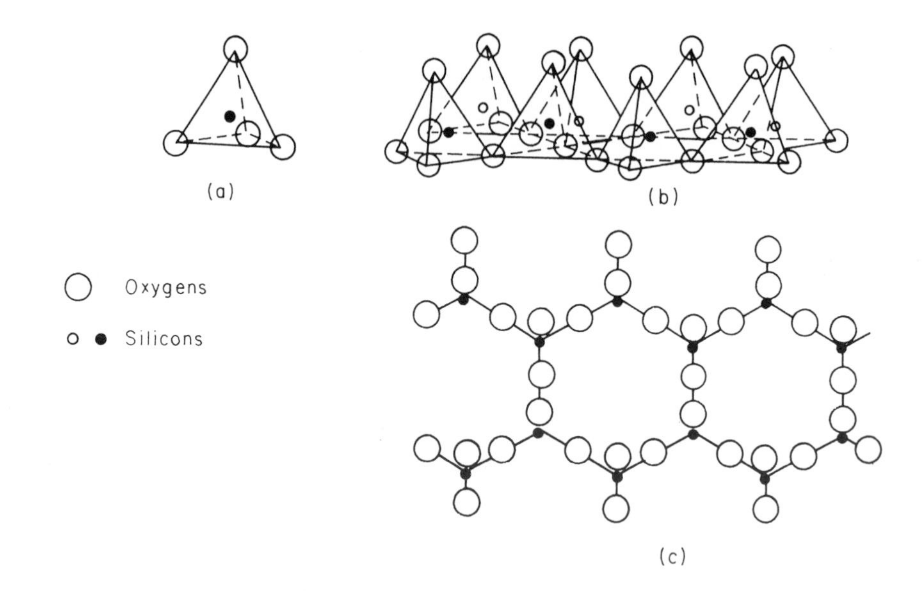

Fig. 10.2. Diagrammatic structure of the octahedral sheet. (a) Octahedral arrangement of Al or Mg with O or OH. (b) Sheet structure of octahedral units. (c) Model of the octahedral sheet, O shaded, OH light

However, the list of the varieties of clays is long and the deposits usually consist of more than one type of clay, often mixed with sand and silt as well.

The *kaolinites* (kaolin group) have a two-layer structure, a Si_2O_3 tetrahedral sheet combined with the $OHO_2Al_2(OH)_3$ octahedral sheet, so that the tips of the silica tetrahedrons form a common layer with one of the layers of the octahedral sheet from which the oxygens satisfy the fourth valency of the silicon (Fig. 10.3). It is thought that the hydrogen atoms of the $(OH)_3$ sheet form partial valency bonds (hydrogen bonds) with oxygen in the silicon sheet and thus provide stronger links between the sheets than those by van der Waals forces. Some of the other members of the group, apart from kaolinite, are dickite, nacrite and halloysite ($2H_2O$) and ($4H_2O$). The halloysites tend to occur in tubular rolled-up form, like scrolls. Kaolinites are more recent sediments and occur mainly in sediments deposited after the Devonian period. They are also associated with advanced stage of weathering. Many intensely weathered soils of the warm and humid tropical regions have clay fractions dominated by kaolinite and gibbsite, and are notoriously infertile.

Fig. 10.3. Model of the 1 : 1 clay mineral kaolinite. The large black spheres represent OH, large white O, the small black spheres represent Si, and the small white Al. Top black layer is hydroxyls of the octahedral layer, the middle layer is hydroxyls and oxygens where the oxygens are common to both layers, and the bottom layer is basal oxygens of the tetrahedral layer. The photo supplied and used with permission, from "Fundamentals of Soil Science", H.D. Foth and L.M. Turk, 5th Ed. J. Wiley & Sons Inc., New York, 1972.

The *montmorillonites* are known as the expanding three-layer clays and are derived from the neutral three-layered pyrophyllite (Al) and talc (Mg) by isomorphous substitution of other atoms. The tetravalent Si atom may be replaced at some locations by trivalent Al. In the octahedral sheet replacement of Al by divalent Mg may take place. Al atoms could also be replaced by Fe, Cr, Zn, Li and other small atoms which fit in the space of Si and Al atoms. A replacement by an atom of lower valance leads to a deficit of positive charge. The excess negative lattice charge leads to absorption on the surface of cations. The latter are too large and do not fit into the lattice. The amount of these cations is expressed in milliequivalent (meq) per 100 grams of dry clay and is known as the cation exchange capacity or the base exchange capacity of the clay. In contact with water the water molecules penetrate between unit layers and this leads to swelling. The structure of montmorillonite is illustrated in Fig. 10.4. Montmorillonite is the main clay mineral in bentonite rock, which originates from volcanic ash. Other members of the group are hectorite, saponite, sauconite, vermiculite, volchonskonite, nontronite. Montmorillonites are less common in sediments older than Mesozoic and represent an intermediate stage of weathering. They are common clay fractions in silty soils of temperate regions which have developed under trees and grass.

Illites (hydrous mica group) are the non-expanding three-layer clays. They exhibit no interlayer swelling with water or organic compounds, and are in structural characteristics similar to the micas. The basic structure of illite unit is the same as of the montmorillonites but here some of the silicons in the tetrahedral sheet are replaced by aluminiums and the charge deficiency is balanced by potassium ions (Fig. 10.5). These potassium ions

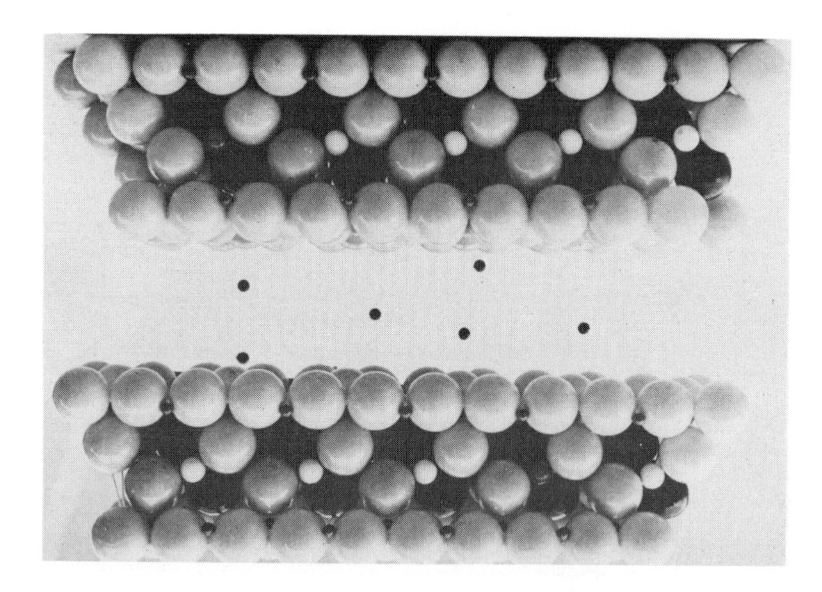

Fig. 10.4. Model of the 2:1 clay mineral montmorillonite. The large white spheres represent O, large black OH. The medium-sized grey spheres represent Mg and white Al. The small grey spheres represent Si and the small black between the two stacks represent exchangeable cations. The gap between the stacks expands and contracts on wetting and drying, and the surfaces of the stacks are available for adsorption of water molecules and exchangeable cations. The cations between the stacks may be hydrated. (The photograph supplied and used with permission, from *Fundamentals of Soil Science*, H.D. Foth and L.M. Turk, 5th ed. J. Wiley & Sons Inc., New York, 1972).

are located on the unit layer surface and are assumed to be the cause of the linking of the layers and of the non-swelling nature of these clays. The non-swelling of illites, as compared to montmorillonites which also have potassium ions on surfaces, is thought to be due to the larger number of potassium ions. In montmorillonites the negative lattice charge is concentrated more in the octahedral sheet and has a greater distance from the potassium ions, resulting in weaker links. However, Radoslovich and Norrish (1962) show that the silica sheet can, by slight rotation, reduce the size of the holes of the hexagonal oxygen rings and thereby adapt their dimensions to those of the octahedral sheet. In illite the unit layers do not open up on contact with water and these ions are not available for exchange, except on external surfaces. Therefore, the cation exchange capacity of illites is of the order of 20-40 meq per 100 g compared to 70-100 meq for montmorillonite, whereas the total amount of charge compensating cations may be 150 meq for 100 g. The muscovite and phlogopite are the base minerals for illites. The

Fig. 10.5. Model of the 2:1 clay mineral illite. Substitution of some lattice silicon +4 by aluminium +3 in the tetrahedral layer produces negative charge which is largely balanced by potassium ions. These fix the units to each other and prevent swelling; the distance between stacks is fixed. Large grey spheres between stacks represent K, large white O, large black OH. Smaller white spheres represent Al and the small grey spheres Si. (The photograph supplied and used with permission, from *Fundamentals of Soil Science*, H.D. Foth and L.M. Turk, 5th ed. J. Wiley & Sons Inc., New York, 1972.)

boundary between illites and montmorillonites is ill-defined and both are in the intermediate stage of weathering. Some of the other types are glancomite, vermiculite, hydrobiotites, betavite and allvardite. Illites frequently occur in extremely fine particle size mixed with other clays. The very old argillaceous (clay-like) sediments are largely composed of illite and chlorite.

The *chlorites* are structurally related to the three-layered clays. In chlorites the charge-compensating cations between the montmorillonite-type unit layer are replaced by a sheet of brucite (octahedral magnesium hydroxide) in which some of the Mg is replaced by Al. This layer now has a net positive charge which appears to compensate for the net negative charge of the unit layers because the cation exchange capacity of chlorites is very low.

The clay deposits do not consist of single crystals, like sand, but of aggregation of basic units in a variety of sizes which have separated along surfaces of weakness, Fig. 10.6. Numerous terms exist to define these such as peds, crumbs, aggregates are applied to units visible by eye; clusters, flocks for microscopic

Fig. 10.6. Kaolinite clay "particles". The stack in the middle of left hand side is about 5 μm wide.

size and domains, packets, etc to those which are identifiable in electron microscopes. The deposits are said to have *texture* and *structure*. Texture refers to the distribution of the size of components. Structure or *fabric* is used to describe the spacial distribution of the units of various sizes. The basic structural arrangements in a clay suspension are illustrated in Fig. 10.7. The structural features are carried over to deposits as the aggregates and flocks settle. Details of these aspects are given by Yong and Warkentin (1975) and Mitchell (1976).

For further reading on clay mineralogy a start could be made with Grim (1962), van Olphen (1966), Marshall (1964) and Low (1961, 1968).

10.2 Clay-Water Electrolyte System

The cohesive properties of soils are the result of the clay-water system in which the interactions between the clay minerals and the water determine the properties. The interactions depend on the clay minerals and on the ion or cation content of water. These affect the properties of clay in every state from suspensions through sedimentation to compacted clay. The structure of the clay, for example, depends strongly on the properties of the electrolyte (water) during sedimentation. It should be noted that the clay properties are influenced not only by the properties and concentration of ions in pore water but also by molecules of organic substances in pore water.

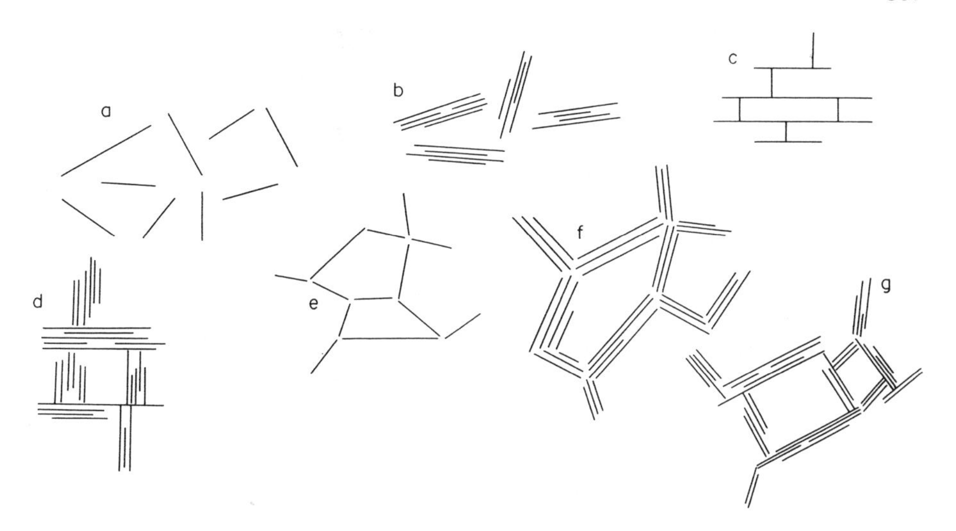

Fig. 10.7. Illustration of particle associations in clay suspensions, after van Olphen (1977):
(a) Dispersed and deflocculated.
(b) Face to face association (parallel orientation) aggregated but deflocculated.
(c) Edge to face flocculated but dispersed.
(d) Edge to face flocculated and aggregated.
(e) Edge to face flocculated and dispersed.
(f) Edge to face flocculated and aggregated.
(g) Edge to face and edge to edge flocculated and aggregated.

In an electrolyte solution the ions are, on the average, uniformly distributed and move about in a continuous random movement. In the vicinity of a boundary, like a clay mineral, however, a distribution of the ion concentration develops, giving rise to what has become known as the double layer, Fig. 10.8. For any two phases of substances in contact an inner or Galvani potential, ϕ, develops which can be subdivided into the volta potential, ψ, caused by charges on the phase and the chi potential, χ, arising from the surface dipoles, i.e.

$$\phi = \psi + \chi$$

Between two different phases, 1 and 2, there is an electric potential difference

$$\Delta\phi = \phi_1 - \phi_2 = \Delta\psi + \Delta\chi \qquad (10.1)$$

The surface of the crystal has an electrical potential, ϕ_s. The plane through the centre of the adsorbed dehydrated ions is called the inner Helmholtz plane (IHP) or the Grahame layer with potential ϕ_1 and the plane through the centres of closest possible hydrated ions marks the outer Helmholtz plane (OHP) or the Stern layer with potential ϕ_d. The plane at which slippage is regarded

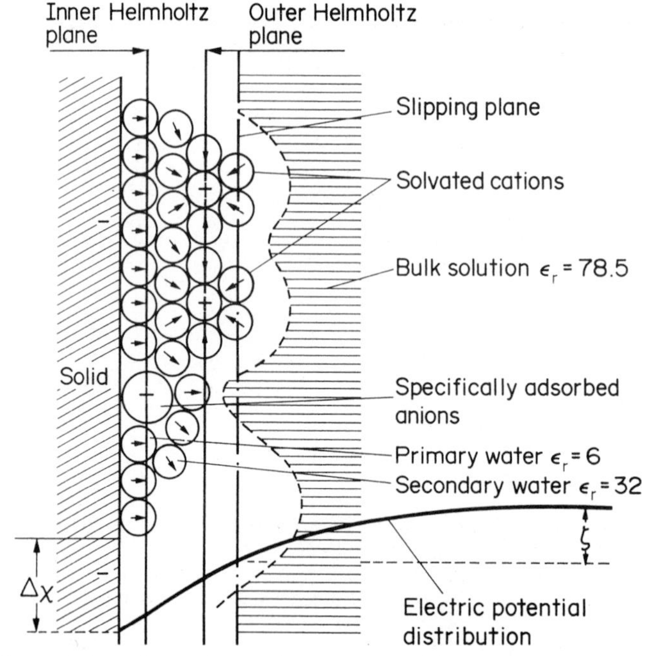

Fig. 10.8. Charged solid surface and double layer (drawing adapted from MacKenzie, 1971).

to occur between the crystal and fluid phases is at a small distance out from the OHP. The potential at this distance is referred to as the zeta (ζ)-potential and the approximation $\zeta \cong \phi_d$ is regarded to be the best estimate.

The condition when the charge on the crystal is zero, i.e. $\Delta\psi = 0$ and $\Delta\phi = \Delta\phi_o = \Delta\chi$ is known as the isoelectric point (iep) and $\Delta\phi_o$ at zero charge as the Lippman potential.

There are two important sources of charge on clay minerals:

1. The isomorphous substitution where the substitution of Al^3 for Si^{4+} or Mg^{2+} for Al^{3+}, for instance, within the clay lattice produces a net negative charge. This charge dominates the negative sheet (basal surface) charge.

2. In mineral oxides the ionization of broken bond surfaces leads to a charge, i.e.

$$M] - OH + H_2O \rightleftarrows M] - O^- + H_3O^+ \qquad (10.2)$$

$$M] - OH + H_3O^+ \rightleftarrows M] - OH_2^+ + H_2O \qquad (10.3)$$

where M] is the surface species of metal M. Reaction eqn (10.2) results in a negative lattice charge and reaction eqn (10.3) in a positive lattice charge. The edges of the lattice structure in clay minerals are considered to behave in the same manner as the broken bond surfaces of simple oxides. However, the basal surfaces of clay minerals are not regarded to be reactive in the same way. Under mildly acidic conditions kaolinite has negatively charged basal surfaces but positively charged edge surfaces. The lower or basal surface iep for kaolinite occurs at pH $\cong$ 3. The iep for the edge surfaces occurs at pH-values between 7 and 8, although 5.8 to 9 have been reported. Montmorillonites do not show a clear iep in the presence of hydroxyl (OH^-) or hydronium (H_3O^+) ions but a strong iep exists in salt solutions.

Under a given set of conditions of temperature and pH a clay mineral has an ion exchange capacity where ions swap position from the pore water to within or on the surface of the clay crystal lattice. The causes are

1. the adsorption of cations (positively charged ions) to balance negative charges due to broken bonds

2. to balance charges due to isomorphous substitution and

3. the exchange of cations with the hydrogen of exposed hydroxyls.

Exchangeable cations from lattice substitutions (isomorphous substitutions) are mostly found on cleavage surfaces, e.g. the basal cleavage surface of the layer clay minerals. This accounts for about 80% of the total cation exchange capacity in montmorillonite and vermiculite. The exchange cations do not necessarily shed their solvation sheaths of orientated water molecules and this gives rise to the high swelling potential of these clays. Adsorption of anions is also observed, known as super-equivalent adsorption. Although electrostatic forces oppose the adsorption of anions onto a negatively charged surface the chemical driving potential is stronger than the electrostatic repulsion.

The charges on clay particles also lead to an interaction between particles. For example, the edges of kaolinite particles are positively charged when the pH-value is less than that for iep-value. Then the positive edge and negative surface charges lead to a "cardhouse" structure of clay. For pH-values corresponding to the iep-value of edges the edge-edge bonds are preferred, governed by the van der Waal forces while the repulsion forces have a minimum. The van der Waal forces arise from interaction of atoms in the particles when these approach each other.

Further reading, Bockris and Reddy (1970).

10.3 Erosion of Cohesive Soils

The subdivision of erosion of cohesive soils and erosion of muds is arbitrary and made here for convenience. Muds refer to soft deposits, particularly in estuarine environments. The fluid-mud interactions usually involve repeated cycles of deposition and resuspension. The term soil is used to identify firm consolidated deposits of cohesive sediments.

The literature on the erosion of cohesive sediments reflects by its diversity the embrionic state of knowledge in this field. The erosion characteristics have been described almost exclusively with the aid of parameters used in soil mechanics which themselves are bulk characteristics and not definable in terms of the parameters of soil physics or chemistry.

The common parameters used are *grain size*, *dispersion ratio*, *clay content*, *Atterberg limits*, *shear* and *tensile strength*, *water content*, *salt content*, *temperature*, *sodium adsorption ratio* and *ion* or *cation exhange capacity*. These have not yielded results which could be used readily for predictive purposes and one is still forced to use empirical data like the permissible velocities quoted by Fortier and Scobey (1926), Table 10.1, or the Mirtskhoulava (1981), Table 10.2; or Fig. 10.9(a), by Garbrecht (1961) or values of critical bed shear stress recommended in the USSR, Fig. 10.9(b).

The *grain size* in noncohesive soils has a dominant influence on erosion since the weight is proportional to diameter cubed. In cohesive soils, in contrast, the grain size (if it can be defined at all) and its weight are quite insignificant in comparison to the elctro-chemical forces. Studies relating to aerosols indicate that the critical shear stress for incipient motion is proportional to d^{-1} to $d^{-4/3}$ power. Figure 10.10 by Croad (1981) shows these tendencies, which are also in keeping with the plot by Sundborg (1956), Fig. 10.11.

The U.S. Bureau of Reclamation has used the *dispersion ratio* as the basis for prediction of erodibility of cohesive soils, but, at the best, it only indicates trends. The dispersion ratio is the ratio of percentage finer than clay sized particles (set arbitrarily at 5 μm) of the non-dispersed and dispersed clays.

The particle size distribution is measured by the standard hydrometer test. For the dispersed sample the clay is dispersed both mechanically and chemically as much as possible. A second test on a sample without chemical dispersant and minimum of agitation yields the other value. Figure 10.12 shows measured data. It shows as expected that clays which disperse more readily have higher erosion rates. Clays which have particularly high levels of dissolved sodium in the pore water are dispersive, e.g. Na-montmorillonite. However, even small amounts of dissolved salts in the eroding water cause a reduction in dispersion. Clays with total salt content in pore water of less than 1 milli-equivalent/ℓ are generally non-dispersive.

The effect of *clay content* was readily emphasised. With increasing clay content the soil-becomes more plastic, its swelling and shrinkage increases and compressibility increases, while permeability and angle of internal friction decrease. The fraction of clay, c, necessary to fill the pores of a soil at a given water content, w, so that the soil grains lose contact is approximately

$$C = 48.4 - 1.42\ w \tag{10.4}$$

where w is the weight percentage of water in terms of dry weight

Table 10.1. Permissible canal velocities, according to Fortier and Scobey (1926)

Canal excavated in	Velocity, after ageing, of canals* carrying: Clear water (m/s)	Clear water (ft/s)	Water-transporting colloidal silts (m/s)	Water-transporting colloidal silts (ft/s)	Water-transporting non-colloidal silt, sand, gravels, or rock fragments (m/s)	Water-transporting non-colloidal silt, sand, gravels, or rock fragments (ft/s)
Fine sand (non-colloidal)	0.46	1.50	0.76	2.50	0.46	1.50
Sand loam (non-colloidal)	0.53	1.75	0.76	2.50	0.61	2.00
Silt loam (non-colloidal)	0.61	2.00	0.92	3.00	0.61	2.00
Alluvial silts when non-colloidal	0.61	2.00	1.07	3.50	0.61	2.00
Ordinary firm loam	0.76	2.50	1.07	3.50	0.69	2.25
Volcanic ash	0.76	2.50	1.07	3.50	0.61	2.00
Fine gravel	0.76	2.50	1.52	5.00	1.52	3.75
Stiff clay (very colloidal)	1.14	3.75	1.52	5.00	0.92	3.00
Graded, loam to cobbles, when non-colloidal	1.14	3.75	1.52	5.00	1.52	5.00
Alluvial silts when colloidal	1.14	3.75	1.68	5.50	0.92	3.00
Graded, silt to cobbles when colloidal	1.22	4.00	1.83	6.00	1.52	5.00
Coarse gravel (non-colloidal)	1.22	4.00	1.68	5.50	1.83	6.50
Cobbles and shingles	1.52	5.00	1.68	5.50	1.98	6.50
Shales and hard pans	1.83	6.00	1.83	6.00	1.52	5.00

*Depth of 1 m or less.

Table 10.2. Permissible non-scouring mean velocities of flow for clay soils

Clay soils at calculated specific cohesion C, kg/cm² (Multiply with 9.81×10^4 for Nm^{-2})	Permissible non-scouring mean velocities in m/s for flow depth in m							
	0.5		1		3		5	
	For contents of easily soluble salts ($CaCl_2$, $MgCl_2$, NaCl, Na_2SO_4, Na_2CO_3, $NaHCO_3$) in % according to dense residue of absolutely dry soil weight							
	less than 0.2	0.2-3	less than 0.2	0.2-3	less than 0.2	0.2-3	less than 0.2	0.2-3
0.005	0.39	0.36	0.43	0.40	0.49	0.46	0.52	0.49
0.01	0.44	0.39	0.48	0.43	0.55	0.49	0.58	0.52
0.02	0.52	0.41	0.57	0.45	0.65	0.52	0.69	0.55
0.03	0.59	0.43	0.64	0.48	0.74	0.55	0.78	0.59
0.04	0.65	0.46	0.71	0.51	0.81	0.58	0.86	0.62
0.05	0.71	0.48	0.77	0.53	0.89	0.61	0.98	0.65
0.075	0.83	0.51	0.91	0.56	1.04	0.64	1.10	0.69
0.125	1.03	0.60	1.13	0.67	1.30	0.76	1.37	0.81
0.15	1.21	0.65	1.33	0.72	1.52	0.82	1.60	0.88
0.20	1.28	0.75	1.40	0.82	1.60	0.93	1.69	1.00
0.225	1.36	0.80	1.48	0.88	1.70	1.00	1.80	1.07
0.25	1.42	0.82	1.55	0.91	1.78	1.04	1.88	1.10
0.30	1.54	0.90	1.69	0.99	1.94	1.12	2.04	1.20
0.35	1.67	0.97	1.83	1.06	2.09	1.22	2.21	1.30
0.40	1.79	1.03	1.96	1.15	2.25	1.31	2.38	1.40
0.45	1.88	1.09	2.06	1.20	2.35	1.39	2.49	2.48
0.50	1.99	1.26	2.17	1.28	2.50	1.46	2.63	1.56
0.60	2.16	1.27	2.38	1.38	2.72	1.60	2.88	1.70

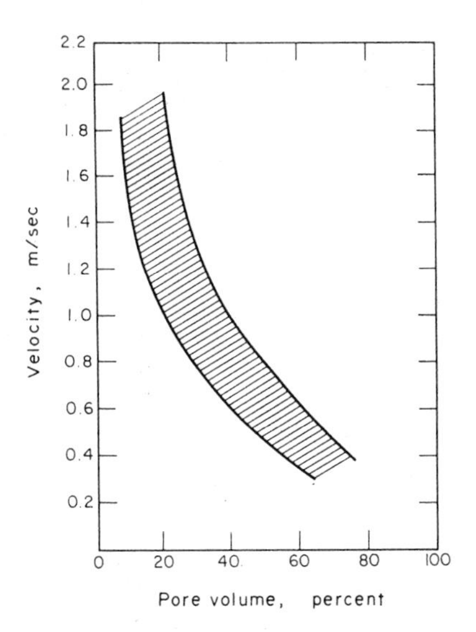

Fig. 10.9(a). Permissible velocity versus pore volume, according to Garbrecht (1961).

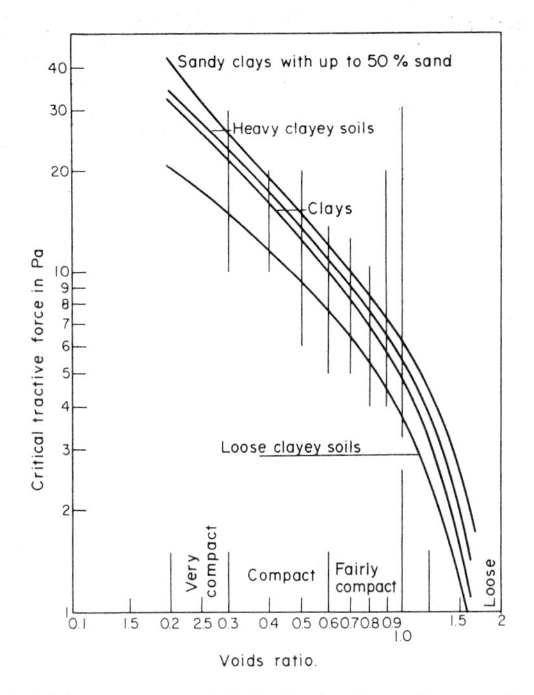

Fig. 10.9(b). Recommended design values for canals in cohesive material, according to Hydro-technical Construction, Moscow, May 1936.

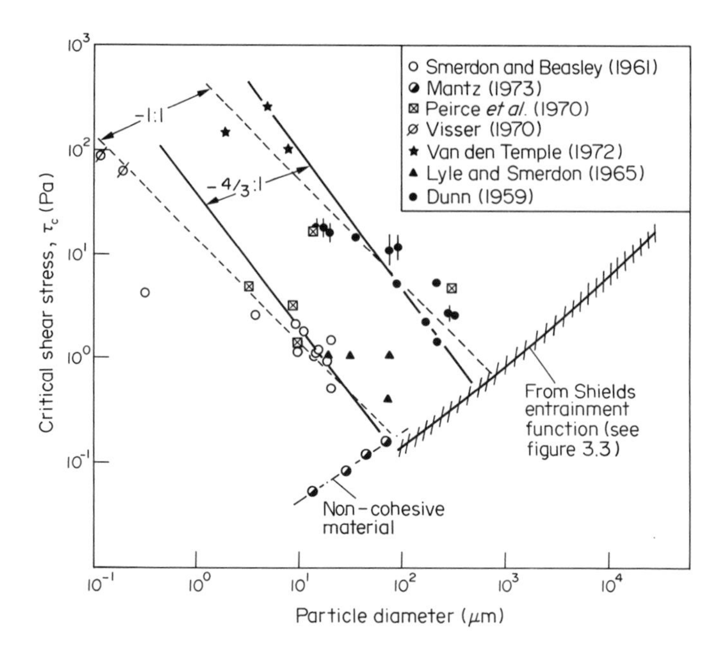

Fig. 10.10. Critical shear stress versus particle diameter.

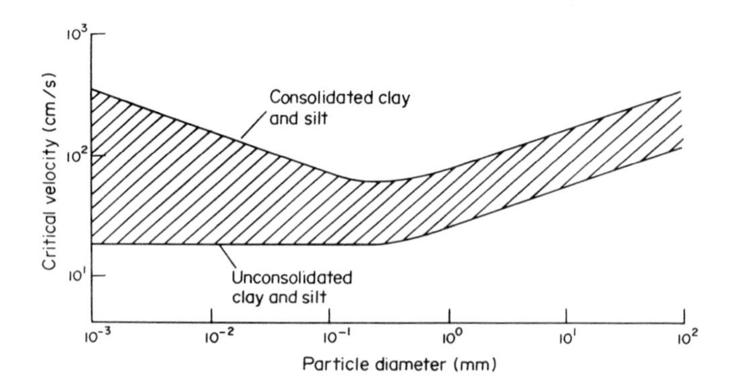

Fig. 10.11. Critical velocity versus particle diameter (Sundborg, 1956).

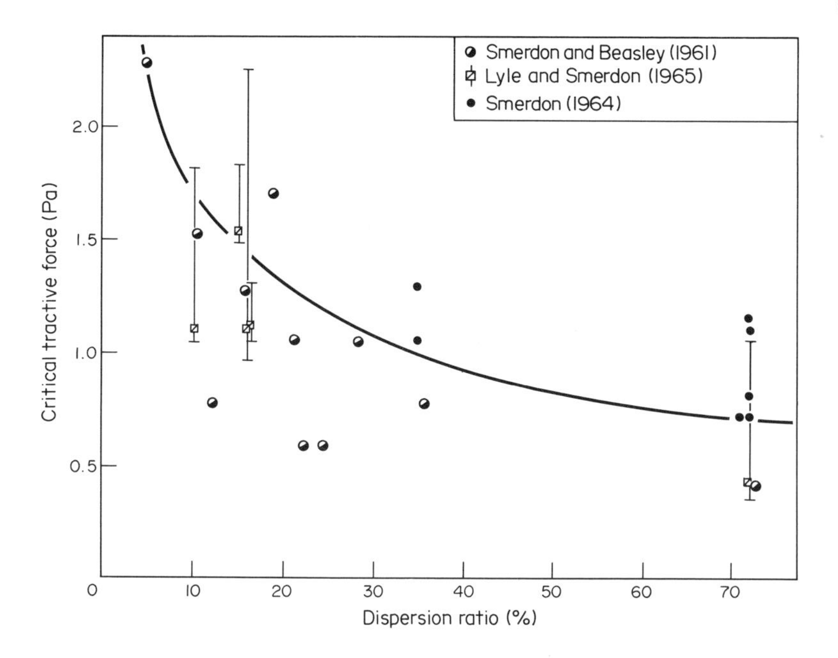

Fig. 10.12. Critical tractive force versus dispersion ratio.

of soil. About 10% of clay will suffice for the clay to assume complete control of soil properties. Figure 10.13 shows data for critical shear stress versus proportion of clay. For comparison the critical shear stress of a noncohesive material of the order of size of 10 μm is about 0.1 Pa.

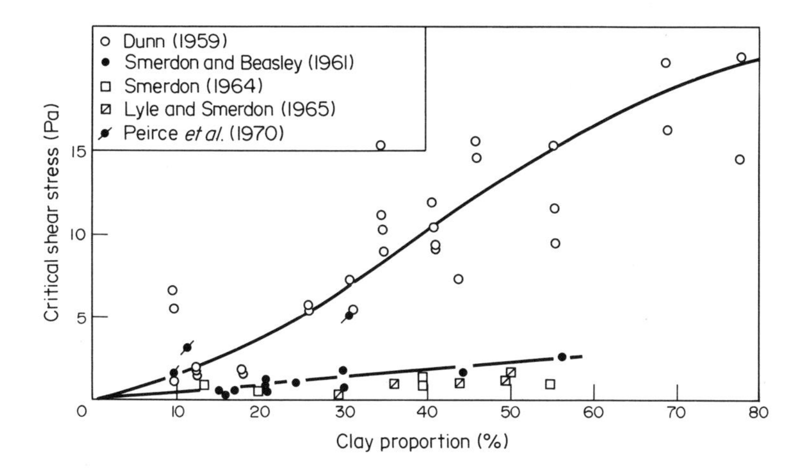

Fig. 10.13. Critical shear stress versus clay proportion.

Atterberg limits, plasticity index and liquid limit, have been widely used to predict the erodibility of clays. Figures 10.14 to 10.16 show that clear trends are discernible but clearly no unique relationship exists.

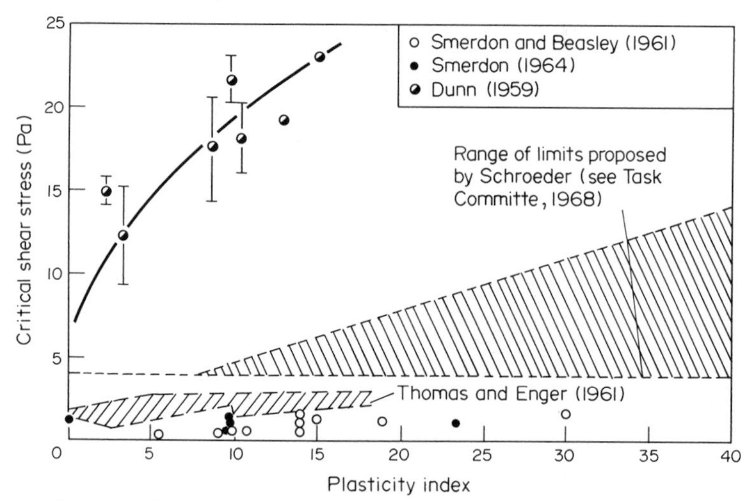

Fig. 10.14. Critical shear stress versus plasticity index.

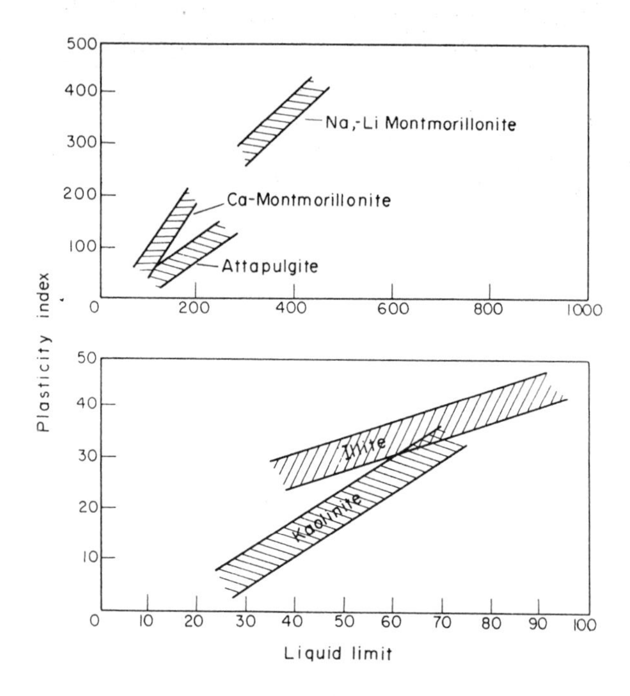

Fig. 10.15. Liquid limit versus plasticity index for the common clay minerals (Grim, 1962).

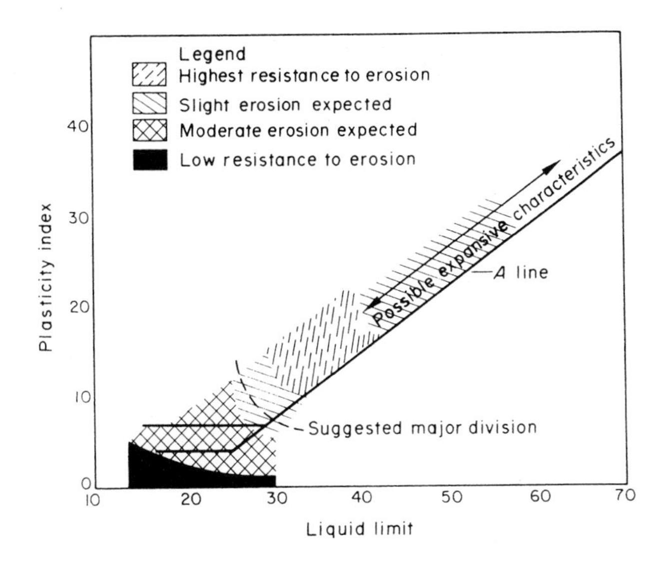

Fig. 10.16. Suggested trend of erosion characteristics for fine graded cohesive soils with respect to plasticity. (After Gibbs, 1962).

Shear and tensile strength of a clay have been extensively used to correlate with critical shear stress for erosion of clays. A plot of data from various sources of this type is shown in Fig. 10.17. Most of the results (e.g. Dunn (1959), Flaxman (1963), Task Committee (1968), Mirtskhulava (1966)) show a positive correlation but no single relationship between shear strength and critical shear stress. The shear strength measurements of Dunn (1959) and Task Committee (1968) are vane shear strengths. The measurements by Flaxman (1963) are unconfined compressive shear strength. Although some researchers have described the erosion of clays as a plucking of clumps from the surface, rather than a wearing type of erosion, only the study by Dash (1968) has been found where the erosion is related to the tensile strength of clay. The results indicate that erosion decreases with increasing tensile strength.

The trends of experimental data on *water content* versus critical shear stress indicate that the critical shear stress decreases with increasing water content. This is in keeping with the concept that the strength (inter particle bonding) of a clay decreases with increasing particle spacing. The results of Grissinger (1966) indicate that it is important whether the soil was saturated during compaction stages. For clays with an orieted structure the after consolidation erosion rate decreases with increasing water content.

The addition of *salt* to pore water leads to a reduction of the thickness of the Helmholtz double layer and as a consequence to a reduction of the repulsive force between the particles. Hence, an increase in erosion resistance could be expected with

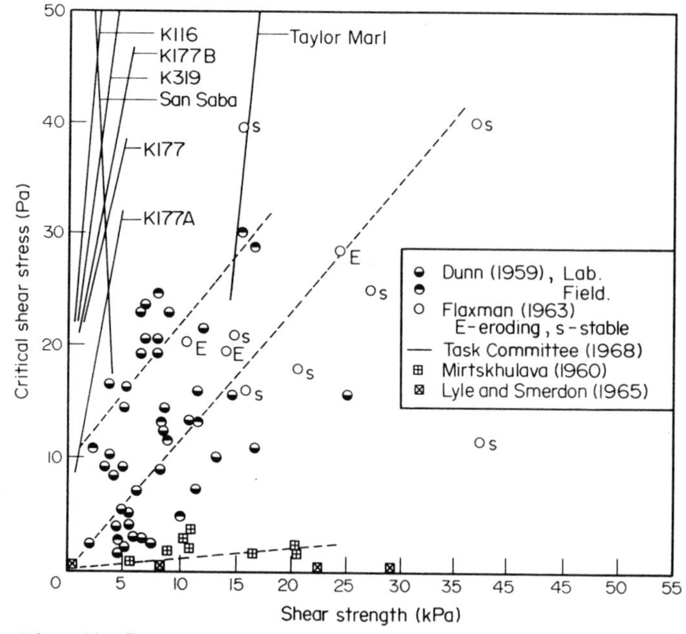

Fig. 10.17. Critical shear stress versus shear strength.

increasing salt concentration in pore water. Such a trend has been illustrated for Na Cl salt by several researchers, e.g. Liou (1970), Sargunam et al. (1973) and Arulanandan et al. (1975). The effect of salt depends on its valency. Smaller concentrations of a 2-valent salt ($Ca\ Cl_2$) are required than of 1-valent salt (Na Cl) to produce the same erosion resistance. Hutchison (1972) also found that the effect of salt concentration increases with the decrease in grain size, i.e., with increasing specific surface area. The effect of salts has been expressed in terms of the *sodium adsorption ratio* (SAR). The SAR is defined as

$$\frac{Na^{+}}{[(Ca^{++} + Mg^{++})/2]^{\frac{1}{2}}} = \text{SAR } (me/\ell)^{\frac{1}{2}} \tag{10.5}$$

and is a measure of the relative abundance of Na^{+} to the other two most common cations. It reflects the concentration of salt adsorbed into the clay.

The erodibility of clays in terms of the SAR has been studied by several researchers. The SAR is determined by chemical analysis of a sample of pore water. Generally low SAR values are associated with interparticle attraction and flocculation. High values of SAR cause the particles to repel each other and dispersion. In consequence, addition of salt increases erosion resistance markedly at low values of SAR and has no effect at high SAR-values. The transition occurs at SAR in the range of 10 to 30, depending on types and concentrations of ions. Differences in the pore and eroding fluid create an osmotic potential that produces local swelling. The ionization of the silicic group in kaolinite causes positive charges on the edges of the kaolinite particles and is pH dependent. At low pH-values kaolinite forms flocculated structures; positive edge to negative face charge attraction. This leads to high erosion resistance. At high pH-values dispersed face-to-face structures predominate, electrostatic repulsion forces dominate and resistance to erosion is low.

The *ion-exchange-capacity* has been linked with erodibility but it gives more information on particle size distribution. It does also vary with the pH-value but is not strictly a function of salt concentration. Ion exchange cannot explain the observation that erosion resistance varies with concentration of neutral salt in water (Hutchison, 1972, Raudkivi and Hutchison, 1974). Montmorillonite with a relatively high ion-exchange capacity can have both very low and very high erodibility. The ion or cation exchange capacity is also a property of the sediment only and not of the sediment-water system which controls the erosion properties. There is evidence to show that the bonds between the plates of which the clay is composed, are stronger for Ca or Mg forms of clay than for Na or Li types. In montmorillonite about 80% of the ion exchange capacity is due to ions exchanged with the clay lattice. Consequently, the swelling potential of Na or Li montmorillonites is high since the interparticle bonds are primarily due to plate-ion-plate bonds and to secondary valence bonds. In kaolinites the swelling is limited because of the ion exchange capacity of kaolinite is mainly with the broken surface bonds. The interplate bonds are due to the strong hydrogen bonds.

The erosion rate is also a function of *temperature*. In general the erosion rate increases with temperature, although the bed shear stress decreases through decreasing viscosity. The dependence of the erosion rate with temperature was studied by Liou (1970) for bentonite, Rao (1971) for halloysite and Raudkivi and Hutchison (1974) for kaolinite.

10.3.1 The erosion process

Only a few descriptions of the erosion process have been reported in the literature. A reason for this is that the flow becomes rapidly opaque due to the suspended colloidal material and obscures the vision. Consequently attention has been focussed on the surface features of the eroded surface after the experiment has been stopped. Moore and Masch (1962) and Masch et al. (1963) describe a "washing of the surface of the clay at low shear stresses whereby flaking of small soil particles from the surface occurs. This is observed up to a critical shear stress, at which time, appreciable quantities of sediment come loose from the sample and the eroding water becomes cloudy". Photographs of the eroded surface are included in the paper by Masch et al. (1963). Judging from the photographs the clay could be described as brittle. The erosion patterns show sharp angles. Partheniades (1965) with his experiments using San Francisco Bay mud and water of ocean salinity observed that, in general, no visible fragments were eroded from the bed but that the eroding water simply clouded up. However, Karasev (1964) states that clay erosion takes place aggregate by aggregate, a conclusion similar to Mirtskhulava (1966).

The features of the eroded surface depend strongly on the orientation of clay aggregates. Scanning microscope pictures show that where the aggregates are stacked face to face the erosion is a kind of peeling off of loose patches where the interlayer bonds have failed. The eroded surface of stiff clays is pitted as shown in Fig. 10.18 for Georgia kaolinite and Fig. 10.19 for a bentonite. The lumps eroded from the kaolinite surface are shown in Fig. 10.20. When the clay is soft the randomly occurring pressure fluctuations remold the crater left when a lump of clay has been eroded from the surface. The resultant shows much softer features as illustrated in Fig. 10.21.

The expressions for the rate of erosion are basically in terms of applied shear stress excess ($\tau_o - \tau_c$) and more recently modelled in terms of analogy to rate process theory.

10.3.2 The rate process theory

Hutchison (1972) and Raudkivi and Hutchison (1974) introduced the analogy to the rate process theory for description of erosion of cohesive soils. This analogy was extensively developed by Croad (1981) and was also used by Kelly and Gularte (1981). Several earlier applications of this concept to soil mechanics have been published of which, according to Kelly and Gularte, the study of the rheological properties of clay by Murayama and Shibata dates back to 1958.

Fig. 10.18. View of the eroded surface of kaolinite.

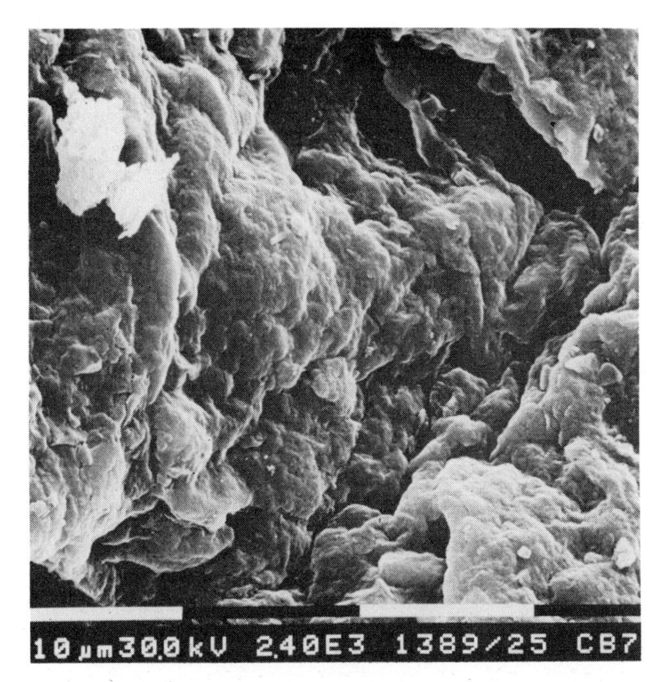

Fig. 10.19. Eroded bentonite surface at pH = 7.

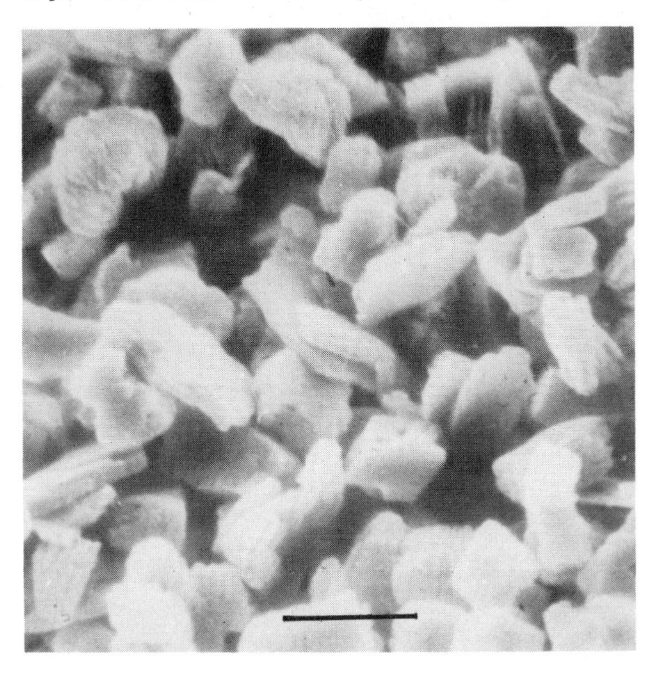

Fig. 10.20. Erosion products from kaolinite surface. Bar scale 10 µm.

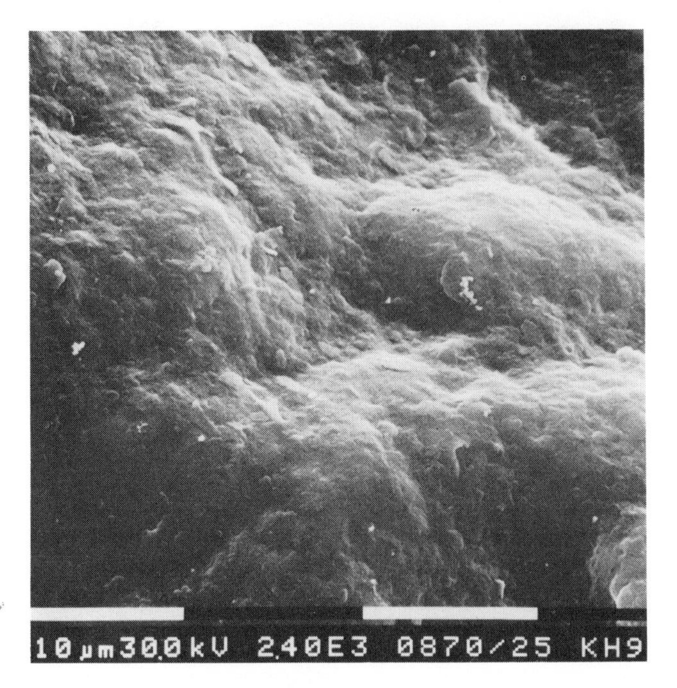

Fig. 10.21. Eroded surface of a soft kaolinite bed.

In analogy to chemical reactions, the erosion process could be expressed as an interaction between a soil module [S] and a fluid module [F]

$$[S] + [F] \rightarrow \text{Erosion product} \tag{10.6}$$

The soil module is the lump of soil removed from the surface by the passing fluid module which provides the activation energy, i.e. breaks the interparticle bonds. Therefore, the erosion rate is

$$\dot{e} = k[S][F] \tag{10.7}$$

where k is a rate constant, [S] and [F] are concentration of soil and fluid module, respectively. The concentration of the soil module (kg m^{-2}) was obtained by assuming that the eroded soil particle is hemispherical, radius a, and the soil fabric is homogeneous,

$$[S] = \frac{\rho_s a}{1 + v} \tag{10.8}$$

where ρ_s is the solid density of the soil, v is the void ratio, defined as the ratio of volume of void to volume of solid; and a is interpreted as the dimension of the soil module perpendicular to the plane of erosion.

The fluid module, i.e. the applied force, was assumed to arise from the boundary pressure fluctuations, the turbulence bursts. Figure 3.8 shows a measured distribution of pressures at a given instant on the boundary surface of a turbulent flow and the pressure fluctuations at a point with time.

The centres of the turbulence bursts on the boundary are spaced on the average at λ_1 and λ_3 from each other and the individual bursts have dimensions of ξ_1 and ξ_3. Thus, the concentration of the fluid module, [F] is

$$[F] = \frac{\xi_1 \xi_3}{\lambda_1 \lambda_3} \tag{10.9}$$

where the subscripts 1 and 3 refer to the streamwise and spanwise direction, respectively. By using published data on the mean sizes of $\xi_1 \xi_3$, λ_1 and λ_3, for example, Hinze (1975), Clark and Markland (1970) and Willmarth (1975), Croad (1981) estimated that $\xi_1 = 30\ \nu/u_*$, $\xi_3 = 20\ \nu/u_*$, $\lambda_1 = 500\ \nu/u_*$ and $\lambda_3 = 100\ \nu/u_*$, where ν is the kinematic viscosity of water and u_* is the shear velocity. Thus, from eqn (10.9)

$$[F] \cong 0.01 \tag{10.10}$$

The rate process theory rests on the empirical Arrhenius equation

$$\dot{e} = A \exp(-\Delta E/RT) \tag{10.11}$$

where $\dot{e}$ is the rate of erosion (reactions), A is a pre-exponential factor, ΔE is the activation energy, R is universal gas content, and T the absolute temperature. The activated complex [SF], with the terminology of eqn (10.6) can be expressed as (Moore, 1972)

$$[SF] = [S][F][Z/Z_S Z_F] \exp(-\Delta E/RT) \tag{10.12}$$

where Z are the partition functions for the particular species. From the rates process theory, eqn (10.11) can also be written as

$$\dot{e} = v^*[SF]$$

$$k = v^* P \exp(-\Delta E/RT) \tag{10.13}$$

where v* is the frequency of passage of the activated complexes, and the partition functions in eqn (10.12) are grouped into the term P. Here, v* is taken to be the frequency of turbulence bursts $1/\overline{T}_B$.

For erosion (reaction) to occur the energy barrier, Fig. 10.22 has to be exceeded. The influence of the entraining force modifies the potential energy profile in favour of the erosion process. Thus, the energy barrier is reduced by an amount $f\lambda L$(J mol^{-1}), the strain energy, where f is the force per bond, λ is the displacement of each bond and L is the Avogadro number.

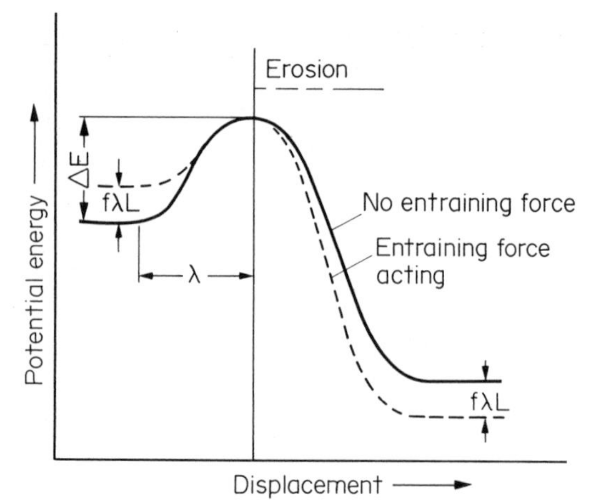

Fig. 10.22. Potential energy profile for the erosion process.

Therefore, the rate constant k becomes

$$k = (1/\overline{T}_B) P \exp(-\Delta E/RT + f\lambda L/RT) \tag{10.14}$$

Croad (1981) argued that the crack formed in the bed, should follow the positive and negative pressure fluctuations and estimated that $a = 35\ \nu/u_*$, which compares favourably with the observed pit size in the eroded bed.

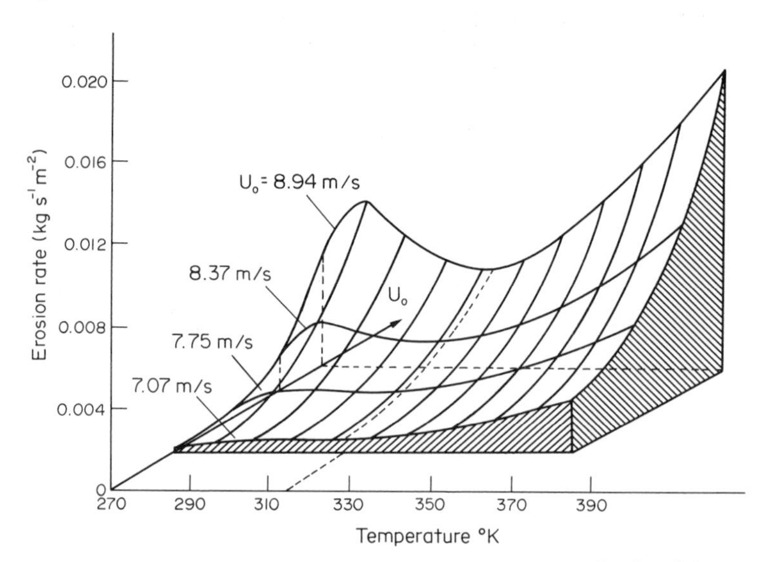

Fig. 10.23. Erosion rate versus temperature relationship according to eqn 10.16 where ΔE = 50 kJ mol^{-1}; $9\ \lambda L/n_B R = 100$; and $[0.07\ \rho_s/(1 + v)]P = 2 \times 10^5$ kgm^{-2}.

The term f was estimated from the general pressure fluctuation characteristics of a two-dimensional channel flow. The pressure peaks, p', appear to have the range of $\pm 6\bar{p}$, where $\bar{p}$ is the RMS value of pressure, and is empirically $\bar{p} \cong 3\tau$, where τ is the bed shear stress. Thus, the value of $p' = \pm 18\tau$ was selected. By assuming that the soil module is hemispherical, f is

$$f = 0.5|p'|/n_B = 9\tau/n_B \tag{10.15}$$

where n_B is the number of bonds per square metre. By substituting eqns (10.8), (10.10), (10.14) and (10.15) in eqn (10.7), and using the empirical value $U_oT_B/\delta = 5$, and $a = 35\ \nu/u_*$, the erosion rate of soil in kg s^{-1} m^{-2} becomes

$$\dot{e} = [0.07\ \rho_s/(1 + v)](U_o/u_*)(\nu/\delta)P\ \exp(9\tau\lambda L/n_BRT - \Delta E/RT) \tag{10.16}$$

where U_o is the velocity of flow outside the boundary layer, δ is the boundary layer thickness and the terms are measured in kg, m, s, °K and Mole.

Equation (10.16) can be expressed in a dimensionless form as

$$\dot{E} = \dot{e}/A_1 = P\ \exp(9\tau\lambda L/n_BRT - \Delta E/RT) \tag{10.17}$$

where $A_1 = [0.07\ \rho_s/(1 + v)](U_o/u_*)(\nu/\delta)$. For constant temperature, eqn (10.17) can be expressed as

$$\ln(\dot{E}) = A_2 + \beta\tau/n_B \tag{10.18}$$

where $A_2 = \ln(P) - \Delta E/RT$ and $\beta = 9\lambda L/RT$.

Equation (10.18), by plotting data as $\ln(\dot{E})$ versus applied shear stress, yields the value of $n_B(m^{-2})$ from the gradient of the plot. Similarly, be rewriting eqn (10.18) as

$$\ln(\dot{E}) - 9\tau\lambda L/n_BRT = \ln(P) - \Delta E/RT \tag{10.19}$$

and plotting the terms on the LHS versus 1/T, the value of the activation energy can be estimated from the gradient of the plot. A reasonably large range of temperatures is required for this method of estimation of ΔE and there are numerous pitfalls as described by Blandamer et al. (1982).

Equation (10.18)

$$\dot{e} = \left\{\frac{0.07\ \rho_s}{1 + v}\ \frac{U_o}{u_*}\ \frac{\nu}{\delta}\ P\ \exp\left[-\frac{\Delta E}{RT}\right]\right\} \exp\left[\frac{\beta}{n_B}\ \tau\right] \tag{10.18a}$$

has the form of $\dot{e} \propto \alpha\ \exp(\beta\tau/n_B)$ and Croad found that as a first approximation this was

$$\dot{e} \propto u_*^n\ \exp(\beta\tau/n_B) \tag{10.20}$$

where n is in the range of 0.14 to 0.25. For small values of τ or β/n_B when $\exp(\beta\tau/n_B) \cong 1$ the $\dot{e}$ versus τ-function is upwards convex and for large values downwards. Croad (1981) concluded from his experimental data that

(i) For acitic conditions the $\dot{e}$ versus τ functions are predominantly convex ($d^2\dot{e}/d\tau^2 > 0$) and concave for alkaline conditions

(ii) for a given clay the convex shape of the $\dot{e}$-τ curves tends to be associated with higher erosion rates than the concave shapes

(iii) erosion in saline conditions is predominantly characterised by convex curves but trends are not as strong as for clays in strongly acitic conditions

(iv) for clays with convex $\dot{e}$-τ-functions no critical shear stress exists but extrapolation of data on a concave curve to a zero erosion rate could imply a critical shear stress.

(v) for some clays as the shear stress increases the shape of the $\dot{e}$-τ curve changes from convex to concave but never the other way.

However, exceptions were noted with the variation of pH-values, types of clay and consolidation history.

The form of $\dot{e}$ versus absolute temperature, T, was shown to be a curve of a Ω-shaped trend merging into a U-shaped trend as T increases. By assuming an Arrhenius type dependence of kinematic viscosity on temperature, and by considering a special case where the relationship of shear stress to free stream velocity is known, Croad obtained a surface function for erosion rate with temperature and free stream velocity as the two variables, Fig. 10.23. The joining of the Ω-shaped trend with a U-shaped trend was also observed by Raudkivi and Hutchison (1974).

The erosion model outlined above, describes the erosion rate of soil with two major parameters, the number of bonds per unit area and the activation energy of the soil, for certain pore and eroding water. The model does not directly incorporate the effects of electrolytes on the erosion resistance of soil. The electrolyte will affect the values of n_B, ΔE and P which are a function of the electrolytes and clay concerned, and have to be determined before the erosion rate of soil for a given shear stress can be calculated.

The major feature of the rate process analogy model is that it relates erosion to an internal/external energy system where the erosion rate is a measure of work done on the system.

The number of bonds per square metre appears to be of the order of 10^{12} and available data show a ratio of maximum to minimum value of about six to seven. Croad proposed

$$n_B = \left[\frac{3c}{\pi(1 + v)d^3}\right]^{0\cdot67} \cong \frac{4}{(1 + v)^{0\cdot67}d^2} \qquad (10.21)$$

where v is the void ratio, d the particle (aggregate) size in m and c is a coordination factor assumed to equal about 8. The parameters c and v allow for the effects of the fabric of the clay. The experimental determination of n_B is affected by the turbulence structure of flow and the relationship between the pressure peaks and shear stress. Mitchell (1976) related the number of bonds in illite to water content. The relationship plotted as log n_B versus water content w% yielded a straight line ($n_B = 5 \times 10^{12}$, w = 0; $n_B = 5 \times 10^{10}$; w = 40%). Data (Raudkivi and Tan, 1984) indicate only a slight increase in n_B with pH-values, with a peak at about pH = 9, except for a kaolinite, with more than traces of ferric potassium and magnesium oxide, where the peak occurred at about pH = 5. No single trend for n_B has been observed with salt concentration. Most of the data show that n_B is insensitive to salt. Kaolinite and rheogel bentonite show an increasing trend, except for the kaolinite with the above mentioned impurities. The n_B-value appears to increase slightly with consolidation pressure, particularly when consolidated at high pH-values when face to face particle associations dominate. Mitchell (1976) shows an increasing number of bonds n_B for San Francisco Bay mud with consolidation pressure ($n_B = 3 \times 10^{10}$; 100 kNm^{-2}; $n_B = 10^{11}$, 400 kNm^{-2}). The effect of consolidation pressure on the erosion resistance is a function of the floc structure that forms in the suspension prior to deposition. When a deposit with an edge to face structure is consolidated the internal forces change little, although the structure may be crushed. Deposits with a face to face structure have elements where the interparticle distances are small and van der Walls forces are significant. Here small reductions in particle spacing lead to substantial increases in the van der Walls forces and erosion resistance.

Data available on activation energy ΔE show a range from 11.5 to 132.2 kJ mol^{-1}. These values are of the same order as quoted by Mitchell (1976) for activation energy for soil creep (80 to 180 kJ mol^{-1}) and Dawson (1975) for flow properties of clay suspensions (50 to 190 kJ mol^{-1}). The activation energy varies with the pH-value. For kaolinite ΔE is a maximum at about pH = 4. The activation energy for erosion is a measure of the probability of failure of enough bonds to allow the entrainment of a clump (aggregate) of soil.

The erodibility of clay in terms of eqn (10.8) using experimentally determined values of n_B as a function of pH and major salt concentration, is displayed in Figs 10.24 and 10.25. The clays used were a low swelling calcium bentonite, a fairly pure kaolin (Koclay) of 1.9 μm median aggregate size and a kaolin (ball clay) with 7% of ferric, potassium and magnesium oxides and a median size of 2.8 μm (Raudkivi and Tan, 1984). Figure 10.24 shows $\dot{E} = f(\tau, pH)$ when salt concentration is zero and Fig. 10.25 as a function of shear stress and molar salt concentration at pH = 7, using sodium chloride (both pore and eroding water). The erosion rate of the kaolinites is seen to decrease with increasing salt concentration. The bentonite shows an initial decrease with concentration and then an increase. The increase in salt will eventually lead to dispersive characteristics. The steps in the surfaces shown arise from computer plotting. Noteworthy is the relative insensitivity of erosion rate on shear stress.

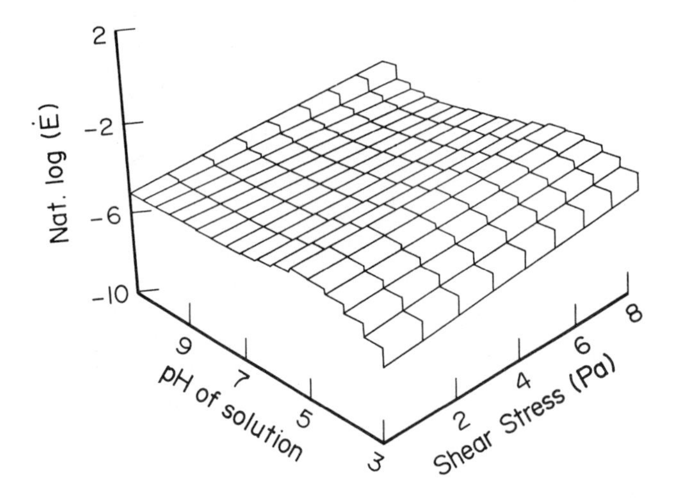

Fig. 10.24(a). The dimensionless erosion rate surface function $\dot{E} = f(\tau, pH)$ at zero salt concentration for bentonite (calben).

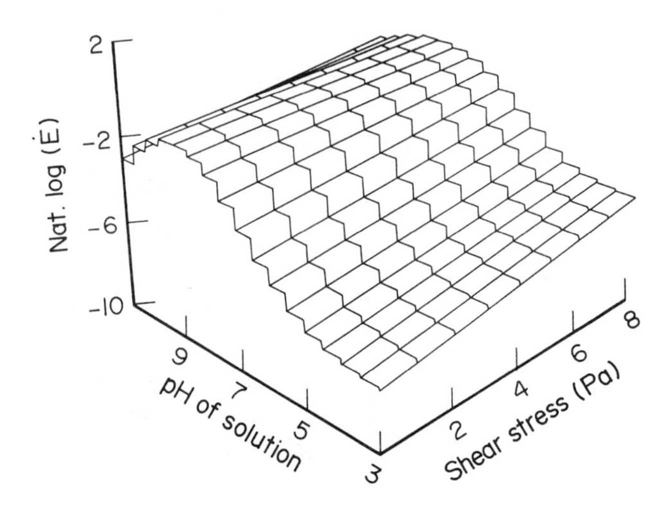

Fig. 10.24(b). The dimensionless erosion rate surface function $\dot{E} = f(\tau, pH)$ at zero salt concentration for kaolin koclay.

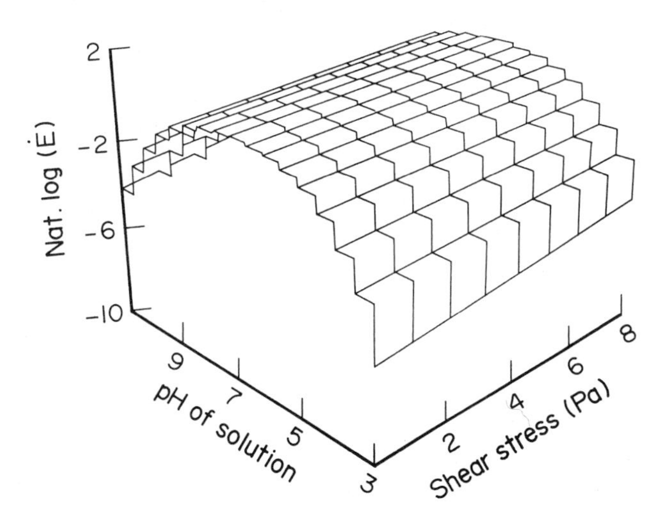

Fig. 10.24(c). The dimensionless erosion rate surface function $\dot{E} = f(\tau, pH)$ at zero salt concentration for kaolin ball clay.

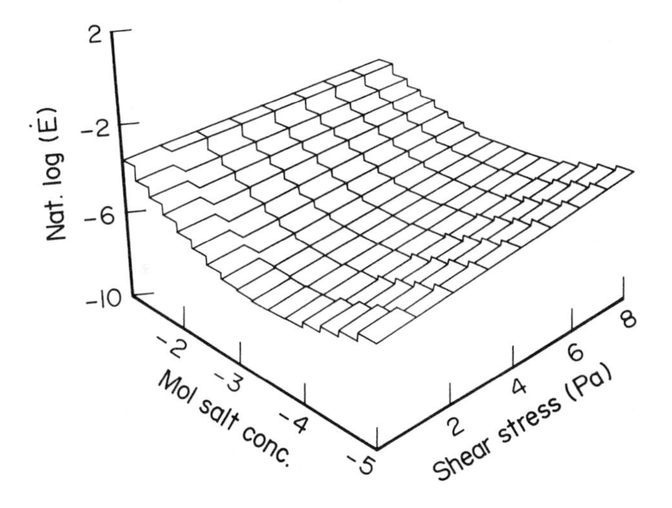

Fig. 10.25(a). The erosion rate surface function $\dot{E} = f(\tau$, molar salt concentration) for bentonite (calben) at pH = 7.

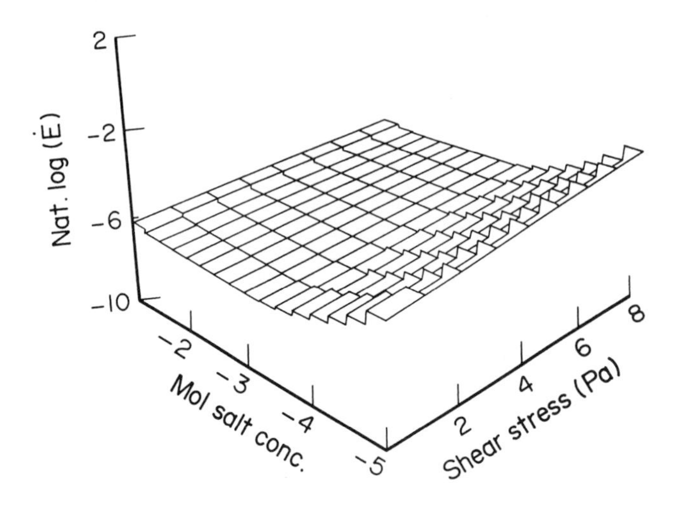

Fig. 10.25(b). The erosion rate surface function $\dot{E} = f(\tau$, molar salt concentration) for kaolin koclay at pH = 7.

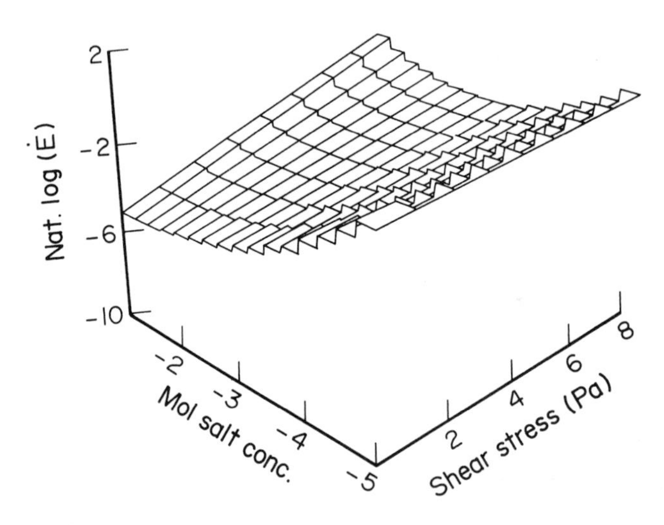

Fig. 10.25(c). The erosion rate surface function $\dot{E} = f(\tau$, molar salt concentration) for kaolin ball clay at pH = 7.

The rate process approach to the erosion of clays is to an extent still an empirical method. It could be compared to the use of Fick's law in solving diffusion problems. However, with a systematic accumulation of data it enables a more rational approach to estimation of erosion characteristics of clays than other available methods.

10.4 Erosion of Cohesive Muds

Fresh and weakly consolidated sediments are a feature of most estuaries and many stretches of coastline. These sediments are usually referred to as marine muds. Their bulk densities are generally less than 1300 kg/m^3. Particularly complex environments are the estuaries. The sediments carried into an estuary by streams and rivers enter a saline environment in which they undergo repeated cycles of erosion, transport and deposition by ebb and flood tide currents and by wave action. Estuaries where fresh water inflow is significant also have a with tides to and fro moving fresh/salt water interface, the salt wedge. The reworking and mixing alters both the characteristics of the sediment entering and sediment present in the estuary. Through the erosion and suspension of the bed material pore water and nutrients are released into the near bed water. The resuspended sediment can also adsorb substances from water. In the bed, oxygen is depleted through decomposition of organic material. This is accompanied by a series of chemical reactions. For example, the sulphate reducing bacteria produce CO_2 which lowers the pH-value and the redox potential becomes negative. Some metals like iron and manganese change to soluble divalent forms, others like lead, copper and cadmium precipitate as solid metal sulphides. When the mud is resuspended, it comes into an oxic environment where metals are oxidized. Metals like iron are precipitated and dissolved metals are adsorbed. Since the release of metal from particulate to dissolved form in the anoxic environment is faster than their removal from water by adsorption and precipitation the net transfer of metal is to dissolved form.

Estuary muds usually have a light brown colour on the surface and dark grey to black only a few cm below the surface. When this black mud is thoroughly stirred in water its colour reverts to brown. Then, when left to stand, the colour slowly changes to greenish grey and back to black. The dark colour arises from ferrous sulphides. These are oxidized by stirring in water to ferric hydroxide which gives the brownish colour. On standing bacterial reduction changes ferric iron to ferrous iron (greenish) and back to ferrous sulphide.

The large specific surface area of mud particles and their mineralogical characteristics lead to a high cation exchange capacity and adsorption capacity. As a consequence muds contain a great variety of chemicals. It has been reported that in the Rhine River muds hardly any element is missing, concentrations ranging from about 0.05 g/m^3 of iridium (Ir) to 50 kg/m^3 of iron (Fe). In addition the muds also adsorbe radionuclides, such as ^{137}Cs, ^{134}Cs, ^{60}Co and ^{210}Pb.

Apart from estuaries, there are large stretches of coastline with muddy bottoms. The washload of rivers like the Amazon, Mississippi and Huanghe transforms into streams of mud along the coastline. An enormous pure mud coast stretches from the Amazon to the Orinoco river. It has a remarkable pattern of wave-like bed features down to a water depth of 20 m and 40 km offshore. The bed waves have 30-60 km spacing, about 5 m height and extended obliquely seaward from the coast. The waves propagate at about 1.5 km per year westward. Extensive fluid-mud banks also extend along the Louisiana coastline, the Gulf of Chihli of the Yellow Sea, the coast of South Korea, and the coast of Kerala southwest India. These soft muds interact with waves and are fluid enough for ships to sail through the fluid-mud bed. A specific feature is that the mud attenuates the water waves to such an extent that they do not reach the shoreline. In Kerala the mud bed areas serve as safe anchorage during southwest monsoons.

The physics and chemistry of estuarine and coastal muds is extremely complex and the formal description of these processes is still one of the challenges to researchers, although significant progress has been made in the last few decades. Of major practical importance are the processes of flocculation, deposition, resuspension, deposition, etc., and the lateral transport. The annual dredging costs and costs of disposal of the dredged mud around the world are of the order of US$1000 million. Hence, even a small reduction of the quantities is worthwhile.

10.4.1 Suspension, flocculation, settling

The colloidal wash load of streams mixes with saline sea water which changes the properties of the clay-electrolyte system. The suspended particles move about due to the Brownian motion, and relative to each other due to different settling velocities and velocity gradients. These movements cause collisions which lead to formation of aggregates (flocs).

The process depends on concentration, interparticle bonding forces and fluid stresses imposed on the particles. Collisions arising from Brownian motion and differential settling are gentle and the aggregates formed weak and fluffy. Rotational collisions due to velocity gradients usually lead to stronger aggregates. The aggregates are much larger than the individual particles. When a floc touches the bed it can bond to the bed surface. The bonding is controlled by the number and strength of inter-mineral contacts between the floc and the bed and the fluid stresses acting there. As a consequence the deposits vary in strength from point to point and with time and depth at a given location.

Fine colloidal sediments are easily suspended and up to about 0.1 g/ℓ do not measurably affect the viscosity or density of water. At these concentrations there is also hardly any flocculation because of the dispersion, i.e., large interparticle distances on the average. Flocculation becomes important at concentrations of 0.1 g/ℓ to 10 g/ℓ. At the latter concentrations the flow becomes decidedly non-newtonian. Its behaviour is then described by the Bingham fluid model, $\tau = \tau_B + \mu \, du/dy$ where τ_B is the Bingham yield stress.

The density of fluid starts to differ perceptibly from that of clear water at concentrations of about 0.5 g/ℓ and viscosity at about 50 g/ℓ. Approximate values of viscosity and bulk density as a function of concentration are as follows:

c (g/ℓ)	50	100	150	200	300	400	500
μ (kg/ms) in fresh water	-	0.001	0.0025	0.01	0.40	7.0	100
μ (kg/ms) in sea water	0.0015	0.011	0.11	0.70	5.0	13.5	100
Bulk density (kg/m³)	1030	1065	1100	1150	1240	1340	1450
	suspension			soft mud		firm mud	

The settling velocity of suspensions is defined in terms of percentage of solids settled out, i.e., w_{50} is the median settling velocity corresponding to 50% of solids by weight settled out. Practically all the data on settling velocities comes from laboratory experiments. The results generally show that salinity increases flocculation. Owen (1970) found an almost linear increase of fall velocity with salinity for the Avonmouth mud. Several researchers have reported that at clay concentrations above a certain value the fall velocity becomes independent of salinity, i.e., there is an increase in fall velocity with salinity at low concentrations. Migniot (1968) observed that the settling velocity remained constant at salinities about 3 $^{o}/oo$ (parts per thousand) for low clay concentrations and at above 10 $^{o}/oo$ for high clay concentrations. He also observed that the settling velocity of flocs can exceed that of the component particles by a factor up to 10^4 when the component particles are in size smaller than 1 μm. The aggregate (floc) and dispersed sediment fall velocities were measured at c = 10 g/ℓ and 30 $^{o}/oo$ salinity for a large number of muds. He introduced a flocculation factor $F = w_a/w_d$, in which w_a is the fall velocity of aggregates and w_d of the dispersed sediment, and related it to the meadian size of the dispersed sediment as $F = kd_{50}^{-m}$, where m = 1.8 and k = 250 for all sediments. The value of F = 1 corresponds to $d_{50} \cong 20$ μm, a size at which aggradation becomes negligible. Field measurements by Burt and Stevenson (1983) show no salinity effect for the Thames estuary mud. They ascribe the difference between the laboratory and field results to the fact that flocculation is a time-dependent process for which the laboratory tests do not allow enough time for full reflocculation to take place. In field the material is held in suspension for long periods. This enables extensive flocculation even at low salinities. The argument is also supported by the laboratory data. For example, Krone (1962) observed initially a period of almost no settling which was followed by a rapid decrease in concentration. The delay time at the beginning depended strongly on the initial concentration and the initial settling rate was that of single particles. The rapid settling of flocs was observed to be proportional to $c^{4/3}$. Burt and Stevenson report that laboratory values of w_{50} were less than half of those measured in the field. They described their field data by

$$w_{50} = 1.341 \times 10^{-4} c^{1.37} \tag{10.22}$$

where w is in mm/s and c in mg/ℓ. The exponent 1.37 is similar to 4/3 by Krone. A more general expression for the Thames estuary mud was given as

$$w_n = c^{1.37}F(n) \tag{10.23}$$

where

$$F(n) = 1.88 \times 10^{-4} n^{2.34}$$

w is in mm/s, c in g/ℓ and n is in percent. The result differs slightly from eqn 10.22 for n = 50%. Their data cover $50 \leq c \leq 5000$ mg/ℓ, salinity 0-30°/oo, temperature 13-20°C, d_{50} = 4 μm, 32% clays by size and 67% silts (Montmorillonite 15%, kaolinite 10%, illite and mica 25%, chlorite 20%, organics 13%, non-clay minerals 17%), and over 200 samples over a 3 year period. However, the exponent and factor in eqn 10.22 have widely varying values from estuary to estuary. The centre of the data band is approximately described by w_{50} = 1.0c. The exponent for c for the Humber estuary data is as low as 0.61 (Delo and Burt, 1986). Relationships of the form of eqn (10.22) apply to the settling of aggregates at relatively low concentrations. At higher concentrations hindered settling, eqn (2.25), occurs as illustrated by the data by Thorn (1981), Fig. 10.26, obtained in laboratory using Severn Estuary mud. The reported limits of onset of hindered settling vary widely, depending on the sediment composition, from about 3 g/ℓ to 15 g/ℓ. For the small particles the exponent in eqn (2.25) is β = 4.65. There are also forms of this equation with an extra coefficient, $w = w_o(1 - kc)^{\beta}$ where k depends on sediment composition and β ≅ 5.

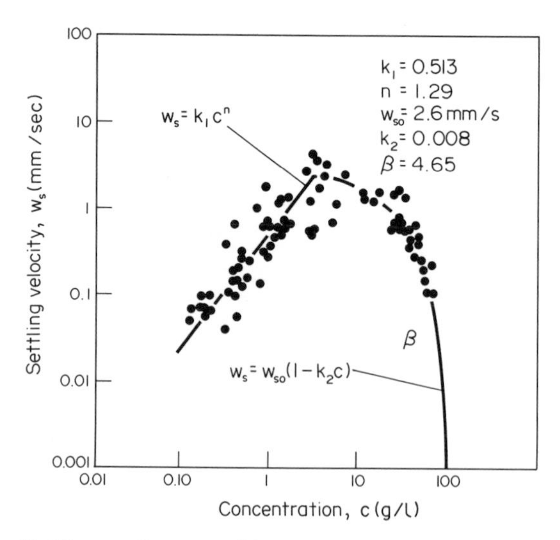

Fig. 10.26. Median settling velocity, w_s, against suspended sediment concentration, based on tests using Severn Estuary (U.K.) mud, after Thorn (1981) where w_{so} is a reference settling velocity.

The settling of sediment leads to a decrease in concentration. Krone found that at concentrations less than 0.3 g/ℓ the concentration varied with time as $c_t = c_o^{-r_1 t}$, where c_o is the initial concentration and t is time from commencement of settling. The decay rate for his flume data was given by $r_1 = (6.6 \times 10^{-4}/h)$ $(1 - \tau_o/0.6)$ where h is water depth and τ_o bed shear stress. For hindered settling (after the initial period) the rate of decrease in concentration was observed too follow log c = r_3 log t + const., where $r_3 = (103/h)(1 - \tau_o/0.78)$.

The vertical distribution of the fine suspended sediments depends on current and wave action. Usually the ε_s = constant model (eqn (7.60)) gives an adequate fit to field data, or eqn 7.65 can be used. The problem is to have enough data on velocity distribution, particularly when strong wave action is present. Frequently the concentration profiles consist of two slopes of exponentially decreasing concentrations, one covering most of the depth and another as a relatively thin layer near the bed within which the concentration increases much more rapidly. When the concentrations generally are very low, the lower zone where the concentration increases more rapidly as bed is approached, is not discernible.

10.4.2 Deposition

In a laboratory settling column the top surface of the suspension falls and a mud layer grows at the bottom. After a certain time the surface of the suspension meets and coalesces with the surface of the mud layer at the bottom, leading to a water/mud interface which from now on will slowly fall as the mud consolidates.

Subjected to currents fine sediments will deposit when the bed shear stress decreases to below a certain critical value, τ_d, which is lower than the critical value for erosion τ_c. In the field the bed shear stresses arise from wave motion and tidal currents. The orbital as well as tidal velocities have to be low for deposition to occur. When sediment concentration is low wave action can inhibit deposition down to appreciable water depth. However, at high concentrations (c > 0.1 g/ℓ) deposition will occur regardless of wave action. Figure 10.27 is a diagrammatic illustration of the variation of bed shear stress over a tidal cycle due to tidal currents. Since generally $\tau_c > \tau_d$ there is a period when neither deposition nor erosion occurs.

The change of mass of suspended sediment per unit area could be written for that part of the time-concentration curve where log c varies linearly with time t as

$$\frac{dM}{dt} = - wc \left[1 - \frac{\tau_o}{\tau_d}\right]' \qquad (10.24)$$

Krone (1962) interpreted the bracket term as the probability that a particle will adhere to the bed. The product $w(1 - \tau_o/\tau_d)$ is the apparent settling velocity. It is the actual settling velocity when $\tau_o = 0$ and zero when $\tau_o = \tau_d$. The value of τ_d depends

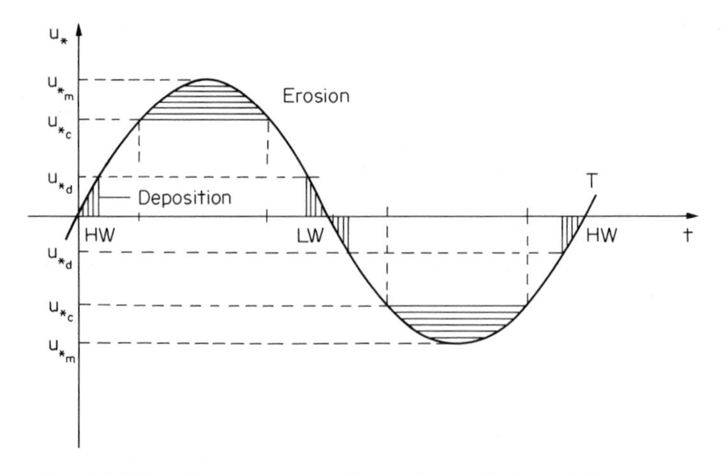

Fig. 10.27. The sequence of erosion and deposition during a tidal cycle in terms of shear velocity or bed shear stress.

on sediment composition. Mehta (1973) found that τ_d ranged from 0.04 N/m² to 0.15 N/m². Since the fall velocity has an initial value and one corresponding to the flocs the rate at which c/c_o decreases will change. Mehta (1973) showed that laboratory data for the initial phase plot as a straight line on log-probability paper with t on log-scale. The second phase plotted as a linear relationship on semi-logarithmic paper with c/c_o on the log-scale. The more important observation was that the fraction of sediment c/c_o which remained in suspension at $\tau_o > \tau_d$ was a function of the bed shear stress, Fig. 10.28. Mehta and Partheniades (1975) demonstrated that for a given sediment-fluid system this ratio c_{eq}/c_o is a function of bed shear stress τ_o only, as shown in Fig. 10.28. The result implies that the amount of sediment which remains in suspension indefinitely at steady state conditions depends, at any given bed shear stress, on the initial quantity of suspended sediment in the flow. Figure 10.28 also shows that there is a minimum bed shear, $\tau_{o\ min}$, below which no sediment will remain is suspension indefinitely. The amount of sediment deposited at any given shear stress τ_o larger than τ_d is $(1 - c_{eq}/c_o)$. The fraction deposited at any given time is $c^* = (c_o - c)/(c_o - c_{eq})$, which for kaolinite in distilled water was found to satisfy

$$c^* = \frac{1}{(2\pi)^{\frac{1}{2}}} \int_{-\infty}^{T} e^{-\frac{1}{2}u^2} du \tag{10.25}$$

where

$$T = \frac{1}{\sigma_1} \log \frac{t}{t_{50}} = \frac{1}{\sigma_2} \log \left[\frac{\tau_o/\tau_{o\ min} - 1}{(\tau_o/\tau_{o\ min} - 1)_{50}} \right] \tag{10.26}$$

σ is the stand. dev. of c versus t plot on log-probability paper,

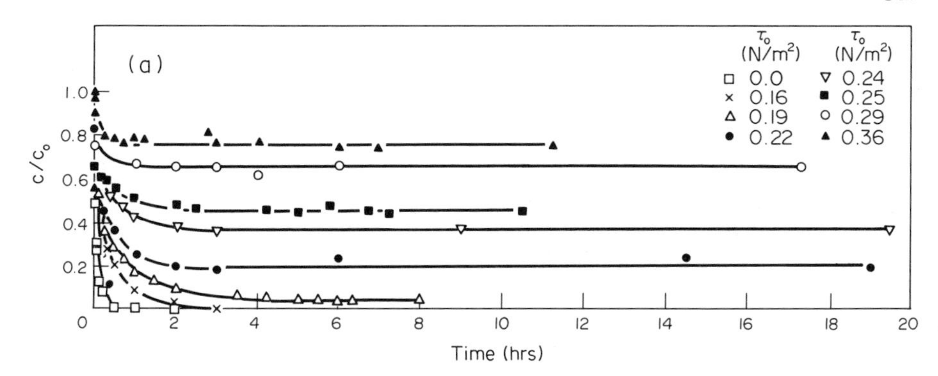

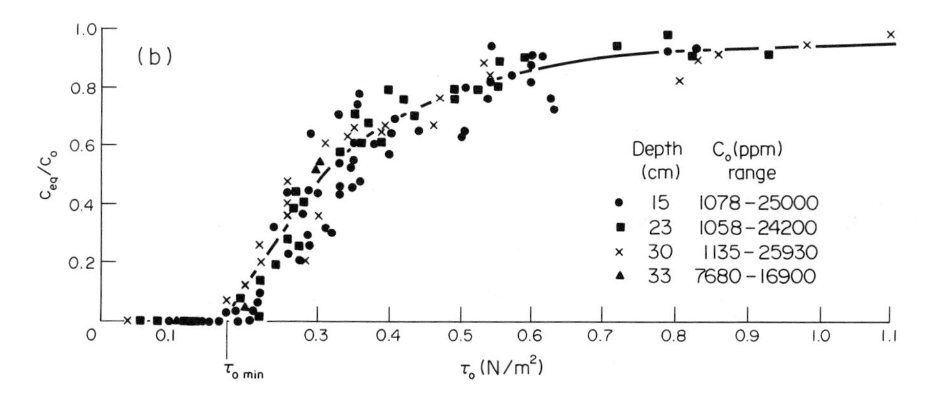

Fig. 10.28. (a) Concentration relative to initial concentration c_o as a function of time for kaolinite in distilled water (Mehta, 1973).

(b) Equilibrium concentration relative to initial concentration c_o as a function of bed shear stress (Mehta and Partheniades, 1975).

a measure of the spread of settling times of flocs, t_{50} is the time for 50% of sediment to be deposited and $(\tau_o/\tau_{o\ min} - 1)_{50}$ the stress for which 50% of the sediment initially in suspension is deposited, and u is a dummy variable. This yielded

$$\frac{dc^*}{dt} = \frac{0.434}{(2\pi)^{\frac{1}{2}}\sigma_1} \frac{1}{t} e^{-T^2/2} \tag{10.27}$$

The parameters σ and t_{50} vary with $\tau_o/\tau_{o\ min}$ which characterises the sediment/water system, with given c_o and flow depth. The values of σ_1 and t_{50} were observed to have a maximum in the range of $\tau_o/\tau_{o\ min}$ of 1 to 1.5. A value of $\sigma_2 = 0.49$ was found to satisfy data from several sources. Partheniades (1986) proposed that

$$\left[\frac{\tau_o}{\tau_{o\,min}} - 1\right]_{50} = 4\,\exp(-0.127\,\tau_{o\,min}) \qquad (10.28)$$

where τ_o is in N/m^2.

If the rate of change of concentration is expressed as the difference between erosion and deposition

$$\frac{\partial c}{\partial t} = E - D \qquad (10.29)$$

then the equilibrium condition for noncohesive sediments is characterised by E = D. Mehta observed with cohesive sediments that there was no resuspension or deposition of sediment. It was hypothesised that with cohesive sediments E and D go to zero. That this is so was shown by Parchure (1984) who slowly replaced the suspension with clear water without stopping the test. The implication of this is that initially larger flocs deposit and adhere to the bed. The particles with weak bonding forces do not form strong enough flocs for these to remain intact near the bed where the velocity gradients are high. Hence, there is initially a sorting process after which the remainder will remain in suspension indefinitely at the given shear stress. Corollary of this is that if sediment is being eroded (resuspended) and no redeposition occurs, steady state can only be reached when τ_c increases with depth of erosion and becomes equal to τ_o.

It has been discussed already that flocs which touch the bed may bond to the surface. The strength of the bonding depends on the number and strength of the mineral to mineral contacts made. The strength of the flocs themselves is affected by the fluid forces, the collisions. Hence, over a tidal cycle there will be deposits of differing strength, a layering.

From eqn (10.27) the change of mass of suspended sediment is

$$\frac{dM}{dt} = \frac{0.434}{(2\pi)^{\frac{1}{2}}\sigma_1}\,\frac{c_o h}{t}\,e^{-T^2/2} \qquad (10.30)$$

At no flow conditions ($\tau_o \sim 0$) and advanced state of flocculation, when the fall velocity is expressed by an equation of the form of eqn (10.22),

$$\frac{dM}{dt} = -\,kc^{n+1} \qquad (10.31)$$

and when the concentration is low

$$\frac{dM}{dt} \cong -\,wc \qquad (10.32)$$

where w is a constant.

The top zone of the mud has a with depth from the mud surface increasing bulk density and shear strength profile, governed by the density and strength of the deposited flocs and subsequent consolidation conditions. Usually there is also a certain amount of layering in the density and strength profile arising from periods of deposition when the flocs are weak or strong. The increase in bulk density of the deposit is initially very rapid. About 80% of the final value is reached after 48 hours and changes in density after 6 days are small.

Typically the bulk density increases from a value close to that of water on the surface to about 1400 kg/m³ within the first 50-60 mm. Mehta et al. (1982) expressed the variation of local dry bulk density as

$$\rho_{sb}/\bar{\rho}_{sb} = a(y/H)^b \tag{10.33}$$

where $\bar{\rho}_{sb}$ is average density over the depth of deposit H, a = 0.794 and b = 0.288. Field data by Owen (1975) and Thorn and Parsons (1980) yielded a = 0.660 and b = -0.347. Migniot (1968) expressed the density of the deposited material, ρ_{sb}, as

$$\rho_{sb}/\rho = 1 + 0.001(S_s - 1)c_w \tag{10.34}$$

where

$$c_w = \alpha \log t + \beta$$

is concentration in terms of weight per volume, t is in hours, α is a factor which depends on the diameter of "elementary" particles (dispersed) and β is a factor which depends on the fluid medium.

Generally, thin deposited layers of mud are stronger and denser than thick layers.

The surface layers are also affected by organic binding of particles. This arises from mucal secretions of bacteria and algae. The strength of such bonding can be substantial but no methods exist to predict it. For further reading on these effects reference is made to Montague (1986), Reineck and Singh (1973) and Führböter (1983).

10.4.3 Erosion

The bed surface erodes and material is *resuspended* when the applied stresses exceed the erosion resistance due to interparticle bonds. The bonding, as discussed earlier, can be due to electrochemical forces or organic bonding or both. Apart from the very complex nature of the bonding the bed is also subject to very complex fluid forces. The waves produce in direction reversing velocities which in shallow water create turbulent boundary layers. The waves in shallow water also subject the mud to substantial periodic pressure fluctuations which force water into and out of the mud. The orbital velocities of long period waves in shallow water can be high and when superimposed on tidal current velocities the resultant peaks may be substantial. The periodic forces from wave motion, plus the associated turbulent forces, subject the mud to continuously varying stresses and fatigue. Under moderate wave and current conditions the erosion is particle by particle where the particle may be a floc, a conglomerate of flocs or an "elementary" particle broken off a floc. The term elementary is used here to indicate the particle size measured in a chemically and mechanically dispersed suspension. The deposition process usually leads to beds with a certain amount of layering, i.e., layers with stronger and weaker bonding. Under more severe wave conditions the deposit may fail in one of these weaker layers and lift off in sheet-like chunks. This process is aided by the pressure fluctuations which affect a decrease in bed density, i.e., the water forced in and out of the mud leads to the rupture of some of the bonds. The wave action produces a degree of liquefaction. Under these conditions, and in very soft muds in general, the "fluid-mud" interface may become unstable and this can lead to substantial mixing, resuspension, of sediment.

Usually the erosion rate is related to the shear stress excess. The critical shear stress, τ_c, for threshold of erosion is assumed to be known and to cater for turbulent fluctuations. Migniot (1968) expressed his experimental results as

$$\dot{e} = M(\tau_o - \tau_c) \tag{10.35}$$

and related the yield shear stress of the mud, τ_c, to bulk density of solids as $\tau_c = n\rho_{sb}{}^m$, where τ is in N/m² and ρ_{sb} in g/ℓ. The exponent m was found to be nearly constant for all his muds at $m = 5$ but n varied from 10^{-12} to 10^{-15}. The constant M is the erosion rate constant with dimensions (s/m). Mignot expressed the critical shear velocity for the top layer of mud as

$$u_{*c} = a\tau_c{}^{0\cdot25}; \quad \tau_c \leq 1.5 \text{ N/m}^2; \quad a \cong 0.01778$$

$$u_{*c} = b\tau_c{}^{0\cdot5}\ ; \quad \tau_c \geq 1.5 \text{ N/m}^2; \quad b \cong 0.016 \tag{10.36}$$

Thorn and Parsons (1980) expressed the critical shear stress of Grangemouth, Brisbane and Belawan muds as

$$\tau_c = 5.42\text{x}10^{-6}\rho_s^{2\cdot28} \tag{10.37}$$

where ρ_s is the dry density of mud. The erosion in 10, 20 and 30 minutes in kg/m² was expressed as a factor times $(\tau_o - \tau_c)$ with the factor being 1.58, 2.14 and 2.51, respectively, τ_c = 0.07 N/m² for the Brisbane mud and 0.10 N/m² for Belawan and Grangemouth muds.

An alternative form of eqn (10.35) for the erosion rate $\dot{e} = dM/dt$ is

$$\dot{e} = \alpha \left[\frac{\tau_o}{\tau_c} - 1\right] \tag{10.35a}$$

where α is again an empirical constant. In an uniform bed, τ_c =

const., the erosion rate would stay constant. However, usually the shear strength varies for a given mud-water deposit from a very low value, τ_{co}, on the surface to a nearly constant maximum value lower down in the deposit, τ_{cm}. Parshure and Mehta (1985) report for kaolinite (1 μm) in salt water a $\tau_{co} = 0.04\ N/m^2$, independent of consolidation period (1 to 10 days), and for lake mud (montmorillonite, illite, kaolinite and quartz, $d_{50} = 1$ μm) at salinities from 0.5°/oo to 1% after 1.7 days a $\tau_{co} = 0.08$ to $0.17\ N/m^2$. The corresponding τ_{cm} values are 0.57 and $0.67\ N/m^2$ for kaolinite and lake mud, respectively. The thickness of the rapidly consolidating top layer decreased in 10 days from 6 to 1.6 mm for kaolinite. Lake mud was consolidated 1.7 days only and yielded top layer thicknesses of 0.5 to 1.7 mm. The mean shear strength in this top layer of kaolinite increased from 0.15 to $0.37\ N/m^2$. The erosion rate was expressed as.

$$\ln \frac{\dot{e}}{\dot{e}_f} = \alpha(\tau_o - \tau_c)^{\frac{1}{2}} \tag{10.38}$$

The parameters $\dot{e}_f$ (the floc erosion rate when $\tau_o - \tau_c = 0$) and α where evaluated from experiment:

Bed		α $m/N^{\frac{1}{2}}$	$\dot{e}_f \times 10^5$ g/cm^2-min
Kaolinite in fresh water		18.4	0.5
Kaolinite in salt water 35°/oo		17.2	1.4
Lake mud, in s.w. 0.5-10°/oo		13.6	3.2
S.F. Bay mud	Partheniades (1965)	8.3	0.04
Lake mud	Lee (1979)	8.3	0.42
Estuarial mud	Thorn and Parsons (1977)	8.3	0.42
" "	" " (1979)	4.2	1.86
Kaolinite	Dixit (1982)	25.6	0.60

The large differences in the coefficient values reflect sediment-electrolyte effects as well as differences in procedure, temperature, apparatus used, etc.

The quantity of erosion or deposition over a tidal cycle could be estimated by integration of eqns (10.24) and (10.37) between the appropriate shear stress or shear velocity limits. As a first approximation the shear velocity over the tidal cycle could be assumed to vary sinusoidally, $u_* = u_{*m} \sin \omega t$, where u_{*m} is the known maximum value of shear velocity, $\omega = 2\pi/T$ and T is the period of tide. Assuming either no wave action or constant wave effect this yields

$$M = 4\int_0^{u_{*d}} \frac{cwT}{2\pi u_{*m}} \frac{\left[1 - \left[\frac{u_*}{u_{*d}}\right]^2\right]}{\left[1 - \left[\frac{u_*}{u_{*m}}\right]^2\right]^{\frac{1}{2}}} du_* - \int_{u_{*c}}^{u_{*m}} \frac{\alpha T}{2\pi u_{*m}} \frac{\left[\left[\frac{u_*}{u_{*c}}\right]^{\frac{1}{2}} - 1\right]}{\left[1 - \left[\frac{u_*}{u_{*m}}\right]^2\right]^{\frac{1}{2}}} du_* \tag{10.39}$$

where c, w and α have to be inserted as functions of u_*. Then a numerical solution is possible. If these parameters are constant then

$$\frac{M}{T} = \frac{2cw}{\pi}\left\{\sin^{-1}\left[\frac{u_{*s}}{u_{*m}}\right]\left[1 - \frac{1}{2}\left[\frac{u_{*m}}{u_{*s}}\right]^2\right] + \left[\frac{1}{2}\sqrt{\left[\left[\frac{u_{*m}}{u_{*s}}\right]^2 - 1\right]}\right]^{\frac{1}{2}}\right\}$$

$$- \frac{2\alpha}{\pi}\left\{\frac{1}{2}\sqrt{\left[\left[\frac{u_{*m}}{u_{*c}}\right]^2 - 1\right]^{\frac{1}{2}}} - \left[\frac{\pi}{2} - \sin^{-1}\left[\frac{u_{*c}}{u_{*m}}\right]\right]\left[1 - \frac{1}{2}\left[\frac{u_{*m}}{u_{*c}}\right]^2\right]\right\} \tag{10.40}$$

where M/T is the average siltation rate in terms of mass per unit area, or divided by bulk density of the deposit it becomes the rate of change of the depth of sediment, i.e. rate of accretion or erosion. Basically, sediment is eroded during that part of the cycle when u_* is greater than u_{*c} and deposited during slack water, $u_* < u_{*d}$. How much, depends on hydraulic conditions. Silt concentrations vary substantially over the tidal period at any given location as well as along the estuary. There will also be with time varying irregular changes. Fine particles and light organic particles in surface layers of flow are carried out of estuary if conditions do not encourage flocculation. Floods dump large quantities of wash load and fresh water into the estuary which again produce transient conditions. Thus, one should consider separately the phenomena which have vastly differing time scales; time scales ranging from a minute or less to hours, days months.

The first model for erosion or resuspension of marine muds was presented by Partheniades (1965). The model predicts the specific erosion rate of a cohesive mud as a function of the average boundary shear stress. According to the model, Fig. 10.29, a flocculated bed at its loosest state is subject to a randomly varying bed shear stress. The maximum interparticle shear stress, σ_m, is taken to be proportional to the shear stress τ and the interparticle bond strength, σ_p, is taken to be proportional to the macroscopic shear strength c. The erosion rate $\dot{e}(gs^{-1}m^{-2})$ is calculated from

$$\dot{e} = (\rho_f \alpha_v d_f{}^3)\left[\frac{1}{\alpha_a d_f{}^2}\right]\left[\frac{1}{\bar{t}_i}\right]\text{Prob. } \{\sigma_m > \sigma_p\} \tag{10.41}$$

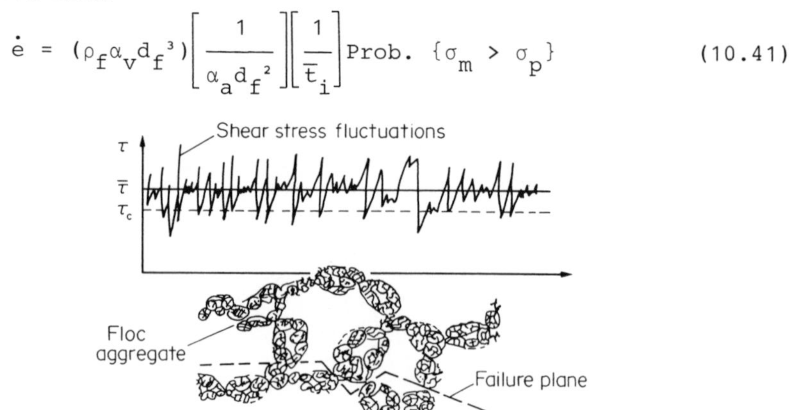

Fig. 10.29. Diagrammatical representation of erosion model of Partheniades (1965).

where the first bracketed term is the mass of a single eroded aggregated floc, ρ_f is the dry density of aggregate, d_f its equivalent diameter and α_v is a volumetric shape factor ($\alpha_v \cong \pi/6 = 0.5$). The second term is the number of aggregates per unit area, where α_a is an areal shape factor ($\alpha_a \cong \pi/4 = 0.8$). The third term is the frequency, where t_i is the average impulse time to break the particle bond. The last term is the probability that for any floc aggregate the eroding forces exceed the resisting forces, assumed to be given by a normal distribution

$$f(r) = \frac{1}{S\sqrt{2\pi}} \exp\left\{-\frac{(r')^2}{2S^2}\right\} \tag{10.42}$$

where $r' = \tau_o'/\bar{\tau}_o = \tau_o/\bar{\tau}_o - 1$; $\tau_o = \bar{\tau}_o + \tau_o'$ and the stand. dev.

$$S = \overline{\{(\tau_o')^2\}}^{0.5}/\bar{\tau}_o \tag{10.43}$$

The probability Prob $(\sigma_m > \sigma_p)$ is taken to be the event

$$\frac{\sigma_m}{\sigma_p} = \frac{k\tau_o}{c} > 1 \tag{10.44}$$

where k is a proportionality factor. This yields for positive values* of instantaneous shear stresses

$$\text{Prob}\,\{\sigma_m > \sigma_p\} = 1 - \frac{1}{S\sqrt{2\pi}} \int_{-[(c/k\tau_o)+1]}^{[(c/k\tau_o)+1]} \exp[-(r')^2/2S^2]\,dr \tag{10.45}$$

and an erosion rate

$$\dot{e} = \frac{\alpha d_f \rho_f}{t_i} \left\{ 1 - \frac{1}{\sqrt{2\pi}} \int_{-[(c/k\tau_o)+1]/S}^{[(c/k\tau_o)+1]/S} \exp(-\frac{t^2}{2})\,dt \right. \tag{10.46}$$

where α is a proportionality factor equal to $\alpha_v/\alpha_a \cong 2/3$ and $t = r'/S$. The integral is a standard integral which can be evaluated from tables. The terms d_f, ρ_f, $\bar{t}/c$ data can be matched graphically with a normalised form of eqn (10.46), i.e.

$$\dot{e}(\bar{t}_i/\alpha d_f \rho_f) = f(k\bar{\tau}_o/c) \tag{10.47}$$

such a plot is shown in Fig. 10.30(a).

Difficulties are encountered with eqn (10.46) for values of $c/k\bar{\tau}_o < 1$ since the integral changes sign. This yields $e > \alpha d_f \rho_f/\bar{\tau}_i$ and implies that the rate of entrainment of aggregates is faster than

*This assumption can be avoided through a change of limits to $[(c/k\tau_o) - 1]/S$ and $-\infty$.

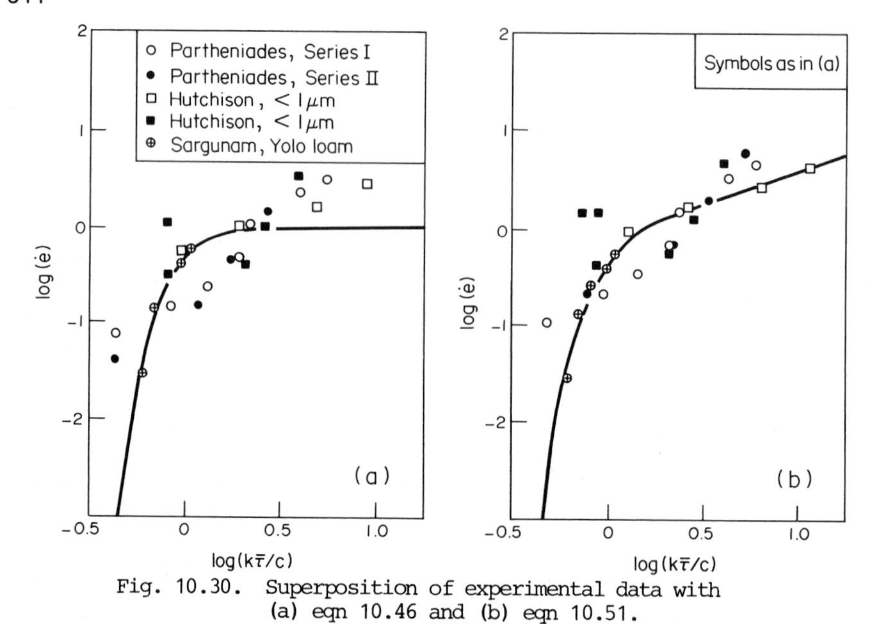

Fig. 10.30. Superposition of experimental data with (a) eqn 10.46 and (b) eqn 10.51.

the fluctuations of shear stress. The equation also tends to a constant erosion rate as $k\tau_o/c$ becomes large which is not in keeping with the trend of data points in Fig. 10.30.

Croad (1981) replaced $\bar{t}_i$ with the period of shear stress fluctuations $\bar{t}_*$, where $\bar{t}_* > \bar{t}_i$, and put $\bar{t}_*$ equal to the frequency of turbulence burst in the boundary layer

$$\frac{U_o \bar{T}_B}{\delta} \cong 5 \tag{10.48}$$

where U_o is the maximum velocity at the outer edge of the boundary layer of thickness δ, here equal to depth of flow. From

$$\frac{\tau_o}{\rho U_o^2} = 0.0128 \left[\frac{U_o \delta}{\nu}\right]^{-1/4} \tag{10.49}$$

with δ = constant

$$\bar{t}_* \propto \bar{\tau}_o^{-4/7} \tag{10.50}$$

This leads to

$$\dot{e} = \beta d_f \rho_f \bar{\tau}_o^{4/7} \left\{ 1 - \frac{1}{\sqrt{2\pi}} \int_{-\infty}^{[(c/k\tau_o)-1]/S} \exp(-\frac{t^2}{2}) dt \right\} \tag{10.51}$$

where β is a proportionality constant and a function of flow depth. According to eqn (10.51) the erosion rate continues to increase as $k\bar{\tau}_o/c$ increases, Fig. 10.30(b).

Mirtskhulava (1966) developed a model similar to those of entrainment of noncohesive material.

Lambermont and Lebon (1978) proposed a model based on analogy between erosion and diffusion.

The approach to modelling of the erosion or deposition of muds, based on the rate process concepts is at present still one of the more promising avenues. The arguments on which it is based do not lose their validity when extended from clays to muds, except that the erosion craters do not remain. The soft surface of mud is rapidly remolded after a cluster of flocs is resuspended.

For numerical modelling of the estuarine mud movements reference is made to Sheng (1986), Hayter (1986) and Parker (1986).

The aspects of transport are discussed in the chapter on Coastal Processes. For the fine suspended sediments the transport is essentially a convection problem, i.e., the translation of the water mass which contains a given amount of sediment.

Chapter 11

Coastal Processes

11.1 Introduction

The term coastal zone describes the shore area, including the coastal dunes, and the sea bed from the water's edge to a depth beyond which the sea bed is seldom disturbed by wave action. Some of the shoreline is rocky, with or without cliffs, some muddy or of fine silts but a substantial part of it is sandy. The sandy coastlines are highly mobile and their features change continuously with changing sea conditions. These changes are referred to as coastal processes. Changes in rocky shorelines are very gradual and will not be considered here. Figure 11.1 illustrates the widely used terminology for description of various parts of the coastal zone.

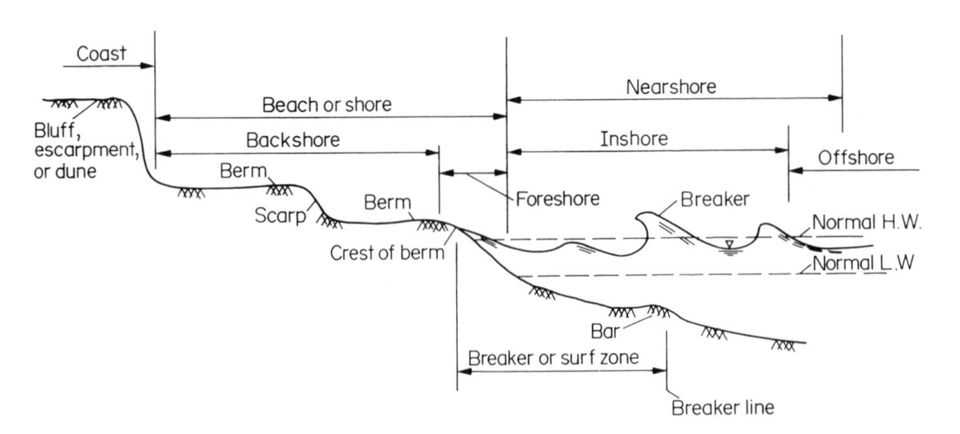

Fig. 11.1. Illustration of terminology in use to describe parts of coastal zone.

The cause of the dynamic state of shoreline is the concentration of energy from wave action on a relatively narrow zone near the water's edge. In this zone the wave energy is transformed into turbulence and is dissipated. The power of waves varies with their height. Near the breaking point it is about 4.3 $H^{2.5}$ kW per metre wave crest or about 240 kW per metre for a 5 m wave. This shows that the energy input during stormy conditions into the narrow zone at water's edge is enormous.

The geometry of a sandy coastal zone changes continuously, the beach may erode or build up, sand banks may form and disappear. Many of these changes are small and may go unnoticed but every now and then large changes occur and we speak of a coastal problem. All changes must be related to a time scale. The beach may generally be eroding or building up, i.e., there may be a long term *trend*. Superimposed on this may be short term changes. The beach may be drastically changed during a storm but could be totally restored by the following sea conditions. These temporal changes of a stable beach are part of a dynamic equilibrium. The short term oscillations of the shoreline with sea conditions may amount to ± 30 m in a few months. These are changes in the human time scale, whereas when a geologist speaks of recent changes these may have occurred 10 000 years ago. Likewise, changes have to be viewed in spatial scale; a sand bank, a bay, the continental shelf, etc.

The physical processes which in complex interactions create the geometry of the coastal zone are diagrammatically illustrated in Fig. 11.2. These processes include one of the most complex sets of hydrodynamic problems. The prediction of waves, current and orbital velocities, the spatial distribution of wave energy, etc. The transport of sediment due to wave action differs from that by unidirectional currents essentially only in the entrainment of sediment. In a current only situation the current has to entrain and to transport the sediment. Under wave action the sediment can be entrained and suspended by the orbital movements of waves, and this suspended sediment could be transported by the feeblest of a current. The analytical description of the process of sediment movement under wave action is, however, much more difficult because of the periodic nature of the fluid forces and the variability of the wave motion in direction and with time. It is particularly important to realise that the movements of sediments, the coastal processes, cannot be treated separately of the hydrodynamic conditions. Coastal processes are intimately coupled to the local hydrodynamics. Because of the very difficult hydrodynamics, there is a tendency to "forget" about it and talk about sand or sediment movement. Whatever the complexities, the coasts are one of the most attractive and valuable land areas. The interface between land and sea is also a most important zone from the environmental, recreational, aesthetic, etc. point of view. The complexities of the detail make it particularly important to acquire a sound understanding of the physics of the processes which shape the shoreline and cause sediment to accumulate or to erode. The subject is extensive and commands a literature in its own right. For introductory reading reference is made to Zenkovich (1967), King (1972) and Bascom (1980). The aim here is to introduce the basic aspects of the hydrodynamics, sediment movement and coastal processes. No attempt is made to introduce the biological, chemical or environmental aspects. A summary on the wave mechanics and wave induced currents is included to give the reader ready access and reference to the more important formulae.

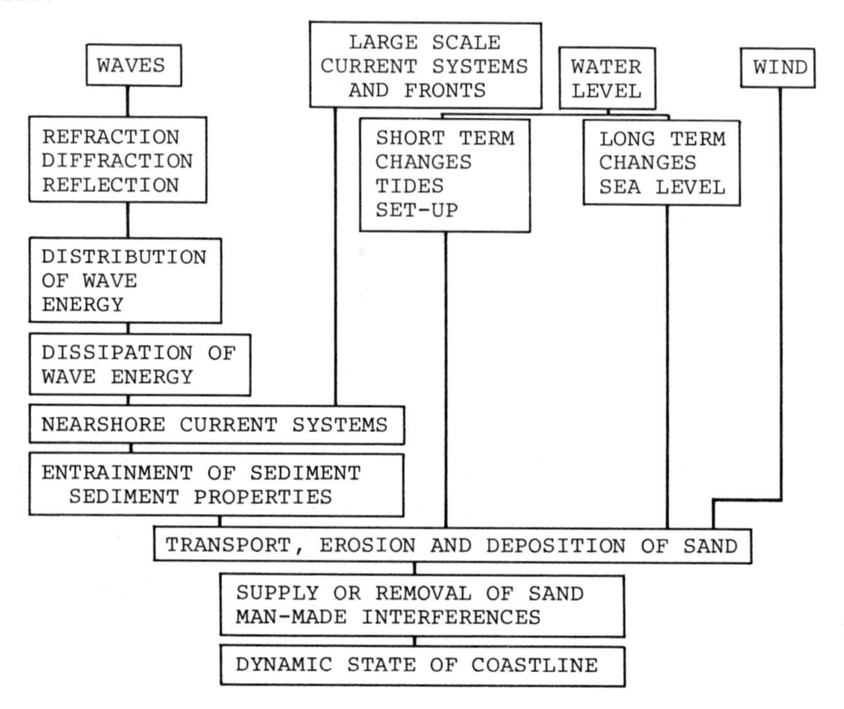

Fig. 11.2. Diagrammatic illustration of factors involved in coastal processes.

From the coastal engineering point of view the density of sea water is its most important property. It is a function of salinity, temperature and pressure, of which the last can be neglected in depth of less than about 500 m. The relationship between density, salinity and temperature is complex and details can be found in Fisher et al. (1970). For many practical purposes, where changes in water temperature are not large, the density can be expressed as $\rho = 1000 + 0.75\,S$ where S is salinity in parts per thousand. The term 0.75 S is usually known as σ_t, i.e. $\sigma_t = \rho - 1000$. Values of $\sigma_t = f(S, {}^\circ T)$ are listed in Table 11.1.

The values of dynamic viscosity as a function of temperature and salinity are shown in Table 11.2.

11.2 Elements of Wave Theory

Waves, through the ages, have fascinated the layman and the applied mathematician alike. The lay observer may be aware only of the short period wind waves and the tidal rise and fall of

TEMPERATURE IN °C

S 0/00	0	2	4	6	8	10	12	14	16	18	20	22	24	26	28	30	40	S 0/00
0	-0.16	-0.06	-0.03	-0.06	-0.15	-0.30	-0.50	-0.75	-1.06	-1.40	-1.79	-2.23	-2.70	-3.21	-3.76	-4.35	-7.78	0
2	1.48	1.57	1.58	1.54	1.44	1.28	1.06	0.80	+0.49	+0.14	-0.26	-0.70	-1.18	-1.70	-2.26	-2.85	-6.30	1
4	3.12	3.19	3.19	3.14	3.02	2.85	2.63	2.36	2.04	1.67	1.27	0.82	+0.33	-0.19	-0.75	-1.35	-4.83	4
6	4.76	4.81	4.80	4.73	4.61	4.42	4.19	3.91	3.58	3.21	2.80	2.34	1.85	1.32	+0.75	+0.15	-3.35	6
8	6.40	6.44	6.41	6.33	6.19	6.00	5.75	5.46	5.13	4.75	4.32	3.86	3.36	2.83	2.25	1.65	-1.87	8
10	8.03	8.05	8.02	7.92	7.77	7.57	7.31	7.01	6.67	6.28	5.85	5.38	4.88	4.34	3.76	3.15	-0.39	10
12	9.66	9.67	9.62	9.51	9.35	9.13	8.87	8.56	8.21	7.82	7.38	6.90	6.39	5.84	5.26	4.64	1.08	12
14	11.29	11.28	11.22	11.10	10.93	10.70	10.43	10.11	9.75	9.35	8.91	8.42	7.91	7.35	6.76	6.14	2.56	14
16	12.92	12.90	12.82	12.69	12.50	12.27	11.99	11.66	11.29	10.88	10.43	9.94	9.42	8.86	8.27	7.64	4.04	16
18	14.54	14.50	14.41	14.27	14.08	13.83	13.54	13.21	12.83	12.41	11.96	11.46	10.93	10.37	9.77	9.14	5.52	18
20	16.16	16.11	16.01	15.85	15.65	15.39	15.10	14.75	14.37	13.94	13.48	12.98	12.45	11.88	11.27	10.64	7.00	20
22	17.78	17.72	17.60	17.44	17.22	16.96	16.65	16.30	15.91	15.47	15.00	14.50	13.96	13.38	12.77	12.13	8.47	22
24	19.40	19.32	19.19	19.02	18.79	18.52	18.20	17.84	17.44	17.00	16.53	16.02	15.47	14.89	14.28	13.63	9.95	24
26	21.01	20.92	20.78	20.59	20.36	20.08	19.75	19.38	18.98	18.53	18.05	17.53	16.98	16.39	15.78	15.13	11.43	26
28	22.62	22.52	22.37	22.17	21.92	21.63	21.30	20.93	20.51	20.06	19.57	19.05	18.49	17.90	17.28	16.62	12.91	28
30	24.23	24.12	23.95	23.75	23.49	23.19	22.85	22.47	22.05	21.59	21.09	20.56	20.00	19.41	18.78	18.12	14.39	30
32	25.84	25.71	25.54	25.32	25.05	24.75	24.40	24.01	23.58	23.11	22.61	22.08	21.51	20.91	20.28	19.62	15.87	32
34	27.44	27.30	27.12	26.89	26.62	26.30	25.94	25.54	25.11	24.64	24.13	23.59	23.02	22.42	21.78	21.11	17.35	34
36	29.04	28.90	28.70	28.46	28.18	27.85	27.49	27.08	26.64	26.16	25.65	25.11	24.53	23.92	23.28	22.61	18.83	36
38	30.64	30.48	30.28	30.03	29.74	29.74	29.41	29.03	28.62	28.17	27.17	26.62	26.04	25.43	24.78	24.11	20.31	38
40	32.24	32.07	31.86	31.60	31.30	30.96	30.58	30.16	29.70	29.21	28.69	28.14	27.55	26.93	26.28	25.60	21.79	40

Table 11.1. Density $\sigma_t = \rho - 1000$ as a function of temperature and salinity

C° \ S°/oo	μ x 10³ kg m⁻¹ s⁻¹ 0	20	40
0	1.78	1.837	1.885
10	1.30	1.349	1.397
20	1.00	1.043	1.088
30	0.806	0.837	0.874

Table 11.2: Dynamic viscosity as a function of temperature and salinity.

water level but nature provides a very broad spectrum of waves. Starting from capillary waves with periods of a fraction of a second they range over the short period wind waves and swell to long period waves like tsunamis, surf beat, infragravity waves, tides, storm surges, gyroscopic and planetary waves with periods increasing from seconds to half a minute to days. In addition, the wave motion is not confined to water surface layers only. Large amplitude long period movements occur in oceans at great depths. The primary source of energy for creation of waves up to about 10 minute periods is the wind. The primary restoring force for waves with 0.05 to about 30 minute periods is the gravity. For very short capillary waves the primary restoring force is the surface tension and for long period waves with $T >$ 5 min. Coriolis force becomes dominant as the restoring force whereas storm systems, tsunamis, sun and moon provide the disturbing force.

In the context of coastal processes the most important waves are the short period wind waves but even here the picture is complex. Wind generates simultaneously waves of many periods and heights - a wave spectrum - and waves from neighbouring wind fields may be superimposed. Kinsman (1965) therefore defined the wave as "a lump on the ocean surface".

Wave theories provide analytical models and descriptions for most of the above wave types but there are still many unsolved problems and waves are still one of the "playgrounds" for applied mathematicians.

11.2.1 The linear or Airy waves

The simplest of the wave theories is the linearised or Airy wave theory, named after Sir George B. Airy who published the theory in 1845. These waves are also referred to as the first order Stokes' waves. The relative simplicity of the theory accounts for its widespread use. The theory postulates that the wave motion is started on the surface of an incompressible frictionless fluid by an external force which is thereafter removed. The wave motion is then subject to gravity only. The resulting waves are two-dimensional, sinusoidal and of very low amplitude.

Linearisation of the surface boundary condition, conservation of energy requirement, leads over a potential and stream function to expressions for the components of velocity vector u and v. The horizontal component is given by

$$u = \frac{\pi H}{T} \frac{\cosh[k(h + y)]}{\sinh kh} \cos(kx - \omega t) \tag{11.1}$$

and the vertical component

$$v = \frac{\pi H}{T} \frac{\sinh[k(y + h)]}{\sinh kh} \sin(kx - \omega t) \tag{11.2}$$

where $k = 2\pi/L$; $\omega = 2\pi/T$; H, T and L are wave height (double amplitude), period and length, respectively, and h is still water depth. The coordinates are in still water surface with y vertically upwards. The associated accelerations are given by du/dt and dv/dt. The movement of water particles in these waves is along closed eliptical paths which for deep water become circles. Hence, u and v are known as orbital velocities. The horizontal half axis of the orbits is

$$A_x = \frac{H}{2} \frac{\cosh k(h + y)}{\sinh kh} \tag{11.3}$$

which at bed where $y = -h$ becomes

$$A = \frac{H}{2} \frac{1}{\sinh kh} \tag{11.4}$$

The maximum horizontal velocities occur under the wave crest and trough, in the direction of wave propagation and opposite, respectively. The distribution of u_{max} under the wave crest is

$$u_{max} = \frac{\pi H}{T} \frac{\cosh k(h + y)}{\sinh kh} \tag{11.5}$$

which at the bed is

$$u_m = \frac{\pi H}{T} \frac{1}{\sinh kh} \tag{11.6}$$

i.e.,

$$u_m = \omega A \tag{11.7}$$

The mean velocity over half period is given by multiplying u_{max} with $2/\pi$.

The celerity of the wave form (speed of translation) is given by

$$c = \frac{\omega}{k} = \left[\frac{gL}{2\pi}\tanh kh\right]^{\frac{1}{2}} \tag{11.8}$$

which in water shallow relative to wave length, $h/L \rightarrow 0$, yields

$$c = c_s = \sqrt{gh} \tag{11.9}$$

and when $h/L \rightarrow \infty$, i.e., in relatively deep water

$$c = c_o = \sqrt{\frac{gL_o}{2\pi}} \tag{11.10}$$

where the subscript o is conventionally used to denote deep water values. Usually water is assumed to be deep when $h > L/2$. However, only 5% error in c results from assuming water to be deep for $h/L_o > 0.25$ and shallow for $h/L_o < 0.05$. Between c_s and c_o the full equation for c has to be used. The deep water values reduce to

$$c_o = 1.25\sqrt{L_o} \quad (ms^{-1}) \tag{11.11}$$

and

$$L_o = 1.56\ T^2 \quad (m) \tag{11.12}$$

since $c_o = L_o/T$.

The u_{max}-distribution leads to dynamic pressure distribution, in excess of the hydrostatic,

$$\Delta p = \tfrac{1}{2}\rho gH \frac{\cosh k(h + y)}{\cosh kh} \cos(kx - \omega t) \tag{11.13}$$

The potential and kinetic energies in Airy wave are equal and the power per unit surface area is

$$P = (\tfrac{1}{2}\rho gH^2)c\left[\tfrac{1}{2}\left[1 + \frac{2\ kh}{\sinh 2\ kh}\right]\right] = Ecn \tag{11.14}$$

where the term $n = n_o = \frac{1}{2}$ in deep water and $n = 1$ in shallow water, and ρ is density of water. The term

$$E = \tfrac{1}{2}\rho gH^2 \tag{11.15}$$

represents wave energy per unit area of water surface. In deep water wave energy is dissipated by viscous action and by breaking of waves, the white caps, but dissipation by viscous action is very minimal and can be ignored.

In shoaling water bottom friction will dissipate energy, in addition to breaking. On a very permeable bed some energy will be consumed by percolation. A special case here are the coastlines with substantial soft mud beds.

The wave height and length change as the deep water waves run into shallowing water:

$$\frac{H}{H_o} = \left[\frac{n_o c_o}{nc}\right]^{\frac{1}{2}} \tag{11.16}$$

and

$$L = L_o \tanh kh \tag{11.17}$$

where the bracket term in eqn (11.16) is known as the *shoaling coefficient*. The increase in wave height is limited by breaking (instability) conditions which are discussed later.

Equation (11.17) can be solved by an iterative method using

$$L_{2i+1} = L_o \tanh \frac{2\pi}{L_{2i}} h$$

$$L_{2i+2} = \frac{2L_{2i+1} + L_{2i}}{3}$$

where i = 0,1,2,... For example, T = 10 s, h = 20 m and L_o = 156 m:

i	L_{2i+1}	L_{2i+2}
0	104.1	121.4
1	121.0	121.2
2		

An appropriate result is given by

$$L = L_o \left[\tanh \frac{2\pi}{L_o} h\right]^{\frac{1}{2}} \tag{11.18}$$

which is only a few percent larger. The Shore Protection Manual, US Army Corps of Engineers, includes tables in terms of h/L_0, h/L and $\tanh 2\pi h/L$ which enable a direct solution of eqn (11.17).

In nature a low amplitude ocean swell approximates to an Airy wave. The major shortcoming of the linearised wave theory is that it does not allow for a net transport of water in the direction of wave propagation. The orbits of water particles of an Airy wave are closed ellipses or circles in deep water. In nature the water particle does not return exactly to the starting point but a little further in the direction of wave propagation. This average over a wave period leads to a net mass transport. When this transport is intercepted by a coastline a build-up of mean water level occurs, the wind or storm tide.

When the waves run into shallowing water the wave form also changes from sinusoidal. The crests become more pointed and the troughs flatter and longer.

11.2.2 Higher order wave theories

In order to describe mass transport and the shape of waves of finite height higher order solutions have been developed. Here the Stokes' wave theory is well known. It is a series solution where the first order approximation yields the linear solution. The horizontal component of the orbital velocity by the second order solution is given by

$$u = \left[\frac{\pi H}{T}\right] \frac{\cosh k(h + y)}{\sinh kh} \cos(kx - \omega t) + \frac{3}{4}\left[\frac{\pi H}{T}\right]\left[\frac{\pi H}{L}\right] \frac{\cosh 2k(h + y)}{\sinh^4 kh} \cos 2(kx - \omega t) \qquad (11.19)$$

The first term is the same as for Airy waves but the second term is seen to have twice the frequency of the first. The orbital paths of this motion are not closed and when eqn (11.19) is integrated over depth and wave period a net transport results. If this transport rate through a vertical section is made zero eqn (11.19) obtains an additional subtractive term $-(^1/_8)gH^2/(ch)$. The distribution of net particle movement over a wave period becomes

$$\overline{U}_y = \frac{\left[\frac{\pi H}{T}\right]\left[\frac{\pi H}{L}\right]}{2\sinh^2 kh}\left[\cosh 2k(h + y) - \frac{1}{2kh}\sinh 2kh\right] \qquad (11.20)$$

which in deep water when $kh >> 1$ becomes

$$\overline{U}_{yo} = \left[\frac{\pi H}{T}\right]\left[\frac{\pi H}{L}\right]\left[e^{-2ky} - \frac{L}{4\pi h}\right] \qquad (11.21)$$

Equation (11.20) defines the distribution over depth of net mass transport velocities. These are in the direction of wave propagation in and near the surface and change to velocities in opposite direction near the bed, i.e. at the top of the boundary layer.

The expressions for wave celerity and wave length by the second order theory are the same as for Airy waves but the third order theory shows that the celerity is both frequency (wave length) and amplitude dependent (dispersive). For deep water

$$c_o = (g/k)^{\frac{1}{2}}[1 + (kH/2)^2]^{\frac{1}{2}} \qquad (11.22)$$

The second order theory also shows that the mean wave energy level is Δh above the mean water level

$$\Delta h = \frac{1}{16}\frac{H^2}{h}\frac{2kh}{\sinh 2kh} \qquad (11.23)$$

The Longuet-Higgins (1953) expression for mass transport is

$$\overline{U}_\eta = \frac{a^2\omega k}{4\sinh^2 kh}\left[2\cosh 2kh(\eta-1) + 3 + kh \sinh 2kh(3\eta^2 - 4\eta + 1) + 3\left[\frac{\sinh 2kh}{2kh} + \frac{3}{2}\right](\eta^2 - 1)\right] \qquad (11.24)$$

where $\eta = y/h$ and $a = H/2$. Equation (11.24) is illustrated in Fig. 11.3. At the bed eqn (11.24) reduces to

$$\overline{U}_o = \frac{5}{4}\left[\frac{\pi H}{L}\right]^{\frac{1}{2}} \frac{c}{\sinh^2 kh} \qquad (11.24a)$$

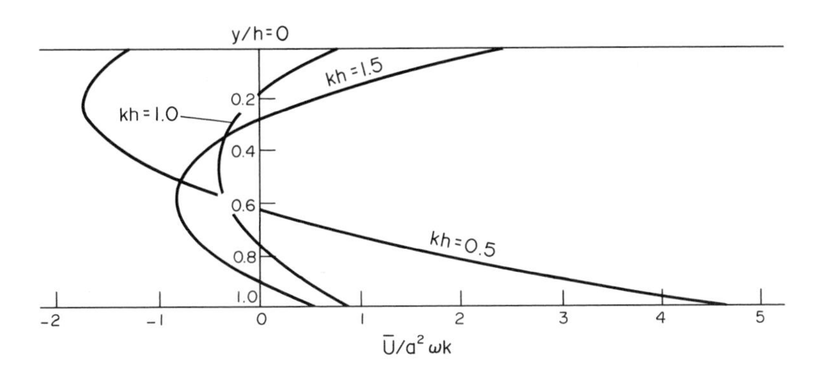

Fig. 11.3. Distribution of mass transport velocities according to eqn 11.24.

Measurements agree well with predictions, particularly near the bed, when $0.7 < kh < 0.13$. In deeper water the Stokes' solution appears to be better. Bijker et al. (1974) studied the influence of bottom slope and bed roughness on the near bottom drift (mass transport) velocities and found trends indicated by eqn (11.24). However, as kh decreases on approaching the shore the return velocity profile changes. At small values of H/L and h/L the return velocity moved up near the surface. Near the breakers, the mass transport velocity profile over a smooth bed has the same shape for all waves with strong forward velocities over the bed and near the surface, and the return flow between these. In the breaker zone the net bottom velocities were always towards the sea and the effect of bed slope was small. Similar observations were made by Kajima et al. (1982) in a large wave flume. Sand roughness slightly reduced the velocities near the bed but the profiles remained essentially as predicted. The presence of (artificial) ripples reduced the initial forward velocity to zero and indiced a consistent shoreward flow above the bed under flat waves and seaward under steep waves. Sleath (1984) concluded that bed roughness significantly increases the time average mean drift velocity near the bed at small values of A/k_s. At large values of A/k_s no clear difference could be observed. The term k_s refers to effective roughness height.

When long waves propagate into shallow water the Stokes' description become inadequate. Here the *cnoidal wave* theory, initiated by Boussinesq (1872) and Korteweg and De Vries (1895), yields better results. Details of the theory are given by Svendsen (1974). The name cnoidal arises from description of the wave profile in terms of the Jacobian elliptical cosine function, denoted by cn. The approximate range of validity is $h/L < 0.1$ and the Ursell (Stokes) parameter $U = L^2H/h^3 > 26$, or on the H_oL_o - h/L_o-plane the area between breaking of waves ($H/h \sim 0.8$, approx. defined by a line $H_o/L_o = 0$, $h/L_o = 0.02$ and $H_o/L_o = 0.14$, $H/L_o = 0.14$) and a line joining $h/L_o = 0.12$, $H_o/L_o = 0$ to $H_o/L_o = 0.14$, $h/L_o = 0.14$.

The wave form is given by

$$\eta = \eta_t + H\,\mathrm{cn}^2(\theta,m) \tag{11.25}$$

where η, η_t and H are shown in Fig. 11.4, $\theta = kx - \omega t$ and

$$\eta_t = \left[\frac{1}{m}\left[1 - \frac{E}{K}\right] - 1\right]H \tag{11.26}$$

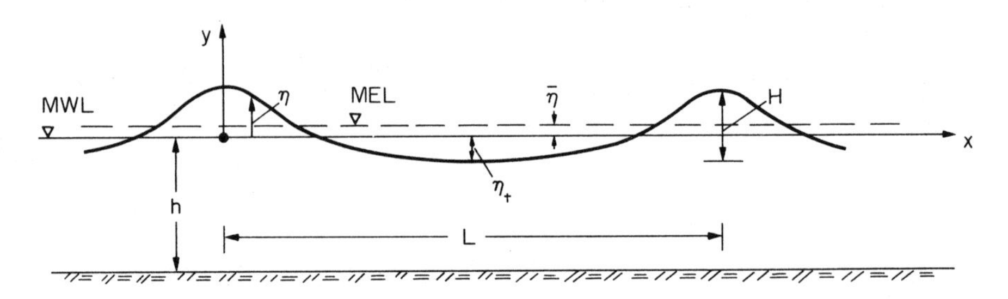

Fig. 11.4. Definition sketch.

The $K = K(m)$ and $E = E(m)$ are complete elliptic integrals of the first and second kind, respectively and the parameter m is related to the Ursell number by

$$U = \frac{HL^2}{h^3} = \frac{16}{3}mK^2 \tag{11.27}$$

The values of m, K and E are tabulated in Table 11.3. The horizontal velocity distributions under wave crest

$$u_c = c\left\{\frac{\eta_c}{h} - \left[\frac{\eta_c}{h}\right]^2 - \left[\frac{1}{3} - \left[\frac{y+h}{h}\right]^2\right]\frac{4HK^2}{gT^2}\right\} \tag{11.28}$$

and trough

$$u_t = c\left\{\frac{\eta_t}{h} - \left[\frac{\eta_t}{h}\right]^2 + \left[\frac{1}{3} - \left[\frac{y+h}{h}\right]^2\right]\frac{4m_1HK^2}{gT^2}\right\} \tag{11.29}$$

U	m	m_1	K	E	η_{min}/H	A	B
1	(-2)7.317	(-1)9.268	1.601	1.542	-0.495	-13.152	0.1250
2	(-1)1.410	(-1)8.590	1.631	1.514	-0.491	-6.565	0.1250
3	(-1)2.038	(-1)7.962	1.662	1.487	-0.486	-4.365	0.1249
4	(-1)2.619	(-1)7.381	1.692	1.462	-0.481	-3.261	0.1249
5	(-1)3.157	(-1)6.843	1.723	1.438	-0.476	-2.596	0.1249
6	(-1)3.655	(-1)6.345	1.754	1.416	-0.472	-2.151	0.1248
7	(-1)4.116	(-1)5.884	1.786	1.394	-0.467	-1.830	0.1247
8	(-1)4.543	(-1)5.457	1.817	1.373	-0.462	-1.588	0.1246
9	(-1)4.937	(-1)5.063	1.849	1.354	-0.458	-1.399	0.1245
10	(-1)5.302	(-1)4.698	1.881	1.335	-0.453	-1.245	0.1244
11	(-1)5.639	(-1)4.361	1.912	1.318	-0.449	-1.119	0.1243
12	(-1)5.952	(-1)4.048	1.944	1.301	-0.444	-1.012	0.1242
13	(-1)6.240	(-1)3.760	1.976	1.285	-0.440	-0.921	0.1241
14	(-1)6.507	(-1)3.493	2.008	1.270	-0.435	-0.842	0.1239
15	(-1)6.754	(-1)3.246	2.041	1.256	-0.431	-0.773	0.1238
16	(-1)6.982	(-1)3.018	2.073	1.243	-0.426	-0.712	0.1236
17	(-1)7.194	(-1)2.806	2.105	1.230	-0.422	-0.657	0.1234
18	(-1)7.389	(-1)2.611	2.137	1.218	-0.418	-0.607	0.1233
19	(-1)7.570	(-1)2.430	2.169	1.207	-0.414	-0.562	0.1231
20	(-1)7.738	(-1)2.262	2.201	1.196	-0.410	-0.521	0.1229
22	(-1)8.036	(-1)1.964	2.266	1.176	-0.402	-0.449	0.1225
24	(-1)8.293	(-1)1.707	2.329	1.158	-0.394	-0.387	0.1220
26	(-1)8.513	(-1)1.487	2.393	1.142	-0.386	-0.333	-0.1216
28	(-1)8.702	(-1)1.298	2.456	1.128	-0.379	-0.285	0.1211
30	(-1)8.866	(-1)1.134	2.519	1.116	-0.372	-0.243	0.1206
32	(-1)9.006	(-2)9.937	2.581	1.104	-0.365	-0.204	0.1201
34	(-1)9.128	(-2)8.720	2.643	1.094	-0.358	-0.170	0.1195
36	(-1)9.233	(-2)7.666	2.704	1.085	-0.352	-0.138	0.1190
38	(-1)9.325	(-2)6.753	2.764	1.077	-0.345	-0.109	0.1184
40	(-1)9.404	(-2)5.959	2.824	1.070	-0.339	-0.082	0.1179
42	(-1)9.473	(-2)5.268	2.883	1.063	-0.334	-0.056	0.1173
44	(-1)9.533	(-2)4.665	2.942	1.057	-0.328	-0.033	0.1167
46	(-1)9.586	(-2)4.139	3.000	1.052	-0.323	-0.011	0.1161
48	(-1)9.632	(-2)3.678	3.057	1.047	-0.317	0.009	0.1155
50	(-1)9.673	(-2)3.274	3.113	1.043	-0.312	0.029	0.1149
50	(-1)9.673	(-2)3.673	3.113	1.043	-0.312	0.029	0.1149
55	(-1)9.753	(-2)2.466	3.252	1.034	-0.301	0.072	0.1134
60	(-1)9.813	(-2)1.874	3.386	1.027	-0.290	0.111	0.1119
65	(-1)9.856	(-2)1.438	3.516	1.022	-0.280	0.145	0.1104
70	(-1)9.889	(-2)1.112	3.643	1.017	-0.271	0.175	0.1090
75	(-1)9.13	(-3)8.668	3.766	1.014	-0.263	0.203	0.1075
80	(-1)9.932	(-3)6.805	3.886	1.012	-0.255	0.227	0.1061
85	(-1)9.946	(-3)5.379	4.003	1.009	-0.248	0.250	0.1048
90	(-1)9.957	(-3)4.278	4.117	1.008	-0.242	0.271	0.1034
95	(-1)9.966	(-3)3.423	4.228	1.006	-0.235	0.290	0.1021
100	(-1)9.972	(-3)2.753	4.336	1.005	-0.230	0.308	0.1009
150	(-1)9.996	(-4)3.955	5.304	1.001	-0.188	0.434	0.0902
200	(-1)9.999	(-5)7.674	6.124	1.000	-0.163	0.510	0.0822
250	1.000	(-5)1.808	6.847	1.000	-0.146	0.562	0.0760
300	1.000	(-6)4.894	7.500	1.000	-0.133	0.600	0.0711
350	1.000	(-6)1.471	8.101	1.000	-0.123	0.630	0.0671
400	1.000	(-7)4.807	8.660	1.000	-0.115	0.654	0.0636
450	1.000	(-7)1.681	9.186	1.000	-0.109	0.673	0.0607
500	1.000	(-8)6.224	9.682	1.000	-0.103	0.690	0.0582
550	1.000	(-8)2.419	10.155	1.000	-0.098	0.705	0.0560
600	1.000	(-9)9.803	10.607	1.000	-0.094	0.717	0.0540
650	1.000	(-9)4.122	11.040	1.000	-0.091	0.728	0.0522
700	1.000	(-9)1.791	11.456	1.000	-0.087	0.738	0.0506
750	1.000	(-10)8.015	11.859	1.000	-0.084	0.747	0.0491
800	1.000	(-10)3.682	12.247	1.000	-0.082	0.755	0.0478
850	1.000	(-10)1.733	12.624	1.000	-0.079	0.762	0.0465
900	1.000	(-11)8.333	12.990	1.000	-0.077	0.769	0.0454
950	1.000	(-11)4.089	13.346	1.000	-0.075	0.775	0.0443
1000	1.000	(-11)2.044	13.693	1.000	-0.073	0.781	0.0434
2000	1.000	(-16)2.421	19.365	1.000	-0.052	0.845	0.0318
3000	1.000	(-20)4.015	23.717	1.000	-0.042	0.874	0.0263
4000	1.000	(-23)2.611	27.386	1.000	-0.037	0.890	0.0230
5000	1.000	(-26)4.066	30.619	1.000	-0.033	0.902	0.0207
6000	1.000	(-28)1.177	33.541	1.000	-0.030	0.911	0.0190
7000	1.000	(-31)5.451	36.228	1.000	-0.028	0.917	0.0176
8000	1.000	(-33)3.663	38.730	1.000	-0.026	0.923	0.0165
9000	1.000	(-35)3.336	41.079	1.000	-0.024	0.927	0.0156
10000	1.000	(-37)3.918	43.301	1.000	-0.023	0.931	0.0149
∞	1.000	0.000	∞	1.000	-0.000	1.000	0.0000

Integers in parentheses indicate powers of 10 by which the following numbers are to be multiplied

Table 11.3. Cnoidal wave functions (Skovgaard et al., 1974).

where

$$c = \{gh[1 + A(m)H/h]\} \cong \sqrt{gh}[1 + A(m)H/2h] \quad (11.30)$$

$$A(m) = \frac{2}{m} - 1 - \frac{3E}{mK} \; ; \quad m_1 = 1 - m$$

$$\eta_c = \eta_t + H \; ; \quad \eta_t = \left[\frac{1}{m}\left[1 - \frac{E}{K}\right] - 1\right]H \quad (11.31)$$

When H, L and h are known U can be calculated and Table 11.3 gives the other terms. When H, T and h are given neither the Ursell number U nor L can be calculated directly since $L = T[gh(1 + AH/h)]^{\frac{1}{2}}$ and $U = (T\sqrt{g/h})^2(H/H)(1 + AH/h)$. The problem can be solved with the aid of Table 11.4 which yields L/h in terms of H/h and $T\sqrt{g/h}$.

The mean drift velocity in cnoidal waves to the second order approximation (Fenton, 1979) with $\varepsilon = H/h$ is

$$\overline{U} = (gh)^{\frac{1}{2}}\left[\frac{\varepsilon}{m}\right]^2\left[-\frac{1}{3} + \frac{m}{3} + \frac{4}{3}\frac{E(m)}{K(m)} - \frac{2}{3}m\frac{E(m)}{K(m)}\right] + \text{Current}$$

The $\overline{U}$ distribution with y differs from uniform in third and higher order terms.
An estimate of the potential energy, E_p, of nonlinear waves at the long wave limit of cnoidal waves (Fenton, 1979) for U > 250 when both m = 1 and E = 1 is

$$E_p = \rho g H^2 \left[\frac{1}{3K(m)} - \frac{1}{2K(m)^2}\right]$$

and of the kinetic energy

$$E_K = (4/3\sqrt{3})(H/h)^{3/2}\rho g h^3$$

where h is depth under trough (or mean depth less 0.22H).
In general, the Stokes theory breaks down in shallow water and cnoidal in deep water and both for high waves.

The cnoidal wave profile approaches a sinusoidal wave in deep water, as the Ursell number U → 0, and a *solitary wave* when U → ∞. A solitary wave is defined as a wave crest only propagating on still water, i.e. no trough. When natural waves run into shallow water their crests become more pointed, wave heights increase and the troughs become longer and flatter. These waves in rather shallow water, near breaking conditions, can then be treated as solitary waves relative to the trough water level. The solitary wave profile is given by

$$\eta = H\,\text{sech}^2[(3H/4h^3)^{\frac{1}{2}}(x - ct)] \quad (11.32)$$

and

$$c = [gh(1 + H/h)]^{\frac{1}{2}} \quad (11.33)$$

The horizontal velocities under the wave crest are

$T\sqrt{\frac{g}{h}}$ \ $\frac{H}{h}$	0.01	0.02	0.03	0.04	0.05	0.10	0.15	0.20	0.25	0.30	0.35	0.40	0.45	0.50	0.60	0.70	0.80
	6.7	6.7	6.7	6.7	6.8	6.8	6.8	6.8	6.9	7.0	7.1	7.1	7.2	7.4	7.6	7.9	8.2
9	8.0	8.0	8.0	8.0	8.0	8.1	8.1	8.2	8.2	8.3	8.4	8.5	8.6	8.8	9.1	9.4	9.7
10	9.2	9.2	9.2	9.2	9.2	9.2	9.3	9.3	9.4	9.5	9.7	9.8	9.9	10.1	10.4	10.8	11.2
11	10.3	10.3	10.3	10.3	10.3	10.3	10.4	10.5	10.6	10.7	10.9	11.0	11.2	11.4	11.8	12.2	12.6
12	11.4	11.4	11.4	11.4	11.4	11.4	11.5	11.6	11.8	11.9	12.1	12.3	12.5	12.7	13.1	13.5	14.0
13	12.4	12.4	12.4	12.4	12.5	12.5	12.6	12.7	12.9	13.1	13.3	13.5	13.7	13.9	14.4	14.9	15.4
14	13.5	13.5	13.5	13.5	13.5	13.6	13.7	13.9	14.0	14.2	14.5	14.7	14.9	15.2	15.7	16.2	16.7
15	14.5	14.5	14.5	14.5	14.6	14.6	14.8	15.0	15.2	15.4	15.7	15.9	16.2	16.4	17.0	17.6	18.1
16	15.6	15.6	15.6	15.6	15.6	15.7	15.9	16.1	16.3	16.6	16.8	17.1	17.4	17.7	18.3	18.9	19.5
17	16.6	16.6	16.6	16.6	16.6	16.8	16.9	17.2	17.4	17.7	18.0	18.3	18.6	18.9	19.6	20.2	20.8
18	17.6	17.6	17.6	17.6	17.7	17.8	18.0	18.3	18.6	18.9	19.2	19.5	19.8	20.2	20.9	21.5	22.2
19	18.6	18.6	18.7	18.7	18.7	18.9	19.1	19.4	19.7	20.0	20.4	20.7	21.1	21.4	22.1	22.9	23.6
20	19.7	19.7	19.7	19.7	19.7	19.9	20.2	20.5	20.8	21.2	21.5	21.9	22.3	22.7	23.4	24.2	24.9
21	20.7	20.7	20.7	20.7	20.8	21.0	21.3	21.6	21.9	22.3	22.7	23.1	23.5	23.9	24.7	25.5	26.3
22	21.7	21.7	21.7	21.7	21.8	22.0	22.3	22.7	23.1	23.5	23.9	24.3	24.7	25.1	26.0	26.8	27.6
23	22.7	22.7	22.7	22.8	22.8	23.1	23.4	23.8	24.2	24.6	25.1	25.5	25.9	26.4	27.2	28.1	29.0
24	23.7	23.7	23.8	23.8	23.8	24.1	24.5	24.9	25.3	25.8	26.2	26.7	27.1	27.6	28.5	29.4	30.3
25	24.7	24.8	24.8	24.8	24.9	25.2	25.6	26.0	26.4	26.9	27.4	27.9	28.4	28.8	29.8	30.7	31.7
26	25.7	25.8	25.8	25.8	25.9	26.2	26.6	27.1	27.6	28.1	28.6	29.1	29.6	30.1	31.1	32.0	33.0
27	26.8	26.8	26.8	26.9	26.9	27.1	27.7	28.7	29.2	29.7	30.2	30.8	31.3	32.3		33.4	34.4
28	27.8	27.8	27.8	27.9	27.9	28.3	28.8	29.3	29.8	30.4	30.9	31.4	32.0	32.5	33.6	34.7	35.7
29	28.8	28.8	28.8	28.9	29.0	29.4	29.9	30.4	30.9	31.5	32.1	32.6	33.2	33.8	34.9	36.0	37.1
30	29.8	29.8	29.9	29.9	30.0	30.4	30.9	31.5	32.1	32.6	33.2	33.8	34.4	35.0	36.1	37.3	38.4
31	30.8	30.8	30.9	30.9	31.0	31.5	32.0	32.6	33.2	33.8	34.4	35.0	35.6	36.2	37.4	38.6	39.7
32	31.8	31.8	31.9	32.0	32.0	32.5	33.1	33.7	34.3	34.9	35.6	36.2	36.8	37.4	38.7	39.9	41.1
33	32.8	32.8	32.9	33.0	33.1	33.6	34.2	34.8	35.4	36.1	36.7	37.4	38.0	38.7	40.0	41.2	42.4
34	33.8	33.9	33.9	34.0	34.1	34.6	35.2	35.9	36.5	37.2	37.9	38.6	39.2	39.9	41.2	42.5	43.8
35	34.8	34.9	34.9	35.0	35.1	35.7	36.3	37.0	37.7	38.4	39.1	39.7	40.4	41.1	42.5	43.8	45.1
36	35.8	35.9	35.9	36.0	36.1	36.7	37.4	38.1	38.8	39.5	40.2	40.9	41.6	42.4	43.8	45.1	46.5
37	36.8	36.9	37.0	37.1	37.2	37.8	38.5	39.2	39.9	40.6	41.4	42.1	42.9	43.6	45.0	46.4	47.8
38	37.8	37.9	38.0	38.1	38.2	38.8	39.5	40.3	41.0	41.8	42.5	43.3	44.1	44.8	46.3	47.7	49.2
39	38.9	38.9	39.0	39.1	39.2	39.9	40.6	41.4	42.1	42.9	43.7	44.5	45.3	46.0	47.6	49.1	50.5
40	39.9	39.9	40.0	40.1	40.2	40.9	41.7	42.5	43.3	44.1	44.9	45.7	46.5	47.3	48.8	50.4	51.9

Table 11.4. L/H as a function of H/h and $T(g/h)^{\frac{1}{2}}$, (Skovgaard et al., 1974).

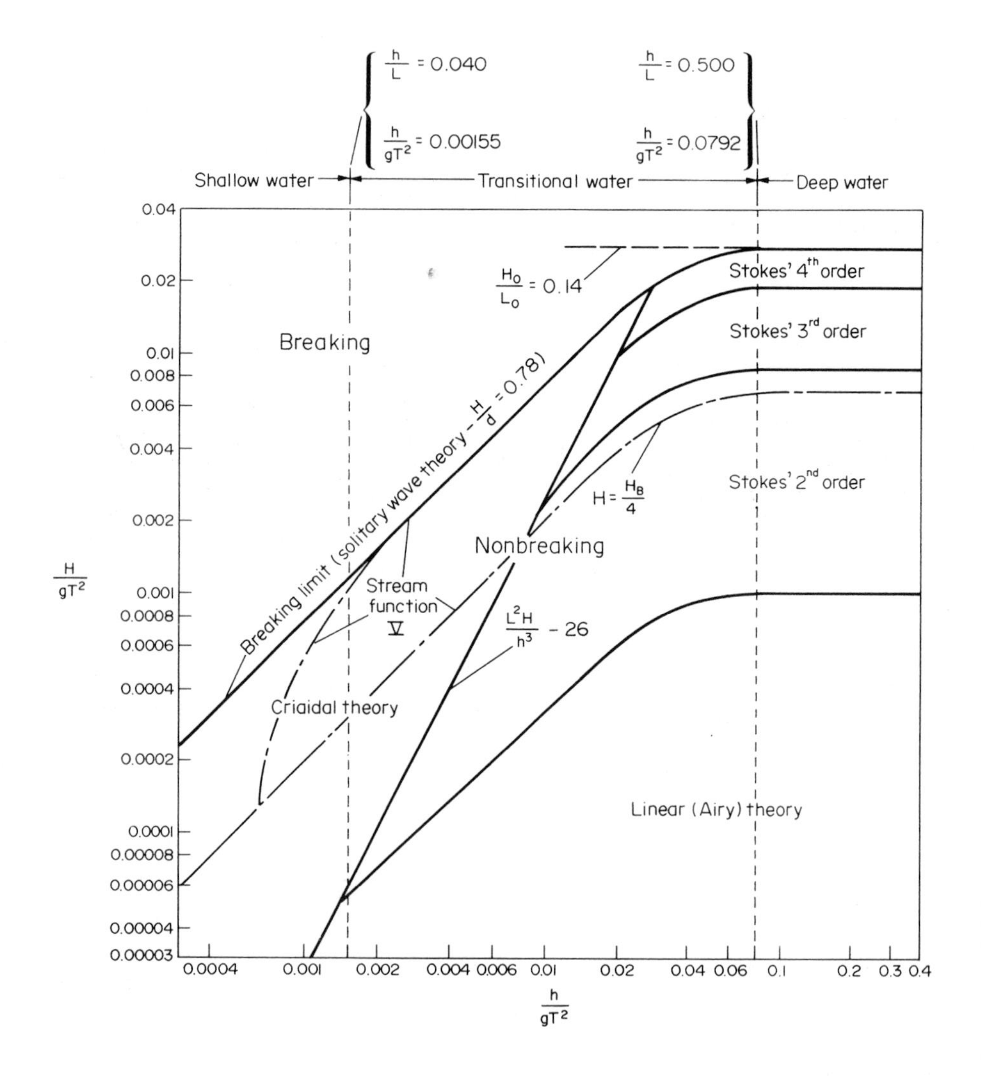

Fig. 11.5. Regions of validity for various wave theories.

$$u_c = \sqrt{gh}\left\{\frac{H}{h} - \frac{3}{4}\left[\frac{H}{h}\right]^2\left[1 - \left[\frac{y+h}{h}\right]^{\frac{1}{2}}\right]\right\} \qquad (11.34)$$

The domains of applicability of the various wave theories are illustrated in Fig. 11.5 from the SPM.

Generally, waves should be considered to be nonlinear for $U \gtrsim 16$.

11.2.3 Wave groups

By superposition of the basic monochromatic waves an infinite number of wave patterns can be formed. For example, superposition of two wave trains with the same amplitude and propagating in the same direction but with slightly different values of k and ω, leads to groups of waves within which the amplitude varies, Fig. 11.6.

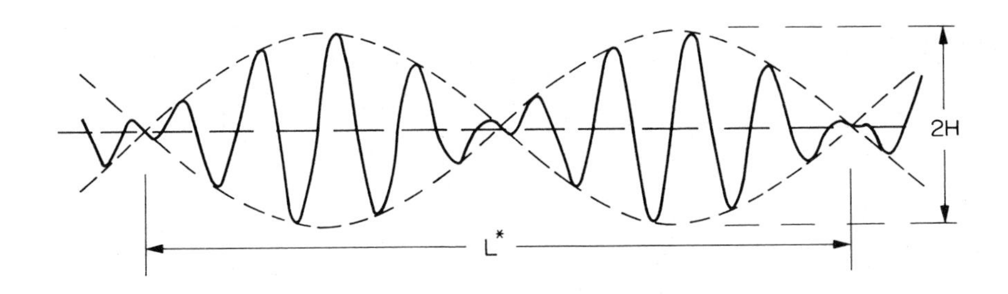

Fig. 11.6. Illustration of wave groups.

The period T*, length L* and the celerity c* of the group are

$$T^* = \frac{4\pi}{\omega_1 - \omega_2} \qquad (11.35)$$

$$L^* = \frac{4\pi}{k_1 - k_2} \qquad (11.36)$$

$$c^* = \frac{\omega_1 - \omega_2}{k_1 - k_2} = \frac{d\omega}{dk} \qquad (11.37)$$

where the last equation leads to

$$c^* = \frac{1}{2}\, c \left[1 - \frac{2kh}{\sinh 2kh}\right] \qquad (11.38)$$

The bracket term in eqn (11.38) tends to one in deep water and to two in shallow water, i.e. in deep water the wave group propagates at half the celerity of the individual waves or the waves travel through the group.

A special case of superposition is that of two identical monochromatic wave trains travelling in opposite directions. The result is a standing wave where the water surface moves up and down between nodal lines in the mean water level surface. Such waves are readily produced by reflection of the incident wave train from a vertical wall.
When the wave trains are angled to each other the result is a diamond pattern of the water surface, three-dimensional wave crests and troughs.
In principle, complex wave patterns, such as created for example by wind, can be decomposed into individual components by Fourier analysis. The wave fields can then be described by a sum of sinusoidal terms

$$a(t) = \sum_{i=1}^{n} a_i \cos(k_i x - \omega_i t)$$

where a is the amplitude of waves (H/2). This leads to descriptions of wave heights, frequencies and energy by methods of spectral analysis and statistics.

11.2.4 Wind waves

The processes by which wind waves are created are complex and will not be discussed. Only a number of the basic characteristics of wind waves will be summarised below. Firstly, wind produces a sepctrum of wave heights and lengths or frequencies. The representative wave height (e.g. H_s or H_{rms}) produced is a function of the wind *speed*, *duration* and the length of the *fetch*, i.e., the length of water surface on which the wind acts. When the wind ceases the waves propagate out of the region at speeds corresponding to their wave lengths and are referred to as *swell*. The wave height, duration and fetch relationship is diagrammatically illustrated in Fig. 11.7.

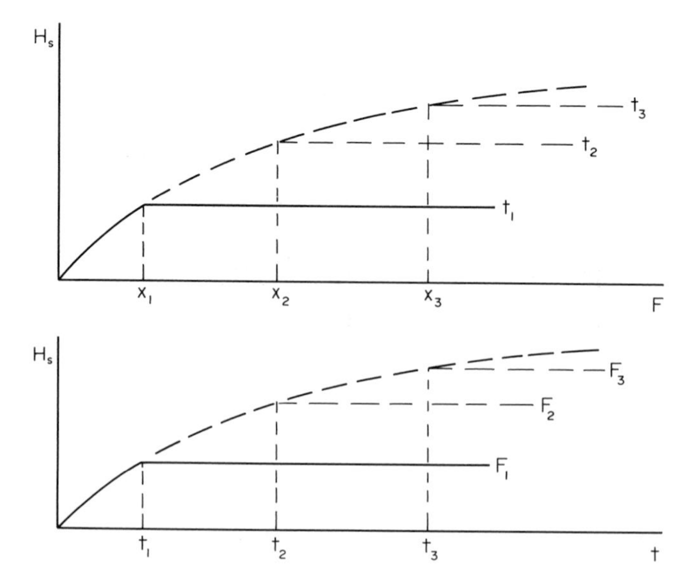

Fig. 11.7. Illustration of wind wave development.

White caps appear on the wave crests fairly soon. These indicate that the given wave is breaking (spilling, i.e. the velocity of water in the wave crest $u \geq c$, the celerity of the wave of that length) and some of the energy is dissipated. If more energy is supplied to this wave element more will be dissipated, i.e. white caps signify a saturation or equilibrium state for the given frequency. This is first reached at the high frequency end of the spectrum but energy is also being transferred to lower frequencies, i.e., the larger waves keep on growing. Thus, there is a gradual saturation process and when all frequencies reach equilibrium we have the state known as *fully arisen sea*. This is not the common case because it requires long periods of wind from the same direction when the fetch is long. The fully arisen sea state is described by a spectrum function, e.g., the Pierson-Moskowitz spectrum or the Jonswap spectrum, and others (an arbitrary choice). The Pierson-Moskowitz spectrum is defined by the distribution function

$$S_H^2(f) = \frac{\alpha g^2}{(2\pi)^4 f^5} \exp\left[-\beta\left[\frac{g}{2\pi V_{19.5} f}\right]^4\right] \tag{11.39}$$

where $S_H^2(f)$ is the energy per unit frequency f or df; $f = 1/T$ or $\omega = 2\pi f$; $\alpha = 8.10 \times 10^{-3}$; $\beta = 0.74$; and $V_{19.5}$ is the wind speed at 19.5 m height. The frequency, f_m, of the peak of the distribution is

$$\frac{g}{2\pi V f_m} = 1.14 \tag{11.40}$$

and the area under the distribution curve is

$$A = \frac{\alpha V^4}{4\beta g^2} = \left[\frac{1}{4} H_s\right]^2 \tag{11.41}$$

or the significant wave height

$$H_s = \frac{2V^2}{g}\sqrt{\frac{\alpha}{\beta}} \tag{11.42}$$

The average wave length of wind waves in shallow water is approximately

$$\bar{L} \cong \bar{T}^2 \tag{11.43}$$

11.2.5 Statistical definitions of wave parameters

Simple definitions of wave height are the root-mean-square value

$$H_{rms} = \left[\frac{1}{N}\sum_{i=1}^{N} H_i^2\right]^{\frac{1}{2}} \tag{11.44}$$

and the *significant wave height* H_s which is the average of the top third of wave heights in a sequence of N waves. These are related by

$$H_s \cong \sqrt{2}\, H_{rms} \tag{11.45}$$

or *average wave height*

$$\overline{H} = 0.886\, H_{rms} \tag{11.46}$$

based on the Rayleigh distribution of waves

$$\text{Prob}(H_n > H) = \exp[-(H/H_{rms})^2] = \exp[-2(H/H_s)^2] \tag{11.47}$$

where $\text{Prob}(H_n > H)$ is the number of waves n larger than H in a total of N waves, i.e., n/N. The standard deviation is

$$\sigma_H = 0.463\, H_{rms} \tag{11.48}$$

and

$$H_s = 1.596\, \overline{H} \tag{11.49}$$

The wave height for any other probability, e.g. 1%, can be calculated from eqn (11.47). The average wave energy per unit surface area is

$$\overline{E} = \frac{1}{8} \rho g \frac{1}{N} \sum_{i=1}^{N} H_i^2 \tag{11.50}$$

which shows that H_{rms} is a measure of average wave energy. The period of a regular wave train of the same energy is related to that of the average period $\overline{T}$ of the irregular waves by

$$T \cong 1.23\, \overline{T} \tag{11.51}$$

The description of waves must also include the variability of wave directions. This can be done by a directional spectrum but data on wave directions is rare. The usual wave gauge data supply only H_s and T. The picture is further complicated by waves from more than one direction occurring simultaneously. For a series of H_s (from different records) the probability P = n/N of H_s greater than H_s' can be estimated (Thompson and Harris, 1972) from $P(H_s > H_s') = \exp[-(H_s' - H_{s\,min})/\sigma_s]$ where σ_s is the st. dev. of H_s. The authors used the approximation $\overline{H}_s = H_{s\,min} + \sigma_s$ and $H_{s\,min} = 0.38$ leading to

$$P(H_s > H_s') \cong \exp[-(1.61\, H_s' - 0.61\, \overline{H}_s)/\overline{H}_s] \tag{11.52}$$

Estimates of wave heights, based on known fetch lengths and assumed wind speeds, can be obtained with the aid of the semi-empirical formulae by Sverdrup-Munk-Bretschneider (SMP), which for deep water are:

$$\frac{gH_s}{U^2} = 0.283 \tanh\left[0.0125\left[\frac{gF}{U^2}\right]^{0.42}\right] \tag{11.53}$$

$$\frac{gT_s}{2\pi U} = 1.20 \tanh\left[0.077\left[\frac{gF}{U^2}\right]^{0.25}\right] \quad (11.54)$$

$$\frac{gt}{U} = 6.5882 \exp\{[A(\ln X)^2 - B\ln X + C]^{\frac{1}{2}} + D\ln X\} \quad (11.55)$$

where A = 0.0161, B = 0.3692, C = 2.2024, D = 0.8798, $X = gF/U^2$, H_s and T_s are significant wave height and period, U is wind speed, F is fetch length and t is the duration of wind required to develop the above H_s wave height.

11.2.6 Wave boundary layers and friction factor

The velocity distributions associated with waves are modified at the bed by the *boundary layer*. When the free stream velocity (orbital) is a maximum, the velocity profile at the bed looks like illustrated in Fig. 11.8. The boundary layer thickness, δ, is defined as the distance from the wall where the velocity is equal the free stream velocity at its maximum. The overshoot by the velocity, just above δ, is due to the small negative velocities near the bed half a period earlier. In terms of the zone of flow where the bed shear stresses are important, the layer thickness is about double the value of δ. The orbital velocity at the bed gives an estimate of the free stream velocity at the outer limit of the boundary layer. In addition to the boundary layer velocity distribution at the bed, the bed shear stress causes a phase shift between the orbital velocities and those in the boundary layer.

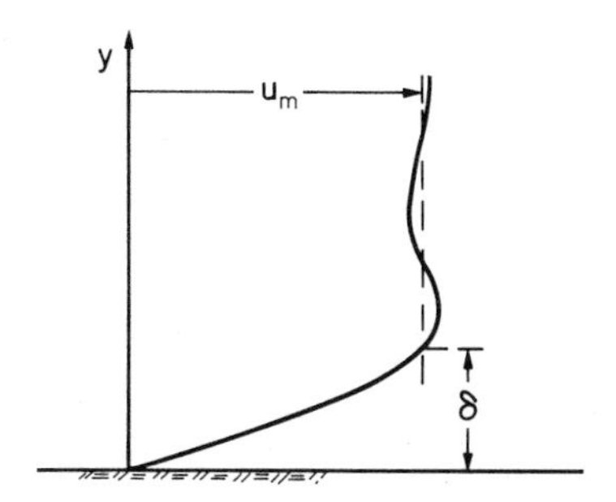

Fig. 11.8. Diagrammatic illustration of velocity distribution at the bed.

The thickness of the laminar boundary layer is defined in terms of the Stokes' length $(2\nu/\omega)^{\frac{1}{2}}$ as

$$\delta = \pi(\nu/2\omega)^{\frac{1}{2}} = (\pi\nu T/4)^{\frac{1}{2}} \quad (11.56)$$

The 1% difference from the mean stream velocity is at about 3δ and Li (1954) expressed the laminar boundary layer thickness as $\delta \cong 6.5(\nu/\omega)^{\frac{1}{2}}$. The boundary layer thickness is also related to the *amplitude* ($A = u_m/\omega$) Reynolds number

$$Re_A = u_m A/\nu \tag{11.57}$$

or boundary layer thickness Reynolds number

$$Re_\delta = u_m \delta/\nu \tag{11.58}$$

The transition from laminar to turbulent boundary layer occurs on a *smooth* bed at about $10^4 \leq Re_A \leq 3 \times 10^5$.

The wave friction factor, f_w, is expressed in terms of the maximum orbital velocity u_m at the bed as

$$\tau_w = \tfrac{1}{2} f_w \rho u_m |u_m| \tag{11.59}$$

and for a laminar boundary layer

$$f_w = 2/\sqrt{Re_A} \tag{11.60}$$

An expression for the turbulent boundary layer thickness on a smooth wall by Jonsson (1967) is

$$\delta/A = 0.0465/Re_A^{0.1} \tag{11.61}$$

and the corresponding

$$f_w = 0.09\ Re_A^{0.2} \tag{11.62}$$

for which Fredsøe (1984) proposed a coefficient 0.065.

The lower limit of turbulence in wave motion over a very rough boundary (Jonsson, 1978) is $Re_A \cong 10^4$ or

$$Re_A = 4130(A/k_s)^{0.45}; \quad Re_A > 2 \times 10^4 \tag{11.63}$$

and on a moderately rough surface

$$Re_A = 223(A/k_s)^{1.17}; \quad Re_A > 2 \times 10^4 \tag{11.64}$$

or the flow is turbulent when $u_m k_s/\nu > 2000$ where k_s is the equivalent roughness height of the surface ($k_s \cong 2.5\ d$ for flat bed).

The thickness of a turbulent boundary layer on a rough wall is given by the empirical relationship by Jonsson and Carlsen (1976) as

$$\delta/A = 0.072(A/k_s)^{-1/4}; \quad 10 < A/k_s < 500 \tag{11.65}$$

and the wave friction factor (Jonsson, 1963) by

$$\frac{1}{4\sqrt{f_w}} + \log \frac{1}{4\sqrt{f_w}} = -0.08 + \log \frac{A}{k_s}; \quad A/k_s > 7 \tag{11.66}$$

which is equivalent to the simpler form by Swart (1974)

$$f_w = \exp[-5.977 + 5.213(A/k_s)^{-0.194}]; \quad 1.57 < A/k_s < 3000 \tag{11.67}$$

where for $A/k_s < 1.57$ the f_w-value is constant, $f_w = 0.30$. Further expressions (Jonsson, 1978) are

$$f_w = \frac{0.0605}{\log^2(22\delta/k_s)} \cong \frac{0.00672}{\log^2(A/12\delta)} \tag{11.68}$$

and Kajiura (1968)

$$f_w = 0.37(A/k_s)^{-2/3}; \quad A/k_s < 50 \tag{11.69}$$

The friction factor definition via shear stress can be linked with boundary layer displacement thickness as

$$\delta = 0.5\ f_w A \tag{11.70}$$

Jonsson used roughness heights obtained from experiment with the aid of logarithmic velocity distribution assuming $k_s \cong 30\ y'$, where y' is the intercept elevation for zero velocity, and the zero datum was set at 0.15 d below the crests. For grain-rough surfaces $k_s \cong 2.5$ d or $k_s \cong 2\ d_{90}$ (Kamphuis, 1975).

For a rough boundary turbulent flow the simple harmonic boundary layer flow is a gross idealisation. With increasing roughness, relative to orbital excursions, the measured variation of $\tau_o = f(t)$ is far from a simple harmonic form. The maximum orbital velocity and shear stress have phase angles to the wave motion which increases with relative roughness, A/k_s, of the bed. Over a flat and almost smooth bed the phase angle is of the order of 10° and about 20° to 25° over very rough bed. The whole problem is further complicated by macro-turbulence, large scale eddies which develop at a rough (rippled) bed.

11.2.7 Waves in shoaling water

In addition to changes in wave height, length and celerity several other processes occur in the gradually shallowing water. Conditions where the depth variation is so graduate that no or only negligible reflection of short periodic waves occurs are referred to as *shoaling water*. Sandy coasts usually satisfy the gentle slope condition. Here the velocities, etc., are calculated as for horizontal bed at any given depth using the appropriate wave theory. The shoaling coefficient in eqn (11.16)

$$K_s = \frac{H}{H_o} = \left[\frac{n_o\ c_o}{n\ c}\right]^{\frac{1}{2}}$$

depends both on h/L_o and wave height. In deep water $K_s = 1$ and in shallow water

$$K_s = \left[\frac{c_o}{\sqrt{gh}}\frac{1}{2}\right]^{\frac{1}{2}} = \left[\frac{1}{4\pi}\frac{L}{h}\right]^{\frac{1}{2}} = \left[\frac{1}{8\pi}\frac{L_o}{h}\right]^{1/4} \quad (11.71)$$

Figure 11.9 from Svendsen and Brink-Kjaer (1972) shows the variation of wave height or K_S and wave length for Airy and cnoidal waves. At the solitary wave limit K_S becomes inversely proportional to water depth as compared to K_S proportional to $h^{-1/4}$ for Airy waves.

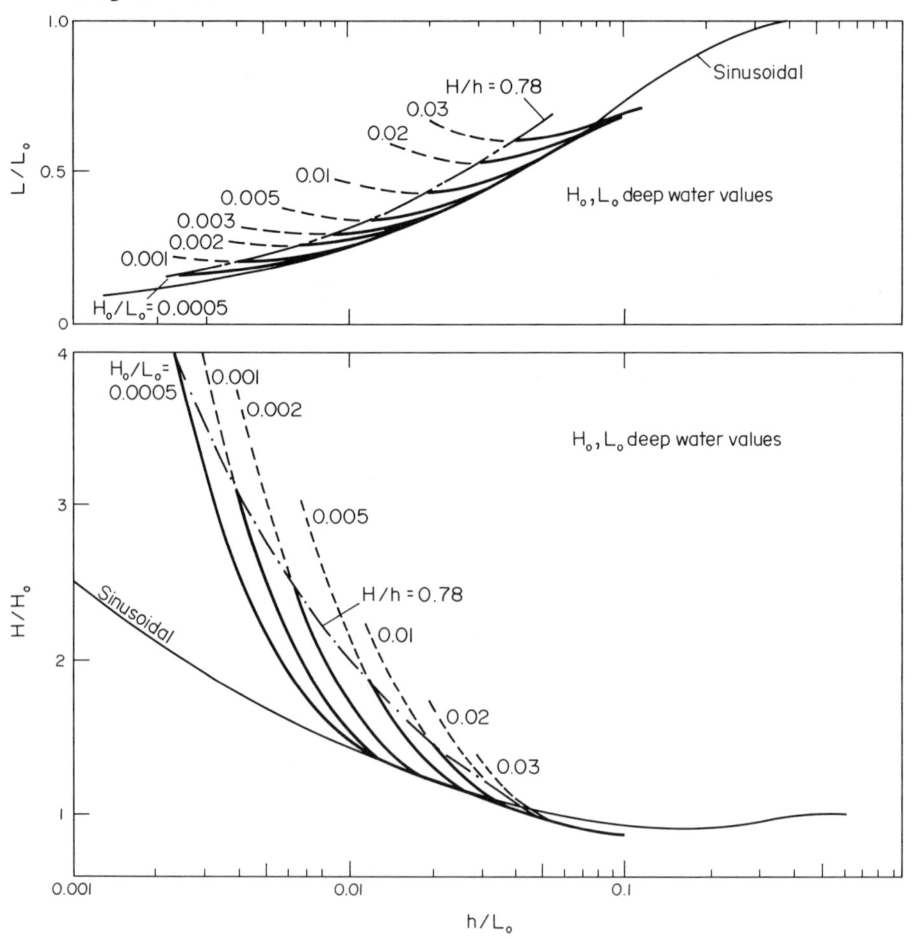

Fig. 11.9. Variation of wave height and length according to linear and cnoidal wave theories in shoaling water, Svendsen and Brink-Kjaer (1972).

Waves running into shallowing water at an angle refract as illustrated in Fig. 11.10. Refraction through redistribution of wave energy and consequential currents plays a very important role in all coastal processes. The energy confined between the wave orthogonals (lines perpendicular to wave crests and tangential to

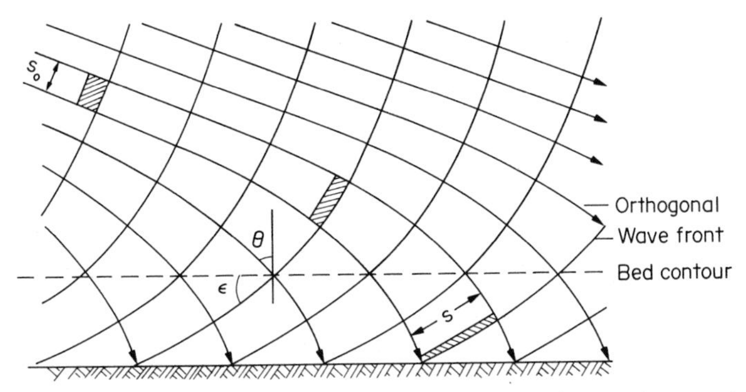

Fig. 11.10. Wave crests and wave orthogonals at the coast. The hatched areas are equal and indicate variation of wave energy per unit length of wave crest.

wave propagation) is constant until dissipated by breakers. Thus, where the orthogonals converge, Fig. 11.11, the wave energy per unit length of wave crest increases, i.e., $H_o s_o = Hs$, and vice versa. Since the temporal mean water level is proportional to H^2 a focussing of wave energy can lead to a longshore current from regions of high waves (higher water level) to regions of low waves.

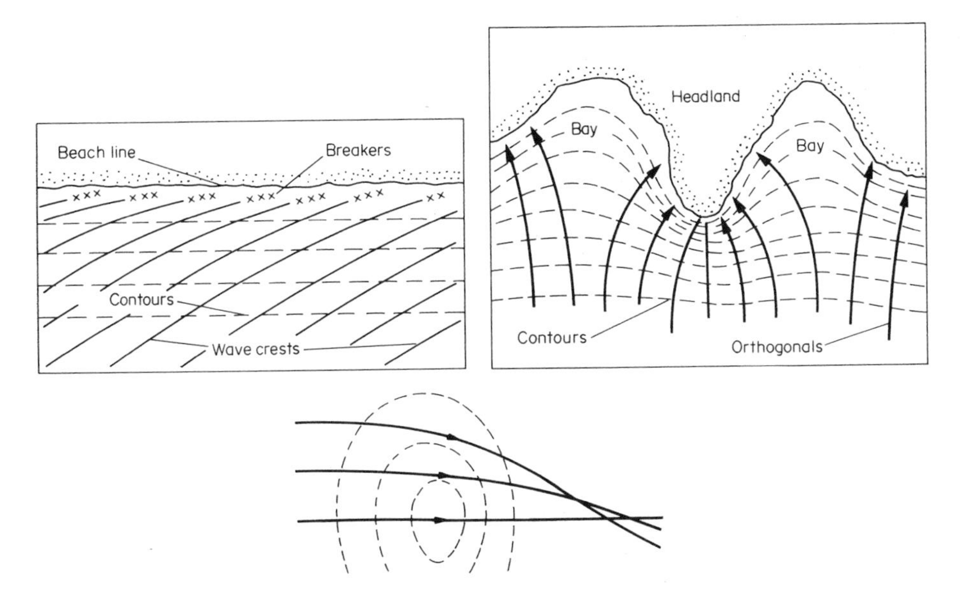

Fig. 11.11. Illustrations of basic refraction patterns.

The methods for solving refraction problems and construction of refraction diagrams are based on Snell's law

$$\frac{\sin\theta_o}{\sin\theta} = \frac{c_o}{c} \tag{11.72}$$

The requirement that $E_o n_o c_o s_o$ = Encs is constant yields

$$\frac{H}{H_o} = \left[\frac{1}{2n}\,\frac{c_o}{c}\,\frac{s_o}{s}\right]^{\frac{1}{2}} = K_s K_r \tag{11.73}$$

where $K_r = (s_o/s)^{\frac{1}{2}}$ is the refraction coefficient. For elementary methods of solution of refraction problems reference is made to the Shore Protection Manual, US Army Corps of Engineers, Coastal Engineering Research Center. For discussion of numerical solutions reference is made to Skovgaard et al. (1975) and references given in it. It should be noted that a solution can lead to crossing of the orthogonals, which implies an infinite wave height. Special techniques have been evolved to cope with these problems. The methods of refraction analysis do not allow for lateral flow of energy from regions of high to low breakers but the effects of currents can be allowed for. In the presence of a current the wave number k is modified. In the direction of propagation $kc = \omega$ becomes $k(c + V \sin\theta) = \omega$ with $k_x = k \cos\theta$ and $k_y = k \sin\theta$, where V is the current velocity in y-directions.

Other processes of importance are *diffraction*, *reflection*, percolation, transmission and trapping of energy.

The flow of energy lateral to the direction of wave propagation is called *diffraction*. It is responsible for the existance of waves in sheltered regions, Fig. 11.12. In most problems diffraction and refraction combine, exceptions being harbour basins of near constant depth. For methods of calculation of wave height distribution due to diffraction reference is again made to Shore Protection Manual.

Any barrier in the path of wave propagation reflects wave energy but the amount depends on its slope. A very flat sand beach reflects only a negligible amount since the wave energy is dissipated in breakers. Vertical walls and cliffs can reflect the total energy and give rise to standing waves of double the incident wave height if the waves approach at right angles. Reflection at an angle leads to three-dimensional surface wave patterns and a high level of turbulence production at the bed where the orbital velocities are likewise angled. Such conditions are particularly conducive to undermining of beach walls by erosion. In principle, reflection problems are solved by superposition of waves. The excess pressure Δp as obtained from second or higher order wave theories has three components. The increase in pressure due to rise of mean water surface Δh (eqn 11.23) and reduction of pressure with depth (eqn 11.13) are counteracting but the third component is independent of depth and does not vanish when $kh \to \infty$. Longuet-Higgins (1950) showed that these pressure fluctuations, produced by reflection from the shore, gave rise to the so-called micro-seisms.

The ratio of reflected to incident wave height is the reflection

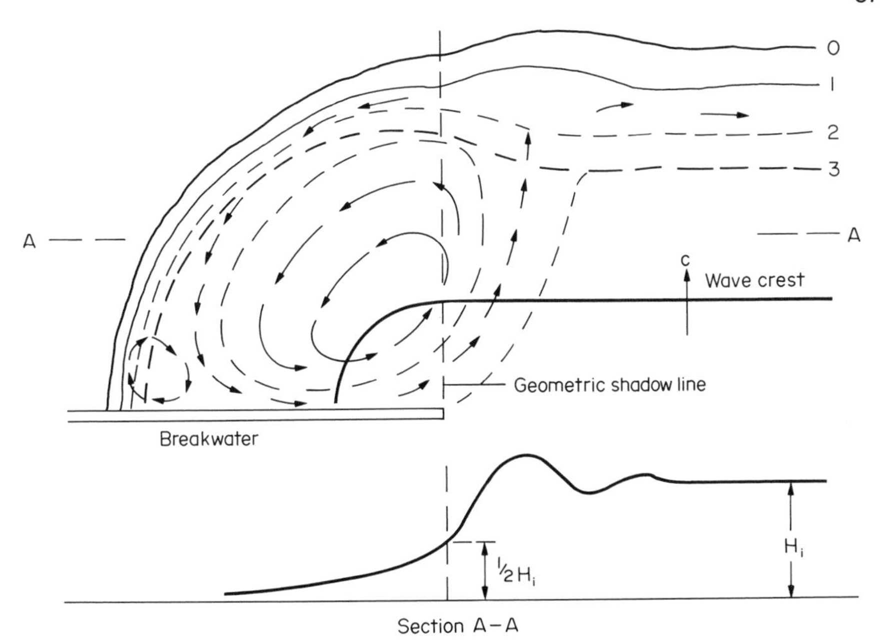

Fig. 11.12. Diagrammatic illustration of diffraction. Notice the induced eddy and the band of high waves (Section A-A). 1 - SW line, 2 - plunge point, 3 - breaking point.

coefficient $r = H_r/H_i$. It is a function of the slope, roughness and permeability of the surface, wave steepness and angle of approach. Breaking of waves on the slope substantially reduces reflection. Miche (1944) derived an expression for the critical slope, separating breaking and non-breaking wave conditions, as

$$(H_o/L_o)_{max} = (2\beta/\pi)^{\frac{1}{2}}(\sin^2\beta)/\pi \tag{11.74}$$

where β is the angle from horizontal (radians). At flatter slopes waves break, at steeper they reflect.

Iribarren and Nogales (1949) proposed a parameter

$$\xi = \frac{\tan\beta}{(H/L_o)^{\frac{1}{2}}} = \frac{1}{\sqrt{2\pi}}\,\frac{\tan\beta}{(H/gT^2)^{\frac{1}{2}}} \tag{11.75}$$

and assigned $\xi_c \cong 2.3$ as the critical value for conditions about halfway between complete breaking and complete reflection. Breaking occurs when $\xi < \xi_c$. Battjes (1974) called it the "surf similarity parameter" and it is also referred to as the Battjes parameter. Figure 11.13 from the SPM shows the reflection coefficient as a function of the similarity parameter. The reflection coefficient is further multiplied by a constant ≤ 1 to allow for the roughness and breaker height at the toe. Reflection from a beach is shown by the graph by Battjes, based on data by Moraes (1970), Fig. 11.14.

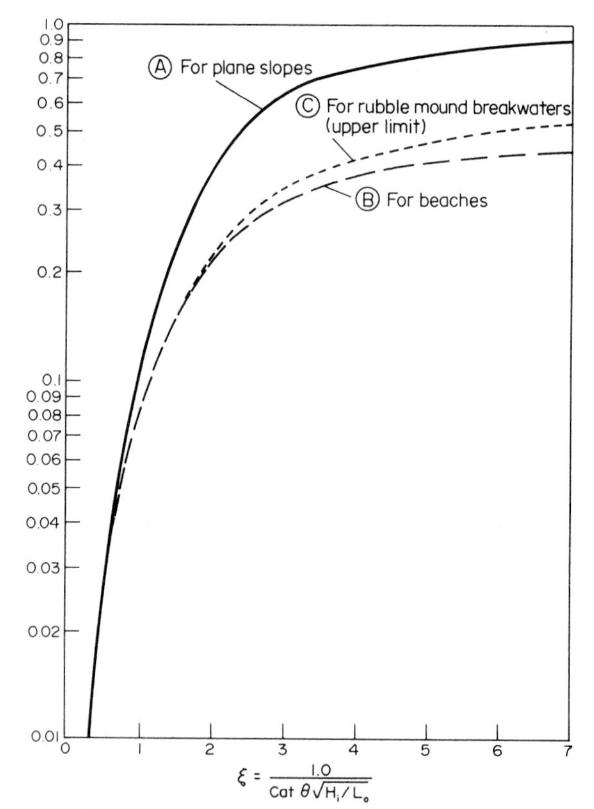

Fig. 11.13. Wave reflection coefficients for slopes, beaches, and rubble-mound breakwaters as a function of the surf similarity parameter ξ.

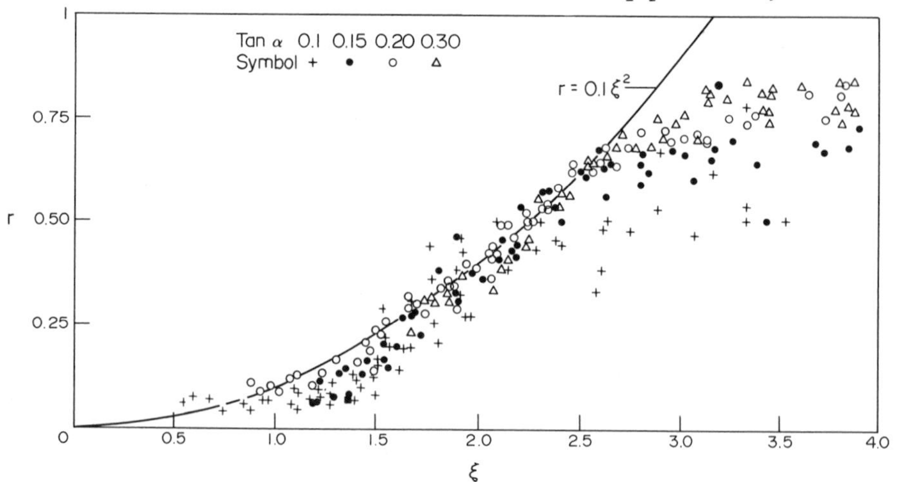

Fig. 11.14. Reflection coefficient from Beaches versus the similarity parameter.

The trapping of reflected energy in the nearshore zone affects several features of the surf zone. In principle, trapping is a phenomenon known as *edge waves*. Ursell (1952) derived the dispersion relationship

$$\omega^2 = g\, k_e \sin(2n + 1)\beta \qquad (11.76)$$

for edge waves, where k_e is the longshore wave number, n is mode and β is beach slope. The trapping condition is $(2n + 1)\beta \leq \pi/2$ which with $\beta = \pi/6$ gives n = 0; the Stokes' edge wave. If $(2n + 1)\beta > \pi/2$ there is no trapping of energy, a condition known as leaky modes. The regions on k_e-ω plane between the modes (resonance conditions) represent weaker forced wave motion. Superposition of edge waves can produce standing edge waves which have been used to explain the generation of beach cusps and cresentic bars of half the wave length of edge waves. Trapping by refraction can be demonstrated with the aid of refraction diagrams. An interesting particular case arising from a cliff or structure along a section of coast, was discussed by Camfield (1982), Fig. 11.15, which can lead to increased wave heights along the coast. He derived for

$$X = \frac{h_s}{S \sin^2\alpha} - \frac{h_s}{S} \text{ or } \frac{XS}{h_s} = \frac{1}{\sin^2\alpha} - 1 \qquad (11.77)$$

and

$$Y = \frac{4\, h_s}{S \sin^2\alpha}\left[\frac{\pi}{4} - \frac{\alpha}{2} + \frac{\sin 2\alpha}{4}\right] \text{ or } \frac{YS}{h_s} = \frac{1}{\sin^2\alpha}(\pi - 2\alpha + \sin 2\alpha) \qquad (11.78)$$

where α is the reflected wave angle in radians, S is the uniform bottom slope and X and Y are defined in Fig. 11.15.

11.2.8 Breaking waves

The limit of wave stability is reached when the maximum orbital velocity at wave crest, u, equals the wave celerity. Michell (1893) showed that in deep water this limit was given by the steepness H/L = 0.142 and Miche (1944) derived for progressive waves in any depth

$$H_b/L_b = 0.142 \tanh(2\pi h_b/L_b) \qquad (11.79)$$

A solitary wave will become unstable when

$$H_b/h_b = 0.78 \qquad (11.80)$$

In the ocean the breaking occurs mainly in the higher frequency part of the wave spectrum, i.e., the shorter waves spill or have white caps. The long waves of the spectrum seldom reach the "fully arisen sea state". A 10 second wave in ocean has to reach a height of about 22 m before the necessary steepness is reached.

The limiting steepness of waves running into shoaling water decreases and becomes a function of both h/L and the beach slope m

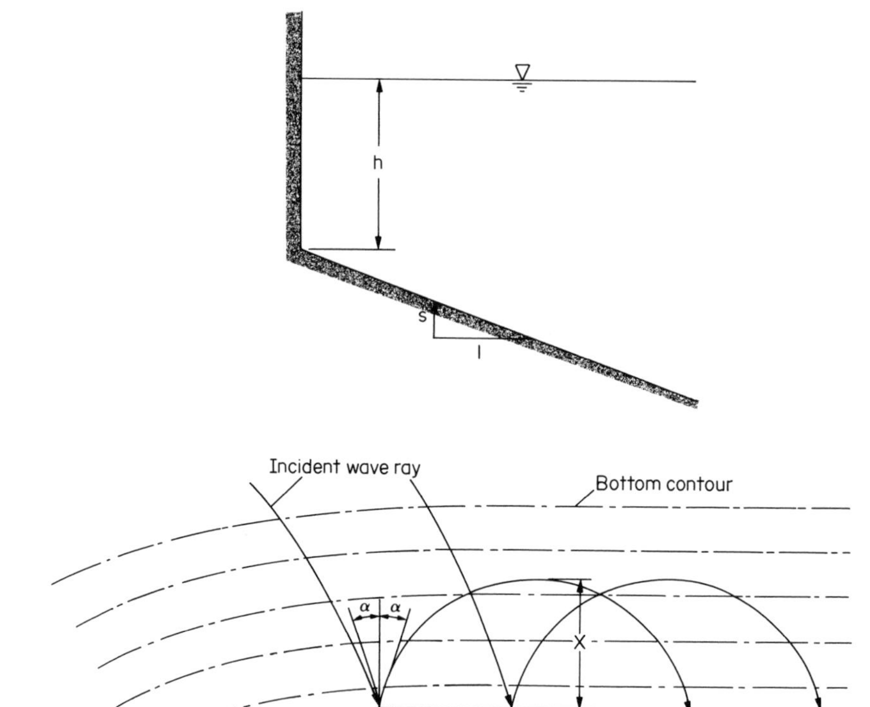

Fig. 11.15. Definition sketch for trapped wave rays, Camfield (1982).

(conventionally slope is expressed as 1:m). Munk (1949) derived the relationship

$$H_b/H_o' = 1/[3.3(H_o'/L_o)^{1/3}] \tag{11.81}$$

for breaker height where H_O' is the deep water wave height not affected by refraction. The SPM gives

$$h_b/H_b = 1/[b - (aH_b/gT^2)] \tag{11.82}$$

$$a = 43.75(1 - e^{-19m})$$

$$b = 1.56/(1 + e^{-19.5m})$$

A descriptive summary of the breaker height versus steepness was presented by Weggel (1972), Fig. 11.16. The wave height in the surf zone is limited by local depth, a saturation condition which Thornton and Guza (1982) described by

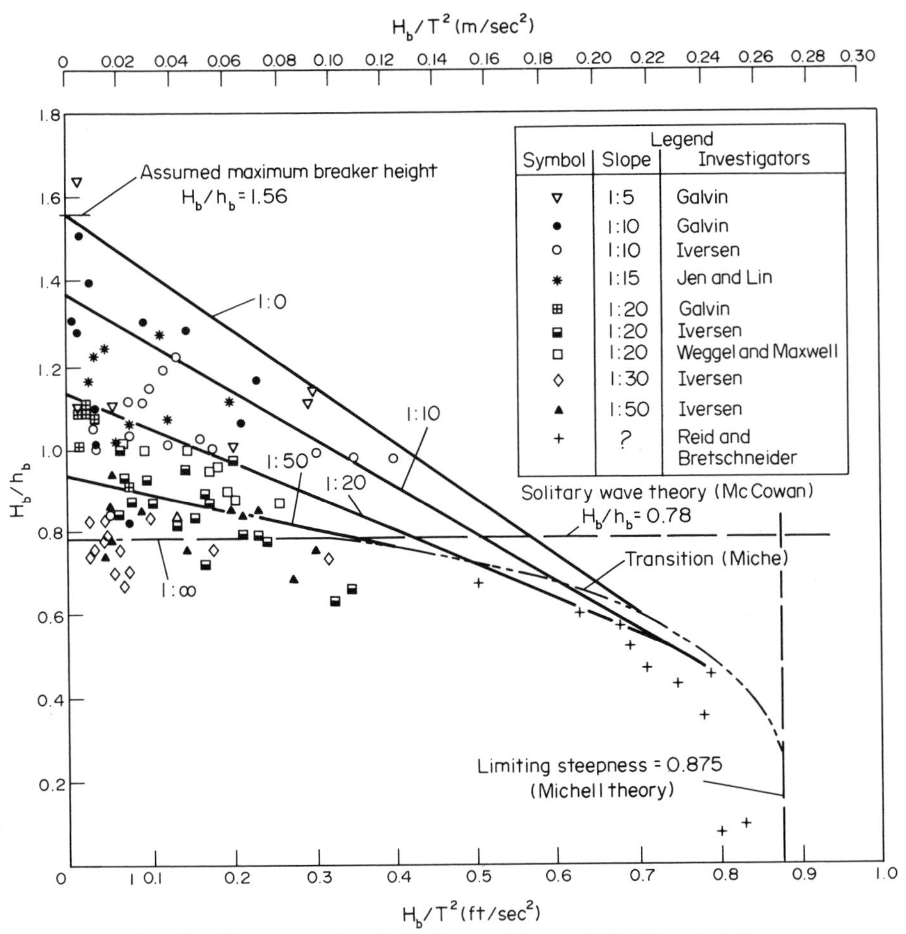

Fig. 11.16. Experimental observations of h_b/H_b versus breaker steepness, H_b/T^2, Weggel (1972).

$$H_{rms} = 0.42\ h \tag{11.83}$$

However, Horikawa and Kuo (1966) stated that after breaking the local wave height is close to half the local water depth. The transition from breaking to broken state follows the relationship

$$H/h = 0.5 + 0.3\ \exp[-\ 0.11\ \Delta x/h_b] \tag{11.84}$$

where Δx is distance from breaker line.

Breakers vary extensively in shape. The basic types are *spilling*, *plunging* and *surging* breakers.

Spilling breakers originate from relatively steep waves running onto very flat beaches. The foamy spill volume in front of the crest grows as the wave runs in and the wave height decreases fairly uniformly over about half the shallow water wave length. Very little of its momentum is reflected back to the sea.

Plunging breakers form when very flat waves run onto fairly steep beaches. The wave crest sharpens and then curls over and plunges in front of the wave. The plunging process produces large amounts of turbulence and entrains air. The reduction in breaker height is rapid and substantial. The reformed waves are usually less than a third of the original breaker height. The reformed wave usually breaks close inshore. Little of the wave momentum is reflected back to sea. The travel distance of the plunging breaker, i.e., the distance between breaker and plunge lines is according to Galvin (1969)

$$x_p = (4.0 - 9.25\ m)H_b \qquad (11.85)$$

Surging breakers occur at steep beaches. The surf zone is usually very narrow and about half or more of the wave momentum is reflected back to sea. The waves peak as if to plunge but then the toe of the waves runs up the beach and the wave collapses.

Frequently the waves between spilling and plunging conditions are referred to as *transitional breakers*. Weggel classified the breakers as shown in Fig. 11.17 and a similar graph by Goda in terms of H_o'/gT^2 is shown in SPM. Battjes (1974) used the surf similarity parameter ξ (eqn 11.75) to differentiate between the breaker types:

surging	$\xi_b > 2.0$	$\xi_o > 3.3$
plunging	$0.4 < \xi_b < 2.0$	$0.5 < \xi_o < 3.3$
spilling	$\xi_b < 0.4$	$\xi_o < 0.5$

where ξ_b and ξ_o refer to breaker height H_b or deep water wave height H_o in eqn (11.75). Best correlation is obtained with the slope just seaward of the breaker line. The breaker index

$$\gamma = H_b/h_b \qquad (11.86)$$

was found to be about 0.8 ± 0.1 for $\xi_o < 0.2$ and increased from $\xi_o = 0.2$ to 2 as $\gamma = 1.06\ \xi_o^{0.176}$ but the scatter in data points is substantial ($\sim \pm 0.15$). The delineation in terms of ξ is also rather tentative. Kajima et al. (1982) concluded from large wave flume tests that the critical value between spilling and plunging breakers was $\xi_o = 0.15$. Kana (1979) defined plunging breakers by $\gamma > 1.1$, transitional breakers $0.93 < \gamma < 1.1$ and spilling breakers $\gamma < 0.93$ where the trough water depth in front of the wave was used as h_b (solitary wave definition) rather than mean water depth. Goda (1975) related the breaker index to bed slope and deep water wave length as

$$(H/h)_b = A\{1 - \exp[-1.5(\pi h_b/L_o)(1+15\ S^{4/3})]\}/(h_b/L_o) \qquad (11.86a)$$

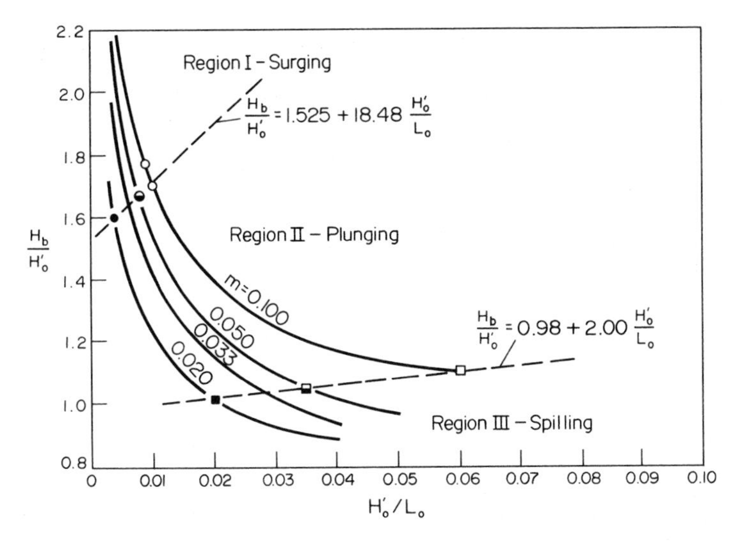

Fig. 11.17. Breaker height index versus deepwater steepness and regions of surging, plunging and spilling breakers.

where the slope of a nonuniformly sloping bed can be expressed by the average slope between h_b and $5h_b$ locations (Izumia and Isobe, 1986).

11.2.9 Radiation stress, wave set-up

Waves running ashore over a gently sloping bed induce a slight drop in mean water level outside the breaker line, the *wave set-down*, and a much larger rise in the surf zone, the *wave set-up*. These are related to the *radiation stress*, a term introduced by Longuet-Higgins and Stewart (1964). The radiation stress may be defined as the excess flow of momentum due to the presence of waves. It is expressed as a momentum flux or stress tensor S with components S_{xx}, S_{yy} and S_{xy}, S_{yx}, where S_{xx} is momentum in the x-direction normal to a yz-plane, etc., i.e. S_{xx} and S_{yy} are normal stresses or pressures and S_{xy} and S_{yx} are shear stresses. The radiation stress is calculated by integrating vertically through the water column the momentum flux and excess pressure force which leads to a force per unit length but the properties of stress are maintained. Distinction has to be made between mean water level, MWL, or mean energy level, MEL, and still water level SWL. The set-up or set down is the difference, h', between MWL and SWL, Fig. 11.18.

For the simplest case of *normal incident* periodic waves and *no net current*, the integration at constant water depth yields the principal components of stress S_{xx} and S_{yy} as

$$S_{xx} = \frac{1}{8} \rho g H^2 \left[\frac{2kh}{\sinh 2kh} + \frac{1}{2} \right] = E \left[2n - \frac{1}{2} \right] \tag{11.87}$$

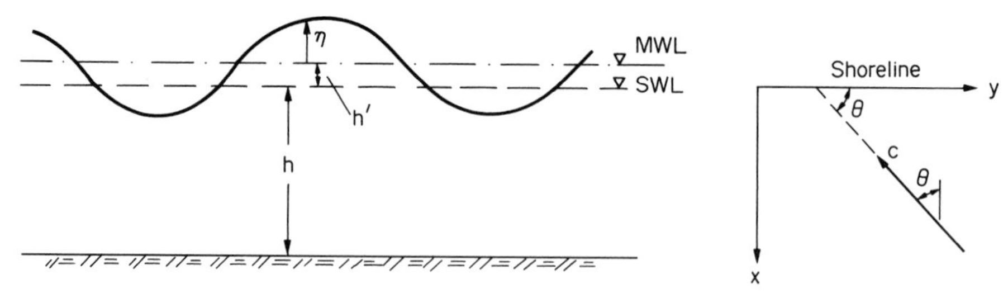

Fig. 11.18. Definition sketch of water levels and directions.

$$S_{yy} = \frac{1}{8}\rho g H^2 \frac{kh}{\sinh 2kh} = E\left[n - \frac{1}{2}\right] \tag{11.88}$$

where n is introduced from eqn (11.14). In deep water $S_{xx} \to (1/2)E$ and $S_{yy} \to 0$ whereas in shallow water $S_{xx} \to (3/2)E$ and $S_{yy} \to (1/2)E$.

The forces due to principal stresses on a control volume $\Delta x \Delta y \Delta z$ in water of constant depth balance, i.e. the radiation stresses have an influence only when depth and wave conditions change. The radiation stress components acting at a given location on planes other than the principal planes can be evaluated with the aid of Mohr's circle, Fig. 11.19. The stresses on a plane at angle θ are

$$S_{xx} = \frac{S_{xx} + S_{yy}}{2} + \frac{S_{xx} - S_{yy}}{2}\cos\theta \tag{11.89}$$

$$S_{yy} = \frac{S_{xx} + S_{yy}}{2} - \frac{S_{xx} - S_{yy}}{2}\cos 2\theta \tag{11.90}$$

$$S_{xy} = \frac{S_{xx} - S_{yy}}{2}\sin 2\theta \tag{11.91}$$

For more detailed and computer based calculation of the radiation stress reference is made to Copeland (1985).

When waves approach the coast, wave parameters change due to shoaling, refraction, diffraction and breaking and consequently radiation stress changes. For the simplest case of waves approaching normal to the coast the momentum equation for the no current case could be written as

$$\frac{d}{dx}(M + F_p) + \rho g(h + h')\tan\alpha = 0 \tag{11.92}$$

where the momentum and excess pressure force are integrated through the depth and the second term arises from projection of the pressure force on the bed boundary of the control volume. This leads to a relationship between the radiation stress and water level

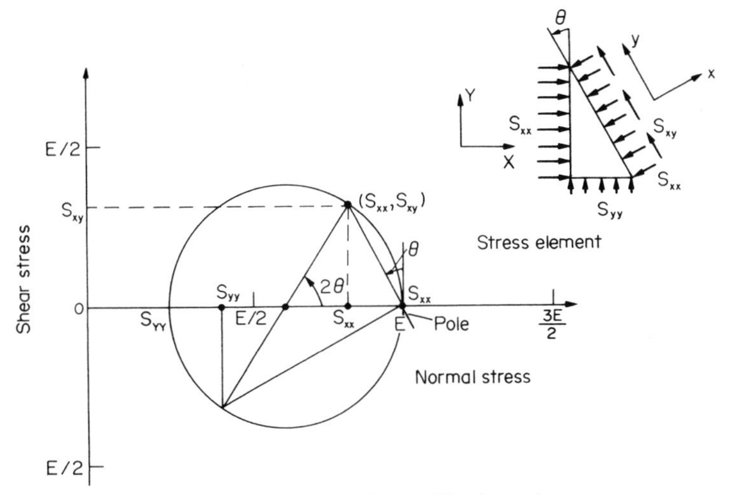

Fig. 11.19. Mohr's circle for radiation stresses.

$$\frac{dS_{xx}}{dx} = -\ \rho g(h + h')\ \frac{dh'}{dx} = 0 \tag{11.93}$$

from which for known S_{xx} the MWL, i.e. h', could be determined. The slope of the mean water level is seen to be negative (downwards in shore direction) when the radiation stress increases, i.e. when the wave height increases in approaching the shore. Inside the breakers, in the surf zone, the wave height decreases rapidly and the reduction in radiation stress must be compensated by an increase in hydrostatic force, i.e. the water surface slopes upwards. Eqn (11.93) can be solved, using the conservation of energy equation, yielding

$$h' = -\ \frac{H^2}{8h}\ \frac{kh}{\sinh 2kh} \tag{11.94}$$

Observing that the energy flux is proportional to the spacing of wave orthogonals, $s_1/\cos\varepsilon = s_2/\cos\varepsilon_2$, eqn (11.16) or (11.73) can be written as

$$H_2 = H_1\left[\frac{\cos\varepsilon_1}{\cos\varepsilon_2}\ \frac{(1 + 2k_1h_1/\sinh 2k_1h_1)\tanh k_1h_1}{(1 + 2k_2h_2/\sinh 2k_2h_2)\tanh k_2h_2}\right] \tag{11.95}$$

and

$$h' = -\ \frac{H_o^2}{16h}\ \frac{2kh/\sinh 2kh}{(1 + 2kh/\sinh 2kh)}\ \frac{\cos\varepsilon_o}{\cos\varepsilon} \tag{11.96}$$

The set-down is generally small but the set-up increases the MWL at shore about 1/3 to 1/2 of the breaker height. The set-up is given by

$$h'_{max} = \frac{3\gamma^2}{8}\ h_b = \frac{3\gamma}{8}\ H_b \tag{11.97}$$

or

$$\frac{h'_{max}}{H_o} = 0.129\gamma^2 (H_o/L_o)^{-1/3} \tag{11.98}$$

where h'_{max} is the elevation at water's edge and $H_b = \gamma h_b$. Eqn (11.98) was derived by Svendsen (1974) for a solitary wave.

Since wave set-up depends on the breaker height, which varies as the wave groups arrive, the mean water level at the shoreline fluctuates. These periodic movements were given the name of *surf beat*, which has now become a name for all movements in surf zone with periods appreciably longer than that of the incident wave, particularly due to edge waves. The velocities in the surf zone, at frequencies of the surf-beat, are dominant for shoreline processes. The amplitude of surf-beat increases approximately linearly with incident wave height in deep water and surf-beat velocities are of the order of 1 m/s.

The effects of currents on the wave set-up are discussed by Jonsson (1978) and Jonsson et al. (1978).

Additional to the wave set-up is the *wave runup*, R_u. It is the vertical distance above mean water level to which water runs on the beach (or slope of a structure) due to wave action. The runup depends on wave steepness, beach slope and roughness of the surface. On flat slopes, typical for beaches, the empirical formula by Hunt (1959) has been found to give satisfactory predictions for short wave conditions.

$$\frac{R_u}{H} = \frac{0.405 \tan\beta}{(H/gT^2)^{\frac{1}{2}}} \quad ; \quad \frac{H}{gT^2} \geq \frac{0.031}{\tan^2\beta} \tag{11.99}$$

which is seen to be equivalent to

$$\frac{R_u}{H} = \xi \text{ for } 0.1 < \xi < 2.3 \tag{11.99a}$$

or

$$R_u = 0.4T(gH)^{\frac{1}{2}}\tan\beta = (HL_o)^2\tan\beta \tag{11.99b}$$

The runup of waves angled to the beach is usually assumed to be proportional to the cosine of the angle of incidence. When used with the significant wave height then the R_u, which is exceeded 2% of events, is the calculated value multiplied by 1.5 to 2. The Delft formula

$$R_u = 8H_s\tan\beta \tag{11.100}$$

for the 2% estimate on slopes less than 16° has been reported to be in good agreement with field observations.

For runup of long waves Togashi and Nakamura (1977) proposed the empirical relationship

$$\log R_u/H_o' = 0.326 + 0.348 \log X - 0.286(\log X)^2 \quad (11.101)$$

for $0.314 \leq X \leq 8.17$, where $X = 2\pi m/(gT^2/h)^{\frac{1}{2}}$ and $1{:}m = \tan\beta$.

11.3 Currents in Coastal Waters

Currents could be subdivided into wave-induced currents, tidal currents, ocean currents, density currents, and fronts.

The wave-induced currents could be further sub-divided into longshore (littoral) currents, refraction currents from regions of high breaker to low breakers, diffraction eddies (Fig. 11.12), rip currents and return current or undertow. Tidal and ocean currents in most regions are well documented and are not discussed here. The density currents arise from differences in the salinity or temperature. Also flows of heavily sediment laden water can lead to density currents, e.g. mud slides on the continental shelf. Density currents usually have a fairly distinct interface to the rest of the water mass and exhibit the features of open channel flow, such as surface waves, hydraulic jump etc. Density interfaces (salt water wedge) are common in river estuaries and affect sedimentation. In regions where the continental shelf is shallow and narrow the colder deep sea water flows landward due to the density difference. The flow is compensated by an outward flow of warmer water. Both currents are deflected by the Coriolis acceleration, giving rise to a three-dimensional movement of water masses on the shelf.

Fronts are prominent in estuary regions of large rivers. The associated up and down currents affect the sediment movement and deposition. Fronts also occur offshore due to movements of water masses in the ocean.

The most important currents in the inshore region are the wave-induced currents but a simple subdivision of these is difficult because of the effects due to the coastline geometry. Along a more or less straight ocean beach the *longshore* current dominates whenever waves approach the shore at an angle. The longshore current is driven by the coastwise component of wave momentum, the S_{xy}-component of radiation stress, and is essentially confined to the surf zone. However, surface wind drag can be an important generating force and can induce additional drift both inside and outside the breaker line. Usually the longshore current velocities are less than 1 m/s but 3.5 m/s has been reported. This longshore current, or most of it, may turn and flow seaward as a *rip current*. The rip currents usually occur at somewhat random spacing along the coast. The currents along coastlines with prominent headlands and bays tend to have well-defined currents from regions of high breakers to regions of low breakers which may be accompanied by diffraction eddies.

The detailed current picture in the surf zone is complex. It is influenced by the extreme variability of wave patterns, the beach slope and coastline topography. Figure 11.20 shows a diagrammatic illustration of the current pattern in the inshore region. In the presence of plunging breakers on a flat beach the longshore

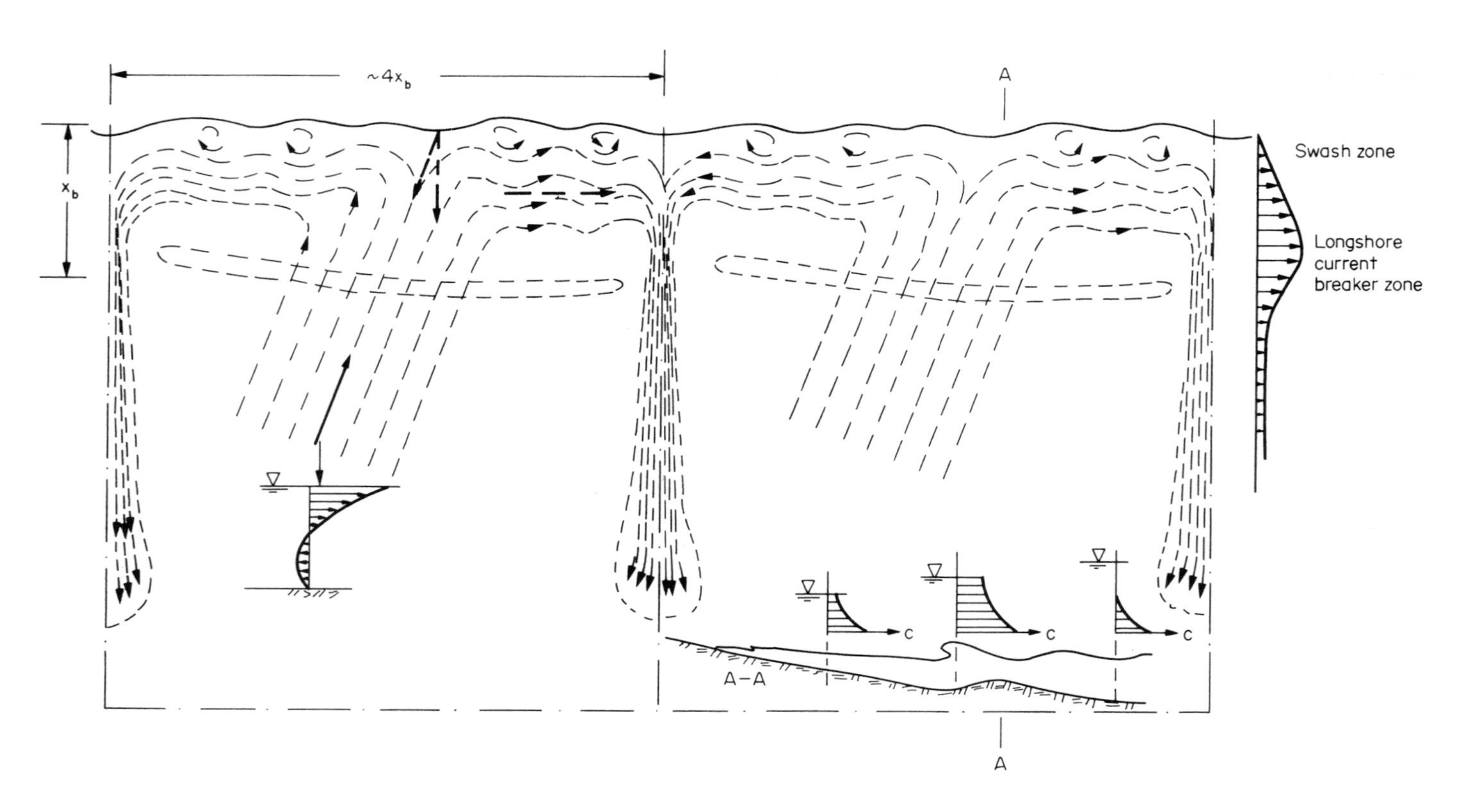

Fig. 11.20. Schematic illustration of the longshore and rip currents.

current tends to display a strongly meandering pattern with cellular eddies between the shoreline and the to shore concave parts of the meanders. On steep beaches and plunging breakers the longshore current tends to be without significant cellular movements.

The least well understood are the *rip currents*. These are well known to lifeguards and people familiar with ocean beaches but no theory exists to predict their location, spacing or current speed. They have been related to edge waves, intersecting wave trains, wave-current interaction and coastal topography (Dalrymple, 1978). Only a few field measurements of rip currents have been reported. The spacing of rip currents clearly depends on the incident wave period and height as well as bed topography. They occur mainly when the incident waves run in almost perpendicular to the beach. At large angles of incidence the longshore current is prominent, with few or no rip currents. The spacing is to some extent random, probably due to the very variable incident wave pattern. Rip current channels left after the storm has abated, and the surf zone circulation appear to affect the rip currents for time afterwards. As a rule of thumb the spacing, with near perpendicular waves, could be taken as four times the width of the surf zone. The variation in spacing is greater with spilling than plunging breakers. The rip current velocity varies extensively. The current may cease altogether when the high waves of a wave group arrive and usually reaches peak values during the arrival of low waves of the wave group. The velocity may reach several metres per second and the stream may extend over a km beyond the breakers. The locations of rip currents are usually marked by breaks in the breaker line, where the opposing current makes the waves steeper, and by a fan of foam on the surface outside the breakers.

The interception of the mass transport by the shoreline leads to a banking up of the mean water surface. This slope drives the *return current* or *undertow* which balances the net mass transport by waves. The undertow is clearly observed in laboratory flumes. Kajima et al. (1982) observed in a large wave flume strong seaward currents throughout the surf zone and beyond the breaker line with maximum velocities of the order of ½ m/s (wave periods 3.1 to 12 s and wave heights 0.46 to 1.76 m).

11.3.1 The longshore current

The longshore current and the associated current patterns are the subject of numerous discussions and classifications. Horikawa (1978), for example, uses the three type classification by Harris, Fig. 11.21. Wright et al. (1984) extended it to six types. At times the actual pattern is very complex, more a random collection of eddies than a steady state system. The pattern is also sensitive to incident wave angle and no criteria exists for linking it to the current patterns. The many and complex interactions of the parameters have to be reduced to idealised models for analytical treatment. Thus, the waves are assumed to be monochromatic, steady two-dimensional gravity waves of constant angle of incidence and usually spilling breakers with constant breaker index. The beach is a gently sloping plane with parallel depth contours, infinitely long, impervious, and reflection of wave

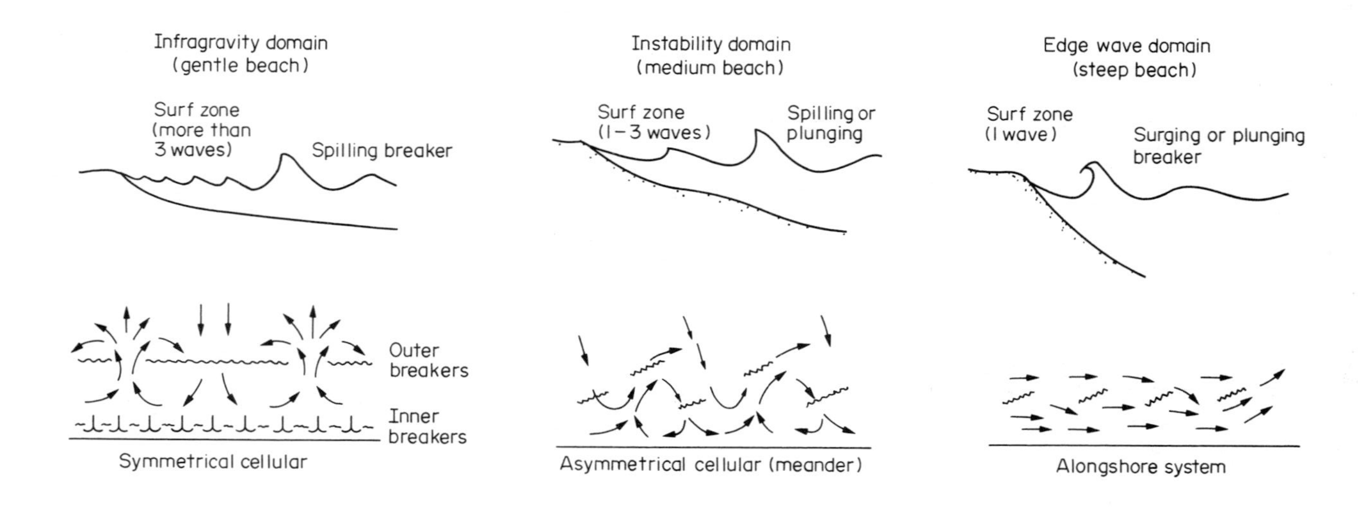

11.21. Beach and inshore current patterns after Sasaki and Horikawa (1978).

energy is negligible. The fluid is incompressible, homogeneous (no air entrainment by breakers) and non-viscous, except in the bottom boundary layer. The longshore current is parallel to the coast and its velocity is averaged over depth and wave period. The models omit wind shear stress, wave-current interaction, rip currents, Coriolis acceleration, etc., and the equations are linearised and approximated.

The analytical treatments of the longshore current date from the introduction of the radiation stress concept by Longuet-Higgins and Stewart (1964). Among the contributions the studies by Bowen (1969), Thornton (1969), Longuet-Higgins (1970), Liu and Dalrymple (1978), Kraus and Sasaki (1979) and Visser (1984) help to mark out the path of development over two decades.

The starting point is the equations of motion which for the x and y directions can be written as

$$\rho h\left(\frac{\partial U}{\partial t} + U\frac{\partial U}{\partial x} + V\frac{\partial U}{\partial y} + g\frac{\partial \bar{\eta}}{\partial x}\right) + \frac{\partial S_{xx}}{\partial x} + \frac{\partial S_{xy}}{\partial y} - \frac{\partial T_{xx}}{\partial x} - \frac{\partial T_{xy}}{\partial y} - P\frac{\partial d}{\partial x} + \bar{\tau}_{ox} = 0 \tag{11.102}$$

$$\rho h\left(\frac{\partial V}{\partial t} + U\frac{\partial V}{\partial x} + V\frac{\partial V}{\partial y} + g\frac{\partial \bar{\eta}}{\partial y}\right) + \frac{\partial S_{xy}}{\partial x} + \frac{\partial S_{yy}}{\partial y} - \frac{\partial T_{xy}}{\partial x} - \frac{\partial T_{yy}}{\partial y} - P\frac{\partial d}{\partial y} + \bar{\tau}_{oy} = 0 \tag{11.103}$$

where U,V are components of mean current, d is still water depth, $h = d + \bar{\eta}$ is the mean water surface above still water level; S is the radiation stress tensor; T is the Reynolds stress tensor; P is the mean dynamic pressure on the bed, and $\bar{\tau}_o$ is the mean fluid shear stress acting on the bed. These equations, with the aid of the various assumptions mentioned above, can be approximated to

$$\rho g h \frac{\partial \bar{\eta}}{dx} + \frac{dS_{xy}}{dx} - \frac{dT_{xy}}{dx} = 0 \tag{11.104}$$

and

$$\frac{dS_{xy}}{dx} - \frac{dT_{xy}}{dx} + \bar{\tau}_{oy} = 0 \tag{11.105}$$

where the last expresses the requirement that the sum of the driving, the lateral friction and bottom friction forces on a control volume must balance.

In the field the bed contours are seldom parallel and refraction will cause wave height as well as its direction of propagation to vary, i.e. $\partial H/\partial y$ and $\partial \theta/\partial y$ terms need to be accounted for. The wave set-up is a function of wave height and angle of incidence which means that there will be gradients of $\partial h/\partial y$ generating coastwise currents from regions of high to low breakers. In applications, the mean set-up is usually added to beach slope. The radiation stress S_{yy} also has a gradient $\partial S_{yy}/\partial y$ which provides an additional driving force.

The energy balance approximates to (Visser, 1984)

$$\frac{dF_x}{dx} + \frac{dF_{tx}}{dx} + (S_{xy} - T_{xy})\frac{dV}{dx} + \overline{v_b \tau_{oy}} = - D_t \tag{11.106}$$

where F_x = Enc cosθ is the wave energy flux towards the shore; F_{tx} is the transport of turbulence energy; v_b is the component of orbital velocity at the bed, and D_t is dissipation of turbulence. The assumption that vertically and time-averaged rates of wave energy dissipation in breakers and turbulence production are equal enables a decomposition of the equation to

$$\frac{dF_x}{dx} + S_{xy}\frac{dV}{dx} + \overline{v_b \tau_{oy}} = - D \tag{11.107}$$

and

$$\frac{dF_{tx}}{dx} - T_{xy}\frac{dV}{dx} = D - D_t \tag{11.108}$$

The terms dF_x/dx and dF_{tx}/dx represent the rates of change of wave power (energy flux) and turbulence power in x-direction; $S_{xy}dV/dx$ and $T_{xy}dV/dx$ the rates of work done by the radiation stress and Reynolds stress against the mean shear; $\overline{v_b \tau_{oy}}$ is the mean rate of work done by bottom friction and D is dissipation of wave energy.

11.3.1.1 The driving force. The driving force for the longshore current, when wind drag is neglected, arises from the gradient of the radiation stress $d/dx(S_{xy})$, where $S = f(x, H_o, T, \theta_o)$. From eqn (11.91)

$$S_{xy} = En \sin\theta \cos\theta \tag{11.109}$$

Between the wave orthogonals outside the breakers

$$Encs = E_o n_o c_o s_o = \text{const} \tag{11.110}$$

The orthogonals of waves, which approach a gently sloping plane beach at an angle, curve into the beach as described by the Snell's law, eqn (11.72). The distance between the orthogonals increases but the distance measured parallel to the depth contours remains constant, i.e.

$$\frac{s_o}{s} = \frac{\cos\theta_o}{\cos\theta} \tag{11.111}$$

Thus, a substitution for sinθ from Snell's law and cosθ from eqn (11.111) in eqn (11.109) and observing eqn (11.110) yields

$$S_{xy} = E_o n_o \sin\theta_o \cos\theta_o = \text{constant} \tag{11.112}$$

This shows that there is no driving force and no longshore current outside the breakers, a result due to Bowen (1969). However, eqn (11.107) which relates the driving force to dissipation of wave energy,

$$\frac{dS_{xy}}{dx} = -\frac{\sin\theta}{c}\left[D + \overline{v_b\tau_{oy}} + S_{xy}\frac{dV}{dx}\right] \qquad (11.113)$$

where $F_x = S_{xy}c/\sin\theta$, indicates that the driving force persists outside the breakers where D = 0, although it is very small in magnitude.

Substitution for $\sin\theta$ from Snell's law, $E = (1/8)\rho g H^2$ from eqn (11.15) and $\gamma = H_b/h_b$ in eqn (11.109) yields

$$S_{xy} = (1/8)\rho g\gamma^2 \frac{\sin\theta_o}{c_o}(h_b{}^2\ nc\cos\theta) \qquad (11.114)$$

and

$$\frac{\partial S_{xy}}{\partial x} = (1/8)\rho g\gamma^2\frac{\sin\theta_o}{c_o}\left[2hnc\cos\theta\frac{dh}{dx} + h^2c\cos\theta\frac{dn}{dx} + h^2n\cos\theta\frac{dc}{dx} - h^2nc\sin\theta\frac{d\theta}{dx}\right] \qquad (11.115)$$

where within the breaker zone n = 1 and dn/dx = 0, $\cos\theta = 1$ and $d\theta/dx = 0$. Introducing from $c = (gh)^{\frac{1}{2}}$, $dc/dx = \frac{1}{2}\sqrt{g}\ h^{-\frac{1}{2}}dh/dx$ yields

$$\frac{\partial S_{xy}}{\partial x} = (1/8)\rho g\gamma^2\frac{\sin\theta_o}{c_o}\left[2h\sqrt{gh}\frac{dh}{dx} + \frac{1}{2}h^2\frac{g}{\sqrt{gh}}\frac{dh}{dx}\right]$$

$$= \frac{5}{16}\rho\gamma^2(gh)^{3/2}\frac{\sin\theta_o}{c_o}m \qquad (11.116)$$

where m = dh/dx is the beach slope. Equation (11.116) evaluated across the surf zone, x = 0 to $x = x_b$, yields the driving force for the longshore current. Note that here the x-axis is perpendicular to shore and positive seawards, y-axis is coastwise because frequently x is used to indicate the longshore direction.

11.3.1.2 Bed friction force. The fluid shear stress on the bed and the current velocity are related by

$$\tau_c = \tfrac{1}{2}f\rho V^2 = \frac{g}{C^2}\rho V^2 = \rho l^2(dV/dz)^2 \qquad (11.117)$$

where the friction factor f is a function of Reynolds number and the relative roughness height; the Chézy coefficient C = 18 log $12z_o/k_s = (g/\tfrac{1}{2}f)^{\frac{1}{2}}$; the mixing length $l = \kappa z$; κ is the Karman constant and z_o is depth of flow or boundary layer thickness. Under wave motion the orbital velocity at the bed varies with time and changes direction periodically. Analogous to eqn (11.117) the bed shear stress could be defined as

$$\tau_w = \tfrac{1}{2} f_e \rho u_b^2 \qquad (11.118)$$

where u_b is the orbital velocity $\underline{u_b} = \underline{u_m} \cos\omega t$ at the bed and f_e is the energy dissipation factor $\overline{E_D} = \overline{\tau_o u_b} = (2/3\pi) f_e \rho u_m^3$ if the phase shift between velocity and bed shear is ignored. Conventionally

$$\tau_w = \tfrac{1}{2} f_w \rho u_m^2 \qquad (11.118a)$$

where the energy dissipation factor and wave friction factor are related as $f_w = f_e/\cos\phi$ if f_e is constant. For rough turbulent boundary layer flow $f_w \cong f_e$ for practical purposes.

When waves and currents are superimposed the interaction produces extra mixing near the bed and the (logarithmic) current profile is shifted by Δu, corresponding to the increase of roughness from $z'_c \cong z'_w$ to z'_{cw}, Fig. 11.22. If data on current measurements are available the shear velocity u_{*c} and z'_{cw} can be determined from the velocity profile. When measurements at one elevation only are avialable the law of the wall

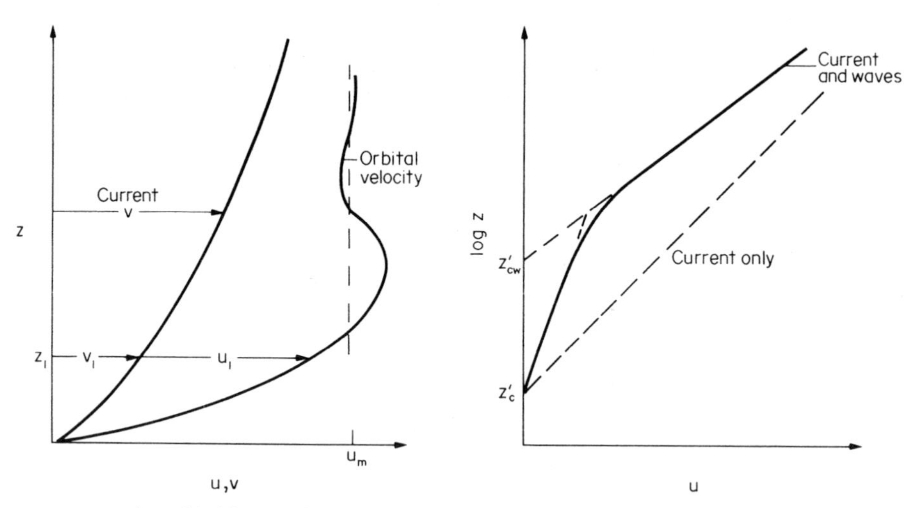

Fig. 11.22. Definition sketch.

$$u_m = \frac{u_{*c}}{\kappa} \ln \frac{z}{z'_{cw}} \qquad (11.119)$$

can be solved by trial and error or with the aid of the approximation

$$z'_{cw}/z'_c = 1 + 0.06 (u_{*w}/u_{*c})^3$$

proposed by Coffey and Nielson (1984) which leads to

$$u_m = \frac{u_{*c}}{\kappa} \ln \frac{z/z'_c}{1 + 0.06(u_{*w}/u_{*c})^3} \qquad (11.119a)$$

where u_m is the measured mean current velocity at elevation z and u_{*w} and u_{*c} are average values. The authors proposed that

$$\frac{z'_{cw}}{z'_c} = F \exp\left[\frac{1}{F} - 1\right]$$

$$F = 1 + (u_{*w}/u_{*c})^3/6$$

Alternatively, the current velocity profile is given by eqn 6.3 for both with or without the presence of waves, only that with waves the equivalent roughness k is increased to an apparent value of k_w due to the effect of the wave boundary layer. The logarithmic form of the velocity distribution of the current remains unaffected because of the assumed linear distribution of shear stress with depth. In the presence of waves the current velocity has a nominal zero at z'_{cw} instead of z'_c as shown in Fig. 11.22. The maximum bed shear stress due to waves is $\tau_o = \frac{1}{2}\rho f_w|u_m|u_m$ and in simple wave motion its time average is zero. When waves and current occur together in the same direction (otherwise vectorial addition is necessary) the mean bed shear stress is

$$\frac{\bar{\tau}_o}{\rho} = \frac{1}{4T} f_w[(u_m + u_\delta)^2 T_1 - (u_m - u_\delta)^2 T_2]$$

assuming that $f_w \simeq f_{wc}$ (wave velocities usually dominate) where u_δ is the current velocity at the outer limit of the wave boundary layer of thickness δ and T_1 and T_2 are the durations of forward and reverse velocities within the wave period T. The factor $\frac{1}{2}$ arises from time averaging (see eqn 11.222).

If $u_m >> u_\delta$ then τ_o approximates to $\bar{\tau}_o = \frac{1}{2}\rho f_w u_m u_\delta$. Combining this with eqn 6.3 and observing that $u_{*c}^2 = \tau_{oc}$ leads to

$$k_w = \delta/\exp\{0.4[u_m/(\tfrac{1}{2}f_w u_m u_\delta)^{\frac{1}{2}} - 8.5]\} \qquad (11.120)$$

The unknown u_δ is related to the mean current speed by eqn 6.17, using $u_* = (\frac{1}{2}f_w u_m u_\delta)^{0.5}$ and k_w.

At the bed the momentum diffusion coefficient (eddy viscosity) ε is affected by ε_w of the wave motion and ε_c of the current. Thus, by analogy to $\varepsilon = \kappa u_* y$ one could write $\varepsilon_{cw} = \kappa(u_{*c} + u_{*w})y$ or since $u_{*c}^2 = \varepsilon du/dy = \kappa(u_{*c} + u_{*w})y\, du/dy$ and this leads to

$$\frac{u}{u_{*c}} = \frac{1}{\kappa} \frac{1}{(u_{*c} + u_{*w})} \ln\left(\frac{y}{k/30}\right)$$

This is an approximation to the lower part of the current velocity profile during wave action, as illustrated in Fig. 11.22. The distribution is approximate only because ε_w decreases rapidly with increasing distance from bed in the upper part of the wave boundary layer, i.e., ε_w is proportional to y only in the lower part. The decrease leads to a transition between the two velocity distributio functions.

The effect of the waves on the current velocity has led to models with an inner and outer region of flow. The treatment by Kajiura (1968) even includes an overlap region. The logarithmic velocity distribution is still assumed to be valid in the inner region but the momentum diffusion coefficient (eddy viscosity) arises now from wave boundary layer turbulence and current. A current aligned with wave propagation may even cancel the opposing orbital velocity, i.e. magnitude fluctuations only without change in direction. When the current is angled complex three-dimensional velocity distributions arise together with higher turbulence production and higher values of momentum diffusion coefficient. For details reference is made to Horikawa and Watanabe (1968), Lundgren (1972), Bakker (1974), Bakker and Doorn (1978), van Kestern and Bakker (1984) Fredsøe (1984) and Jonsson (1980) which also includes a literature survey.

Brijker (1967) assumed a logarithmic current velocity profile and used the elevation $z_1 = ez'_c$ where the tangent from the origin meets the profile to evaluate current velocity. Here z'_c is the elevation at which the logarithmic velocity is zero, $z'_c = k_s/30.2$ to $k_s/33$. Since $dv/dz \cong v_1/z_1$ and $l = \kappa z_1$, eqn (11.117) yields

$$v_1 = \frac{v_*}{\kappa} \tag{11.121}$$

where the current shear velocity is denoted by $v_* = (\tau_c/\rho)^{\frac{1}{2}}$, or the bed shear stress due to current

$$\tau_c = \rho\kappa^2 v_1^2 \tag{11.122}$$

and due to waves

$$\tau_w = \rho\kappa^2 u_1^2 \tag{11.122a}$$

Bijker set

$$u_1 = pu_b \tag{11.123}$$

Between eqns (11.118), (11.123) and (11.124) and observing that $f_e \cong f_w$

$$p = \frac{\sqrt{\frac{1}{2}f_w}}{\kappa} \tag{11.124}$$

and with substitution for v_* from eqn (11.117)

$$\frac{v_1}{u_1} = \frac{v_*}{\kappa}\,\frac{\kappa}{\sqrt{\frac{1}{2}f_w}\,u_b} = \frac{\sqrt{g}}{C\sqrt{\frac{1}{2}f_w}}\,\frac{V}{u_b} = \frac{V}{\xi u_b} \tag{11.125}$$

where V is the mean velocity of the current over depth and ξ is generally greater than one.

By analogy to eqns (11.122)

$$\tau_{cw} = \rho\kappa^2 v_r^{\,2} \tag{11.126}$$

where v_r is the resultant velocity at elevation z_t.

The velocities at any elevation are combined vectorially, Fig. 11.23.

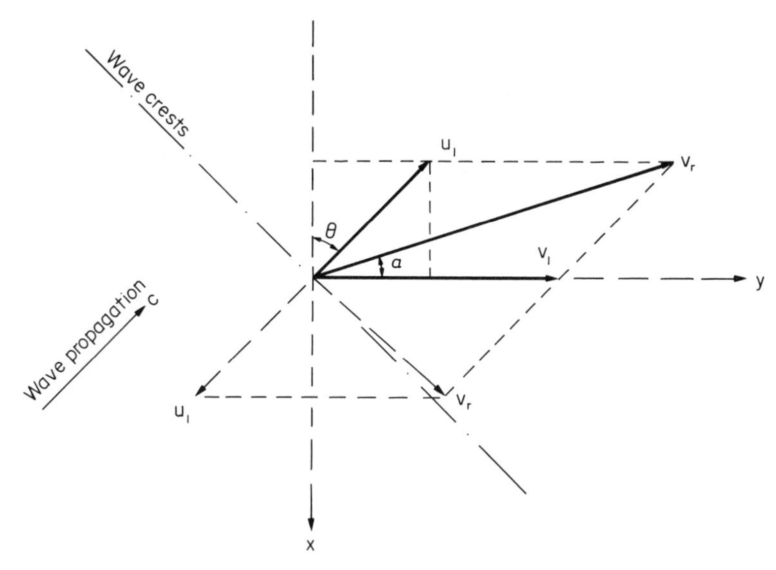

Fig. 11.23. Vector addition of orbital and current velocities at the bed.

The absolute value of the resultant velocity at elevation z_1 is

$$|v_r| = [u_1^{\,2} + 2u_1v_1\sin\theta + v_1^{\,2}] \tag{11.127}$$

or

$$\tau_{cw} = \rho\kappa^2 u_1^{\,2}\left[1 + 2\,\frac{v_1}{u_1}\,\sin\theta + \left[\frac{v_1}{u_1}\right]^2\right] \tag{11.128}$$

or

$$\frac{\tau_{cw}}{\tau_w} = \left[1 + 2\frac{v_1}{u_1}\sin\theta + \left[\frac{v_1}{u_1}\right]^2\right] \qquad \frac{\tau_{cw}}{\tau_c} = \left[\left[\frac{u_1}{v_1}\right]^2 + 2\frac{u_1}{v_1}\sin\theta + 1\right]$$

$$= \left[1 + 2\frac{V}{\xi u_m}\frac{\sin\theta}{\cos\omega t} + \left[\frac{V}{\xi u_m}\right]^2 \frac{1}{\cos^2\omega t}\right] \qquad = \left[\left[\xi\frac{u_m}{V}\right]^2 \cos^2\omega t + 2\xi\frac{u_m}{V}\sin\theta\cos\omega t + 1\right] \tag{11.129}$$

on substitution from eqn (11.125) and (11.119).

The time average of τ_{cw}/τ_w over half a wave period Swart (1974) approximated by

$$\frac{\overline{\tau_{cw}}}{\tau_w} = 1 + M\left[\frac{V}{\xi u_m}\right]^N ; \quad M = 1.91 - 1.32\sin\theta, \quad N = 1.24 - 0.08\sin\theta \tag{11.130}$$

Visser (1984) expressed the longshore component as

$$\frac{\overline{\tau_{cwy}}}{\tau_c} = \frac{1}{T}\int_0^T \left[1 + 2\xi\frac{u_m}{V}\sin\theta\cos\omega t + \left[\xi\frac{u_m}{V}\right]^2\cos^2\omega t\right]^{\frac{1}{2}} \left[1 + \xi\frac{u_m}{V}\sin\theta\cos\omega t\right] dt = F \tag{11.131}$$

where $\cos\alpha = (u_1\sin\theta + v_1)/v_r$. The value of F was approximated by

$$F = \begin{cases} 1 + \frac{1}{4}\left[\xi\frac{u_m}{V}\right]^2\left[1 + \sin^2\theta\right]; & \frac{u_m}{V} < 1 \\ \frac{2}{\pi}\xi\frac{u_m}{V}\left[1 + \sin^2\theta\right]; & \frac{u_m}{V} >> 1 \end{cases}$$

Bijker approximated the expression for $\overline{\tau}_{cwy}$ for $\theta < 20°$ by

$$\overline{\tau}_{cwy} = \tau_c\left[0.75 + 0.45\left[\xi\frac{u_m}{V}\right]^{1 \cdot 13}\right] \tag{11.132}$$

and for small angles between wave crests and currents ($\sin\theta = 0$) and $\xi u_b >> V$

$$\overline{\tau}_{cwy} = \tau_c\frac{2}{\pi}\xi u_m/V = \frac{\rho}{\pi C}(2g_{fw})^{\frac{1}{2}}\, Vu_m \tag{11.133}$$

For shallow water eqn (11.6) reduces to

$$u_m = \frac{\pi H}{T\kappa h} = \tfrac{1}{2}\gamma\sqrt{gh} \tag{11.134}$$

and

$$\overline{\tau}_{cwy} = \frac{\rho g}{\sqrt{2}\,\pi C}\gamma\sqrt{h}\sqrt{f_w}\, V \tag{11.135}$$

These are drastic simplifications and limit the applications of the equations.

Generally, bed friction is important for the force balance but the energy dissipation due to bed friction is small compared to that due to turbulence in the surf zone.

11.3.1.3 Lateral friction force. The most difficult term to estimate is the lateral friction force arising out of the Reynolds shear stress $-\overline{\rho u'v'}$, where u' and v' refer to the turbulence components of velocity. In the shear flow these components are strongly correlated and $\overline{\rho u'v'}$ is of the same order as $\rho q'^2$, where q' is the turbulence velocity vector. In the surf zone the principal source of turbulence energy is the transfer of wave energy. Any turbulence produced by currents are usually ignored in comparison, i.e. transfer of wave energy is assumed to be the sole source of turbulence energy. In the turbulence produced by the breakers the u'v' correlation is weak and only in the presence of a velocity gradient dV/dx is a horizontal anisotropy produced in the turbulence eddies. The Reynolds stress tensor

$$T_{xy} = -\overline{\int_h^\eta \rho u'v' dz} \tag{11.136}$$

is usually in analogy to $\tau = \rho \ell^2 (du/dy)^2 = \rho\varepsilon du/dy$ written as

$$T_{xy} = \rho\varepsilon h \frac{dV}{dx} \tag{11.137}$$

where ε is the momentum diffusion coefficient ($m^2 s^{-1}$), ℓ is the mixing length, and V is the depth averaged mean velocity of longshore current at distance x from the shore. The problem centres around the definition of ε. Clearly, there will be differences in the turbulence structure, and in the nature of the diffusion coefficient, with breaker type. The coefficient will also depend on the distance from breaker line, bed shear stresses etc. However, the biggest single factor will be the type of breakers. Bowen assumed ε to be a constant, Longuet-Higgens put it proportional to the distance from shore, and ε has been related to $H_b x_b/T$, to α, to u_m, etc. Battjes (1975) argued that $\overline{u'v'}$ is proportional to $\varepsilon \ell dV/dx$ and ε is proportional to $q\ell$ where q is a characteristic turbulence velocity vector and put ℓ equal to h, the water depth, i.e.

$$T_{xy} = K\rho q h^2 \frac{dV}{dx} \tag{11.138}$$

The production and dissipation balance over the surf zone. Battjes also assumed that they balance locally. Then the vertically averaged mean rate of energy dissipation, D_t, per unit area is

$$D_t \sim \rho h \partial \tag{11.139}$$

where ∂ is the mean rate of turbulence energy dissipation per unit mass averaged over depth. The kinetic energy of turbulence per unit mass is $\frac{1}{2}\rho q^2$ and its dissipation rate can be expressed through a characteristic velocity and characteristic length, e.g., q/h. This is based on the concept of cascade of turbulence where the eddies break down to smaller and smaller ones without a significant loss of energy until the dissipation size is reached where viscous action converts the energy into heat. The dissipation rate per unit mass is then $\rho q^3/h$ or ρq^3 over depth, i.e.

$$D_t \sim \rho q^3 \tag{11.140}$$

and this is proportional to the dissipation of wave energy. The flux of wave energy towards the shore F_x = Enc cosθ and the dissipation

$$D = -\frac{dF_x}{dx} = \frac{5}{16}\rho\gamma^2 g^{3/2}\left[\frac{h_b}{x_b}\right]^{5/2} x^{3/2} \quad ; \quad x \leq x_b \tag{11.141}$$

At the breaker line

$$D_b = \frac{5}{2}\frac{1}{x_b}(E_o n_o c_o \cos\theta_o) = \frac{5}{32}\frac{1}{x_b}\rho g H_o{}^2 c_o \cos\theta_o \tag{11.142}$$

Comparison with eqn (11.116) shows that

$$D = \frac{c_o}{\sin\theta_o}\frac{dS_{xy}}{dx} \tag{11.143}$$

Thus, the characteristic turbulence velocity at the breaker line from eqn (11.140) is

$$q_b \sim (D_b/\rho)^{1/3} \tag{11.144}$$

Visser utilised eqn (11.108), in which he neglected $T_{xy}dV/dx$ as small, to obtain a solution for distribution of q

$$\frac{dF_{tx}}{dx} D_y + D = 0 \tag{11.145}$$

where the power density of turbulence ($Nms^{-1}m^{-1}$) per unit area is expressed as

$$F_{tx} = \varepsilon h \frac{d}{dx}(\tfrac{1}{2}\rho q^2) \tag{11.146}$$

and

$$\varepsilon = Kqh \tag{11.147}$$

Togehter with eqn (11.140)

$$q^3 + \frac{K}{3}\frac{d}{dx}\left[h^2\frac{dq^3}{dx}\right] + \frac{D}{\rho} = 0 \tag{11.148}$$

where the two coefficients are combined in one.

The variation of $\tilde{q} = q/q_b$ is shown in Fig. 11.24 with the value of the constant K = 2.5 the maximum turbulence velocity is

$$q_m = 0.8(D_b/\rho)^{1/3} \tag{11.149}$$

where q_b is from eqn (11.144) with the proportionality constant equal to one, or

$$q_m = 0.8\left[\frac{5}{32}\,\frac{1}{px_b}\,gH_o{}^2c_o\,\cos\theta_o\right]^{1/3} \tag{11.150}$$

where $p = x_p/x_b$ and x_p is the location of the plunge line (of plunging breakers), i.e. the solution can allow for the width of the breaker zone but does not differentiate for breaker form.

Introduction of eqns (11.147) and (11.150) in eqn (11.137) yields

$$\frac{d}{dx}\,T_{xy} = K\rho q_b v_o\left[\frac{h}{x}\right]_b^2 \frac{d}{d\tilde{x}}\left[\tilde{q}\tilde{h}^2\,\frac{d\tilde{V}}{d\tilde{x}}\right] \tag{11.151}$$

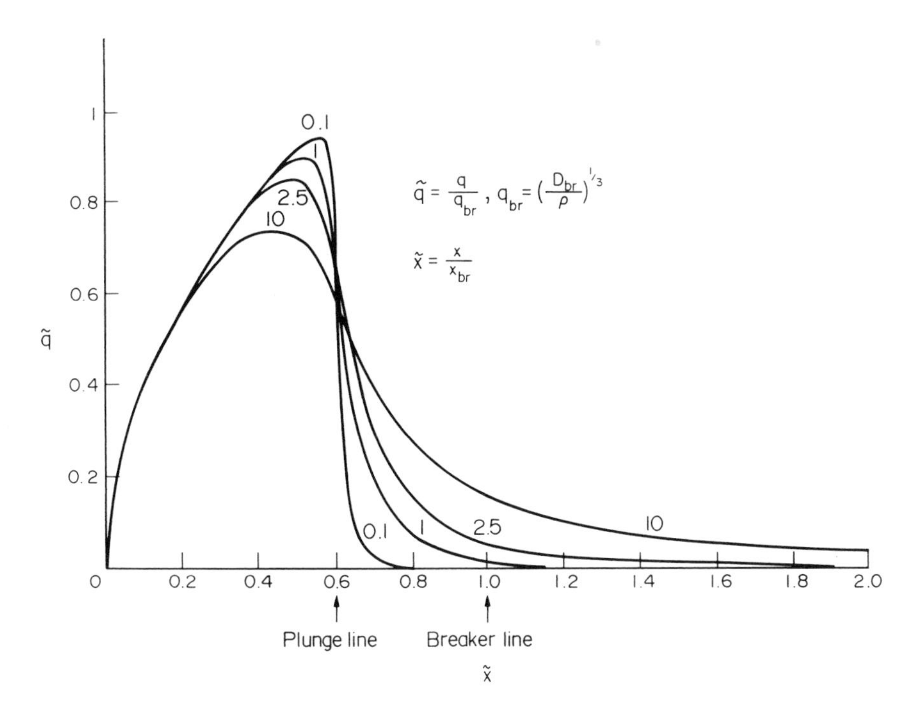

Fig. 11.24. Computed dimensionless turbulent velocity $\tilde{q}$ as function of $\tilde{x}$ for different M, Visser (1984).

where

$$\tilde{h} = h/h_b, \quad \tilde{q} = q/q_b, \quad \tilde{x} = x/x_b, \quad \tilde{V} = v/v_o$$

$$v_o = \left[\frac{2}{f_b}\frac{\sin\theta_o}{c_o}\frac{D_b}{\rho}\right]^{\frac{1}{2}} \quad ; \quad u_m/\bar{V} < 1$$

$$v_o = \frac{\pi}{f_b}\frac{\sin\theta_o}{c_o}\frac{D_b/\rho}{(u_m)_b} \quad ; \quad u_m/\bar{V} >> 1,\ \sin^2\theta << 1$$

The v_o expressions are characteristic longshore velocities for a strong and a weak longshore current, respectively.

11.3.1.4 Calculation of longshore current. The solutions of the longshore current problem centre on eqn (11.105)

$$\frac{dS_{xy}}{dx} - \frac{dT_{xy}}{dx} + \bar{\tau}_{oy} = 0$$

where the driving force $d/dx(S_{xy}) = -(\partial F_x/\partial x)(\sin\theta/c)$ and $F_x = Ecn\cos\theta$. As indicated earlier, the second term (lateral mixing force) is the most difficult to define. Longuet-Higgins (1970) put it equal to $\partial/\partial x(\rho\varepsilon_L h\partial v/\partial x)$ and ε_L proportional to the distance from shore. His solution for the distribution of the mean longshore velocity, $\tilde{V} = V/V_o$, for small wave angles is shown in Fig. 11.25 in terms of X and P, where $X = x/x_b$ is the distance from water's edge divided by the breaker distance and P is a mixing parameter which incorporates the beach slope β, the momentum diffusion coefficient in the surf zone, the bed friction coefficient c_b (defined as $\tau_o = \rho c_f u^2$) and the breaker index $\gamma = H/h$, i.e.

$$P = \pi N\tan\beta/c_f\gamma \tag{11.152}$$

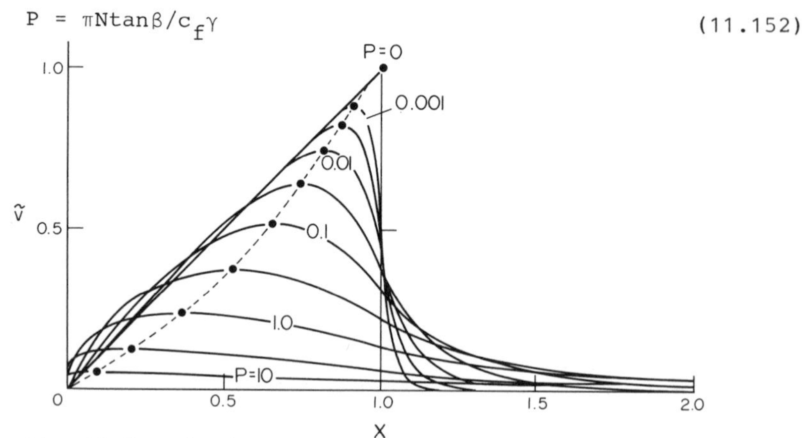

Fig. 11.25. The form of longshore current profiles, according to Longuet-Higgins as a function of the mixing parameter P.

where N is a dimensionless constant in the range of 0 to 0.016 in $\varepsilon_L = Nx(gh)^{\frac{1}{2}}$, and $\tan\beta$ usually includes wave setup, i.e.,

$$-\frac{\partial h}{\partial x} = \tan\beta' = \frac{\tan\beta}{1 + (3/8)\gamma^2} \qquad (11.153)$$

The normalising velocity V_o is the theoretical value of V at the breaker line for a triangular distribution (P = 0) and not a physically measurable quantity

$$V_o = \frac{5\pi}{16}\gamma\frac{\tan\beta'}{c_f}(gh_b)^{\frac{1}{2}}\sin\theta_b \qquad (11.154)$$

or

$$\frac{V_o}{u_m} \cong 2\frac{\tan\beta'}{c_f}\sin\theta_b \qquad (11.155)$$

where $5\pi/8 \cong 2$ and $u_m = (\gamma/2)(gh_b)^{\frac{1}{2}}$ at the breaker line.

Kraus and Sasaki (1979) refined the Longuet-Higgins' solution, in particular by avoiding the small wave angle assumption. As seen in Fig. 11.26 the maximum velocity V_{max} and its position X_{max}, the velocity at the breaker line V_B, the mean velocity inside the breaker line $\bar{V}$, and the mid-surf velocity $V_{\frac{1}{2}}$ all decrease with increasing P and increasing breaker angle θ, except $V_{\frac{1}{2}}$ which remains approximately constant for $P > 0.1$. The latter is in keeping with the empirical value (Komar, 1976) for

$$V_{\frac{1}{2}} = 2.7\, u_m\sin\theta_b\cos\theta_b \qquad (11.156)$$

For constant P the maximum velocity moves with increasing breaker angle towards the shoreline. Figure 11.27 shows examples of longshore current distribution as measured in laboratory and Fig. 11.28 from field ($\theta_b = 9°$, $h_b = 1$ m). Note the apparent influence of the beach profile.

Visser (1984) derived for the longshore current distribution inshore of the plunge line

$$\tilde{V} = (\tilde{x}/p)^{3/4} \qquad (11.157)$$

and

$$\tilde{V} = \tilde{x}/P \qquad (11.158)$$

for strong and weak currents, respectively. (For notation see eqn (11.151).) From a numerical solution of eqn (11.105) with eqn (11.151) and from laboratory experiments on a *fixed beach* at 1:10 slope he proposed a longshore current distribution as shown in Fig. 11.29.

All the models refer to a well defined breaker line and a plane beach of constant slope. In nature the waves are mostly irregular and the distance of the breaker line varies substantially as the

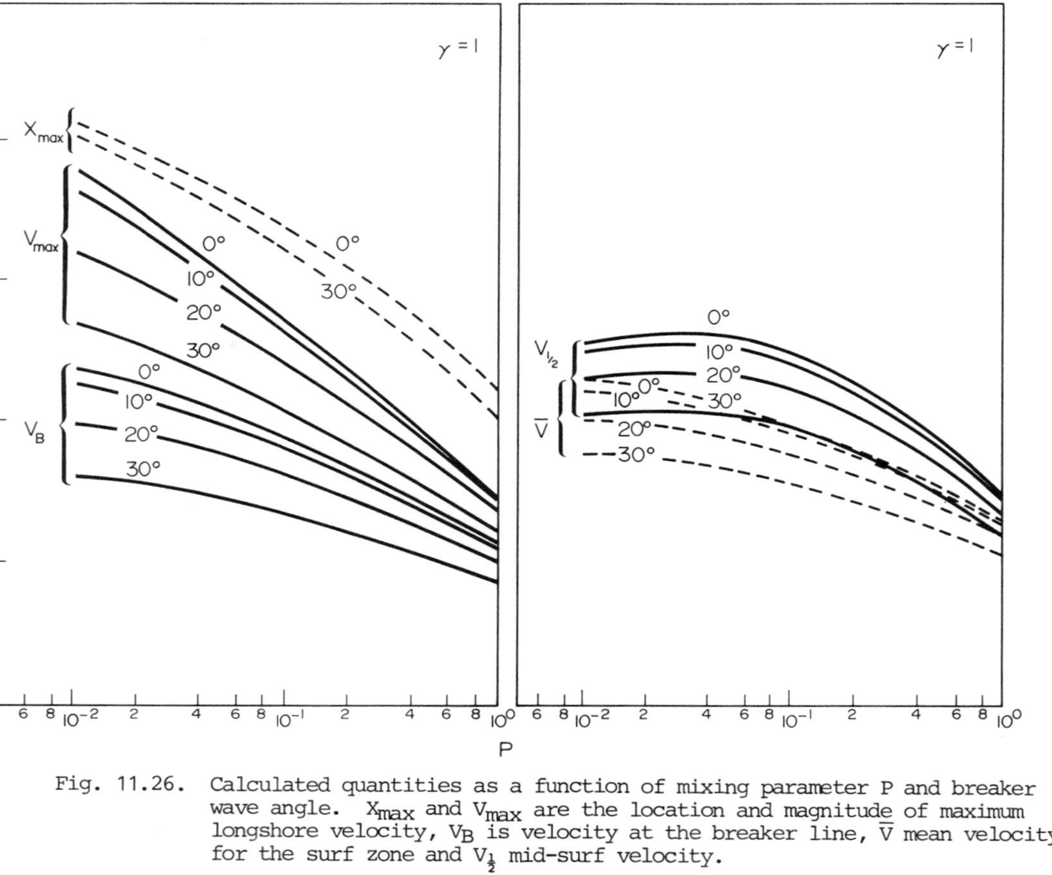

Fig. 11.26. Calculated quantities as a function of mixing parameter P and breaker wave angle. X_{max} and V_{max} are the location and magnitude of maximum longshore velocity, V_B is velocity at the breaker line, $\overline{V}$ mean velocity for the surf zone and $V_{\frac{1}{2}}$ mid-surf velocity.

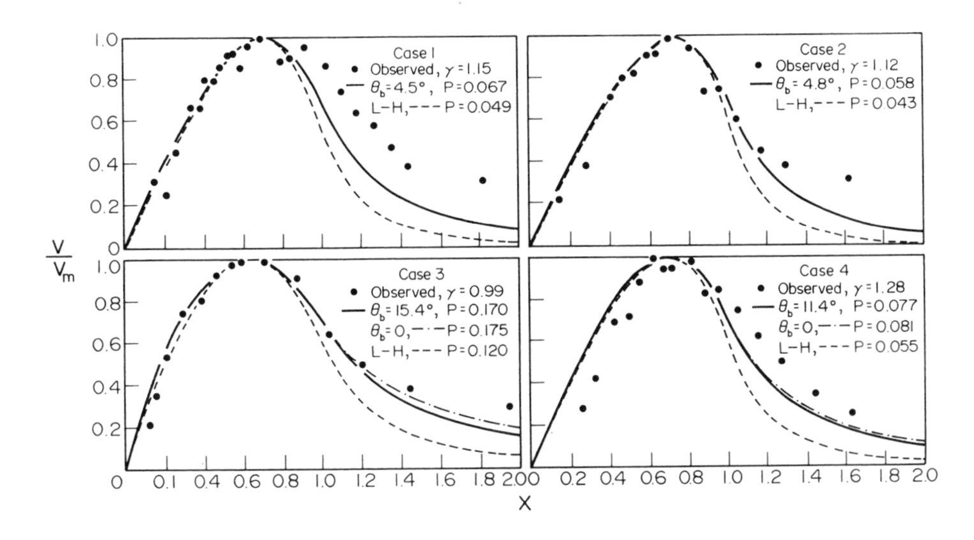

Fig. 11.27. The laboratory longshore current measurements and theoretical profiles determined by equating the maximum velocity V_m, according to Kraus and Sasaki (1979).

waves of a wave spectrum arrive. The major effect of this is that the decay of the longshore velocity seaward of the breakers is much more gradual than indicated by the theoretical longshore current profiles. The wave height used to describe a random sea is H_{rms}. The beach profiles are seldom flat and this has a strong influence on the actual longshore current distribution across the surf zone.

It should also be noted that the models give an average longshore current velocity. In nature the current is not steady but fluctuates wildly, about ± 100%, and the periods of these fluctuations can be much longer than the average incident wave period. Dette and Führböfer (1974) recorded periods of velocity fluctuations up to nine times the average period of incident waves.

The simplest approximation of eqn (11.105) is

$$\frac{d}{dx} S_{xy} + \overline{\tau}_{Dy} = 0 \tag{11.159}$$

which ignores the Reynolds stresses. Thus, from eqn (11.116) and (11.131)

$$\frac{5}{16} \rho \gamma^2 (gh)^{3/2} \frac{\sin\theta_o}{c_o} m = \tfrac{1}{2} f_c \rho V^2 \begin{cases} 1 + \frac{1}{4}\left[\xi \frac{u_m}{V}\right]^2 (1 + \sin^2\theta); & \frac{u_m}{V} < 1 \\ \frac{2}{\pi} \xi \frac{u_m}{V} (1 + \sin^2\theta); & \frac{u_m}{V} >> 1 \end{cases} \tag{11.160}$$

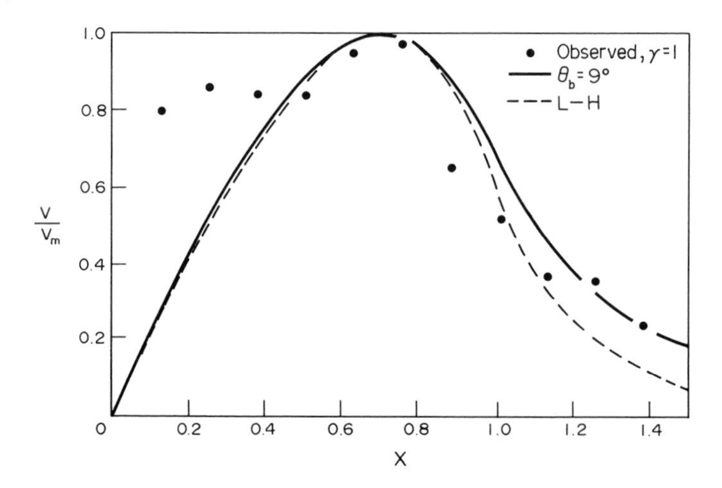

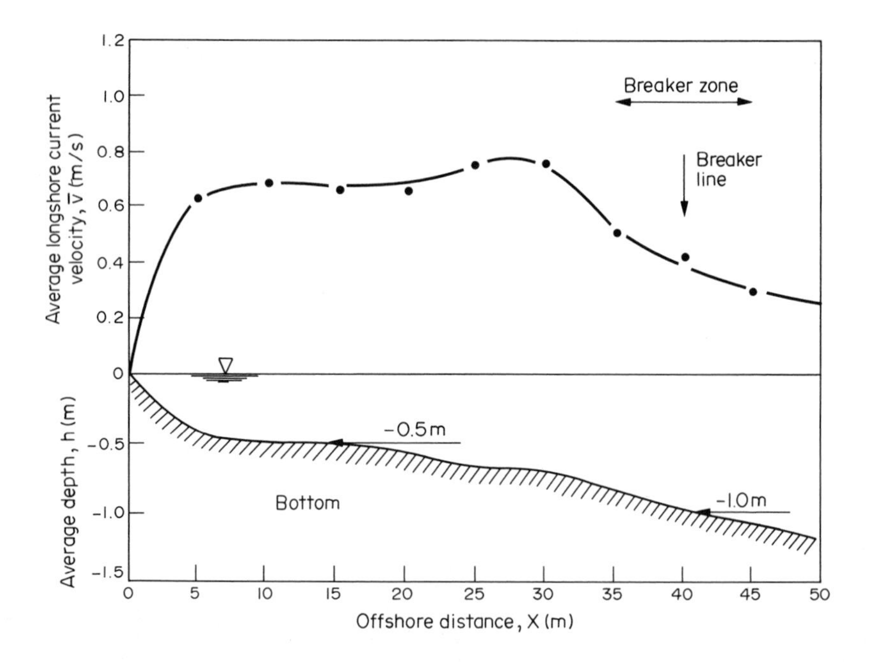

Fig. 11.28. Theoretical longshore current profile fitted to field data at the location of maximum velocity (above) and measured values (below) together with surf zone profile. Note that the step profile leads to almost constant current velocity in the surf zone, after Kraus and Sasaki (1979).

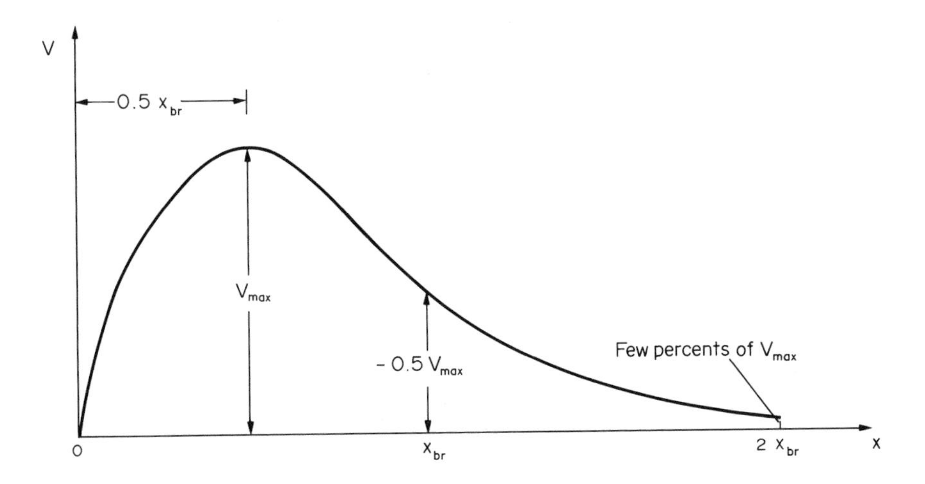

Fig. 11.29. Schematization of measured longshore current profiles, Visser (1984).

$$= \frac{g}{C^2} \rho V^2 [0.75 + 0.45(\xi u_m/V)^{1.13}] \tag{11.161}$$

with eqn (11.132). The former is easier to solve. The latter can be solved using the Runge-Kutta iteration. However, if eqn (11.135) is used

$$= \frac{\rho g}{\sqrt{2}\pi C} \gamma \sqrt{h} \sqrt{f_w} V \tag{11.162}$$

then

$$V = \frac{5\sqrt{2}\pi}{16} \sqrt{g} \frac{\sin\theta_o}{c_o} \gamma \frac{C}{\sqrt{f_w}} hm \tag{11.163}$$

If γ, m and $C/\sqrt{f_w}$ are constant eqn (11.163) yields a longshore current velocity distribution which increases linearly from zero at the water's edge to V_{max} at the breaker line and zero outside. Up to about $\frac{1}{2}x_b$ from shore the calculated velocities are of the same order as measured but seaward from there the observed velocities differ appreciably. The observed velocities reach a maximum in the region of $(^2/_3)x_b$ and fade out at about $x = 2x_b$. The velocity distribution depends strongly on the breaker type and on the actual beach profile.

An approximation in use represents the velocity profile by a triangle from shore to $x = 1.6\ x_b$ with its peak at $x = (^2/_3)x_b$. The peak value of the longshore current velocity at the breaker line V_b is calculated using

$$\int_0^{1.6x_b} \overline{\tau}_{cwy} dx = S_{xy}\Big|_{x_b} \tag{11.164}$$

and eqn (11.135).

11.4 Incipient Motion and Bed Features

11.4.1 Incipient sediment motion under waves

The threshold of motion of noncohesive particles in steady current was discussed in Chapter 3. In a flow where the velocity vector changes periodically in magnitude and direction the solution of the threshold problem is appreciably more complex. Apart from the definitions of drag and lift forces and their points of application, inertia forces, virtual mass forces, forces due to pressure gradients as well as phase shifts between velocities, accelerations and forces now enter into the problem. In principle, the forces present are weight force form drag force, lift force, inertia and virtual mass forces, pressure gradient force, friction and rolling resistance forces, and forces exerted by support points. The definition of the hydrodynamic forces and their points of application are complicated by the oscillating boundary layer flow and the flow of water into and out of the bed as the waves pass over.

Eagleson et al. (1958,1963) developed an analytical model for grain motion under waves, considering all but the flow in and out of the bed. They solved it for the incipient conditions, i.e., the sum of moments is zero $\Sigma M = 0$, and the established sediment motion condition, i.e., the balance of forces $\Sigma F_x = 0$. Other authors have developed similar models. However, many assumptions remain which render the use of such complex models unattractive. This has led to developments of simplified semi-empirical models, utilizing dimensional analysis and experimental data. In general, the forces which arise from pressure gradients and inertia effects are important only with very short steep waves, e.g. in laboratory work. The effect of water penetration into the bed is also likely to be small. Pressure measurements by the Research Station Lubiatowo (Police Academy of Sciences, 1980) showed, for example, an average double amplitude of pressure fluctuations of 0.743 m of water column 0.55 m above the bed surface, and 0.718 m at 0.55 m below the bed, under wind waves with an average wave height $\overline{H}$ = 1.125 m in 7.30 m of water. However, due to the effective incompressibility of water the associated volumes of water movement would be very small.

For practical application the Shields diagram for threshold in unidirectional flow, has been adapted to describe the incipient conditions under wave motion. This approach effectively ignores the very significant differences in the structure of the boundary layer as well as in the surface condition. Under waves grains may be rolled to and fro and consequently there will always be grains which have come to rest in very unstable positions. Among those grains is frequently a larger fraction of coarse grains than in the general sample of bed material. Those grains are through their exposure more mobile on a flat sand bed. Komar and Miller (1974) and Madsen and Grant (1975) showed that the various experimental data, for incipient motion due to waves, agree reasonably well with the Shields diagram if the shear stress is calculated with the bed friction factor, f_w, by Jonsson. The data, however, fall in the transition region where the f_w-curve is ill-defined due to lack of data. The method requires the estimation of the roughness height of the bed, instead of the shear velocity u_*. If the velocity distribution is known then u_* can be obtained from the momentum equation

$$- u_*^2 = - \int_{y'}^{\delta+y'} \frac{\partial}{\partial t} (u_b - u)\,dy$$

where u_b is the orbital velocity at the bed.

Figure 11.30 by Komar and Miller shows a θ_c versus Re = $u_m d/\nu$ presentation. To make it easier to use the graph can be replotted for grains of constant density and liquid of constant viscosity, as was done by Bagnold (1963), Fig. 11.31.

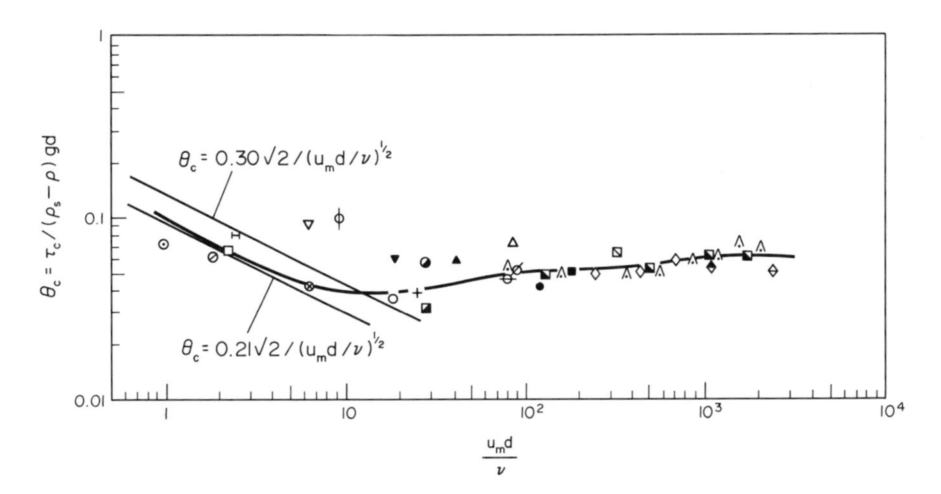

Fig. 11.30. Threshold θ_c versus Reynolds number $u_m d/\nu$, after Komar and Miller (1974). + steel d = 0.6 mm, all round symbols relate to quartz sands, square symbols to coal (S_s = 1.30), ▽,▼ glass beads, △,▲ plastic pellets (S_s = 1.28, 1.052), Λ limestone chips, ◇ glass spheres, ⬘ concrete cubes, ⬙ perspex cubes.

Sleath (1978) expressed the shear stress in the Shields parameter as $\tau_c = \frac{1}{2} f_1 \rho u_m^2$ and plotted f_1/f_w against $f_w u_m/(\omega\nu)^{\frac{1}{2}}$; the idea being that if the data were affected by drag only then $f_1 = f_w$ and the plot of f_1/f_w made the experimental curves for various values of A/d collapse. The f_1/f_w-function is shown in Fig. 11. 32 where f_1=f_w for large values of the abscissa and f_w is defined by equation (11.66) using d as the roughness height. His plot of $\theta_c = \frac{1}{2} f_1 \rho u_m^2/(\rho_s - \rho)gd$ against the nondimensional grain size $D_* = (\Delta g/\nu^2)^{1/3} d$ is shown in Fig. 11.33 which is an equivalent

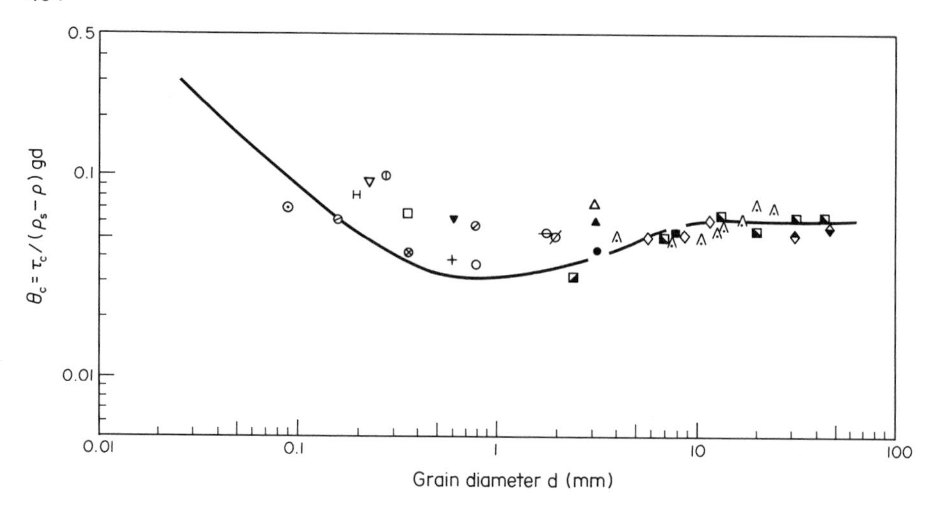

Fig. 11.31. Comparison between the Bagnold (1963) curve for threshold under unidirectional flow and the threshold data for oscillatory wave motions (Komar and Miller, 1974). Symbols as for Fig. 11.30 except for H which signifies data of Horikawa and Watanabe (1968).

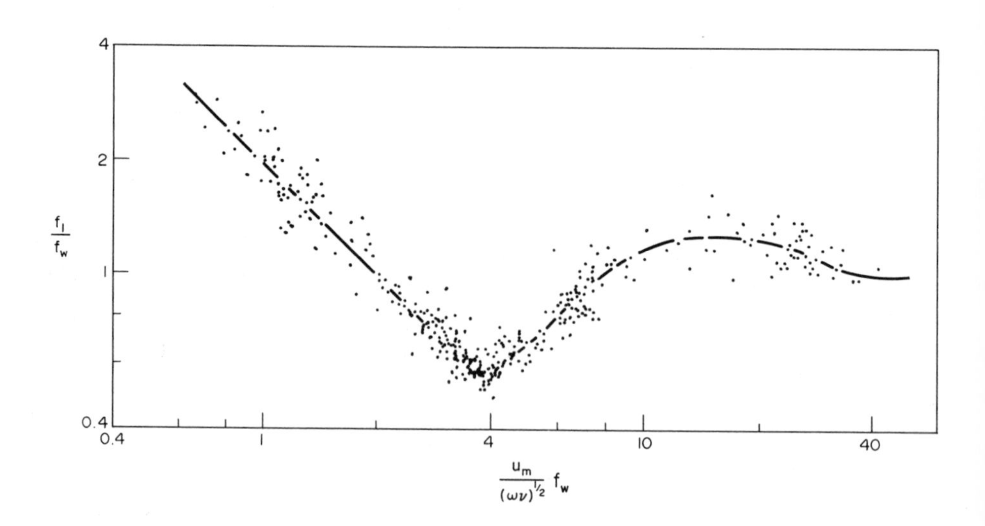

Fig. 11.32. Variation of f_1 with $u_m/(\omega\nu)^{\frac{1}{2}}$ and f_w, according to Sleath (1978).

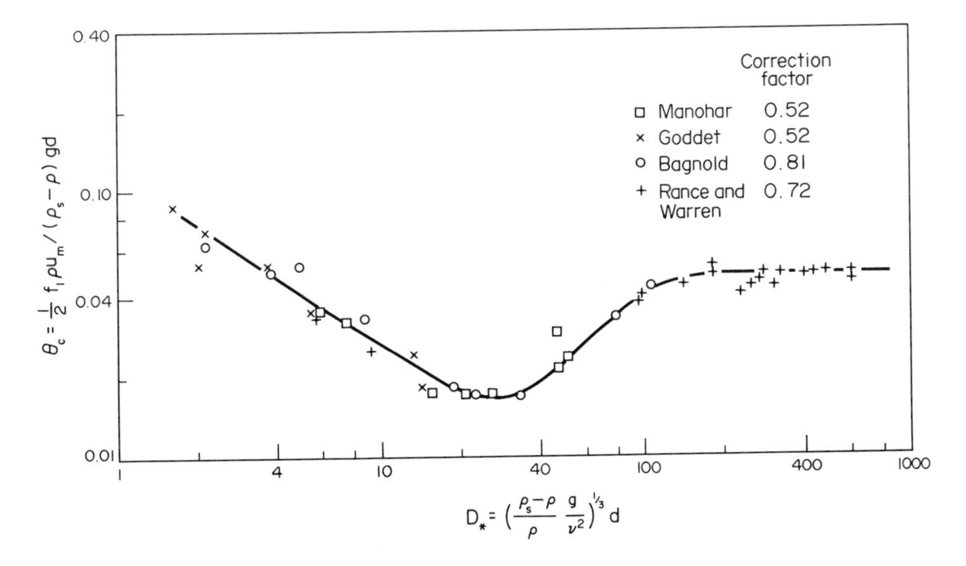

Fig. 11.33. Modified Shields curve for unsteady flow, after Sleath (1978).

presentation to the Shields diagram. Noteworthy is the lower minimum value of θ_c than in other presentations. The minimum corresponds to quartz grains of d ≅ 1-1.5 mm at usual conditions.

Komar and Miller also proposed empirical equations for threshold which define the laminar and turbulent rough boundary asymptotes

$$\frac{\rho u_m^2}{(\rho_s - \rho) gd} = 0.21 \left[\frac{d_o}{d}\right]^{1/2} ; \quad d < 0.5 \text{ mm} \tag{11.165a}$$

$$\frac{\rho u_m^2}{(\rho_s - \rho) gd} = 0.46\pi \left[\frac{d_o}{d}\right]^{1/4} ; \quad d > 0.5 \text{ mm} \tag{11.165b}$$

where d_o = 2A, the double amplitude of orbital movement. These equations are very convenient for use. For quartz sand in water these reduce to

$$u_m = 0.337 (g^2 Td)^{1/3} ; \quad d < 0.5 \text{ mm} \tag{11.166a}$$

and

$$u_m = 1.395 (g^4 Td)^{1/7} ; \quad d > 0.5 \text{ mm} \tag{11.166b}$$

respectively. The functions, defined by eqn (11.165), join well for waves with periods greater than 10 seconds. At shorter periods they have a small offset in the vicinity of d = 0.5 mm.

A similar expression for initiation of motion was proposed by Dingler (1979) for $0.18 \le d \le 1.454$ mm

$$\left[\frac{\rho_s - \rho}{\rho}\right]\frac{gT^2}{d} = 240\left[\frac{d_o}{d}\right]\left[\frac{\rho(\rho_s - \rho)gd^3}{\mu^2}\right]^{-1/9} \tag{11.167}$$

which in terms of maximum orbital velocity is

$$u_m = 0.052\ [g^5(S_s - 1)^5/\nu]^{1/6}(Td)^{\frac{1}{2}} \tag{11.168}$$

where u is the dynamic viscosity of water. For $\mu \cong 1.25 \times 10^{-3}$ at 15°C, 4% salinity and quartz grains

$$u_m \cong 5.1\ \sqrt{Td} \tag{11.169}$$

where u_m is in m/s and d in m. Figure 11.34 by Clifton and Dingler (1984) shows a comparison of the two sets of equations.

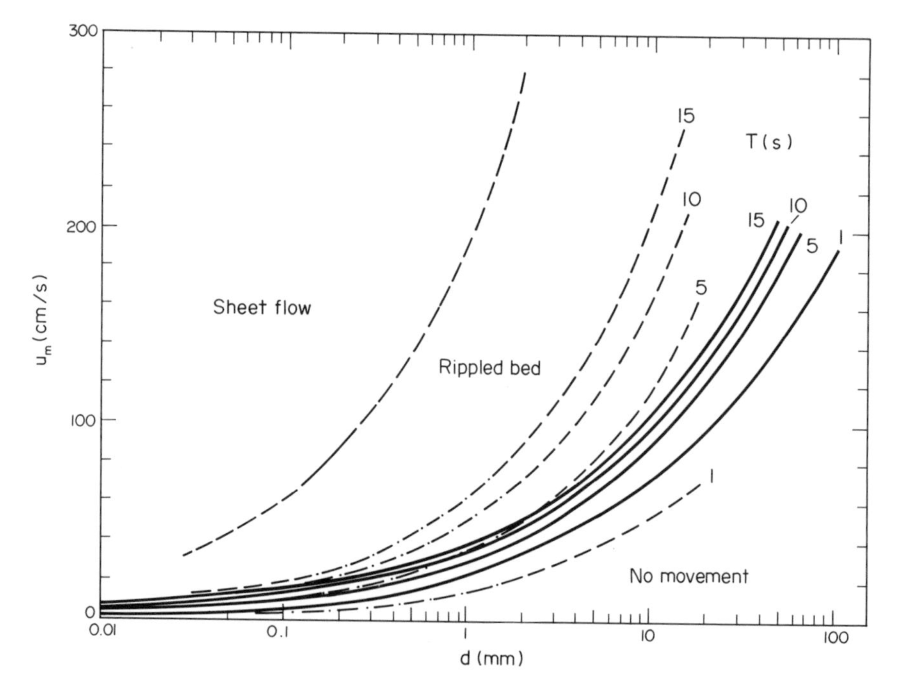

Fig. 11.34. Velocity threshold for grain movement and sheet flow of quartz sand in water, after Clifton and Dingler (1984).
——— Komar and Miller (1974),
—— . —— . —— Dingler (1979),
— — — — — extrapolated beyond experimental data range.

A set of semi-empirical equations for the incipient motion conditions was also proposed by Hallermeier (1980), and there are several expressions of the form $A/d \propto (S_s - 1)^m(g/\omega^2 A)^n$ or $A/d \propto (Re_A/D_*)^p$. For example

$$\frac{A}{d} = C \frac{(S_s - 1) g^2}{\omega^2 \alpha} = C' \frac{Re_A}{D_*}^{6/5} \qquad (11.170)$$

where $C \sim 0.1$ and $C' \sim 1.7$. Sleath (1984) lists over two dozen formulae for incipient motion.

The conditions of incipient motion relate to peak velocity at the bed which is a function of wave period, height and water depth. Thus, the movement occurs briefly under wave crest and trough only. The mobility of the grains is, apart from size, also affected by exposure. Thus, a large grain on the surface can actually have a very low threshold, as discussed in Chapter 3.

When comparing predicted values of threshold it is necessary to reflect that the values in nature of u_m or A at the bed under wave action are not as well defined as might appear from the linearised wave theory. The conditions at the bed are strongly affected by the roughness and in areas of major interest the linear wave theory is no longer valid. Even estimates of velocities from the cnoidal wave theory do not represent the true conditions because of the asymmetry of wind driven waves, not to mention complications which arise from wave spectrum and multi-directional seas. Additional uncertainties arise from armouring effects of the bed, from biological effects, shells and shell fragments, algae growth, etc. Particularly on very flat tidal coasts the effects of biological activity can be very significant (Führböter, 1983). In some coastal areas small amounts of fine silt and clay are mixed with the sand and these can make the deposit cohesive and much more resistant to movement.

11.4.2 Bed features and roughness

The ripple patterns on the beach possess an unusual fascination and are described already in ancient literature. The beginning of the modern investigations of ripples under waves could be dated from the studies by Bagnold (1946) although reports of studies from more than a century ago exist. Bagnold produced ripples on a sand tray which he oscillated in its own plane under water. When the velocity slightly exceeded the threshold value a few grains were carried along by the peaks of the velocity and low ridges formed. The flow did not separate at these ridges which remained stable up to about twice the threshold velocity. These he called the *rolling ripples*. At higher velocities the ridges grew in height and shed vortices from their crests. These were called the *vortex ripples*. They are symmetrical in shape in the absence of net sediment transport and asymmetric in the direction of net sediment transport. Beyond a certain velocity the ripple height starts to decrease leading again to rolling ripples (where flow does not separate) and ultimately to *sheet flow* conditions over a flat bed (equivalent to the transition flat bed in unidirectional flow). Bagnold also observed that at small values of the ratio of orbital half axis A at the bed to wave length L, $A/L < \sim 0.5$, ripples formed which he named "brick pattern". In this case the lee vortex breaks up into a series of horseshoe vortices, i.e., it does not remain as a continuous transverse vortex. The ripple troughs now show a transverse wave pattern. The brick pattern disappears when significant amounts of sediment move over the crests.

The beautiful long-crested (two-dimensional) ripple patterns are mostly found in shallow water where the waves tend to become more regular and the orbital diameter of water movement at the bed, $d_o = 2A$, is large, or under pure swell conditions. Usually, the waves in deeper water are irregular, almost never sinusoidal and often three-dimensional. These waves produce bed features which are more three-dimensional and smaller in size.

Very irregular patterns of highly mobile bed features are frequently found in the surf zone. These are up to a few hundred mm high, a few metres in "wave length" as well as in crest length, that is, the pattern is generally three-dimensional. The crests are well rounded and appear to shed no vortices. The three-dimensionality of these features is linked with three-dimensional water movements near the bed giving rise to intermittent vortices with vertical axes which lift sediment to appreciable heights above the bed.

The vortex ripples are the dominant form in coastal regions and they have a major effect on bed roughness and on suspension of sediment. Sand is carried during each half cycle of the orbital velocity over the ripple crest and into the vortex which has velocities of the same order of magnitude as the maximum orbital velocity at the bed. Some of the sand settles in the lee of the ripple, some can be observed (in laboratory) to pass over the ripple crest and vortex as a thin sheet of sediment laden flow and some is *trapped in the vortex*. Nielsen (1984,1985) deduced that the effectiveness of trapping increases with u_m/w until after $u_m/w > 10$ the trapping is fully effective, where w is the fall velocity of grains. On reversal of the flow direction the sediment laden vortices are carried into the flow, Fig. 11.35. These vortices, with high concentrations of sand, can rise a few hundred mm before breaking up.

The analytical description of ripples under wave motion is complex and incomplete as can be seen, for example, from papers by Sleath (1975,1976). A detailed discussion of theories of wave generated ripples is given by Hedegaard (1985). Because of the complexities most of the reported studies concentrate on empirical or semi-empirical descriptions, usually starting with dimensional analysis. By assembling the variables, which clearly affect the problem, dimensional analysis readily shows that a variable X will be a funciton of dimensionless groups of parameters, of the form

$$X = f\left[\frac{u_m d}{\nu}, \frac{\rho u_m^2}{(\rho_s - \rho)gd}, \frac{\rho_s}{\rho}, \frac{A}{d}\right] \qquad (11.171)$$

where u_m is the maximum value of the orbital velocity and A the half axis of the orbital excursions. The groups are the familiar Reynolds number, the Shields parameter in terms of u_m, density ratio and the length ratio A/d. The second group has become known as the mobility number

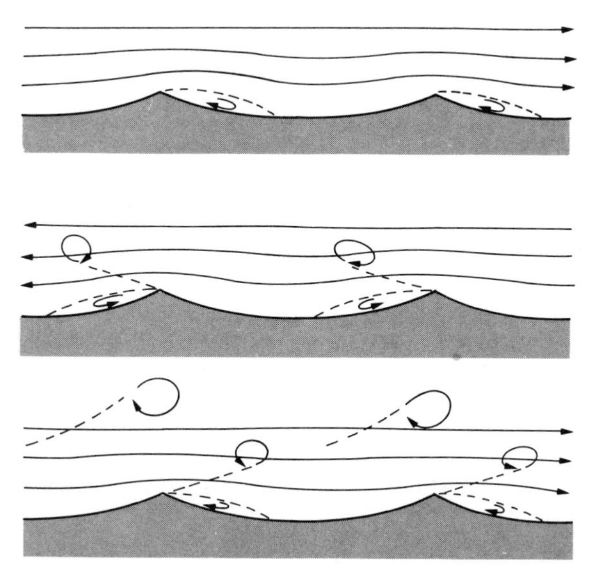

Fig. 11.35. Diagrammatic illustration of vortex ripples.

$$\psi = \frac{\rho u_m^2}{(\rho_s - \rho) g d} = \frac{(A\omega)^2}{\Delta g d} \tag{11.172}$$

where $\omega = 2\pi/T$. If the bed shear stress is expressed as $\tau_o = \frac{1}{2} f_w \rho u_m^2$ then the Shields dimensionless shear stress becomes

$$\theta = \tfrac{1}{2} f_w \psi \tag{11.173}$$

The data on initiation of ripples has been fitted with expressions like $\psi_c = C D_*^a (A/d)^b$ where D_* is the dimensionless grain size, but the exponents and constant vary widely. The proportionality of ψ_c to $(A/d)^a$ is, however, more distinct. The exponent $a \cong 1/3$ at small values of A/d and $a \cong 2/3$ at large value of A/d. Such an approximate relationship can be drawn on log-log paper from ($A/d = 10$, $\psi_c = 2.4$) with a slope of $a = 1/3$ to about $A/d = 5000$ and continued at $a \cong 2/3$ to $A/d = 10^4$, $\psi_c \cong 62$. A flat transition curve extends from $A/d = 100$ to 2000.

Mogridge and Kamphuis (1972) wrote eqn (11.171) as

$$X = \left[f \; \frac{\Delta g d^3}{\nu^2} \, , \; \frac{d}{\Delta g T^2} \, , \; \frac{\rho_s}{\rho} \, , \; \frac{d_o}{d} \right] = f\left[X_1, \; X_2, \; X_3, \; X_4 \right] \tag{11.174}$$

The first term is now the dimensionless grain size D_* but it is difficult to give a physical interpretation for the second group. In the last parameter the authors added the net distance of water movement during the wave period to $d_o = 2A$. They concluded from their data that viscous effects were negligible but that density ratio was significant with short period waves. They presented their data as λ/d and η/d versus X_2 as shown in Fig. 11.36 and Fig. 11.37 and as design curves λ/d and η/d against d_o/d with X_2

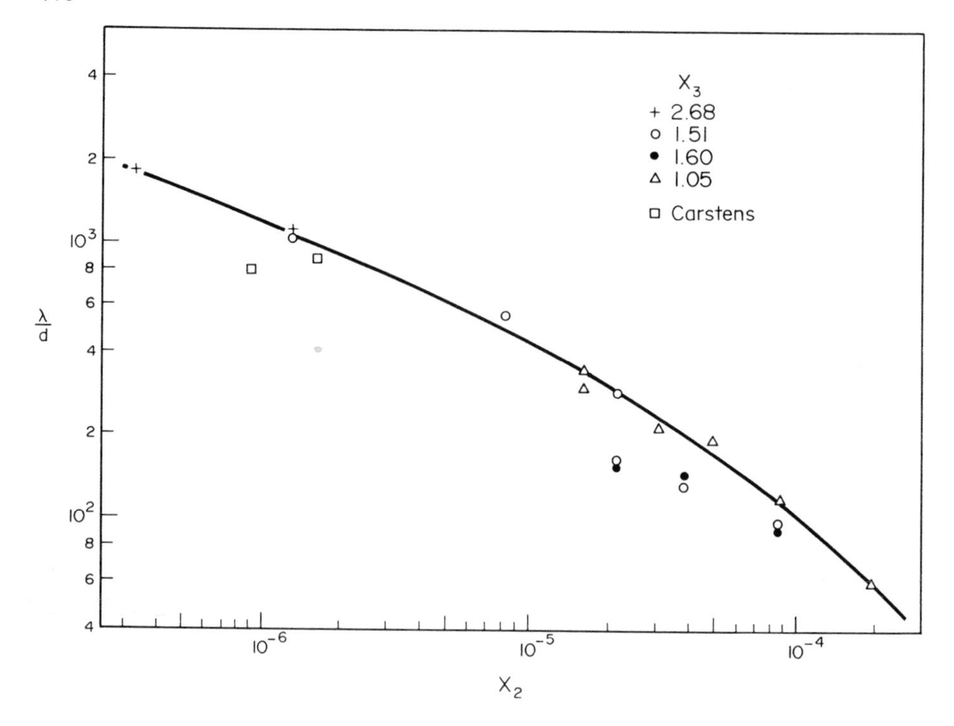

Fig. 11.36. Maximum values of ripple length λ/d versus $X_2 = d/\Delta gT^2$, after Mogridge and Kamphuis (1972).

as parameter, Fig. 11.38 and Fig. 11.39, where λ and η are the ripple length and height, respectively.

Generally, ripple length and height increase initially with length of orbital path at the bed up to a certain value. The ripple length then becomes proportional to grain size and independent of orbital path length. These are conditions as in unidirectional flow where $\lambda \cong 1000\ d$ is used as a first estimate. The reported data on ripple length as a function of orbital diameter fall within the band $0.65 \leq \lambda/d_o \leq 1.0$ but where the functions curve away (Fig. 11.38) they show a maximum and then fall back to a lower, more or less constant (horizontal) value for large values of d_o/d. However, around the maximum region the scatter in data is large, particularly for fine sands where the data band width can vary by a factor up to five. The scatter is less for coarse sands.

With increasing orbital velocity the ripple height gradually decreases to post vortex rolling ripples with $0.036 < \lambda/d_o < 0.059$ and disappear altogether. This occurs according to Dingler and Inman at $\psi \cong 240$. Chan et al. (1972) expressed the onset of sheet flow by

$$\psi_s = 23.2\ D_*^{-1/3}(A/d)^{1/3} = 18.4\ D_*^{-1/3}(d_o/d)^{1/3} \quad (11.175)$$

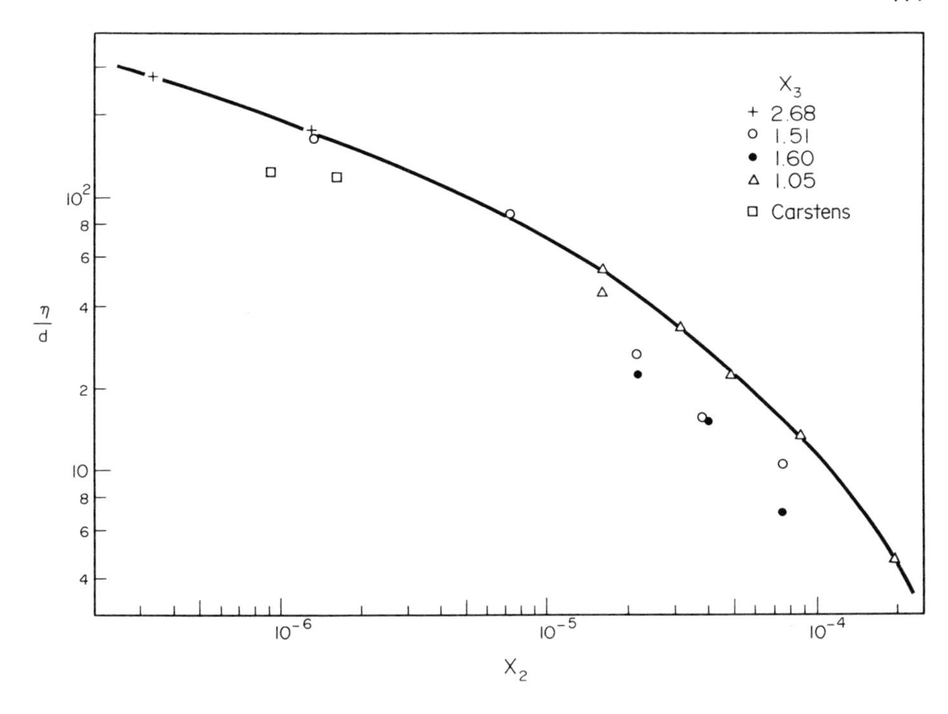

Fig. 11.37. Maximum values of ripple height η/d versus $X_2 = d/\Delta gT^2$, after Mogridge and Kamphuis (1972).

for $1.04 < \rho_s/\rho < 5.1$ and $0.16 < D_* < 160$. The criterion for inception of sheet flow conditions by Manohar (1955) is

$$\frac{u_m^2}{(S_s - 1)gd} = \frac{2 \times 10^3}{(u_m d/\nu)^{\frac{1}{2}}} \tag{11.176}$$

and by Komar and Miller (1974)

$$\frac{f_w u_m^2}{2(S_s - 1)gd} = \frac{4.4}{(u_m d/\nu)^{1/3}} \tag{11.177}$$

Lofquist (1978) reported from analysis of available data that stable two-dimensional ripples with nearly constant λ/A and η/A, independent of u_m or T form for $\psi < 30$, with

$$0.55 < \lambda/d_o < 0.8 \qquad \overline{\lambda/d_o} \cong 2/3 \tag{11.178a}$$

$$0.125 < \eta/\lambda < 0.22 \qquad \overline{\eta/\lambda} \cong 0.16 \tag{11.178b}$$

His own experiments in a water tunnel yielded with 0.18 mm sand stable two-dimensional features when $\psi < 21.3$, $u_m \leq 0.252$ m/s or

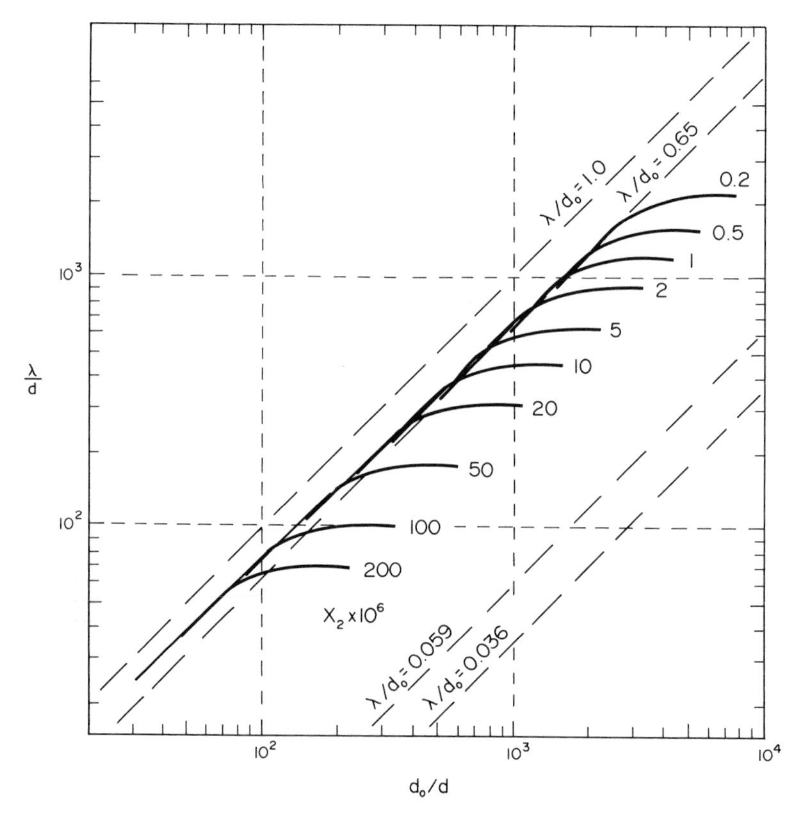

Fig. 11.38. Design curves for ripple length, after Mogridge and Kamphuis (1972).

$d_o/d \leq 0.444$ T, and $\eta/A = 0.18$ or $\eta/d_o = 0.09$. The final ripple form was always stable with $d = 0.55$ mm sand and $\eta/d_o = 0.12$.

Miller and Komar (1980) proposed from laboratory data that the maximum ripple length, λ_m, when the proportionality to orbital diameter ceases to be valid, is a function of grain size only

$$\lambda_m = 3.07\, d_{(mm)}{}^{1.68} \qquad (11.179)$$

but the expression tends to underestimate.

It should be noted that ripples have been studied with essentially uniform sands. The relationships between ripple geometry and grain size will become more complex as the grain-size distribution broadens.

According to Fig. 11.38 the ripple length λ/d is a function of $d/\Delta gT^2$ and d_o/d. Yalin and Karahan (1978) combined the parameters X_1 and X_2 of eqn (11.174) into $\nu T/d^2$ and interpreted it to represent the ratio of laminar and turbulent boundary layer thicknesses. Accordingly,

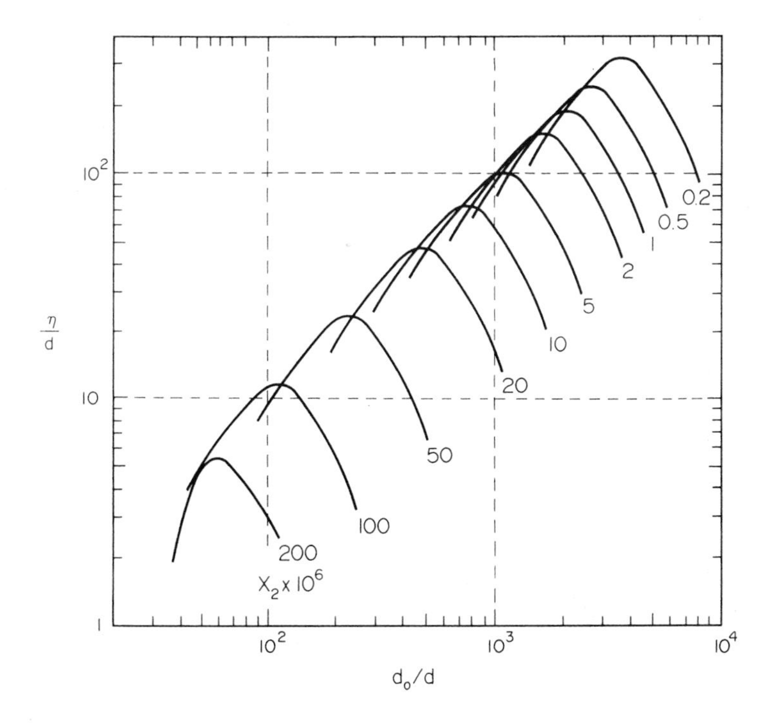

Fig. 11.39. Design curves for ripple height, after Mogridge and Kamphuis (1972).

$$\lambda/d = f(\nu T/d^2,\ d_o/d) = f(Re) \tag{11.180}$$

since d_o/T is a velocity, i.e. $\lambda/d = f(Re)$ as for ripples in unidirectional flow. This is in keeping with the constant ripple length under long waves, i.e., where the conditions approach those of unidirectional flow. They further normalised the family of curves in Fig. 11.38 with ∇ and Λ_o, which are the values of d_o and λ at the point where the individual curves start to deviate from the λ proportional to d_o line. Figure 11.40 shows their replot of data. They then replotted the data as $\Lambda_o/\nabla_o = f(\nu T/d^2)$, Fig. 11.41, and produced a single function relationship. Thus, with known ν, R and d Fig. 11.41 yields Λ_o and the relationship

$$\nabla = \text{constant}\ \Lambda_o \tag{11.181}$$

the value of ∇, where the constant is in the range of 1.6 to 1.0 and the authors used 1.0, although 1.5, i.e. $\overline{\lambda/d_o} \cong 2/3$, appears to be more widely supported. With known d_o/∇_o Fig. 11.40 yields λ/Λ_o and λ for given conditions.

The data from field and laboratory show distinct differences. The ripples in the field are invariably smaller. This has been explained by the irregular nature of waves in the sea. Nielsen

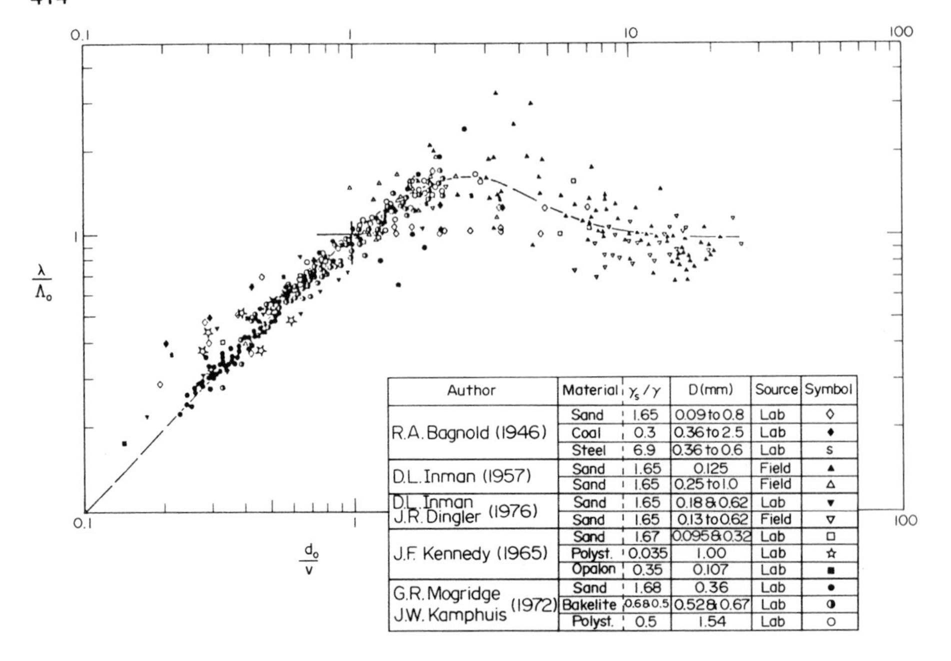

Author	Material	γ_s/γ	D(mm)	Source	Symbol
R.A. Bagnold (1946)	Sand	1.65	0.09 to 0.8	Lab	◇
	Coal	0.3	0.36 to 2.5	Lab	◆
	Steel	6.9	0.36 to 0.6	Lab	s
D.L. Inman (1957)	Sand	1.65	0.125	Field	▲
	Sand	1.65	0.25 to 1.0	Field	△
D.L. Inman J.R. Dingler (1976)	Sand	1.65	0.18 & 0.62	Lab	▼
	Sand	1.65	0.13 to 0.62	Field	▽
J.F. Kennedy (1965)	Sand	1.67	0.095 & 0.32	Lab	□
	Polyst.	0.035	1.00	Lab	☆
	Opalon	0.35	0.107	Lab	■
G.R. Mogridge J.W. Kamphuis (1972)	Sand	1.68	0.36	Lab	●
	Bakelite	0.68 0.5	0.52 & 0.67	Lab	◑
	Polyst.	0.5	1.54	Lab	○

Fig. 11.40. Normalised ripple length, λ/Λ_o according to Yalin and Karahan (1978).

(1981) analysed the available data and fitted functions to the data sets which are reasonably convenient to use. He plotted the data on ripple steepness, η/λ, against grain roughness Shields parameter, $\theta' = \frac{1}{2}f_w\psi$, and concluded that this reduced scatter compared to a plot in terms of ψ. The functions fitted are

Laboratory Data (a)	Field Data (b)	
$\lambda/A = 2.2 - 0.345\ \psi^{0.34}$	$\lambda/A = \exp\left[\dfrac{693 - 0.37\ \ln^8 \psi}{1000 + 0.75\ \ln^7 \psi}\right]$	(11.182)
$\eta/A = 0.182 - 0.24(\theta')^{1.5}$	$\eta/A = 0.342 - 0.34(\theta')^{0.25}$	(11.183)
$\eta/A = 0.275 - 0.022\ \psi^{\frac{1}{2}}$	$\eta/A = 21\ \psi^{-1.85}$	(11.184)

where η/λ is constant and independent of θ' for $\theta' < 0.20$ and λ/A is constant for $\psi < 20$. According to eqn (11.183(a) the sheet flow conditions are reached at $\theta' \cong 0.83$ which is generally a little low. Laboratory observations suggest $\theta' > \sim 1$, i.e. the factor 0.24 could be reduced. For field data A was evaluated in terms of the significant wave height $H_s \cong \sqrt{2}\ H_{rms}$.

The differences can be illustrated with the aid of a numerical example: T = 6 s, A = 0.5 m, u_m = 0.524 m/s, d = 0.2 mm, ψ =

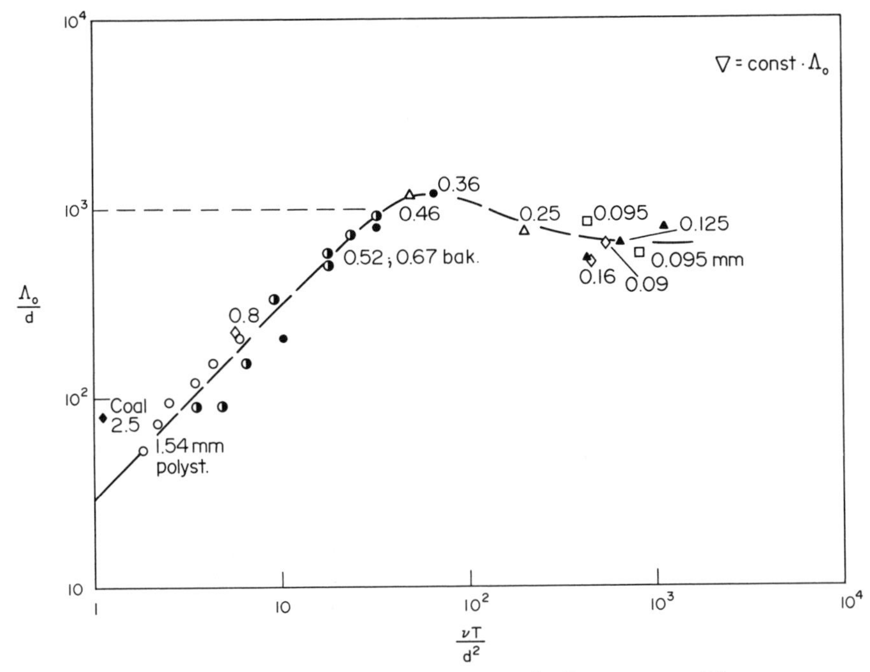

Fig. 11.41. Ripple data plotted as Λ_o/d versus $\nu T/d^2$, according to Yalin and Karahan (1978).

84.8, $f_w' = 0.0099$, $\theta' = 0.42$, $X_2 = 0.34 \times 10^{-6}$. Thus from Figs 11.38 and 11.39 $\lambda/d \sim 1700$, $\eta/d \sim 130$, $\lambda \cong 0.34$ m, $\eta \cong 0.026$ m and $\eta/\lambda \cong 0.0765$.

From eqns (11.182) to (11.184):

Laboratory:		Field:	
$\lambda/A = 0.6386$	$\lambda = 0.319$ m	$\lambda/A = 0.1247$	$\lambda = 0.0623$ m
$\eta/\lambda = 0.1164$	$\eta = 0.037$ m	$\eta/\lambda = 0.0681$	$\eta = 0.0042$ m
$\eta/A = 0.0724$	$\eta = 0.036$ m	$\eta/A = 0.0057$	$\eta = 0.0028$ m

Ripples are the dominant but not the only bed features present under wave motion. Large regular features are called, *dunes*, *megaripples*, and *sand banks*. Dunes or megaripples are features created by currents, analogous to those in unidirectional flow. The main difference in the formation of these large features on the sea bed from those in rivers, is that in the sea the sediment is made mobile, or its threshold is lowered, by the agitation arising from wave motion. Thus, quite weak currents can transport it and build dune-like features. A special case are the dunes in tidal channels, in estuaries and their *off*shore extensions. In tidal channels the dunes look and behave like giant vortex ripples, with the orbital motion fixed by the period of the tide, as illustrated by the sonar record by Stehr (1975), Fig. 11.42. The sediment flow over the crests is clearly discernible. These dunes vary is size with flow conditions. Giant

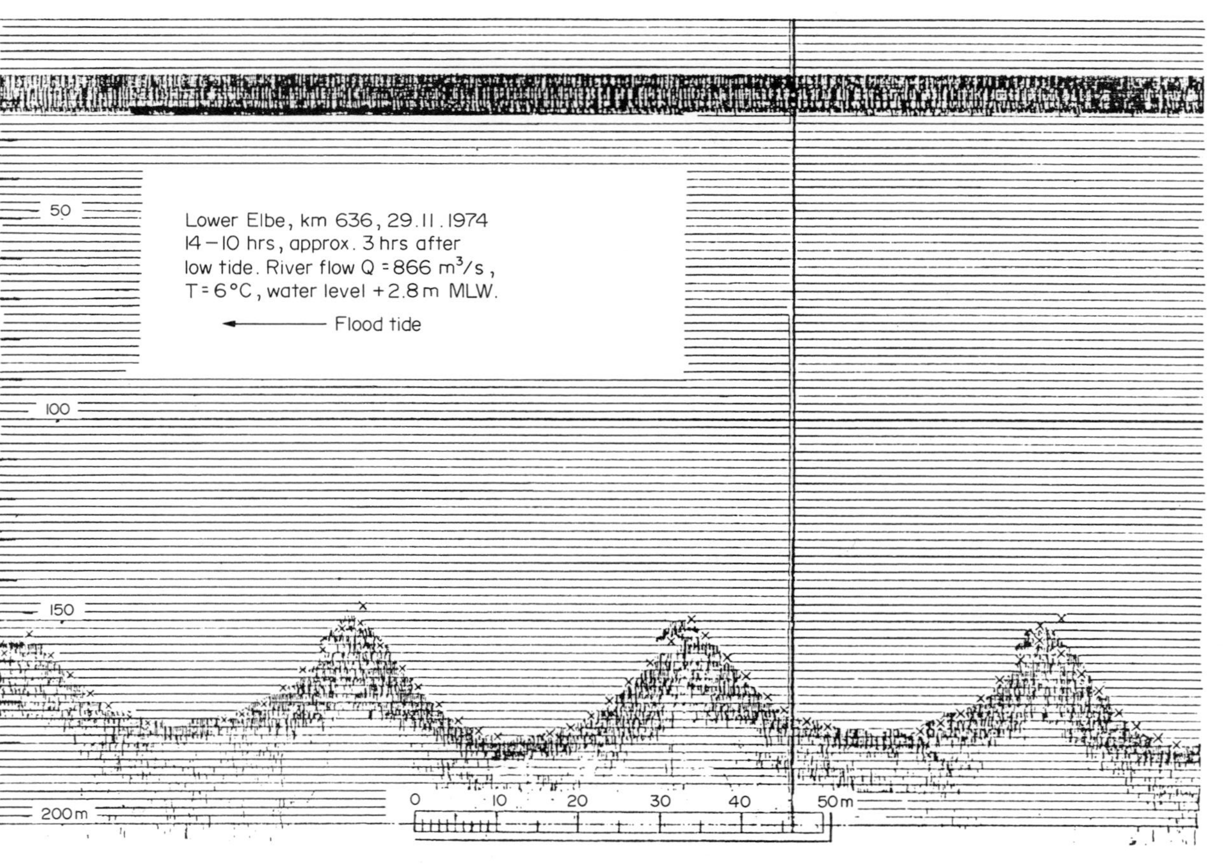

Fig. 11.42. Echogram from the Elbe River, Stehr (1975).

ripple patterns in offshore tidal channels are also reported by Pasenau and Ulrich (1974), St Andrew Bay, Florida, etc. About 25 m high features were reported in Sofala Bay, East Africa and very large sand waves have been measured in the Irish Sea. The banks are not just confined to sands. Moving gravel banks in more than 12 m of water at the North Sea end of the Straits of Dover off Deal are well known.

Most coastlines display bars which run parallel to the coast with spacings of about three wave lengths of the dominant sea conditions. These have been related to the interaction of the sea bed with the higher order components of wave energy. Such longshore bars are most prominent on coasts with negligible tidal range (Baltic and Black Sea, Great Lakes) and are discussed by Boczar-Karakiewicz et al. (1981,1983) and Boczar-Karakiewicz and Bona (1981).

Bed features with less regular patterns are referred to as sand banks. These are related to the distribution of wave energy and currents and, in principle, indicate areas of deposition. Particularly complex topographical patterns are a feature of shallow coastal areas, like the coast of the North Sea from Denmark to Holland, but similar tidal or near tidal areas can be found in most parts of the world. These have sand banks which translate towards the shore. The banks are initiated at the outer surf zone of the shallow area by a sand supply from the sea. The sand is eroded at the seaward face and deposited at the lee of the bank which leads to the slow landward translation. The complex interactions of waves, refraction and wave induced currents lead to intricate shapes. Some details of such banks in the Wattensee are given by Göhren (1975).

The bed features from a hydrodynamic point of view are roughness elements and therefore have an important influence on current patterns. They also are an important source of turbulence and eddy production and hence play a significant part in mobilizing and suspending of sediment. The estimating of wave friction factor f_w requires A/k_s where k_s is the effective roughness height of the bed. Various expressions have been proposed for this, for example, Swart (1976) proposed the empirical relationship

$$k_s = 25\ \eta^2/\lambda \tag{11.185}$$

Grant and Madsen (1982)

$$k_s = 27.7\ \eta^2/\lambda + (S_s + c_M)160(\sqrt{\theta'} - 0.7\ \sqrt{\theta_c})^2 d \tag{11.186}$$

where S_s is the specific gravity of grains and c_M is the mass coefficient (≅0.5). Nielsen (1981,1983) proposed

$$k_s = 8\eta/\lambda + 190(\theta' - 0.05)^{\frac{1}{2}} d \tag{11.187}$$

where in eqns (11.186) and (11.187) the additive terms account for roughness due to sediment in motion. Raudkivi (1988) fitted the data by Carstens et al. (1969) with

$$k_s = 16\eta^2/\lambda + 0.16\ u_m^{2.25} \tag{11.188}$$

where ripple height and length could be expressed with eqns (11.182) to (11.184), u_m is in m/s and k_s, η and λ are in m. In eqn (11.188), as the ripple height goes to zero, the first term goes to the grain roughness height, e.g. 2.5 d, rather than zero.

11.5 Suspension of Sediment

The process of sediment suspension by wave action differs substantially from that in unidirectional flow, particularly in the surf zone. In unidirectional flow suspension is the result of diffusion of turbulence from the bed. Under wave action, in addition to diffusion of turbulence, convection of sediment by vortices (macro turbulence) plays a major part. In the surf zone the input of turbulence energy from the dissipation of wave energy can overshadow the turbulence production in the bottom boundary layers.

It is customary to talk of suspended sediment concentration profiles but these refer only to long term average values. Due to the high levels of macro turbulence short term measurements differ substantially. Average values have to cover at least 100 waves or about 15 minutes. In connection with sediment concentration the expression "average over a wave period" is meaningless. Apart from the average with respect to time, spatial averaging is necessary whenever bed features are present. The spatial average should cover an area representative of a few bed features. In addition, the problems with measurement are manifold. The instrumentation available to measure sediment concentrations is cumbersome and most of these cannot be used in heavy seas, that is, in conditions for which the data are needed most.

11.5.1 Suspension offshore of breakers

Offshore of the breakers the oscillatory movement of water is mostly over a bed covered with vortex ripples. The vortices swept upwards from the crests of the ripples at each reversal of velocity are laden with sediment and the instantaneous concentration distribution over the bed varies substantially with time and location, Fig. 11.43. The sediment is put in suspension by the convective transport by vortices. The effectiveness of trapping of sand in the vortices increases with u_m/w and the trapping becomes fully effective for $u_m/w > 10$ (Nielsen, 1984) where w is the fall velocity of grains. The convection process of sand upwards is rapid, whereas the falling out is gradual at the fall velocity w. This, according to Fredsøe et al. (1983) leads to a layer of suspension above the bed, about the thickness of the boundary layer, in which the concentration is more uniform. However, measurements close to bed are rare.

The time average suspended sediment distribution outside the breakers is confined to a relatively thin layer over the bed, usually less than half a metre thick. Across this layer the average concentration decreases by three to four orders of magnitude. The suspended matter above this bottom layer consists mainly of fine fractions and is somewhat similar to the wash load in rivers. The lower layer of concentration can usually be fitted with

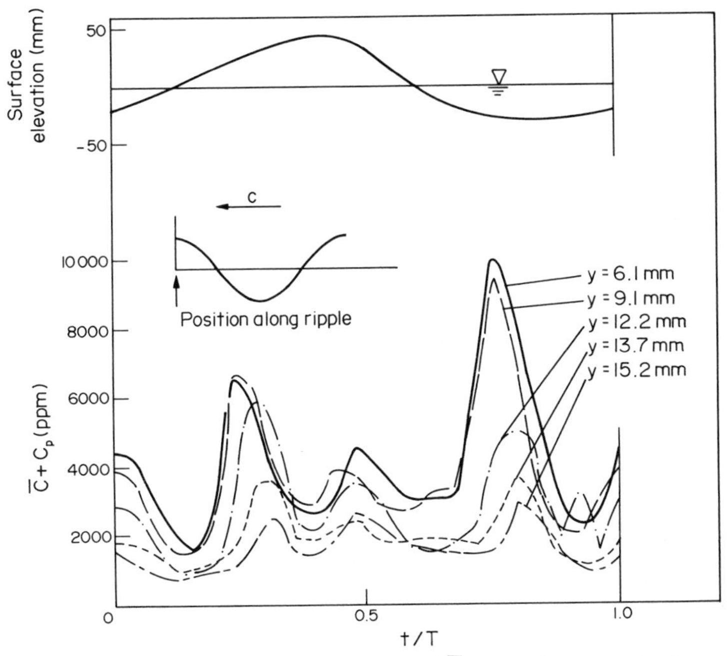

Fig. 11.43. Temporal variation of $\bar{c} + c_p$ above a ripple peak in the horizontal bed experiments (Kennedy and Locher, 1972). (c_p is the periodic time-dependent component of c).

$$\bar{c} = \bar{c}_o e^{-az} \tag{11.189}$$

where z is the vertical height above bed level, a is a decay parameter with dimension m^{-1} (a mixing length) and the bar refers to time-averaged values. This simple expression is very convenient for use in various calculations. MacDonald (1977) concluded from his experimental study that the slope of the concentration profile is approximately the same as that of the rms-values of the vertical velocity fluctuations. Field data as well as wave tunnel measurements indicate that over the usual range of beach sand sizes the average concentration profile is effectively independent of grain size. Field data by Nielson (1984) indicate a decay rate a ≅ 9 over well developed vortex ripples (H = 0.5 m, h = 1.45 m, T = 6.1 s, d = 0.45 mm, θ' = 0.31, η = 0.10 m, λ = 0.55 m, non-breaking waves). Measurements reported by Kossyan et al. (1982) and Antsyferov et al. (1983) in Black Sea show a ≅ 9.8 to 11.5 (h = 3.6 m, $\bar{H}$ = 0.64 m, $\bar{T}$ = 3.8 s). Typical concentration profiles are shown in Fig. 11.44. The profiles were presented at approximately 3 hour intervals over three days and show remarkable consistency in the slopes while the concentrations at a given elevation vary up to about one order of magnitude (factor of 10). The slopes of concentration profiles at stations 3, 4 and 5 and between 2 and 3 from measurements in 1977 and 1978 are all within $7.1 \leq a \leq 14.7$. The grain size in this lower suspension layer decreased from the

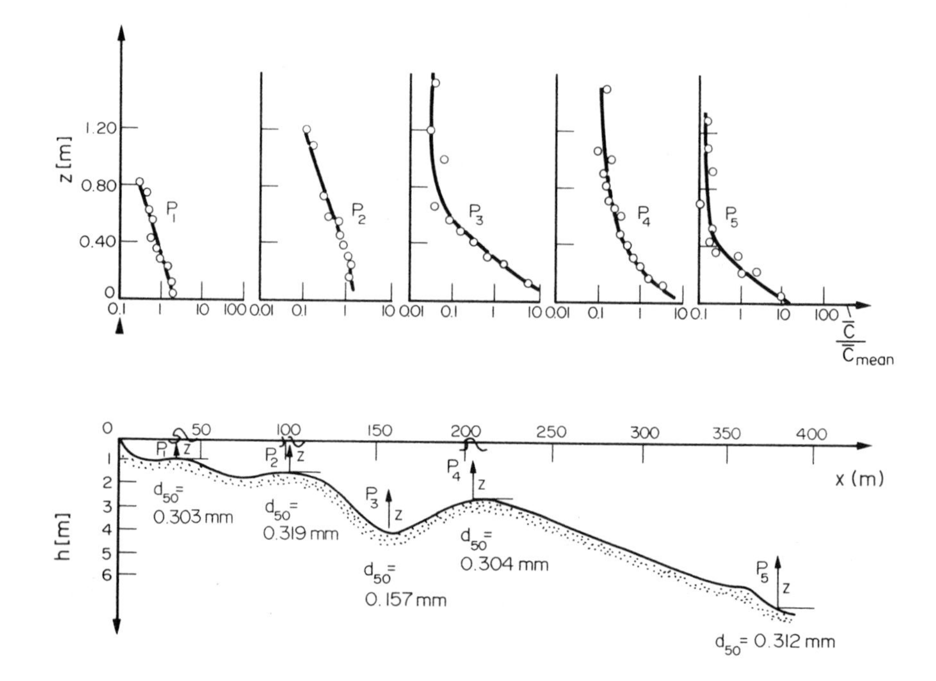

Fig. 11.44. Vertical profiles of relative concentration during a storm at the Black Sea coast (Kamchiya). Some of the highest waves broke at P_4 but frequent breaking was at P_2, Antsyferov et al. (1983).

mean of over 300 μm at the bed to about 70 μm and remained essentially constant throughout the depth. However, if the fine sediments have a broad range of grain-size distribution a size distribution can also occur in this upper layer due to differences in the fall velocity. Like the wash load of rivers the concentration of this upper layer depends primarily on the availability of the fine and colloidal fractions of sediment rather than on hydraulic conditions. Over relatively clean sand beds the concentration of the fines is almost independent of sea conditions whereas over silty beds it is very strongly related to wind and wave conditions. Measurements over the silty beds of the shallow seas off the East Frisian coast, West Germany, correlate well with wind speed. For wind velocities $V > 2$ m/s the data can be described by

$$\log c = -1.9 + 4 \log V \tag{11.190}$$

where c is in mg/ℓ and V in m/s. Notice the fourth power relationship to wind speed. The data scatter at low wind speeds by about ± 50% and at higher speeds about ±20%. Maximum wind speed 16.5 m/s.

In order to separate this suspension from the suspension supported by diffusion of momentum from bottom boundary layer the term *diffused or dispersed sediment layer* is suggested.

At high applied shear stresses ripples are wiped out and conditions known as *sheetflow* develop. Then the concentration distribution at the bed shows two distinct regimes, as illustrated by wave tunnel data in Fig. 11.45 and Table 11.5. Under sheet flow conditions the lower layer of sediment is carried by the bed through momentum exchange from grain to grain and grain to bed (dilatation stress). This layer represents a *layer of mobilized sediment*. Above it is a *layer of suspended sediment* which is maintained by the diffusion of turbulence produced at the bed.

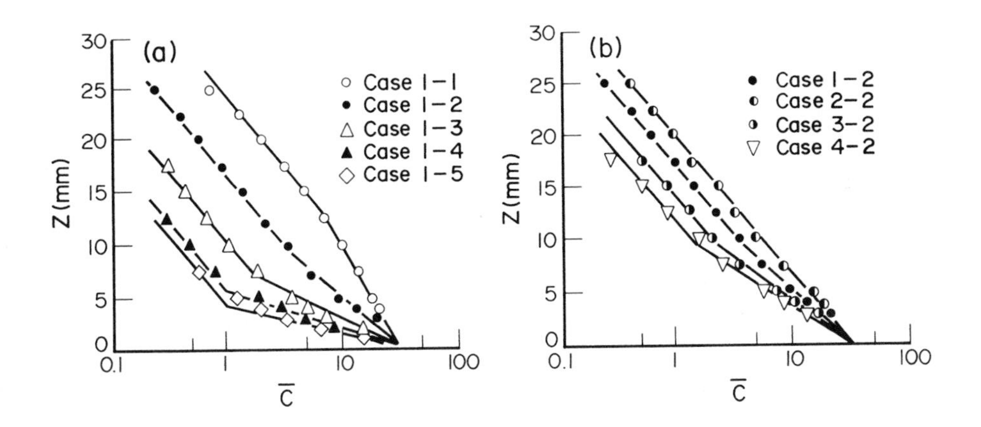

Fig. 11.45. Sediment concentration data in g/ℓ by Horikawa et al. (1982) at sheet flow conditions; (a) constant orbital path length of 0.72 m and 0.2 mm sand, (b) constant maximum value of orbital velocity u_m = 1.08 m/s.

In principle, the concentration distributions of suspended sediments seaward of breakers are related to the production of turbulence at the bed and are correlated with the velocities at the bed. The data by Horikawa et al. (1982) suggest that the decay parameter 'a' (eqn 11.189) in the sheet flow layer is related to the maximum value of the orbital velocity, u_m, at the bed as

$$a = 315\ u_m^{-4} \tag{11.191}$$

i.e., 'a' is inversely proportional to the square of kinetic energy per unit mass of water at the bed. The sheet flow thickness (change of slope), z_1, is given as

$$z_1 = 8 \times 10^{-3} u_m^{2.25} \tag{11.192}$$

where u_m is in m/s, a in m^{-1} and z in m. The reference concentration c_o at the bed remained essentially constant and independent of u_m at a very low value of 33.5 g/l. The distinct discontinuity in the concentration profile from stationary bed to mobilized sediment layer appears to constitute the usual case.

Above the mobilized sediment layer the a-value remains essentially constant and the extrapolated value of reference concentration, c_{ou}, varies with u_m. The above data yield

$$\bar{c}_{os} = 10\, u_m^{7.16} \tag{11.193}$$

TABLE 11.5: Wave tunnel data by Horikawa et al. (1982).

Case No	d mm	T s	u_m m/s	u_m^+ m/s	A m	f_w'	θ'	ψ	a m^{-1}
1-1	0.2	3.6	1.27	1.256	0.72	0.0090	2.24	495.2	122.5
1-2		4.2	1.08	1.077			1.62	358.1	242.4
1-3		4.8	0.95	0.943			1.25	277.1	413.1
1-4		5.4	0.87	0.838			1.05	232.4	638.4
1-5		6.0	0.76	0.754			0.80	177.3	837.3
2-2	0.5	3.4	1.08	1.072	0.58	0.0124	0.92	143.3	182.7
3-2	0.7	2.6	1.08	1.063	0.44	0.0151	0.80	102.3	287.8
4-2	4.0	2.0	1.08	1.068	0.34	0.0352	0.46	165.1	328.9

+values calculated from A and T. Specific gravity 2.66 except 4-2 which was 1.18.

where $\bar{c}_{os}$ is the value of concentration as obtained by extrapolation of the suspended sediment regime to the bed level in g/ℓ and u_m is in m/s. Notice that the concentration is extremely sensitive to u_m.

A similar sensitivity is indicated by the expression proposed by Nielsen (1986) from field data, according to which $\bar{c}_{os} \propto u_m^6$. This explains why field measurements of concentration scatter so excessively, as illustrated by the data by Kossyan et al. (1982), Fig. 11.46. The slopes of the profiles are seen to be essentially constant (9.8<a<11.5) while the magnitude of concentration and the $\bar{c}_{os}$-values vary appreciably under waves of $\bar{H}$ = 0.64 ± 0.1 m. The waves over the about 36 h measuring period at station 4 varied from H_s = 0.54 m, T_s = 4.55 s to H_s = 0.747 m, T_s = 3.56 s. The highest concentration relates to the 4.55 s waves. Notice also the mean grain size distribution.

Thus, outside the surf zone three sediment regimes can be recognized:

(i) The bed load or mobilized sediment layer carried by the dilatation stresses,

(ii) the suspended sediment layer supported by diffusion of turbulence from the bed boundary layer, and

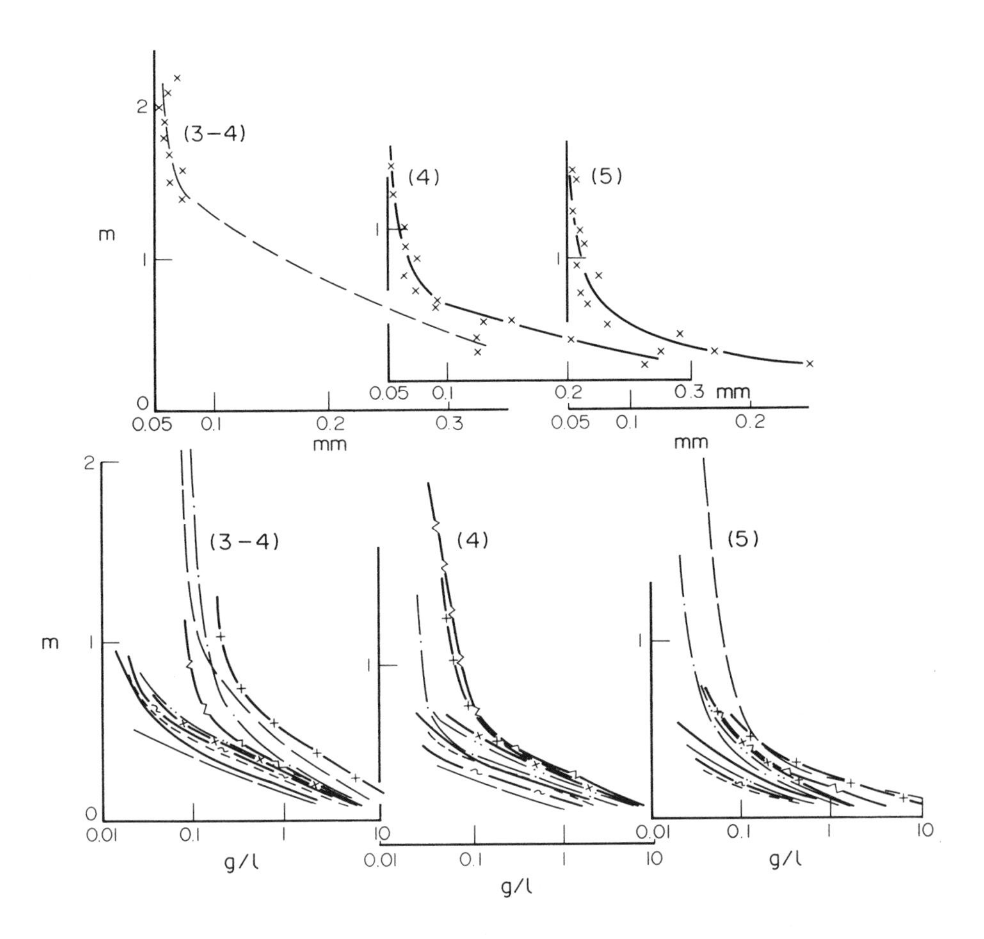

Fig. 11.46. Concentration and mean grain size profiles of suspended sediment at three stations seaward the breakers in the Black Sea. Station [3-4] at 140 m from shoreline, 3 m water depth. Station [4] at 165 m from shoreline and 3.7 m water depth and Station [5] at 220 m and 5.6 m water depth. Sea bed slope approximately 1:40. (Kossyan et al. 1982).

(iii) The dispersed sediment layer of a low concentration of the very fine fractions of sediment.

The mobilized sediment layer is thin and recognizable only at sheet flow conditions. Over a rippled bed the deterministic system of injection of sediment into the fluid above the bed by the vortices and the "raining out" of it at fall velocity dominates the lowest layer. The suspended sediment layer is usually up to about half a metre thick and the rest of the depth is occupied by the dispersed sediment.

It is necessary to establish more reliable relationships for the description of the suspended sediment distributions in terms of the orbital velocity at the bed, i.e., in terms of wave height and period.

11.5.2 Suspension in the surf zone

In the surf zone the dissipation of wave energy injects large amounts of turbulence into the water. This turbulence dominates the suspension process and usually overshadows the contribution of turbulence produced in the boundary layer at the bed. The breakers have a variety of forms and the breaker zone of a random sea covers a significant part of the surf zone width in which additional breaking of the reformed waves takes place. The breaker forms range from pure spilling breakers ($H_b/h_b < \sim 0.96$) to pure plunging breakers ($H_b/h_b) > \sim 1.07$) with transitional breakers ($0.91 < H_b/h_b < 1.12$) between these. The surging breakers are rare. These are confined to steep beaches and very flat waves, although they do occur in form of secondary breakers in bar and trough type surf zones with steep foreshore slope.

In the front of a plunging breaker the seaward velocities at the bed are about 10% of the wave celerity, about ten times greater than in the trough in front of a spilling breaker, and there is a strong upward current under the steep front of the breaker. The plunging sheet of water superimposes a strong downward component on the shoreward movement of water. The bulk of the energy of the plunging breaker is transferred to turbulence at the plunge line which excavates and distributes large quantities of sediment. The plunging process also entrains large amounts of air. At the plunge line the sediment concentration can be very high. Sand, and even gravel, can be seen swirling in the surface layer and sediment laden plumes of water spout at times well above the local water level. These plumes are driven by the escaping air pockets. Plunging breakers generally produce the highest concentrations. These are highest in the vicinity of the plunge line and decrease with distance shoreward until a new boost is given by the inshore breaking of the reformed waves.

In spilling breakers the velocities in the trough ahead of the breaker are much lower and there is little upward movement under the wave front. The spilling breaker action is on the surface and turbulence diffuces downwards. In addition, spilling breakers tend to occur one after another, several over the width of the surf zone. Thus, the turbulence in the surf zone increases in shorewards direction.

Under strong breaker action the concentration distribution is totally governed by the turbulence from wave energy dissipation. The strong eddies swirl clouds of sediment through the depth and overshadow any effects from diffusion of turbulence produced at the bed. However, in broad surf zones the distinct suspended sediment distribution profile features are frequently discernible in the measured concentration profiles. Parts of such broad surf zones may have ripples which contribute to suspension through the transport of sand by the vortices. The surf zone will also have areas where sheet flow conditions prevail with the corresponding mobilized sediment layer but no field measurements so close to the bed are known. A further feature of surf zones are the very mobile irregular three-dimensional bed features with spacing and crest lengths of a few metres and a few hundred mm high. The crests of these features are well rounded and flow does not separate but they do give rise to three dimensional movements and eddies with vertical axes which, like tornadoes, carry sediment high up into the flow.

The decay rate (a-value) as well as the magnitude of concentration should correlate with the dissipation of wave energy, i.e., at a given location

$$a(x) = f[D(x),\ \xi_o,\ h] \text{ and } \bar{c}_{os}(x) = f[D(x),\ \xi_o,\ h] \qquad (11.194)$$

where $D(x)$ is the wave energy dissipation or conversion (W/m²), $\xi_o = \tan\beta/(H_o/L_o)^{\frac{1}{2}}$, with $\tan\beta$ being the bed slope just outside the breakers, and h is the local water depth. The ξ_o-value relates $a(x)$ and $\bar{c}_{os}$ to breaker type. For a steady state the balance of energy flux can be written as $\delta/\delta x(Ec) = -D(x)$, where E is the wave energy density per unit surface area (Nm/m²) and c is the wave group celerity which in shallow water approximates to $(gh)^{\frac{1}{2}}$. As yet empirical data on $a(x)$ and $\bar{c}_{os}(x)$ as a function of wave parameters are lacking and should be accumulated with the aid of the large wave flumes. The energy conversion has been modelled primarily as that of a propagating bore (hydraulic jump). Significant contributions are papers by Le Méhauté (1962), Hwang and Divoky (1970), Svendsen et al. (1978), Battjes and Janssen (1978), Thornton and Guza (1983), Stive (1984), Dally et al. (1985) and Battjes (1986).

Little is known over the intensities or structure of turbulence in the surf zone but local energy concentrations can be very high. For example, measurements in the large wave flume in Hannover showed a 2.4 m high breaker reduced to 0.7 m over a distance of only 6 m, i.e., a conversion of about 32.7 hW/m or 5.45 kW/m². Using a very simplistic relationship, $D(x) = (1/2)\rho q^3$, where q is a representative turbulence velocity vector, implies $q \cong 2.2$mm/s or a velocity about 100 times greater than the fall velocity of a typical beach sand grain. In a spilling breaker energy dissipation occurs mainly in the bore with some diffusion of turbulence downward. Consequently, the turbulence intensity in a surf zone covered with spilling breakers is lower than in one with plunging breakers. Likewise, the suspended sediment content is lower as shown by the field measurements by Nielsen (1984) where for comparable conditions (runs 24 and 70) the following data were recorded:

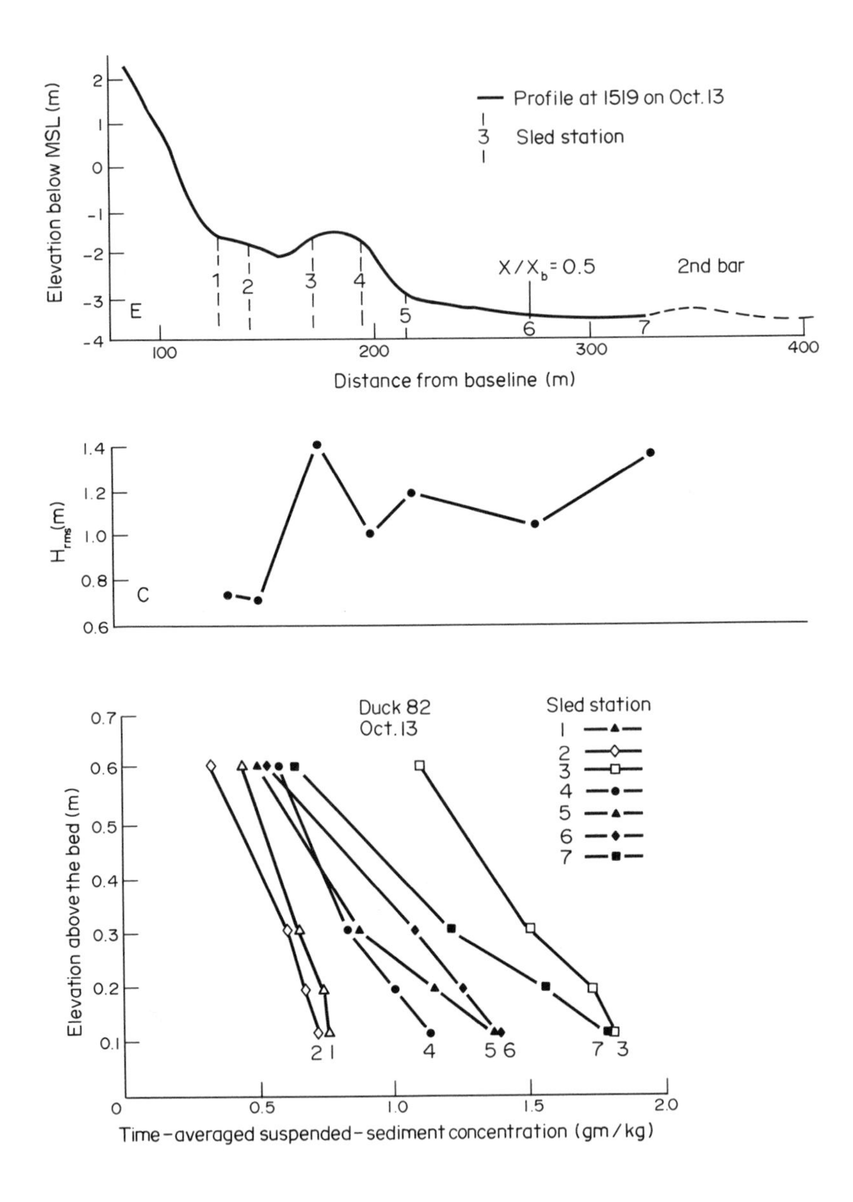

Fig. 11.47. Field data on suspended sediment concentration, Jaffe et al. (1984). First line of breakers was at about 600 m and the dominant breaker line at about 175 m, mean depth 2.4 m. Three or more breakers were present at one time.

24: Plunging breakers, $H_b = 0.70$ m, $T = 9.3$ s, $h = 1.50$ m

70: Spilling breakers, $H_b = 0.82$ m, $T = 9.2$ s, $h = 1.50$ m

and concentrations in ppm at 0.1 m vertical spacing starting with 0.1 m above the bed

24:	1500	655	391	137	64	89	84
70:	-	490	250	86	55	37	39

Air entrainment also plays a major role in lifting of sediment and in particularly high with plunging breakers. The effects of air entrainment on energy dissipation were discussed by Führböter (1971).

The detailed picture of suspension in the surf zone is extremely complex. In addition to the macro-turbulence produced by dissipation of wave energy complex current systems exist in the surf zone which interact with waves to create three-dimensional eddying and vortices. The bed level and bed geometry too can change rapidly. Changes of the order of 2 m in elevation have been reported in a single storm. Likewise, the sediment characteristics vary with wave conditions across the surf zone.

Kana (1979) plotted the measured concentrations in the surf zone against $x = (1-m)^4 h_b/H_b$, where m is the beach slope, e.g. m = 0.1 is 1:10. The concentrations were measured at 0.10, 0.30 and 0.60 m above the bed in the surf zone within 10 m of breakers which ranged up to 1.5 m in height. The sand had a $d_{50} = 0.22$ mm. Kana fitted the mean concentrations at 0.10, 0.30 and 0.60 m above the bed by

$$\bar{c}_1 (g/\ell) = 13.888 \exp[- 4.017(1-m)^4 h_b/H_b] \tag{11.195a}$$

$$\bar{c}_3 (g/\ell) = 4.832 \exp[- 3.597(1-m)^4 h_b/H_b] \tag{11.195b}$$

$$\bar{c}_6 (g/\ell) = 1.501 \exp[- 3.054(1-m)^4 h_b/H_b] \tag{11.195c}$$

Note, that h_b is defined as the water depth in the trough of the wave before breaking. The decay value, a, of the mean concentrati at the measuring station given by the data is

$$a = 4.1 - 1.66\ h_b/H_b \qquad (11.196)$$

This gives with $h_b/H_b \geq 1.28$ and $a \geq 2$ which is a too large a value for more severe wave conditions where $a \rightarrow 1$.

The vertical distributions were based on an average of up to five two litre rapidly drawn samples.

The results by Kana could be compared with field data by Jaffe et al. (1984), obtained during the final stages of a storm which generated waves in excess of 2 m. Initially the periods were 7 seconds but later as the storm moved offshore 12 to 15 seconds. The beach profile, the H_{rms} distribution and the concentration profiles are shown in Fig. 11.47. The data for Stations 5, 6 and 7 can be fitted with $a \cong 2.2$ and Stations 1, 2, 3 and 4 with $a \cong 1.1$. This shows that in the nearshore part of the surf zone the concentration was nearly uniform, one order of magnitude chang over more than 2 m of elevation. The values compare with $a = 1.84$ by Kana.

The variation of the sediment concentration from seaward of breakers to the shoreline is graphically illustrated in Fig. 11. 48. It shows field data collected at about 3 hour intervals over three days on the Black Sea coast. The record starts with the typical suspended and diffused sediment layers outside the breakers. The concentration profile becomes almost uniform over depth at the plunge line and further inshore the effect of diffusion of turbulence from the bed becomes apparent. Notice the mean particle size distribution over depth at the plunge line. For a true description of the sediment distribution within the surf zone the concentration distribution has to be related first to the breaker type and thereafter modified for the effects of currents within the surf zone. These too are dependent on the breaker type but vary also with the direction of wave approach.

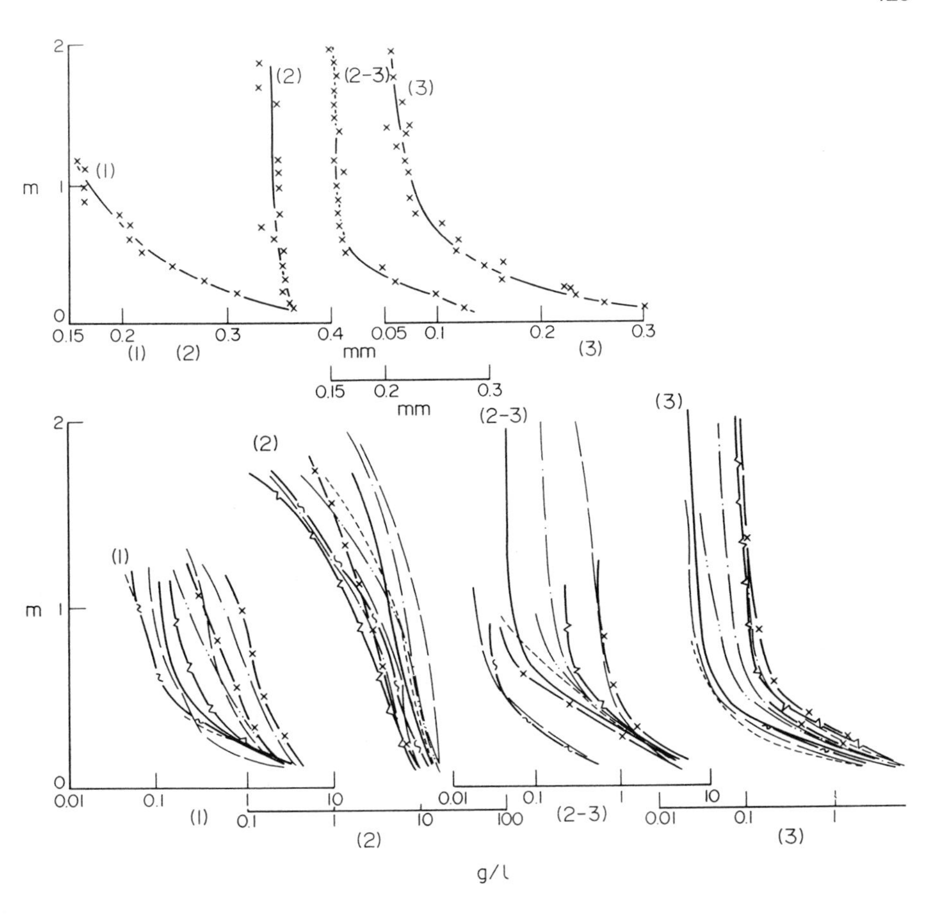

Fig. 11.48. Concentration and mean grain size profiles of suspended sediment at four stations across the breaker zone. Station (3) at 105 m from shoreline and 2.2 m depth is outside breakers. St. (2-3) breaker line 75 m and 2 m depth. St. (2), trough and plunge line area 40 m, 2.7 m depth. St. (1), at 10 m from shoreline and 1.1 m depth. (Kossyan et al., 1982).

11.5.3 Analytical models of concentration distribution

The conventional approaches to calculation of the time averaged vertical sediment concentration distribution, based on the one-dimensional diffusion equation, are even less applicable under wave motion than in unidirectional flow. However, numerous models exist which have adapted this approach to oscillatory flow and assume an expression for the sediment diffusion coefficient ε_s.

In the gradient-diffusion type solutions the coefficient ε_s arises from three effects:

(i) Diffusion due to orbital movement, $\varepsilon_v \sim (\overline{v^2})^{\frac{1}{2}}\ell$, where v is the vertical component of orbital velocity and ℓ is a mixing length, usually related to the vertical half axis of orbital movement $B = (H/2)\sinh k(h+y)/\sinh kh$. The vertical velocity averaged over wave period is

$$\overline{v^2} = \frac{1}{T}\int_0^T v^2 dt = \frac{1}{2}\left[\frac{\pi H}{T}\right]^2 \frac{\sinh^2 k(h + y)}{\sinh^2 kh} \qquad (11.197)$$

and

$$\varepsilon_v \sim \frac{\pi H^2}{2\sqrt{2}T} \frac{\sinh^2 k(h + y)}{\sinh^2 kh} \qquad (11.198)$$

which varies from zero at the bed to $\pi H^2/2\sqrt{2}T$ at the mean water level.

(ii) Diffusion due to boundary layer turbulence. The boundary layer thickness can be estimated with the aid of eqn (11.65) and assuming a logarithmic velocity distribution $u = (u_*/\kappa)\ln y/k_s$ the diffusion coefficient

$$\varepsilon_b = \ell^2 \left|\frac{du}{dy}\right| = \frac{2\kappa}{\pi} u_{*m} y \qquad (11.199)$$

averaged over time, where u_{*m} is the shear velocity based on the maximum orbital velocity at the bed. The turbulence produced decays rapidly with distance from the bed and is insignificant outside the boundary layer. This can be accounted for with an exponential multiplier to ε_b, for example, $[1 + b(y/\delta)\exp(y/\delta)]^{-1}$, where b is a factor which allows matching with observed data. Lundgren (1972) used this approach with $b = 1.34(\frac{1}{2}f_w)^{\frac{1}{2}}$.

(iii) Diffusion due to vertical velocity gradient. The vertical component of orbital velocity varies from zero at the bed to a maximum at water surface but studies of the effect of this gradient have shown it to be negligible.

The combined effects of (i) and (ii) are illustrated in Fig. 11.49. These ignore any effects which may arise from local velocity gradients in a multidirectional random sea as well as any injection of turbulence from breaking of waves.

Bijker (1967) adapted the concentration distribution derived by Rouse (1937) for suspension in a unidirectional flow with logarithmic velocity distribution

$$\frac{\overline{c}}{\overline{c_a}} = \left[\frac{h - y}{y} \frac{a}{h - a}\right]^z \; ; \quad z = \frac{w}{\kappa u_*} \qquad (11.200)$$

for use in the surf zone with longshore current. He replaced the

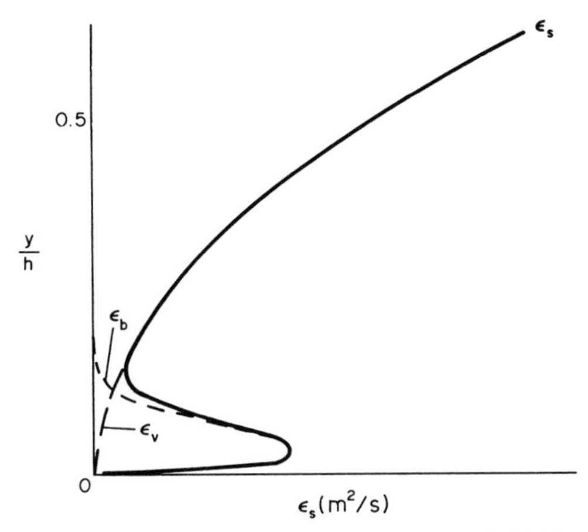

Fig. 11.49. Diagrammatic superposition of diffusion coefficients due to orbital velocity ε_v and momentum diffusion from bed ε_b.

shear velocity u_* with $(\overline{\tau}_{cw}/\rho)^{\frac{1}{2}}$ from eqn (11.129), by observing that the integrals of cosx and cos²x from 0 to 2π divided by 2π are equal to 0 and ½, respectively, i.e.,

$$\overline{\tau}_{cw} = \tau_c[1 + \tfrac{1}{2}(\xi u_m/V)^2] \tag{11.201}$$

Further examples are solutions based on calculations of the diffusion coefficient, e.g. Hom-ma and Horikawa (1962), Kennedy and Locher (1972) and Liang and Wang (1973). Liang and Wang set

$$\varepsilon = H_1 K_2 \frac{H\omega}{2 \sinh kh} \sinh ky \tag{11.202}$$

where $K_1 = \ell$ is a constant with length dimension to be determined experimentally and K_2 is a delay constant introduced by Hattori (1969)

$$K_2 = \frac{3}{2(\rho_s/\rho + \frac{1}{2})} \tag{11.203}$$

Equation (11.202) is basically $\varepsilon \sim \ell(\overline{v^2})^{\frac{1}{2}}$. When inserted in $w\overline{c} + \varepsilon_s d\overline{c}/dy = 0$ the result is

$$\frac{\overline{c}}{\overline{c}_a} = \left[\frac{\tanh \frac{1}{2} ky}{\tanh \frac{1}{2} ky_a}\right]^z ; \quad z = \frac{2w \sinh kh}{K_2 K_1 kH\omega} \tag{11.204}$$

which in shallow water reduces to

$$\frac{\overline{c}}{\overline{c}_a} = \left[\frac{y}{y_a}\right]^z ; \quad z = \frac{2wh}{K_2 K_1 H\omega} \tag{11.205}$$

The K_1 is given by the slope of a log-log plot of data.

Introduction of eqn (11.198) in the diffusion equation yields

$$\ln \bar{c} = -\int_{\varepsilon}^{w} -dy = -\int w \frac{2\sqrt{2T}}{\pi H^2} \frac{\sinh^2 kh}{\sinh^2 ky} dy$$

$$\frac{\bar{c}}{c_a} = \exp\left[w\left[\frac{\sqrt{2LT}}{\pi^2 H^2} \sinh^2 kh \coth ky\right]\right] \tag{11.206}$$

When the boundary diffusion term is included as well, the integral for $\bar{c}$ can be solved only numerically.

The simplest assumption, i.e., ε_s = const. leads to $\bar{c}/\bar{c}_o = \exp(-wy/\varepsilon_s)$ as discussed for unidirectional flow.

The above relationships imply that the concentration profile depends on grain size, i.e., the fall velocity w. The measured profile shapes of beach sands do not support this. Nielsen (1984) separated samples at each elevation to grain size groups and found that the concentration profiles plotted parallel, whereas from the diffusion equation the concentration decay

$a(m^{-1}) = -\left[\frac{1}{c}\frac{dc}{dy}\right]^{-1} = \frac{\varepsilon_s}{w}$. Likewise, the wave tunnel data by

Horikawa et al. (1982) coalesce to single functions when w and d are not considered. Apparently in the surf zone, and outside the breakers during heavy seas, the turbulence levels are so high that fall velocities of beach sands, of the usual size range, are insignificant. The diffusion model has some logical support when applied to suspension outside the breaker line. Within the surf zone the amount of turbulence produced at the bed may be quite insignificant compared to the turbulence energy input from dissipation of wave energy. Only under spilling breakers could one talk of diffusion of turbulence and this is downwards. Serious conceptual difficulties arise when the diffusion model is related to suspension by combined wave and current action.

Field data indicate that the suspended sediment concentration profiles can be fitted by segments of the simple form

$$\bar{c}(y) = \bar{c}_o e^{-ay} \tag{11.189}$$

where a is a decay parameter with dimension m^{-1}, i.e., the inverse of a "mixing length", which can be estimated from the plot of y versus log $\bar{c}$ as

$$a = (\Delta y \log e)^{-1} \tag{11.207}$$

where Δy is the height for one log-cycle change of concentration.

In the absence of field data the reference concentration $\bar{c}_o$ is still difficult to assess. Nielsen (1986) from field data related $\bar{c}_o$ to the grain roughness Shields parameter θ', where $\tau_o = \frac{1}{2} f_w \rho u_m^2$

and f_w is evaluated with $k_s = 2.5$ d. Using data by Du Toit and Sleath (1981) the θ was modified to

$$\theta_r = \frac{\theta'}{(1 - \pi\eta/\lambda)^2}$$

in analogy to the velocity over ripple crests $u_r = u_m/(1 - \pi\eta/\lambda)$. The data plotted as $\bar{c}_o$ versus θ_r showed reduced scatter and yielded

$$\bar{c}_o = 0.005\ \theta_r^3 \tag{11.208}$$

where the refinement $(\theta_r - \theta_c)^3$ was considered to be not warranted. However, the scatter of data may also be due to correlation with θ. Wave tunnel data by Horikawa et al. (1982) when plotted in terms of θ scatter about an order of magnitude, yet plotted against u_m, the scatter is minimal.

11.6 Transport of Sediment by Wave Action

The discussion under this heading is confined to noncohesive sediments. For cohesive sediments reference is made to Ch. 10.

The wave forces which entrain and transport sediment vary periodically, as illustrated in Fig. 11.50, and over parts of the wave period the applied stresses are less than the threshold value for sediment movement. Under an idealised sinusoidal wave the orbital movements are to and fro only and no net mass transport occurs. An actual net transport of sediment arises either from the non-linearity of waves of finite height or a superimposed current or both, provided always that critical values for threshold of grain movement are exceeded. Thus, sediment transport under wave action differs from that due to a uniform current in several important aspects.

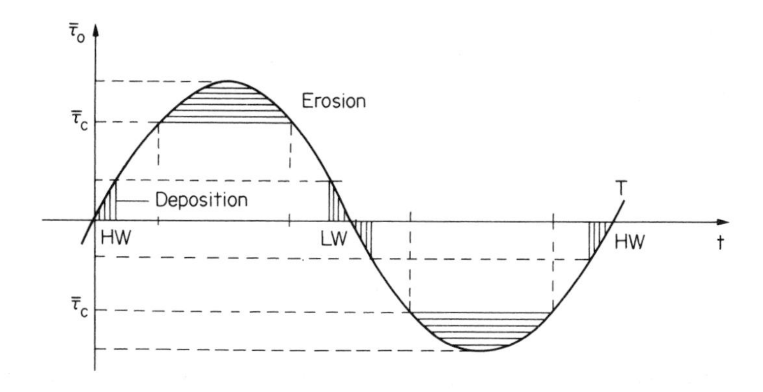

Fig. 11.50. Diagrammatic variation of average bed shear stress over a wave period.

Firstly, there is a difference between *sediment load* and *sediment transport*. The *load* is the sediment in motion, either as *mobilized sediment* or *suspended sediment*. Transport, refers to the net transport and is the difference in the forward and backward translation of the sediment load, i.e. transport is a relatively small difference between two large quantities and this is one of the causes of difficulties associated with prediction of transport rates. Additional problems arise with the estimation of the sediment load under given wave conditions, particularly when a current is superimposed. Then, over a bed covered with vortex ripples, even the direction of net transport may be in doubt. The vortices due to the higher velocity over the crests contain more sand, than those of the weaker flow in reverse direction. On reversal of flow direction these vortices are carried "upstream" and lose their sediment there, i.e. there could be a net sediment transport against the net movement of water. It also follows that *bed load transport* is associated with the translation of mobilized sediment.

A major conflict in philosophy exists over the actual process of sediment transport. The transport models adapted from unidirectional flow, from river engineering, set the transport rate, q_b, of "bed load" proportional to velocity or shear velocity to some power n, i.e., $q_b \sim u^n$, where in open channel flow n ranges from about 3 to 6. As a consequence, the sum of u^n of forward and reverse velocities would differ substantially even by small differences in velocities. Another school of thought is that the sediment load, mobilized and suspended sediment, translates as an integral part of the grain-water mass. The momenta of the grain and the water it displaces differ only by a factor of about 1.6 which would not lead to a significant slip and decoupling of the movements. This implies that a net transport is given by the sum over depth of the product of local net velocity and concentration.

The net sediment transport problems could be subdivided into transport inside and outside the surf zone, transport as bed load and transport as suspended load, transport by wave action and transport by wave and current action.

11.6.1 Transport outside the breaker line

Transport outside the breaker line is responsible for the long term changes in the coastal morphology. The transport rates are usually low and the transport is confined to a layer close to the bed. Seldom is sediment in significant concentrations more than half a metre above the bed.

Under strong wave action, when the grain roughness Shields parameter θ' is one or greater, sheet flow conditions exist. Then the sediment concentration can be approximated by two exponential relationships of the form $\bar{c} = \bar{c}_o \exp(-az)$ describing the bed load or mobilized sediment layer and the suspended load layer. The bed load layer is the sediment mobilized by wave action and carried by the bed through grain to bed and grain to grain momentum exchange, the dilatation stress. The suspended sediment load is maintained by the diffusion of turbulence from the boundary.

Bagnold (1954,1956,1963) introduced the concept of dilatation stress to sediment transport. Already Reynolds showed that a shearing action of granular material must be accompanied by a normal stress (dilatation stress) if the grains are not to be crushed. Bagnold showed that this stress did not disappear in the sheared granular-fluid flow after the grains were mobilized, i.e. it was maintained by the momentum transfer from grain to grain. The second concept used was the Coulomb yield stress criterion. Accordingly, in the plane between the stationary and moving grains the shear stress is equal to the normal load per unit area times the friction factor, tan ϕ.

From the definition sketch, Fig. 11.51

$$\tau_{zx} = w'\sin\beta + \tau_o\cos\alpha = \tau_{gx} + \tau_o\cos\alpha \tag{11.209}$$

where $w' = (\rho_s - \rho)gV$; ρ_s and ρ are the densities of solids and fluid, respectively; g is gravitational acceleration; V is volume of solids mobilized per unit area; τ_o is the applied fluid shear stress on top of the mobilized grain layer of thickness z_1; $\tau_{gx} = w'\sin\beta$ and τ_{zx} is the shear stress on the x-y-plane in x-direction.

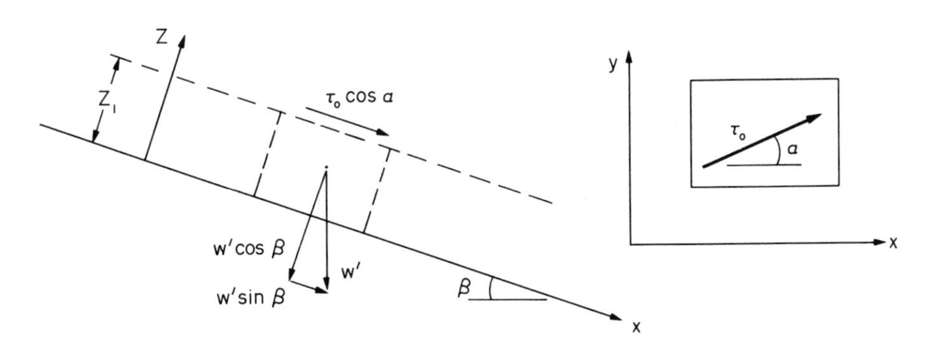

Fig. 11.51. Definition sketch.

The resultant shear stress is from τ_{zx} and $\tau_{zy} = \tau_o\sin\alpha$

$$\tau_{zr} = (\tau^2{}_o + 2\tau_{gx}\tau_o\cos\alpha + \tau^2{}_{gx})^{\frac{1}{2}} \cong \tau_o + \tau_{gx}\cos\alpha \tag{11.210}$$

if $\tau_{gr} << \tau_o$, which is the case when β is small. Thus,

$$w'\cos\beta\tan\phi = \tau_o + w'\sin\beta\cos\alpha$$

or

$$\tau_o = w'\cos\beta(\tan\phi - \tan\beta\cos\alpha) \tag{11.211}$$

Bagnold postulated that the rate of mass transport per unit width was $\rho_s V$ times its mean velocity $\overline{U}_b$ and that this was proportional to the rate of energy dissipation per unit area. The tangential force affecting the transport is τ_o and the work done is

$$\tau_o \bar{U}_b = (\rho_s - \rho) gV \cos\beta (\tan\phi - \tan\beta \cos\alpha) \bar{U}_b \tag{11.212}$$

where he defined the immersed weight rate of transport per unit width as

$$i_b = (\rho_s - \rho) gV\bar{U}_b \cos\beta \tag{11.213}$$

Thus, the fluid power expended, the dissipation rate, is

$$i_b (\tan\phi - \tan\beta \cos\alpha) = P = \varepsilon\tau_o u_m = \tfrac{1}{2}\varepsilon f_w u_m^3 \tag{11.214}$$

or

$$i_b = \frac{\tfrac{1}{2}\varepsilon f_w}{\tan\phi - \tan\beta \cos\alpha} u_m^3 \tag{11.215}$$

where the efficiency factor ε of the transport process hides the actual value of $\bar{U}_b$, the differences arising from the use of $V\bar{U}_b$ instead of $\int c(z)\bar{u}(z)dz$, any effects arising from use of τ_o rather than $(\tau_o - \tau_c)$, and most significantly the proportioning of wave power, $\tau_o u_m$, at the bed between sediment movement and turbulence production.

Madsen and Grant (1976) adapted the Einstein-Brown sediment transport formula to describe the time averaged transport rate, $\bar{q}_b$, during half the period of orbital motion as

$$\frac{\bar{q}_b}{wd^2} = 12.5\ (\theta')^3 \tag{11.216}$$

where w is the fall velocity of grains and $\theta' = \tfrac{1}{2}f_w u_m^2/\Delta gd$ based on grain roughness.

Sleath (1978) described his experimental results by

$$\frac{\bar{q}_b}{\omega d^2} = 47(\theta - \theta_c)^{3/2} \tag{11.217}$$

where $\omega = 2\pi/T$.

The net transport rate by either of the above equations would be the difference in the two $\bar{q}_b$-values for the opposing directions of orbital velocity.

Equations (11.215) to (11.217) can be compared in terms of the volume of sediment mobilized.

According to eqn (11.211) the volume of grains mobilized is

$$V = \frac{\tau_o}{(\rho_s - \rho) g \cos\beta (\tan\phi - \tan\beta \cos\alpha)} \tag{11.218}$$

or if $\beta = \alpha = 0$

$$V(t) = \frac{\tau_o}{(\rho_s - \rho) g \tan\phi} \tag{11.219}$$

or

$$V(t) = \frac{\tau_o - \tau_c}{(\rho_s - \rho) g \tan\phi} = \frac{1}{\tan\phi} (\theta - \theta_c) d$$

$$\cong 1.6(\theta - \theta_c) d \tag{11.220}$$

The instantaneous shear stress is

$$\tau_o(t) = \tau_o |\cos\omega t| \cos\omega t \tag{11.221}$$

where

$$\tau_o = \tfrac{1}{2} f_w \rho |u_m| u_m$$

$$f_w = \exp[-5.977 + 5.213(A/k_s)^{-0.194}]$$

where A is the half axis of orbital movement and k_s is the roughness height of the bed. This is not k_s = d or 2.5 d when a significant volume of sediment is in motion.

The τ_c or θ_c value can be estimated from data given in literature, but its significance is questionable because in the to and fro movement of grains under wave action there are always some grains perched on top of others in very unstable positions.

The time average shear stress is

$$\overline{\tau_o - \tau_c} = \frac{2}{T} \int_{-t'}^{t'} (\tau_o \cos^2 \omega t - \tau_c) dt$$

$$= \frac{2}{T} \left[\frac{\tau_o}{2} (t + \frac{1}{2\omega} \sin 2 \omega t) - \tau_c t \right]_{-t'}^{t'}$$

$$= \frac{2}{T} \left[\left[\frac{\tau_o}{2} - \tau_c \right] 2t' + \frac{\tau_o}{2\omega} \sin 2 \omega t' \right]$$

when $\tau_c << \tau_o$, $t' \cong T/4$ and $\overline{\tau_o - \tau_c} \cong \tfrac{1}{2} \tau_o - \tau_c$ (11.222)

$\tau_c = \tau_o/2$, $t' \cong T/8$ and $\overline{\tau_o - \tau_c} \cong \tau_o/2\pi$ (11.223)

Conditions corresponding to eqn (11.223) are associated with only a few grains in motion and the concept of a mobile layer is not applicable. Numerical experiments show that the "layer thickness" is only a small fraction of the grain diameter, i.e., only an occasional grain moving here and there.

The time averaged volume of grains mobilized according to the Bagnold model is then given by eqn (11.218) or (11.219) and eqn (11.222).

Equations (11.216) and (11.217) can also be expressed in terms of the volume of grains mobilized. The average transport rate over half period could be interpreted as the product of the volume of grains mobilized, V, and the mean orbital velocity $\bar{u} = (2/\pi)u_m$. Thus, the Sleath's expression yields

$$V = 23.5\ \pi(\theta - \theta_c)^{3/2}\omega d^2/u_m \tag{11.224}$$

and that by Madsen and Grant

$$V = 6.25\ \pi d(w/u_m)(\theta')^3 \tag{11.225}$$

The three equations (11.220, 11.224, 11.225) show very different dependence on θ.

Outside the surf zone the true bed load transport is associated with sheet flow conditions. In the presence of vortex ripples the transport and suspension are dominated by the vortices shed from the crests of the ripples. At just past the threshold of motion conditions the movement on a flat bed involves only the occasional movement of a few grains.

At sheet flow conditions a clearly defined bed load layer exists, as shown by the data in Fig. 11.45. This layer is carried by the dilatation stresses. Since the layer is very thin the volume of sediment in it can be estimated from

$$V = \bar{c}_o \int_0^{y_1} e^{-ay} dy = \frac{\bar{c}_o}{a}\left[1 - e^{-ay_1}\right] \tag{11.226}$$

which for $y_1 \to \infty$ becomes

$$V = \bar{c}_o/a \tag{11.226a}$$

Assume, for example, regular waves with T = 8 s, H = 2 m, water depth h = 8 m and sand of d = 0.2 mm, ρ_s = 2650. Then u_m = 0.920 m/s; A = 1.171 m; $\psi = u_m^2/\Delta gd$ = 265.7, i.e., sheet flow conditions. From eqn (11.188), $k_s = 5 \times 10^{-4} + 0.16\ u_m^{2.25}$ = 0.133 m; from eqn (11.67) f_w = 0.0775, $\tau_o = \frac{1}{2} f_w \rho u_m^2$ = 33.676 Nm^{-2}, and θ = 10.297. For d = 0.2 mm $\tau_c \cong$ 0.14 Nm^{-2}, and $\theta_c \cong$ 0.042. From eqn (11.222) $\overline{\tau_o - \tau_c}$ = 16.698 Nm^{-2}. Thus, from eqn (11.219) with $\tan\phi \cong 0.615$ the volume of sediment mobilized per unit area is

$$V = \frac{\overline{\tau_o - \tau_c}}{(\rho_s - \rho)g\tan\phi} = \underline{1.678 \times 10^{-3}\ m^3/m^2}$$

From eqn (11.226a), $V = \bar{c}_o/a$ and with $\bar{c}_o < 0.30$ the decay constant of concentration is a < 178.

From eqn (11.224)

$$V = 23.5\ \pi(10.297 - 0.042)^{1.5}\ \frac{2\pi}{8}\ \frac{4 \times 10^{-8}}{0.920} = \underline{0.083 \times 10^{-3}\ m^3/m^2}$$

and from eqn (11.225) with $w = 0.0211$ m/s, $f_w' = 0.0081$ and $\theta' = 1.0718$

$$V = 6.25\,\pi\, 0.0002\ \frac{0.0211}{0.920}\ 1.0718^3 = \underline{0.11 \times 10^{-3}\ m^3/m^2}$$

The three values are in the ratio of 20.2:1:1.3. If these volumes are expressed in terms of grain roughness $f_w' = 0.0081$, $\theta' = 1.072$, $\tau_o' = 3.520\ Nm^{-2}$ then the volumes are 0.163×10^{-3}, 0.0026×10^{-3} and $0.111 \times 10^{-3}\ m^3/m^2$, respectively, i.e., 63:1:43 with the Bagnold formula giving the highest volume. In both cases the Sleath formula yields the lowest value.

Horikawa et al. (1982), Fig. 11.45, case 1-1 had the following data: $u_m = 1.256$ m/s, A = 0.72 m, T = 3.6 s, d = 0.2 mm, $\rho_s = 2660\ kg/m^3$, $\bar{c}_o = 0.0126$ and $a = 122.5\ m^{-1}$. These yield: $f_w' = 0.0090$; $\theta' = 2.19$; $\tau_o' = 7.14\ Nm^{-2}$; $k_s = 0.269$ m; $f_w = 0.1882$; $\tau_o = 148.43\ Nm^{-2}$; $\theta = 45.6$ and $\tau_o - \tau_o = 74.1\ Nm^{-2}$. The corresponding values of the volumes are

$$V_B = 7.28 \times 10^{-3}\ m^3/m^2 \quad \text{or} \quad V_B' = 0.337 \times 10^{-3}\ m^3/m^2$$

$$V_s = 1.261 \times 10^{-3}\ m^3/m^2 \quad \text{or} \quad V_s' = 0.013 \times 10^{-3}\ m^3/m^2$$

$$V_M' = 0.69 \times 10^{-3}\ m^3/m^2$$

for Bagnold, Sleath and Madsen and Grant, respectively. The volume from test data is

$$V_H = \bar{c}_o/a = 0.103 \times 10^{-3}\ m^3/m^2$$

The results clearly show that estimation of the volume of sediment mobilized with the shear stress based on effective roughness grossly over estimates the amount of sand in motion. This implies that only a small amount of the energy dissipated at the bed goes into moving sediment.

A speculative proposal is that the volume be estimated with τ_o', i.e. V_B' and the concentration gradient from eqn (11.191). Thus, for the first example, $V_B' = 0.163 \times 10^{-3}\ m^3/m^2$, $a = 315\ u_m^{-4} = 439.7\ m^{-1}$ and $\bar{c}_o = V_B'\ a = 439.7 \times 0.163 \times 10^{-3} = 0.0717$. The *net sediment transport as bed load* is then given by the integral of the local product of $\bar{c}$ and U over the thickness of the mobilized layer, where U is the net mass transport (water) velocity. As a first approximation the mean value of concentration $\bar{c} = 0.5\bar{c}_o$ could be multiplied with the mean mass transport velocity at the elevation of $\bar{c}$, i.e.

$$\bar{y} \cong 0.7/a \qquad (11.227)$$

This velocity is estimated by extending the mass transport velocity at the outer edge of the boundary layer down to the bed as a logarithmic distribution. In the absence of currents the net mass transport velocity at the outer limit of the boundary layer is part of the logarithmic distribution within it, i.e.

$$\overline{U}_o = 2.5u_* \ln\delta/k_s + 5.9 \tag{11.228}$$

and

$$\frac{\overline{U}_o - u}{u_*} = 2.5 \ln y/\delta + 3.7 \tag{11.229}$$

Although, functions can be fitted to the measured points of concentration profiles, these tend to be cumbersome and it is doubtful whether at present state of knowledge the extra complications are warranted. It is suggested that a reasonable method for estimation of the net transport rates is to approximate the concentration profile by an exponential function for each of the "layers", that is, the bed load, the suspended load and where warranted also the upper diffused load (wash load) layer. Then the net water transport profiles has to be estimated with the aid of an appropriate wave theory and assuming a logarithmic velocity profile in the boundary layer. Its thickness could be estimated with the aid of eqn (11.65), assuming the datum to be at 0.25 k_s below the top of the nominal k_s. The resultant net sediment transport is then given by

$$q_s = \int_0^{y_1} \overline{c}Udy + \int_{y_1}^{y_2} \overline{c}Udy + \int_{y_2}^{y=|h|} \overline{c}Udy \tag{11.230}$$

where U is the net mass transport velocity at a given elevation; y_1 and y_2 are the elevations of the breaks in concentration profile between bed load and suspended load, and suspended load and diffused or wash load. Over a rippled bed it is considered adequate to omit the bed load layer and take the suspended layer down to nominal bed level. In each of the layers $\overline{c} = \overline{c}_o \exp(-ay)$, where $\overline{c}_o$ is the concentration at the lower boundary. By using an a-value as indicated by field data the thickness of the suspended layer can be estimated. The net mass transport velocity over this thickness outside the boundary layer can be approximated by a linear function.

Appreciable additional complications arise when currents are superimposed on wave motion or the waves are strongly non-linear. Then, expressions of the form

$$q_s = \frac{1}{T}\int_0^T \int_0^y cudydt \tag{11.230a}$$

would have to be evaluated, i.e., in terms of instantaneous velocities and concentrations. In those situations the forward and backward velocities, and shear stresses, can differ substantially and since the absolute concentration increases very rapidly with the velocity at the bed, significant differences in the entrainment rates occur. The fall-out of sediment is almost independent of wave action at fall velocity w, except for changes in turbulence levels. The steady fall-out feature can, when the sediment is lifted high enough, lead to some equalising or averaging effect or it can lead to significant errors due to erroneous correlation between velocity and concentration as discussed in connection with transport over vortex ripples.

The results from the integration using mean values, eqn (11.230), or instantaneous values, eqn (11.230a), are not identical but eqn (11.230) is the usual starting point for practical applications.

For transport under waves and currents Grass (1981) proposed that the transport rate g_s was proportional to that due to current only, expressed as $g_{sc} = \alpha V^n$, where V is depth-averaged mean current velocity. The proportionality factor was related to the total kinetic energy of the fluctuating velocity field near the bed and expressed as

$$M = [1 + \beta(u/V)^2]^{(n-1)/2}$$

where u is the rms-value of the near bottom orbital velocities. Thus,

$$g_s(u,V) = \alpha V^n [1 + \beta(u/V)^2]^{(n-1)/2} \qquad (11.231)$$

The coefficients α and n are obtained by matching the formula at u = 0 (current only) limit with field data or a current only transport formula. Grass obtained from the Maplin Sands (in Outer Thames Estuary) data α = 1380 and n = 4, where u and V are in m/s and g_s in $gs^{-1}m^{-1}$. The value of β is inversely proportional to the drag coefficient, defined by the shear stress at the bed $\tau_o = \rho C_D V^2$. For a logarithmic velocity distribution $C_D = \{\kappa[\ln h/k_s - 1]\}^2$ and $\beta = 0.08/C_D$. The value of M was $M \cong 28$.

For a spectrum of waves Soulsby and Smallman (1986) showed that for the JONSWAP spectrum the near bottom rms-values of orbital velocity can be expressed as

$$\frac{u}{H_s}\left[\frac{h}{g}\right]^{\frac{1}{2}} = \frac{0.25}{(1 + At^2)^3} \; ; \quad 0 < t < 0.55$$

where

$$t = (h/gT_z^2)^{\frac{1}{2}}$$

T_z is the zero crossing period of significant wave H_s, h is water depth and

$$A = [6500 + (0.56 + 15.54t)^6]^{1/6}.$$

11.6.2 Transport within the surf zone

The sediment transport in the surf zone is usually dominated by the turbulence energy from dissipation of wave energy and the wave induced currents. The large turbulence energy input from dissipation of wave energy leads to a sediment distribution which is almost uniform throughout the depth, a diffused sediment layer. However, over parts of the surf zone a suspended sediment layer, carried by diffusion of momentum from the bottom boundary layer, may develop. This depends on the breaker type and width of the surf zone. Sediment transport in the surf zone accounts for the rapid and at times spectacular changes of the beach shape and profile. The surf zone could be subdivided into three transport zones. The first is the breaker region from the breaking point to the plunge point. This is followed by the zone of broken waves to the point of wave reformation and a new breaker region or to the swash zone. In the swash zone or runup zone the flow conditions are very different from the rest of the surf zone, even if small inshore breaking waves are present. Sand transport in the swash zone depends on the characteristics of the sediment and the runup bore and the local slope. Avalanching occurs if the angle of initial yield (28-30°) is exceeded. The residual angle after shearing is usually 10°-15° or less. Such a subdivision is an arbitrary concept.

A subdivision into bed and suspended load transport exists in literature but since the sediment is essentially suspended by macro-turbulence, vortices and whirls, such a subdivision has a limited physical meaning, except over areas of flat bed where a thin layer of mobilized sediment exists. In principle, an estimate of the net transport of sediment is given by the convection of the fluid-grain mass, that is, as described for outside the breakers, by summation throughout the depth of the local products of mean concentration and mean net velocity. The concentration could have three segments: mobilized, suspended and diffused layer. However, as indicated by eqn (11.191) the thickness of the mobilized layer under sheet flow is of the order of 10 mm and the error introduced by ignoring it in the surf zone is small. When the concentration of the suspended or the diffused sediment layer (depending on location) is extrapolated to bed level the differences in concentration over the thin layer will lead to quantities which are small compared to other sources of error. Most formidable are the problems with description of the currents in the surf zone (as already discussed) which do most of the transporting. A further significant problem is the onshore-offshore transport of sediment. Near the breaker line where the waves approximate to solitary waves, the differences in the onshore and offshore velocities are substantial. The amount of sand carried shoreward by the onshore velocity, assisted by the breaker, is larger than that in the return flow. At the same time strong net seaward return currents develop which carry sand offshore, Fig. 11.52. However, it should be noted that these measurements relate to an essentially two-dimensional situation in a 3.4 m wide wave flume. In nature the rip currents return substantial volumes of water offshore and this must affect the return currents shown in Fig. 11.52. A rigorous solution of the sediment transport in the surf zone does not appear to be likely in the foreseeable future.

The sediment transport formulae in use are generally adaptations from river engineering, except the formula by the U.S. Army Coastal Engineering Research Center, the CERC-formula.

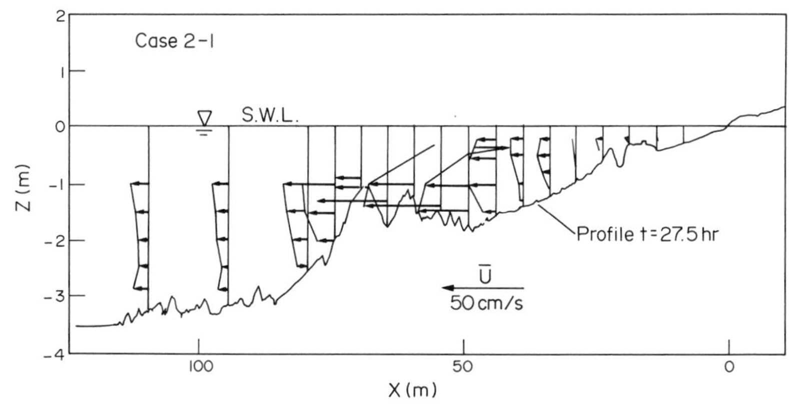

Fig. 11.52. Distribution of time averaged horizontal velocity, Kajima et al. (1982). Case 2-1, d_{50} = 0.47 mm, T = 6.0 s, H = 1.76 m, width of channel 3.4 m, initial slope 0.03.

A widely used transport formula in coastal engineering is the Kalinske-Frijlink formula (Kalinske, 1947; Frijlink, 1952) which Bijker (1967) adapted for coastal use (also referred to as the Bijker formula). The formula expresses the bedload transport rate per unit width as

$$q_b = Bd\,\frac{V}{C}\sqrt{g}\;r\,\exp\left[-\,0.27\,\frac{\Delta C^2 d}{rV^2}\right] \tag{11.232}$$

where B ≅ 5 is a dimensionless constant; d is grain size; V is mean velocity; C is Chézy coefficient; r is an empirical ripple factor accounting for form drag of bed features; and $\Delta = (\rho_s - \rho)/\rho$. Bijker omitted r from the first part, assuming it to be incorporated in C. The ratio $C^2/V^2 = \rho g/\tau_o$ and $V\sqrt{g}/C = u_*$ is the shear velocity. In order to account for waves and currents, C^2/V^2 was replaced by g/u_*^2 and u_* was inserted from eqn (11.129) by observing that

$$\frac{1}{2\pi}\int_0^{2\pi}\cos x\,dx = 0 \text{ and } \frac{1}{2\pi}\int_0^{2\pi}\cos^2 x\,dx = \frac{1}{2}, \text{ i.e.}$$

$$\bar{\tau}_{cw} = \tau_c\left[1 + \frac{1}{2}\left[\xi\,\frac{u_m}{V}\right]^2\right] \tag{11.233}$$

$u_*^2 = \bar{\tau}_{cw}/\rho$ and $\tau_c = \rho g V^2/C^2$. The use of an average value of τ instead of the average of the sum of instantaneous values introduces in the non-linear relationship an error but simplifies the computations substantially. Thus,

$$q_b = Bd\,\frac{V}{C}\sqrt{g}\,\exp -\left\{\frac{0.27\,\Delta d C^2}{rV^2\left[1 + \frac{1}{2}\left[\xi\,\frac{u_m}{V}\right]^2\right]}\right\} \tag{11.234}$$

Bijker also postulated that the bed load occurred in a layer, equal in thickness to the roughness height, and that the sediment concentration in this layer is constant

$$c_b = \frac{q_b}{\int_0^{k_s} v_{(z)}\,dz} = \frac{q_b}{6.34\ u_* k_s} \tag{11.235}$$

where the integral

$$\int_0^{k_s} v_{(z)}\,dz = \tfrac{1}{2} z_1 v_1 + \frac{u_*}{\kappa}\int_{z_1}^{k_s} \ln\frac{z_1}{z'}\,dz = 6.34\ u_* k_s$$

$z_1 = ez'$ is the elevation at which the tangent from $v = 0$ meets the logarithmic velocity profile and $z' \cong 33/k_s$ is the elevation where the logarithmic velocity distribution goes to zero.

The suspended load component was evaluated with the Einstein formula

$$q_s = 11.6\ u_* c_a y_a [\ln(33h/k_s) I_1 + I_2] \tag{11.236}$$

where $c_a \cong q_b/11.6\ u_* y_a$ and $u_* = (\bar{\tau}_{cw}/\rho)^{\frac{1}{2}}$. Using eqn (11.235) Bijker showed that

$$q_s = 1.83\ q_b [\ln(33h/k_s) I_1 + I_2] \tag{11.237}$$

Similarly, the Engelund-Hansen formula for total sediment transport can be re-arranged as

$$q_{Ts} = \frac{0.05\ C\tau_c^{\,2}}{\rho^2 g^{5/2} \Delta^2 d_{50}} V \tag{11.238}$$

where V is the mean current velocity. For the combined current and waves τ_c is replaced by eqn (11.233).

The Ackers-White formula eqn (7.155), expressed in terms of the total transport rate is

$$q_T = Ud \left[\frac{U}{u_*}\right]^n C \left[\frac{F_{gr}}{A} - 1\right]^m \tag{11.239}$$

where the symbols and parameters are defined in Chapter 7. If the deposited or eroded volume is considered then the right hand side has to be divided by one minus porosity.

Any of the formulae for sand transport could be transcribed in a similar manner.

The CERC, also BEB (Beach Erosion Board) *formula* empirically correlates wave energy flux into the surf zone with littoral transport rate. According to SPM (1984)

$$Q(m^3/\text{year}) = 1290 \left[\frac{m^3 s}{N\text{-}yr}\right] P_{ys} \left[\frac{W}{m}\right] \tag{11.240}$$

where

$$P_{ys} = \frac{1}{8} \rho g H_{sb}\, c^*_b \cos\theta_b \sin\theta_b$$

$$= \frac{1}{16} \rho g H_{sb}\, c^*_b \sin 2\theta_b = [\frac{1}{16} \rho g c_o H_o{}^2 K_{rb}{}^2 \cos\theta_b] \sin\theta_b \tag{11.241}$$

where H_{sb} is significant wave height at breaker line, c^*_b the wave group celerity and the shoaling coefficient $K_s = 1$ in deep water. Introduction from solitary wave $c = [gh(1 + H/h)]^{\frac{1}{2}} \cong (2gH_b)^{\frac{1}{2}}$ where $(H/h)_b = 1.28$; and $H_b/H_o = K_{sb}K_{rb} \cong 1.3$; $K_{rb}{}^2 = \cos\theta_o/\cos\theta_b$ and $\cos\theta_b \sim 1$ yields

$$P_{ys} = 0.0884\ \rho g^{3/2}\ H_{sb}{}^{5/2} \sin 2\theta_b$$

$$= 0.05\ \rho g^{3/2}\ H_{so} (\cos\theta_o)^{1/4} \sin 2\theta_b \tag{11.242}$$

The interpretation of P_{ys} as the longshore component of wave energy in terms of the significant wave height is physically meaningless since energy is a scalar quantity without components. The terms within the square bracket represent a constant energy flux between the wave orthogonals and could be compared with the radiation stress

$$S_{xy} = E_o n_o \cos\theta_o \sin\theta_o = E_b n_b \cos\theta_b \sin\theta_b$$

which is also constant outside the breakers. The gradient dS_{xy}/dx is the driving force of the longshore current in the surf zone. It is seen that $S_{xy}c_b$ equals eqn (11.241) which implies that the sediment entrainment is assigned to c_b. That this is reasonable can be seen from orbital velocity at the bed which in shallow water approximates to

$u_m = \frac{\pi H}{T} \frac{1}{kh} = \frac{\lambda}{T} \frac{H}{2h} = c_b \left[\frac{H}{2h}\right]_b$, i.e. c_b incorporates the important

features, velocity at the bed and the breaker height.

Equation (11.240) is displayed in Fig. 11.53. Notwithstanding the explanation of the equation it does incorporate the best field data available.

Whatever method is used to estimate the sediment transport rates in the surf zone the confidence level remains low. The reasons for this are numerous. Firstly, as has been discussed, the estimates of current patterns and velocities are subject to large errors. The adaptation of sediment transport formulae from river engineering is unsatisfactory, particularly for suspended sediment which forms the major part. The suspension mechanisms in the surf zone and a river are not similar. In the surf zone the

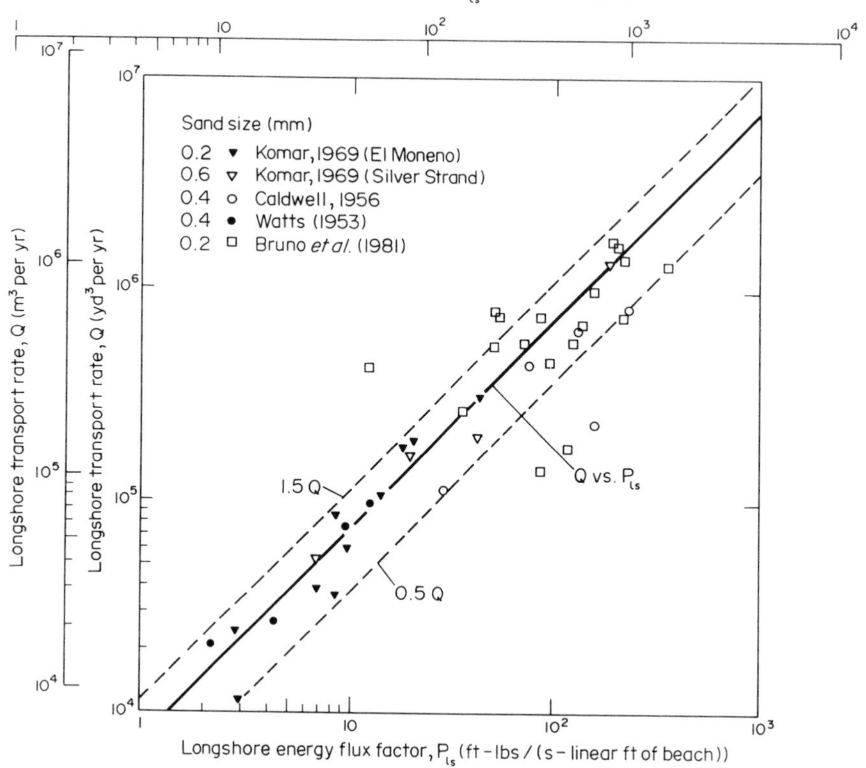

Fig. 11.53. The CERC empirical longshore sediment transport relationship.

sediment is suspended primarily by the turbulence generated by dissipation of wave energy and from the interaction of the long-shore current, return current, rip currents and other numerous eddy systems with wave motion. The wave motion almost always involves a wave spectrum, at times even multi-directional. Hence, a choice has to be made whether to superimpose transport due to component frequency components or use the simpler H_{rms} value to describe the wave field.

For the two-dimensional cross-shore transport numerical models can be established based on measured data, c.f. Kriebel and Dean (1985), but nature is seldom two-dimensional. Additional problems arise due to the concurrent net velocities in both onshore and offshore directions in the vicinity of the breakers, which lead to local sorting of sediment because the grain size of the suspended sediment tends to decrease with elevation above the bed.

The sediment transport across the surf zone is primarily a function of wave energy dissipation, i.e., of turbulence energy distribution. The wave decay in the surf zone could be expressed as $dF/dx = (a/h)(F - F_S)$, where $F = Ec^*$ is the energy flux, F_S is the stable energy flux, a is an empirical decay coefficient equal to about 0.15 and h is the stillwater depth. Since the wave group celerity c^* is shallow water becomes $(gh)^{1/2}$ the energy flux is $F = (1/8)\rho gH^2 (gh)^{1/2}$. The stable energy flux is usually expressed in terms of local stable wave height, H_S, and water depth as $H_S = \alpha h$, where α is a dimensionless coefficient with values in the range of 0.35-0.50, with 0.4 being a typical value. The wave attenuation increases with decreasing bed slope due to the increasing decay distance in the surf zone (see, e.g., Horikawa and Kuo, 1966; Dally et al., 1985). A proper description of the turbuluence energy distribution across the surf zone would lead via erosion or deposition characteristics to the various beach profile forms, development of bars, etc., without the use of arbitrary zones of flow regimes. Only the swash zone would require special treatment.

Since the numerical modelling of coastal zones is aimed at simulation of current patterns, usually depth integrated, it will be simpler to concentrate on description of concentration and to treat the transport as a convection of the fluid-grain medium. However, substantial research effort is required to define the sediment concentration spatially as a function of wave conditions.

11.7 Beach Profile and Shape

Erosion or accretion mean visible changes in beach condition and these may have economic or recreational consequences. Substantial effort has been devoted to the description and prediction of beach profiles and their plan geometry. The rapid and spectacular changes are confined to the inshore region but granular material can be moved by big seas at substantial depths. Thus, the first problem is the definition of the coastal zone. A rule of thumb is that on ocean coasts the sea bed is active down to about 20 m water depth. However, profile changes of significance are confined usually to depth less than that. Hallermeier (1981) concluded from field data that intense onshore/offshore transport of sediment occurs in depths

$$h_i \leq 2H_{s50} + 12\sigma_{Hs} \qquad (11.243)$$

relative to MSL, where H_{s50} is the median value of annual significant wave heights, set equal to $(\overline{H}_s - 0.307\ \sigma_{Hs})$ and σ_{Hs} is their standard deviation, approximately equal to $0.62\ \overline{H}_s$. This was shown to equate approximately to twice the nearshore wave height exceeded 12 hours per year. The seaward limit of the changes was expressed as

$$h_o \cong H_{s50}\overline{T}_s(g/5000\ d_{50})^{\frac{1}{2}} \tag{11.244}$$

where d_{50} is the median grain size at approximately $h = 1.5\ h_i$. Changes in sea bed topography between h_o and h_i are usually small and gradual. However, this simplistic picture can be affected substantially by coastal currents.

The vertical cross-section of the beach normal to the shoreline is called the *beach profile* and the nearshore part of it changes continuously with wave conditions. The nearshore profiles, which effectively cover the surf zone, are subdivided according to wave conditions into *storm or winter profile* and *swell or summer profile* with an approximate demarcation of $H_o/L_o \cong 0.025$, according to Johnson (1949). The storm profile is associated with steep waves which erode the inshore beach and form a bar at about the breaker line. The flat waves (swell) slowly bring this material back. The effect of steep and flat waves on a plane beach is illustrated in Fig. 11.54 and Fig. 11.55 is an illustration of storm wave action on a beach from the SPM. The rapid movement of sand offshore and slow return are illustrated by field measurements shown in Fig. 11.56. As a rule of thumb the depth of the trough on a storm profile is of the order of $0.6\ H_o$ and the bar height above trough about H_o.

Sunamura and Horikawa (1974) classified shoreline changes in terms of a parameter

$$C = (H_o/L_o)(\tan\beta)^{0.27}(d_{50}/L_o)^{-0.67} \tag{11.245}$$

where $\tan\beta$ is the average slope from the initial shore to the depth of sediment movement (20 m), or more logically to the depth h_i of eqn (11.243). The demarcation between recession and accretion of shoreline was C = 4 to 8 in laboratory and 9 to 18 in field. From subsequent studies C = 18 was recommended for field use.

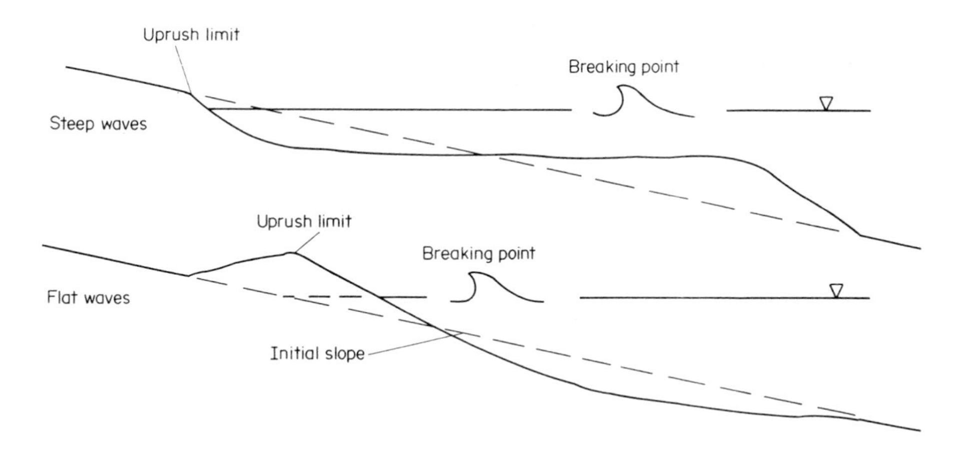

Fig. 11.54. Diagrammatic illustration of the effect of steep and flat waves on an initially plane beach.

Dean (1973) related beach erosion and deposition to wave steepness and the fall velocity of grains through

$$\frac{H}{L} = K \frac{\pi w}{gT} \tag{11.246}$$

where K is the proportionality of landward orbital movement and the fall distance of grain. A value of K = 2.0 to 2.4 was suggested. However, the role of the fall distance in surf during a storm is difficult to interpret since the suspension is then totally governed by the macro-turbulence. A study by Allen (1985) shows that field data on erosion plotted as H_b/L_o versus $\pi w/gT$ scatter above the line defined by eqn (11.246) but data on deposition are well spread over the entire plane.

The shape of the beach profile is further complicated by water level changes due to tides. A storm profile may have several bars, a high tide outer bar and an inner bar where the reformed waves break. At low tide the earlier outer bar may become the inner bar, as illustrated in Fig. 11.57.

Pronounced differences in profile shape exist between sand beaches and gravel beaches. A typical profile of a gravel or shingle beach is shown in Fig. 11.58. For steady state conditions the shelf is about the incident wave height below still water level, SWL, and the step is at the breaker line. The steep slopes of gravel beach profiles lead to the waves breaking in relatively shallower water, c.f. Fig. 11.16. Once equilibrium has been established there is little movement of gravel on the shelf. The material scoured during the formation of the shelf is thrown up on the beach. The inshore slope of a gravel beach is very steep, compared to sand beaches, and the material is strongly graded. Both particle size and beach slope increase up the beach. On

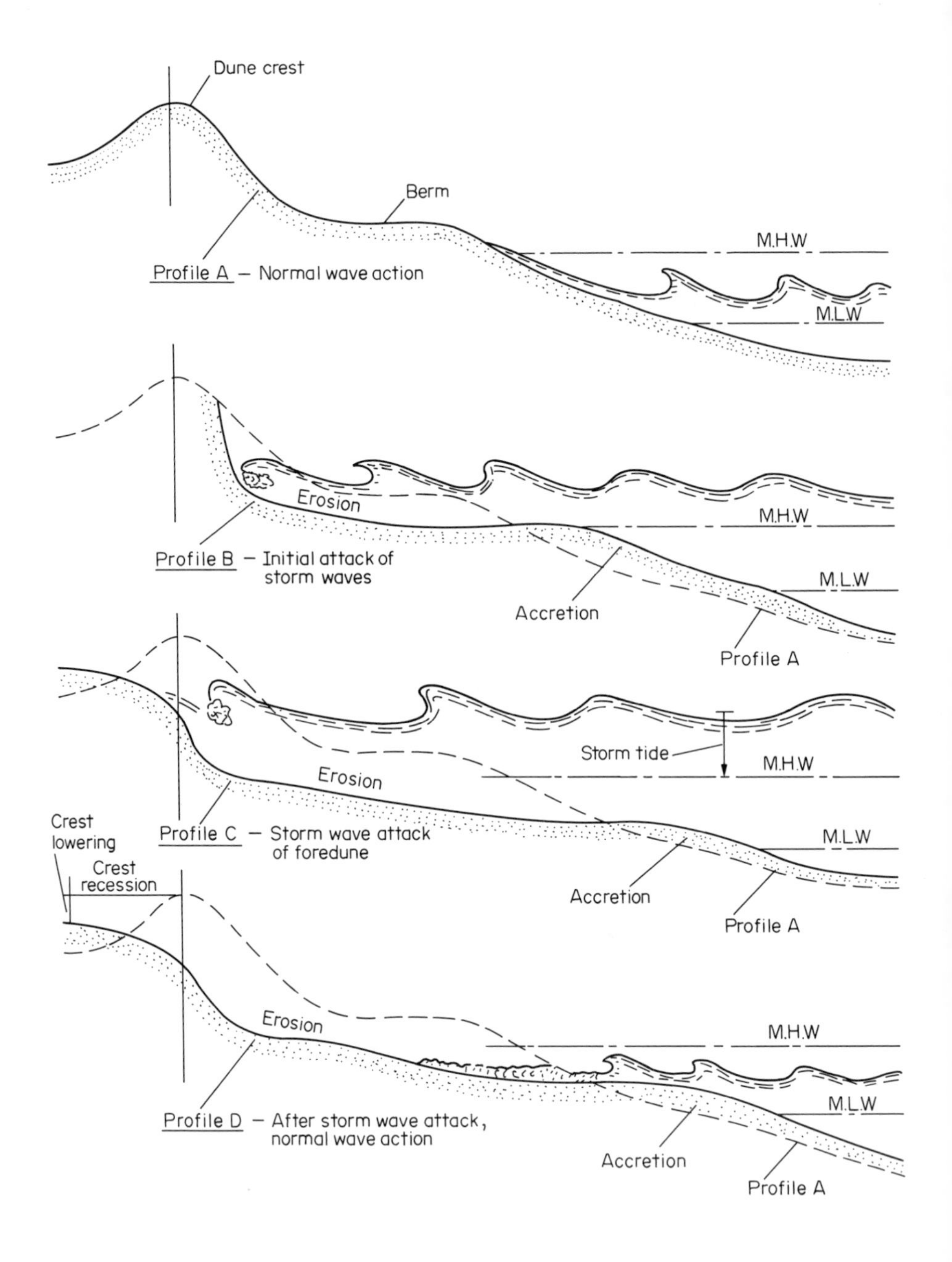

Fig. 11.55. Diagrammatic illustration of beach changes due to erosive storm wave attack, after Shore Protection Manual.

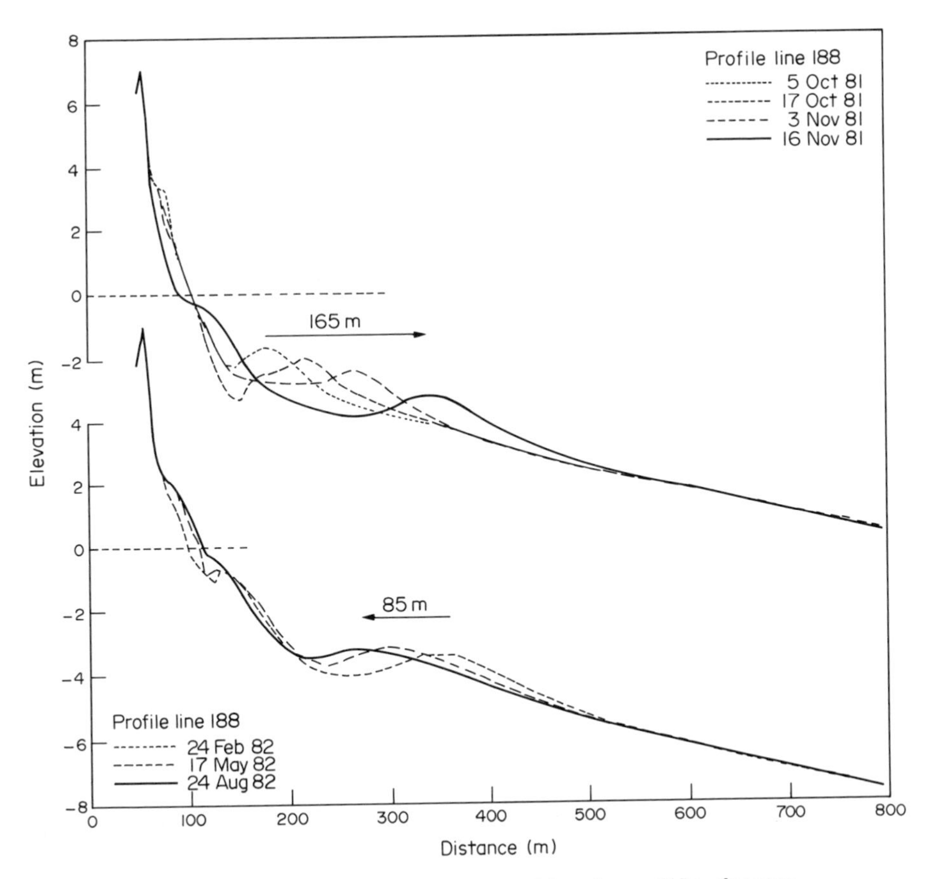

Fig. 11.56. Field measurements of beach profile changes. Rapid erosion due to 1981 autumn storms above and below slow onshore movement of the bar during the following six months of low wave conditions, Birkemeier (1984).

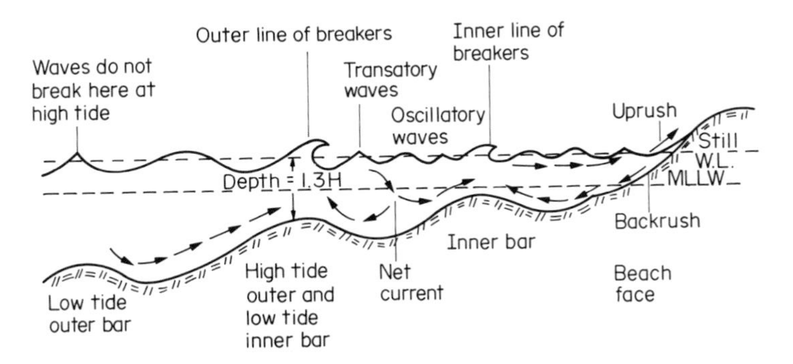

Fig. 11.57. Diagrammatic illustration of the effect of water level changes on beach profile.

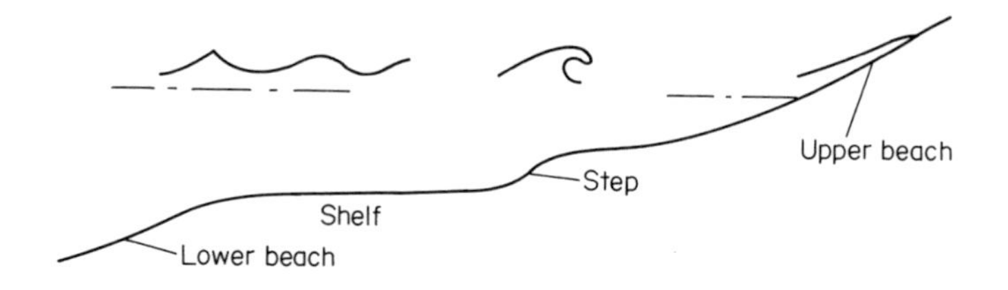

Fig. 11.58. Diagrammatic profile of shingle beach.

steep profiles the wave forces up the slope are greater than those seaward during the wave troughs. The difference in wave forces is balanced by the weight component of gravel and the coarser the gravel the greater the equilibrium slope. The wave force difference is amplified on the upper beach slope through percolation which can substantially reduce the return flow.

A sorting of gravel also occurs along the beach if the beach is angled to the predominant direction of large waves. The large waves move gravel along the beach but the smaller waves from other directions are capable only of returning the smaller fractions. A size grading is also common along curved gravel beaches.

Kajima et al. (1982) expanded the classification of the beach profile types on the basis of test data in a large wave flume, Fig. 11.59. Type I profile is an erosion profile, subdivided to mono-crested and bi-crested profiles. Type II profile corresponds to the separation zone between erosion and accretion, characterised by the C = 18 value in eqn (11.245). Type III, again subdivided into mono-crested and bi-crested profiles, characterises accretion conditions. According to the authors mono-crested profiles occurred when $(d_{50}/L_o)^{0.67}/(\tan\beta)^{0.27} > 10^{-3}$ and bi-crested when smaller.

It has to be kept in mind that in nature the observed beach geometry is the result of accumulation of antecedent processes, each having different scales of space and time. Rapid changes of beach state essentially involve only the surf zone and are related to recent antecedent conditions. Major changes in beach profile, extending beyond the surf zone, involve sediment movement over the entire inshore region and these changes evolve slowly. In recent years the empirical eigenvector analysis has been introduced as an objective method for analysis of profiles. Basically, it is a statistical procedure which separates in a rectangular matrix the variation of data into two sets of orthogonal functions. Each successively higher eigenvector explains, in the least squares sense, the variance remaining in the data. Random uncorrelated variations in the field data are filtered out to higher order eigenvectors. The procedure is somewhat analogous to Fourier analysis. Detailed discussion of the procedure is given by Aubrey (1980) and Vincent and Resio (1977).

A very visible feature of the beach profile is its foreshore slope. It increases with grain size but is also affected by the wave climate, beach porosity and density of the beach material. Wave

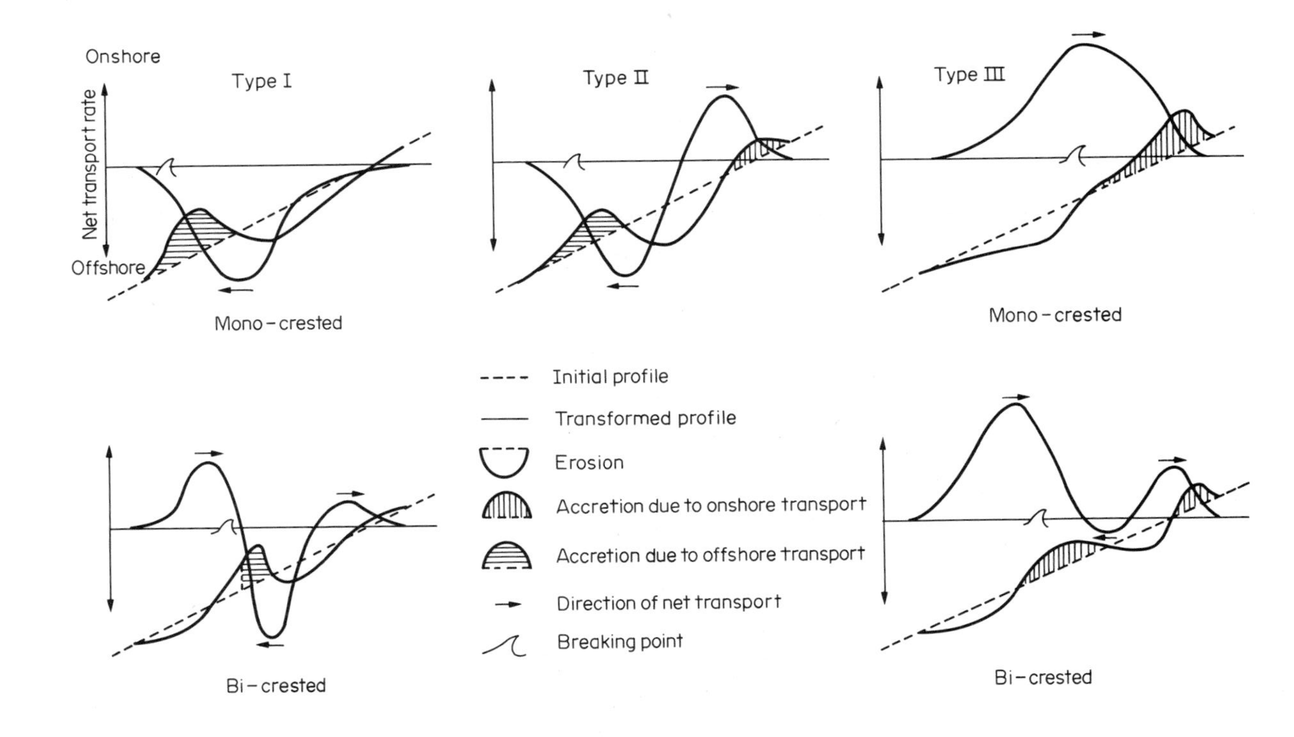

Fig. 11.59. Classification of beach profile types, according to Kajima et al. (1982).

height and steepness have a dominant effect on foreshore slope. The SPM shows some field data from east and west coasts of U.S.A. The data have been fitted with linear functions in terms of tangent of beach slope S and median grain size d in mm as

$S = 0.633(d-0.15)$ for Florida Panhandle
$S = 0.378(d-0.11)$ for New Jersey-North Carolina
$S = 0.236(d-0.008)$ for West Coast

where West Coast has the highest annual mean wave height of about 0.8-1.0 m and Florida the lowest at about 0.3-0.5 m, i.e. the higher the annual mean wave height the flatter is the foreshore slope. A minimum slope envelope to the data by Bascom (1951), Half Moon Bay, California, is given by

$$S = 0.277\, d_{mm}^{2} \tag{11.247}$$

11.7.1 Analytical profile models

Generally the beach profile changes with wave conditions and with water level, which alters the depth - wave height conditions. It always tends towards the most stable condition. An increase in water level narrows the surf zone and increases the energy to be dissipated per unit volume. Observations show that the equilibrium profile can be described, as a first approximation, by

$$h = A\, x^{n} \tag{11.248}$$

For spilling breakers and more or less uniform dissipation of wave energy over the surf zone $n \cong 2/3$. The scaling parameter A depends on fluid and sediment characteristics and is determined from field data. Coarser sediment leads to larger values of A and a steeper beach. In Coastal Sediments '87 Dean quotes $A = 0.067\, w^{0.44}$ in which A is in $m^{1/3}$ and w in cm/s. Kriebel and Dean (1985) related the scaling parameter to dissipation of wave energy D per unit volume by

$$A = \left[\frac{24}{5}\,\frac{D_e}{\gamma^2 \rho g^{3/2}}\right]^{2/3} \tag{11.249}$$

where D_e signifies equilibrium conditions and γ is the breaker index. The concept of equilibrium profile implies that the offshore/onshore transport of sediment q_s is proportional to the difference between the actual energy dissipation rate and the equilibrium dissipation rate

$$q_s = K(D - D_e) \tag{11.250}$$

where an empirical transport rate coefficient $K = 2.2 \times 10^{-6}\ m^4/N$ was used.

$$D = \frac{1}{\bar{h}}\,\frac{\partial P}{\partial x} \tag{11.251}$$

and the energy flux in shallow water

$$P = \frac{1}{8} \gamma^2 \rho g^{3/2} h^{5/2} \qquad (11.252)$$

For non-equilibrium conditions the term $(N/K)dh/dx$ is added in the brackets of eqn 11.250, where N is the transport rate coefficient for the slope term of the order of 0.001 m^2/s.

Thus, in the absence of longshore transport, the continuity for sand is given by

$$\frac{\partial x}{\partial t} = - \frac{\partial q_s}{\partial h} \qquad (11.253)$$

where x and q_s are positive in the offshore direction and the depth h is positive. The authors present a finite differences scheme for solution.

Swart (1974a,b, 1976) used a similar upward concave beach profile, the D-profile, to describe offshore transport conditions for $d_{50} \leq 5$ mm sands. The profile extends from the wave runup limit to the "seaward boundary of intense profile changes". The D-profile essentially caters for the surf zone. A transition profile is used to join the D-profile into the more or less stationary sea bed. Erosion of the backshore above the wave runup limit is also considered separately. The upper limit of the D-profile is based on the runup formula by Hunt (1959)

$$R_u = a_1 H(H/L_o)^{b_1} \tan\beta \qquad (11.254)$$

where a_1 and b_1 are constants and $\tan\beta$ is the foreshore (wetted beach) slope, given by

$$\tan\beta = a_2 d_{50}^{c_2} \qquad (11.255)$$

where the constants a_2 and c_2 depend on sea conditions (c.f. eqn 11.247). Swart prepared a number of empirical correlations, mainly based on small scale laboratory data and some large flume data. With the aid of formulae and graphs presented the shape of a storm profile can be calculated.

A simple formula for estimation of the erosion profile in the surf zone due to a storm surge attack was proposed by Vellinga (1983):

$$7.6 \frac{y}{H_{os}} = 0.47 \left[\left[\frac{7.6}{H_{os}}\right]^{1.28} \left[\frac{w}{0.0268}\right]^{0.56} x + 18 \right] - 2.0 \qquad (11.256)$$

where y is the depth below MWL during the storm surge and x = 0 at the toe of the dune. The limit, at about the breaker line, is given by

$$x = 250 \ (H_{os}/7.6)^{1.28}(0.0268/w)^{0.56}$$

$$y = 5.72(H_{os}/7.6) \qquad (11.257)$$

where H_{os} is the deep water significant wave height and w is the fall velocity of d_{50}-sized grains in still water at 10°C. The dune erosion is given by balancing the areas of erosion and accretion, Fig. 11.60.

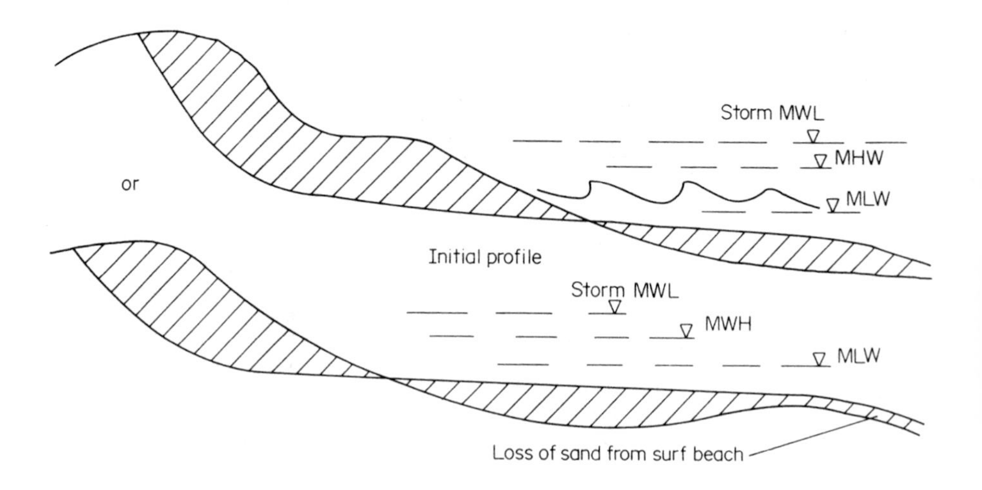

Fig. 11.60. Schematic dune erosion.

These methods essentially predict the erosion profile in the surf zone. However, frequently the eroded and deposited areas are not equal. During heavy storms substantial quantities of sand can be transported seaward of the bar. Although there the changes in bed level are not as spectacular as in the surf zone, the large areas involved can account for large volumes of sand. On the west coast of New Zealand the author measured after a storm flat sand ridges on the earlier profile in depths from 9 to 21 m with a height up to ½ m.

There are no satisfactory methods for prediction of accretion profiles. These develop slowly and the process may be interrupted by periods of steep eroding waves. Estimates can be made using measured shapes of summer profiles of the given beach.

In regions with a trend in the mean sea level, and no longshore sediment transport, the beach profile will move inshore or offshore according to whether the MSL is rising or falling. The rising MSL and lack of lateral sediment inflow lead to beach erosion problems. The extent of shoreward translation (erosion) of the beach profile is estimated by translating the original profile up and landward and matching the volumes of erosion and deposition (to reduce the depth) within the depths of active profile changes, $h < h_i$.

11.7.2 Beach nourishment

Beach nourishment is a method of dealing with a local erosion problem. It means that beach material is imported from an external source and placed to improve a beach or repair erosion damage. It is a visually and environmentally attractive method since no structures, such as groynes, breakwaters or seawalls are involved, but for most problem areas constitutes an ongoing maintenance (sand supply) commitment. Part of the design of beach nourishment is the prediction of the beach profile.

In general the beach fill material should be coarser than the native beach material. This would reduce the amount of sand lost offshore. Usually the fill material is different and some extra material is required to allow for the initial sorting and loss of fines offshore. There are no proven methods for computing the amount of overfill required. The methods of estimation of the beach fill requirement in use are derived from the work by Krumbein (1957) and Krumbein and James (1965). The object is to estimate for each size fraction the proportion which is removed to produce a material with the same size distribution as the original beach sand. This postulates that the original sand is the most stable under the given wave climate. The use of the overfill criterion developed by James (1975) is probably the best estimate available.

The quantity of sand to be placed to obtain the beach profile compatible with the native material is given by multiplying the required volume by the overfill ratio R_A, Fig. 11.61. In it σ_ϕ is the standard deviation of the sand-size distribution in terms of the ϕ-scale (eqn 2.7). A $\sigma_{\phi f} > \sigma_{\phi n}$ means that the fill is more poorly sorted than the native material. M_ϕ is the mean diameter of the grain size distribution and $M_{\phi f} > M_{\phi n}$ means that the fill is finer than the native material. Assuming log normal grain-size distribution

$$\sigma_\phi = \tfrac{1}{2}\,(\phi_{84} - \phi_{16}) \text{ and } M_\phi = \tfrac{1}{2}\,(\phi_{84} + \phi_{16})$$

A second graph shows the renourishment factor R_J, Fig. 11.62, which indicates the rate at which the fill material will erode. The factors R_A and R_J are independent estimates and can give contradictory predictions, e.g., that an overfill is required and that the fill will erode more slowly than the native material. This is the result of a simplistic description of a very complex process.

The onshore/offshore movement of sediment is also an idealised situation. Usually some currents are present which further complicate the analytical description of the geometry and lead to changes in plan geometry of the coastline.

11.7.3 Coastline shape

Coastline changes, in principle, arise from imbalance of coastwise sediment transport. This can arise from changes in input, trapping or geometry of coast. The input of sediment is from rivers and erosion of coastline. Large variations can arise from land

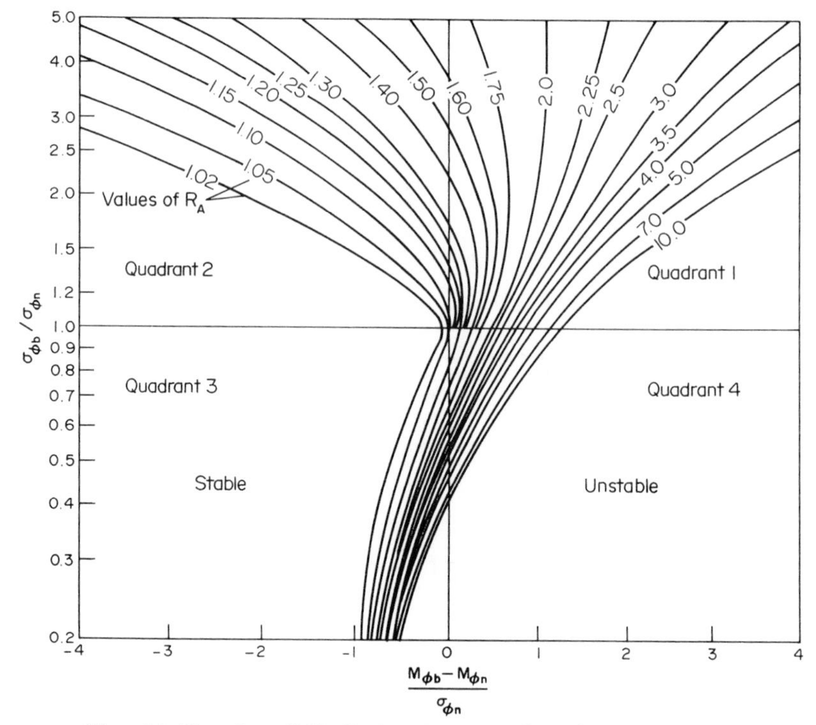

Fig. 11.61. Overfill factor R_A as a function of relative size distribution and mean grain size, after James (1975).

management practices and trapping of sediments in large reservoirs c.f., Aswan dam and Nile River delta. Sediment can also be trapped locally, for example, by man-made structures, such as breakwaters, which trap littoral drift and cause accretion on the updrift side and erosion of the downdrift shore.

A great variety of beach geometries arises from rocky headlands and other more permanent features along the coast. The shapes of the beaches between such fixtures arise from distribution of wave energy and currents due to refraction and diffraction.

In some inlets the small stable beaches may be facing each other and be more or less parallel to the direction of the main waves entering from the ocean.

A curved coastline always has a varying rate of coastwise sediment transport under given wave direction because the angle of wave attack changes. Basically, a seaward convex coast will erode and a concave accrete. In both cases the coastline tries to realign itself so as to become normal to the approaching waves. The conditions along a convex coast are illustrated in Fig. 11.63. If erosion of the shoreline between 1 and 2 is to be prevented, the sand lost could be replaced from time to time or the orientation of the shoreline could be changed to yield a constant transport rate.

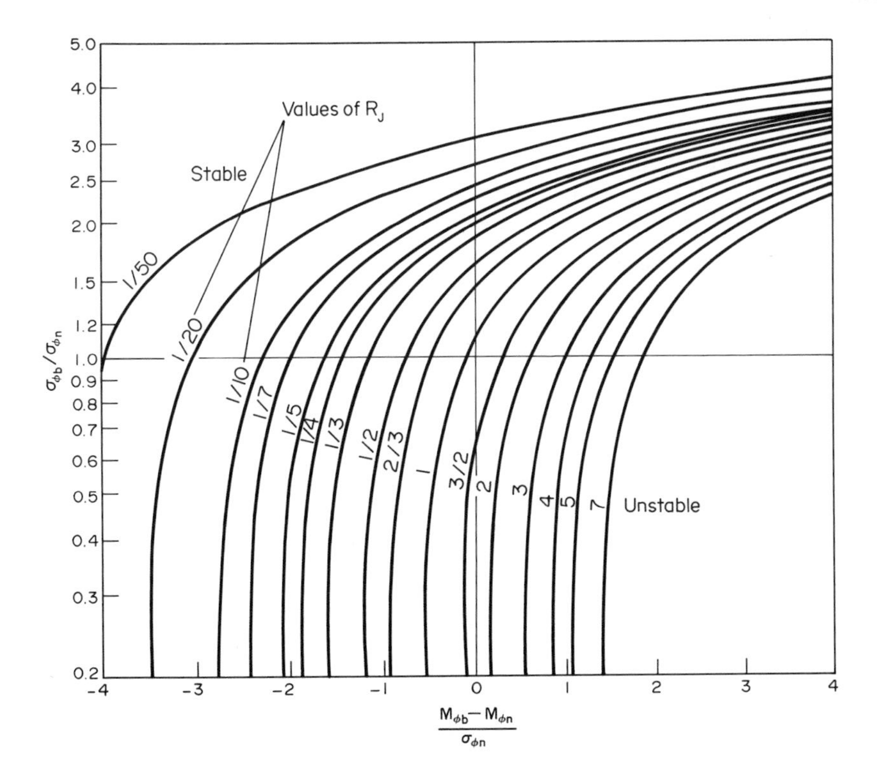

Fig. 11.62. Renourishment factor R_J as a function of relative size distribution and mean grain size, after James (1975).

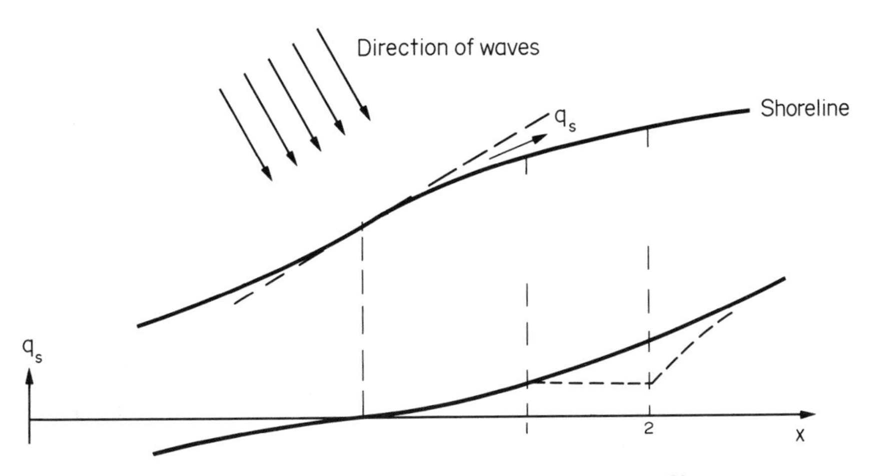

Fig. 11.63. Schematic variation of longshore sediment transport rate.

The replacement of sand lost by sand from an external source (beach nourishment) is an environmentally attractive solution but constitutes also an ongoing commitment. A constant transport rate, $\partial q_s/\partial x = 0$, could be achieved by realigning the segment of beach, for example, at the angle at location 1. If the beach is realigned normal to wave direction then the transport rate goes to zero, which means that sand will accrete at the updrift end. At the downdrift end, 2, the gradient $\partial q_s/\partial x$ is increased. The diminished transport at 1-2 leads to transport deficiency and increased erosion beyond 2. If the resulting changes of shoreline beyond 2 are acceptable the reorientation of the segment 1-2 could be achieved by structural means. In practice the stability has to be analysed for the cumulative effect of all the possible wave directions, wave height, steepness and spectra.

One of the simplest models for calculation of coastline changes is that by Pelnard-Considére (1954). With the symbols of Fig. 11.64 the continuity of sediment transport is given by

$$q_{sy}dt - (q_{sy} - dq_{sy})dt = hdxdy$$

Since

$$dq_{sy} = \frac{\partial q_{sy}}{\partial y} dy \text{ and } dx = \frac{\partial x}{\partial t} dt$$

this becomes

$$\frac{\partial q_{sy}}{\partial y} + h \frac{\partial x}{\partial t} = 0 \tag{11.258}$$

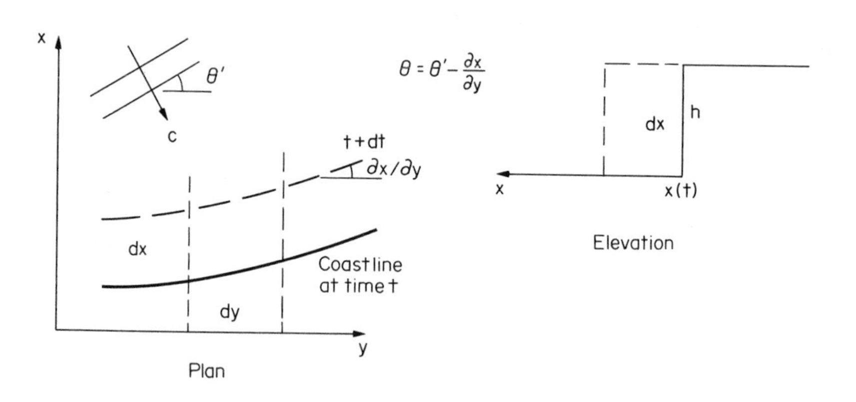

Fig. 11.64. Definition sketch.

For a constant direction of waves along the coast the transport rate, q_{sy}, is a function of the wave angle θ only, which is one of the parameters in the longshore transport formulae, e.g., eqn (11.238). Writing $\partial q_{sy}/\partial_y = (\partial q_{sy}/\partial\theta)\partial\theta/\partial y$ and observing that for small changes of $\partial\theta$ it can be replaced by $-\partial x/\partial y$ (increase in θ) and $\partial\theta/\partial y = -\partial^2 x/\partial y^2$ eqn (11.258) becomes

$$- \frac{\partial q_{sy}}{\partial \theta} \frac{\partial^2 x}{\partial y^2} + h \frac{\partial x}{\partial t} = 0 \qquad (11.259)$$

where $\partial q_{sy}/\partial \theta \cong$ constant $= q_y{}^*$ for small values and changes in θ. The angle θ is measured at depth h where the shoreline modification starts. Dividing by the constant depth h yields

$$C = \frac{q^*_y}{h} = \frac{q_s}{\theta h} \qquad (11.260)$$

and

$$C \frac{\partial^2 x}{\partial y^2} - \frac{\partial x}{\partial t} = 0 \qquad (11.261)$$

Equation (11.261) yields the new coastline location, given the boundary conditions of the coastline, shape at t = 0 and sediment transport as a function of time at two locations. For example, a groyne or breakwater as shown in Fig. 11.65.

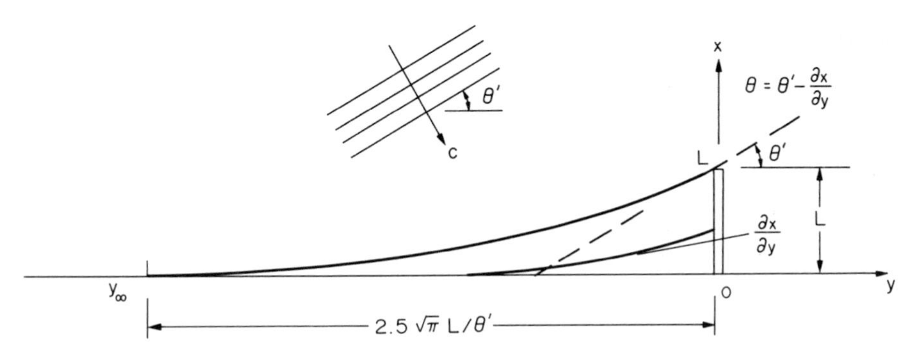

Fig. 11.65. Definition sketch.

At $t = 0$ $\quad x = 0$ for all y

$y = -\infty$, q_{sy} remains constant at q_s

$y = 0$ $\quad q_{sy} = 0$ for $t > 0$

$$\frac{\partial x}{\partial y} = \theta' \text{ for } t > 0.$$

The solution of eqn (11.261) is

$$x = \theta'(4Ct/\pi)^{\frac{1}{2}} \left[e^{-u^2} - u\sqrt{\pi}\ E \right] \qquad (11.262)$$

where

$$u = - \frac{y}{\sqrt{4Ct}} \text{ and } E = \frac{2}{\sqrt{\pi}} \int_0^\infty e^{-u^2} du$$

is the complimentary error function, i.e.,

$$\text{erfc } x = 1 - \text{erf } x = 1 - \frac{2}{\sqrt{\pi}} \int_0^u e^{-u^2} du$$

which is tabulated in mathematical tables. The magnitude of E decreases from 1 at u = 0 to about 4×10^{-4} at u = 2.5 or 2.2×10^{-5} at u = 3. Thus, the influence of the barrier becomes insignificant at $y_\infty \cong -2.5\ (4Ct)^{\frac{1}{2}}$. The advance of the coastline at the breakwater, y = 0, is from eqn (11.262)

$$x(t) = \theta' \left[\frac{4C}{\pi}\right]^{\frac{1}{2}} \sqrt{t} = 2\left[\frac{q_s\theta'}{\pi h}\right]^{\frac{1}{2}} \sqrt{t} \tag{11.263}$$

At the limit of growth when x = L and small angles θ', $x = L = \theta' y$. Generally, from eqn (11.263) $y = 2(Ct/\pi)^{\frac{1}{2}}$ and $y_\infty/y \cong 2.5\sqrt{\pi}$. Thus, y_∞ is about $4.4\ L/\theta'$.

The total area $0Ly_\infty$ of the accretion

$$A = C\theta' t \tag{11.264}$$

The real situation involves a beach slope and sediment will commence to pass around the head of the groyne when the toe of the slope arrives at the head, Fig. 11.66. The amount of by-passing is relatively small as long as the breaker line remains shoreward of the head. Significant by-passing commences when the depth at the head has been reduced to h_b. From eqn (11.263) the time for this is

$$t_L = \frac{\pi}{4} \frac{x^2 h}{q_s \theta'} \tag{11.265}$$

When sand is transported around the head the boundary condition at y = 0 has to be changed to

$$y = 0: \quad x = L + \varepsilon \text{ for all } t \geq t_L$$

where ε is a small length. At $t = t_L$, x is given by eqn (11.263) and the coastline is given by the earlier solution.

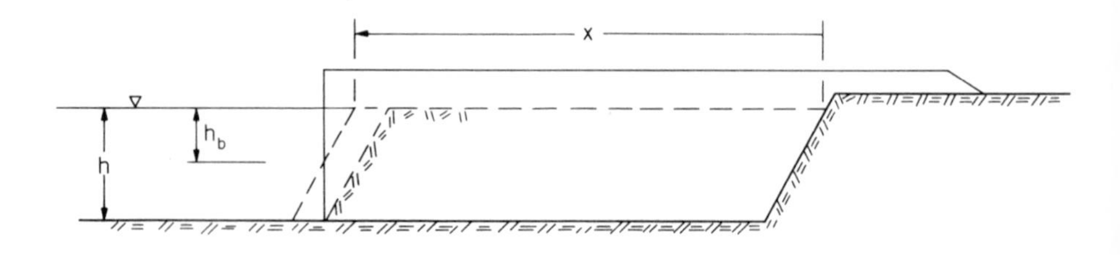

Fig. 11.66. Definition sketch.

This type of change of coastline is associated with groynes, breakwaters, etc., which intercept the littoral transport. Another common type of coastal structure is the seawall, frequently built to protect sea front land. Its effect, however, can lead to the opposite result. A wall from which a substantial fraction of wave energy is reflected leads to an increase in the steepness of the incident waves. In general, the equilibrium beach slope for a given sand is flatter for steeper waves. The only way the beach can flatten its slope is by removing material at the toe of the sea wall. This process is substantially aided by the three-dimensional velocity pattern produced in front of the wall by wave reflections whenever the waves strike the wall at an angle. The end result is a substantial lowering of the beach profile, often associated with undermining of the sea wall.

In conjunction with coastal protection works, it is important to keep in mind that these works cannot be done for one property. These have to work for a bay or a stretch of a coastline which forms a hydrodynamic unit. Nature does not recognize boundaries marked on land.

Little concrete can be said about the *shape of the coastline* in general, other than that a sandy coast is in a continuous state of change, it is always trying to reach equilibrium with the seas running in. Since the waves vary continuously in size, form, direction and spectral composition an equilibrium state might be reached for only short periods. On a longer time scale the coast is also subject to varying sediment supply through natural and man-made causes. On a longer time scale still the variation of mean sea level relative to land affects the coastline.

The basic plan geometry of our coastlines is dominated by geology, headlands, rock outcrop, etc., which provide essentially fixed points on the human time scale. All the rapid changes in geometry occur in the surf zone which is called upon to dissipate the energy of the waves. Many of the changes are small and come and go in quick succession, such as ripples and cusps, which are very eye-catching, or sand banks. However, on many coasts major features exist, such as barriers or barrier islands, spits and tombolos.

11.7.4 Barriers, spits and deltas

Barriers are features created by sediment moving towards the shore. They are associated with relatively shallow and wide coastal zones. Waves which break at the seaward parts of the shallow water gradually build up a ridge which becomes a berm and finally a barrier running parallel to the earlier shoreline. The formation of the barrier (at constant water level) is associated with a flat slope of sea bed. If the coast is steep it will be eroded and material is carried away. At times waves may wash sediment over the barrier or break through it, forming gaps. This leads to the formation of barrier islands. From sand drying on the beach during low water, winds gradually build dunes on the barrier or islands. Extensive developments of barrier islands exist along the north-west coast of the Gulf of Mexico, the east coast of the USA and in the North Sea coast of the Netherlands and Germany. For discussion of barrier islands reference is made to Leatherman (1979). Barriers can also close bays and form brackish lakes or swamps.

Spits are tongue-like formations, which usually continue the coastline in the direction of dominant sediment transport as illustrated in Fig. 11.67. Spits can also lead to closure or near closure of bays and estuaries. The spit heads are generally highly mobile and change in shape with waves and currents. On many coasts littoral transport changes direction with changing sea conditions. This can lead to a spit formation from both sides of the bay or estuary, with a gap between them. Usually the head of the spit in the dominant transport direction protrudes relative to the leeward head. The spit may at times form a long lagoon between itself and the coast, with a gap at or near the downdrift end. Such spits may be breached during major floods of river(s) discharging into the lagoon. This trend can be associated with the littoral transport which for a given wave height was seen to be proportional to sin 2θ, i.e., a maximum at $\theta = 45°$ and two equal lesser rates at $\pm \Delta\theta$. If θ is more or less than 45° erosion of spit head increases the angle but refraction of waves decreases it. Thus, the sediment supply is distributed over the face of the head. However, currents in and out of the lagoon also play an important part in shaping of the head(s).

The gaps in spits or between barrier islands subject to tidal flows, tend to be bridged by arc-shaped bars, Fig. 11.68, usually covered with slowly translating asymmetric features called swash bars or plates. The bar shape is linked to ebb current. The outward flow deflects the littoral transport seaward, tending to form a ridge, which in turn changes refraction. The angle θ is reduced and sediment accumulates on the updrift side of the gap. The initial concave form of the accumulation near the water's edge changes to convex as sediment is carried along the arc and across the tidal gap. The entire sand-bed configuration is referred to as the *ebb-tidal delta*. The head at water's edge is shaped by high water conditions and mostly shows the usual inward curved form at the downdrift end of the island. Figure 11.69 shows an aerial photo of the arc across the Wichter Ee between the islands of Norderney and Baltrum and Fig. 11.70 the sand ridges as these come ashore on Langeoog (c.f. Fig. 11.68), East Frisian Islands. Details are presented by Homeier and Kramer (1959) and Fitzgerald et al. (1984).

The dimensions of the channel connecting an estuary or a tidal area behind the barrier islands to the sea, depend on the volume of water flowing through the gap at each tide. Observations show that the average velocity in the channel is only a little higher than the threshold velocity for the sand involved and usually in the range of 0.6 - 0.8 m/s. This led to the relationship between the volume of tidal prism and the stable cross-sectional area of the channel (the O'Brien equation of 1931, see O'Brien 1969). Since then additional data have been collected. The SPM (1984) quotes the results by Jarrett (1976), according to which the minimum cross-sectional area, A_c, of the channel (below MSL) and the spring tide volume of tidal prism are related by

$$A_c = 5.74 \times 10^{-5} P^{0.95} \qquad (11.266)$$

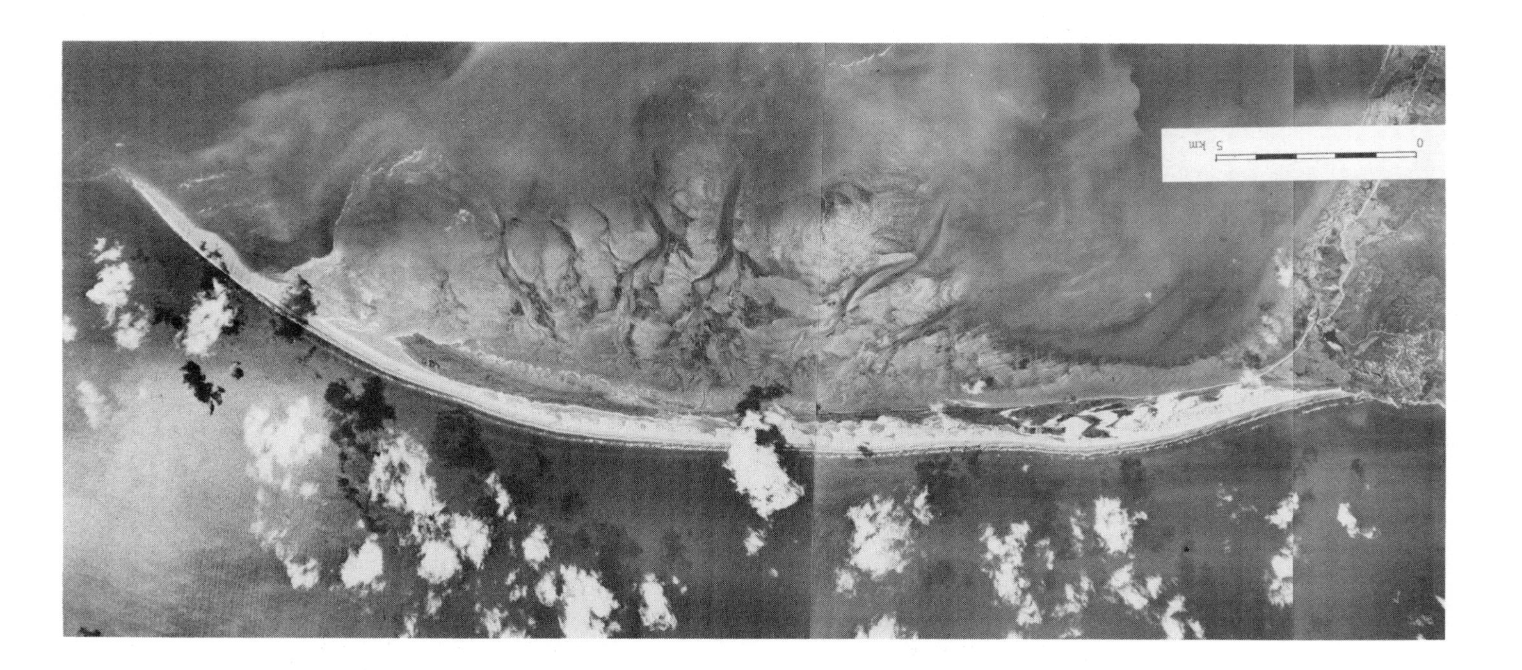

Fig. 11.67. Farewell Spit at the northwestern end of South Island, New Zealand (172°50' E, 40°31' S, approx.), running west to east with the end curving to southeast. (Photo by NZ Aerial Mapping, 2377;A1,A2,A5; 21-11-70).

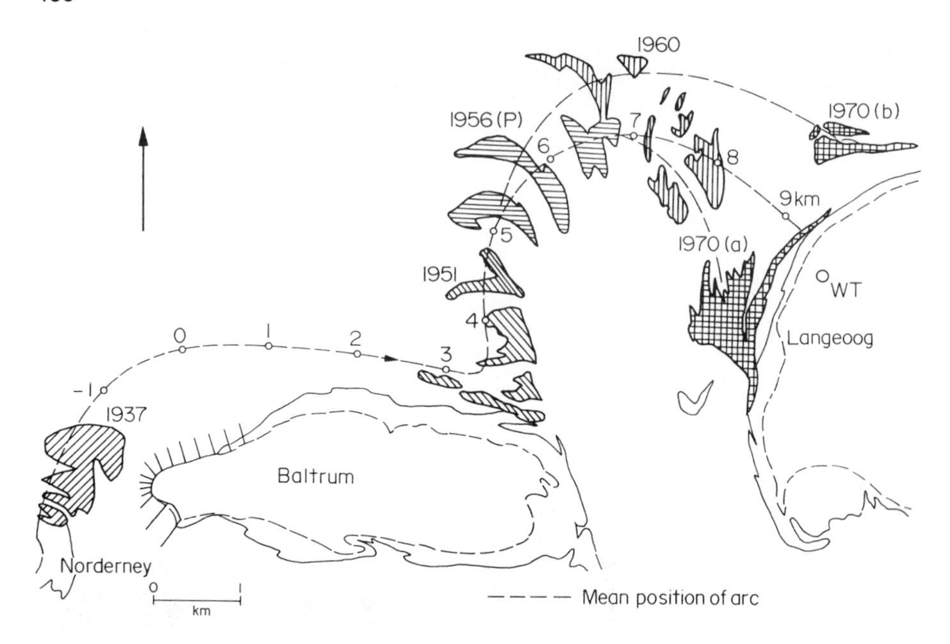

Fig. 11.68. Path of sediment movement between barrier islands, from Norderney to Langeoog 1937/1970. The bars translate about 400 m per annum. Distance between numbered points 1 km. (Courtesy Bauamt für Küstenschutz, Norden).

where A is in m^2 and P in m^3. Small differences were noted between Atlantic, Gulf and Pacific coasts. Vincent and Corson (1980) also related A_C and the channel length. Their data can be described by

$$A_C = 0.046 \, L^{2.1} \tag{11.267}$$

where L is in m.

The volume of the tidal prism, P, is equal to $\bar{U}A_C T/2$ or

$$A_C \cong \frac{2P}{\bar{U}T} \tag{11.268}$$

which with T = 12.4 hours and U = 0.7 m/s yields

$$A_C = 6 \times 10^{-5} P \tag{11.269}$$

The beds of tidal inlet channels are frequently armoured with shell. This tends to reduce depth and increases width.

Fig. 11.69. Aerial view of the arc across the gap between Norderney and Baltrum in the North Sea.

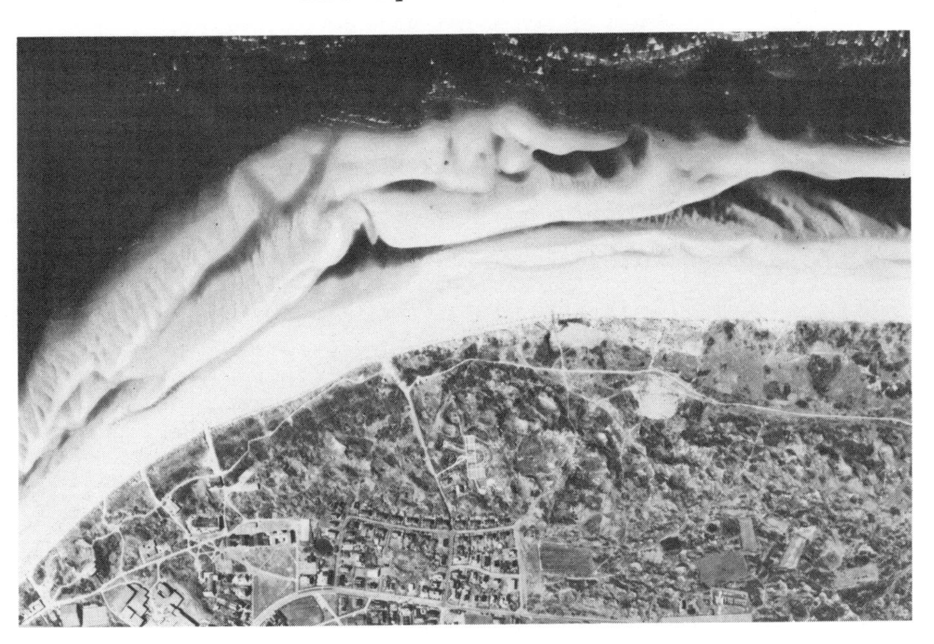

Fig. 11.70. Aerial view of the sand transport "coming ashore" from the arc bridging Baltrum and Langeoog, c.f. Fig. 11.66.

The problem of water level variation in the lagoons due to tides is treated in the 1984 edition of SPM, including graphs for quick estimation.

Rivers discharging into the sea develop localised features known as *deltas*. Their shape and size is dependent on the water and sediment discharge of the river, on coastal currents, wave climate, tidal range and coastal sediment transport. In the absence of wave action or currents a pear-shaped delta develops. As the delta grows in length, resistance to flow increases and this leads to the river "breaking out" sideways. This process is repeated at some other location, as the headloss in this new channel increases, and gradually the typical multi-channel delta evolves. Wave action will spread the sedimant leading to a bulge rather than the pear-shaped delta. Superimposed coastal currents can further modify the shape or restrict the delta formation to the estuary as for the Amazon River. Here the South Equatorial current reaches longshore velocities of up to 4 knots and distributes the enormous sediment load of the Amazon to the northwest. Features of major deltas like the Mississippi, the Nile, the Niger, the Ganges, etc are apparent on maps of the regions.

Offshore features, such as islands or rock outcrops, can by interception of wave energy shelter a stretch of coastline. This leads to some of the littoral transport being deposited in the shadow area. Gradually water depth decreases, refraction pattern changes, and the deposit may grow into a strip of land joining the island to the shore. Such accretions are known as *tombolas*. Offshore breakwaters are used to stimulate this type of accretion and to protect a stretch of a beach. However, the effects on the downdrift coastline have to be carefully evaluated. The breakwater is usually designed to provide only patial protection and maintain a certain littoral transport rate.

Coastal morphology is the manifestation of long term average sediment transports due to waves and currents. Only in the surf zone are the changes rapid and at times visually dramatic. The major contribution to long term average sediment transport is made by the usual wave and current conditions. The contribution of the extreme events is small because these occur too infrequently. The most effective transport conditions arise from the joint action of waves and current.

For a given location the depth averaged mean current speeds, V_i, have a probability distribution $P_c(V_i)$. For a given depth and sediment the sediment transport rate, g_c, due to current only is a function of the velocity and its long term average is $\bar{g}_c = \Sigma P_c(V_i) g_c(V_i)$. The proportion, $S_c(V_i)$, of the long term transport rate, $\bar{g}_c$, contributed by the current in the interval V_i is

$$S_c(V_i) = \frac{P_c(V_i) g_c(V_i)}{\bar{g}_c}$$

Likewise, there is a probability distribution for the occurrence of the rms-value of orbital motion, u_i, due to wave action. The simplest approach to estimation of u_i is to divide the record of significant wave height, H_s, and the corresponding zero crossing wave periods, T_z, into intervals, e.g., 0.5 m and 0.5 s. The u-value corresponding to each increment would then be calculated with the linearised wave theory. An improvement on this is the approach by Soulsby and Smallman (1986) who used linear wave theory at each frequency of the waves in the JONSWAP spectrum, defined by H_s and T_z, and fitted a simple function which describes the near bottom rms-value of orbital velocity (see eqn 11.231). The u_i values produce a scatter diagram on the H_s - T_z plane in terms of the number of occurrences of u_i within a particular velocity increment. From this diagram a probability $P_w(u_i)$ distribution can be prepared.

Soulsby (1987) assumed that these two probabilities are independent. Thus, the joint probability of waves and current is $P_{cw}(V,u) = P_c(V)P_w(u)$. Each such event causes a sediment transport rate $g_{cw}(V_i,u_i)$. The resultant long term mean sediment transport rate is

$$\overline{g}_{cw} = \sum_i \sum_j P_{cw}(V_i,u_j)\, g_{cw}(V_i,u_j)$$

and the proportion S_{cw} of its due to V_i and u_j is

$$S_{cw}(V_i,u_j) = \frac{P_{cw}(V_i,u_j)\, g_{cw}(V_i,u_j)}{\overline{g_{cw}}}$$

Using data from the Dowsing Light Vessel (Fortnum, 1981) Soulsby (1987) prepared a joint probability diagram $P_{cw}(V_i,u_j)$, shown in Fig. 11.71. The probabilities are seen to be concentrated at the lower left hand corner, i.e., at low values of u_i and V_i. He used the formula by Grass (1981), eqn (11.231) to estimate the sediment transport by the combined action of waves and current. The transport rate increases exponentially with u and V leading to a transport rate diagram with the high transports concentrated to the top right hand corner area, i.e., the opposite arrangement to that seen in Fig. 11.71.

The long term mean transport rate is given by the product of the appropriate entries of these two diagrams with the opposing trends. The total time average transport rate $\overline{g_{cw}}$ is given by the sum of these products and it can be used to normalise the individual products. The result of such a procedure applied to the Dowsing data is shown in Fig. 11.72. The main feature of this diagram is the concentration of transport around the larger but not too infrequent wave action. The largest observed waves are seen to make only a minor contribution.

For further reading on coastal zone reference is made to Davis, Jr. (1978) and Stanley and Swift (1976).

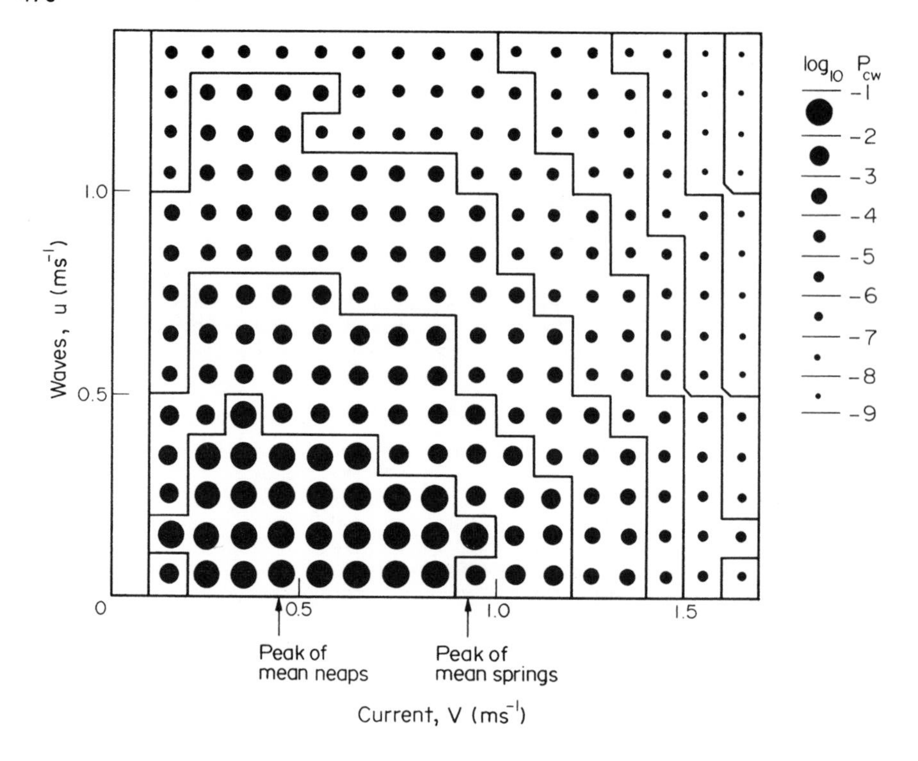

Fig. 11.71. Joint probability distribution $P_{cw}(V_i, u_j)$ for Dowsing data, after Soulsby (1987).

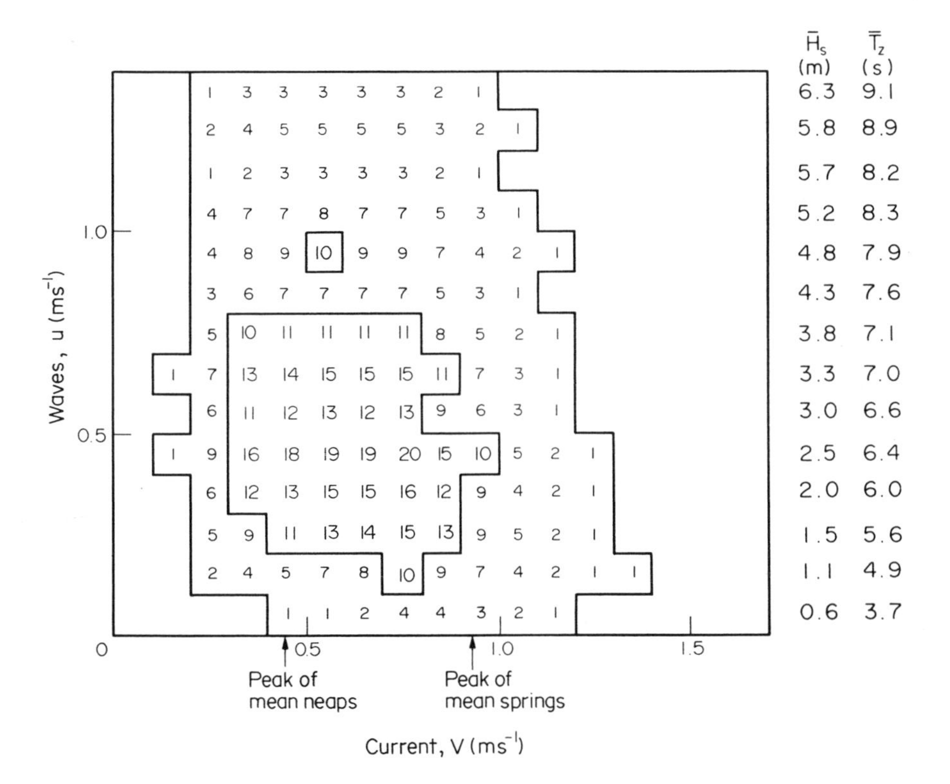

Fig. 11.72. Sediment transport as a function of waves and currents in parts per thousand of total based on Dowsing data. Values less than 1 ppt are not shown, (Soulsby, 1987).

Chapter 12

Transport in Pipelines

Transport in pipelines is not strictly loose boundary hydraulics, since the boundaries usually are fixed. However, because pipeline transport is closely linked to river and coastline engineering problems a brief discussion of its features is included. Transport of solids by fluids in pipelines has a very wide variety of applications: dredging, transport of sand, coal, etc.; pumping of wood-pulp, industrial processes, pneumatic transport of grain, wood chips, etc. Sundeberg (Stockholm) even has a pneumatic solid-waste-removal system.

The very brief discussion below is confined to transport by water. Pneumatic conveyancing is outside the scope here but for discussion of the physics reference is made to the paper by Owen (1969).

Transport of solid-fluid mixtures through pipelines can be subdivided into four characteristic modes (Fig. 12.1). The demarcation is not distinct, and no sudden changes occur at these lines, but the mechanics of flow is sufficiently different to warrant the classification.

Solids finer than about 40 μm form readily into homogeneous suspensions and coarser solids approach this condition with increasing velocity of flow. Heterogeneous suspensions are maintained by turbulence, but with graded solids the smaller particles may be transported as homogeneous suspension while the coarser form a heterogeneous suspension. The flow regimes with bed load and with a stationary bed are normally avoided because of the reduction in cross-section of the pipe and the increased resistance caused by ripples on the stationary bed.

The flow can also be divided into regimes according to concentration as shown diagrammatically in Fig. 12.2.

The primary design parameters for a given system are
 (i) head loss, energy gradient,
(ii) sediment concentration that can be transported,

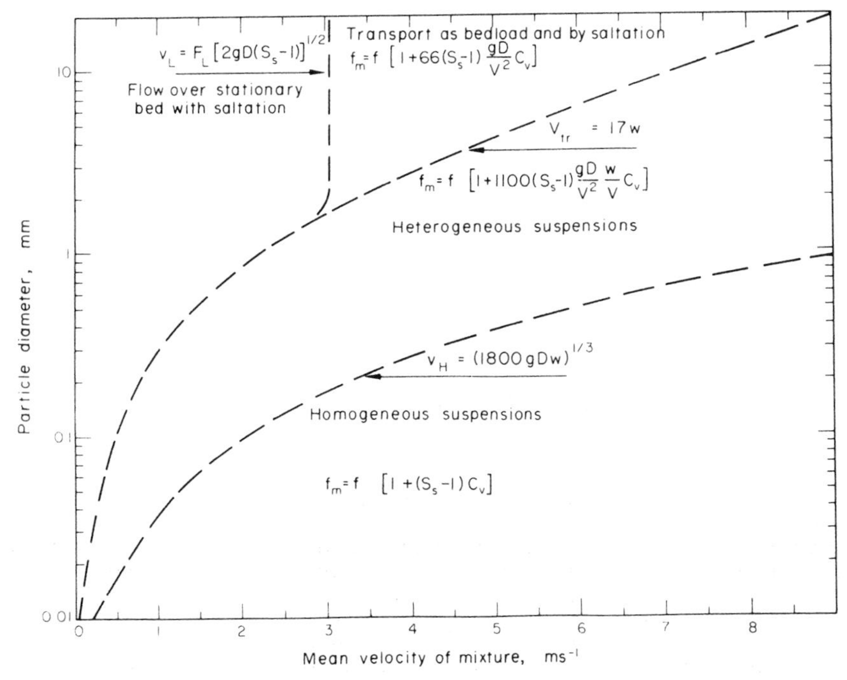

Fig. 12.1. Flow regimes in a 150-mm-dia. pipeline carrying water and solids with specific gravity of 2.65, according to Newitt et al. (1955).

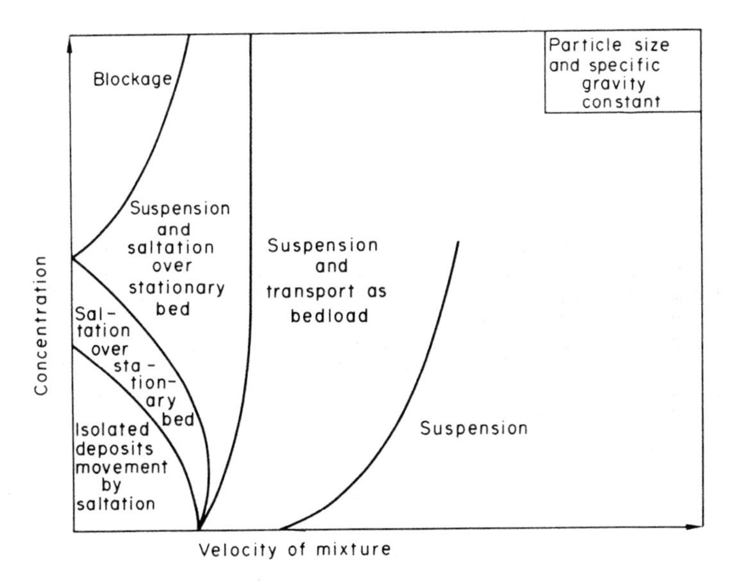

Fig. 12.2. Flow regimes as a function of concentration.

(iii) rate of abrasion of pipe,
(iv) rate of particle attrition.

Only the first two topics will be discussed here. Figure 12.3 shows a typical head loss relationship for a horizontal pipe. However, when a stationary bed develops, the head loss curve at lower velocities falls again, as was shown by Acaroglu and Graf (1969) (Fig. 12.4). For a given size of pipe and sediment the excess pressure gradient varies with concentration, c, as shown in Fig. 12.5. These curves form a single line if the data is plotted with $(i_m - i)/c_v$ as the ordinate. The same type of reduction is affected for the family of curves with the pipe diameter as the parameter by converting the abscissa to $V/\sqrt{(gD)}$. Likewise, some of the data can be condensed by introduction of $(S_s - 1)$ and the drag coefficient C_D. However, a comprehensive solution valid through all the regimes and fluid/solid combinations is still outstanding.

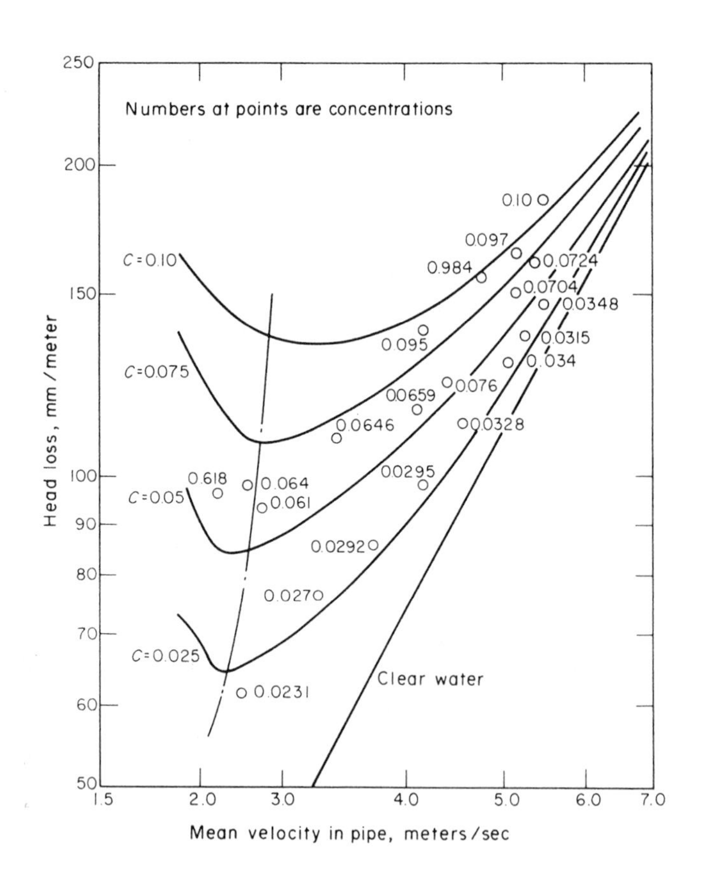

Fig. 12.3. Effect of concentration of suspensions on head loss in pipes. Coarse sand d = 2.04 mm. (After Durand).

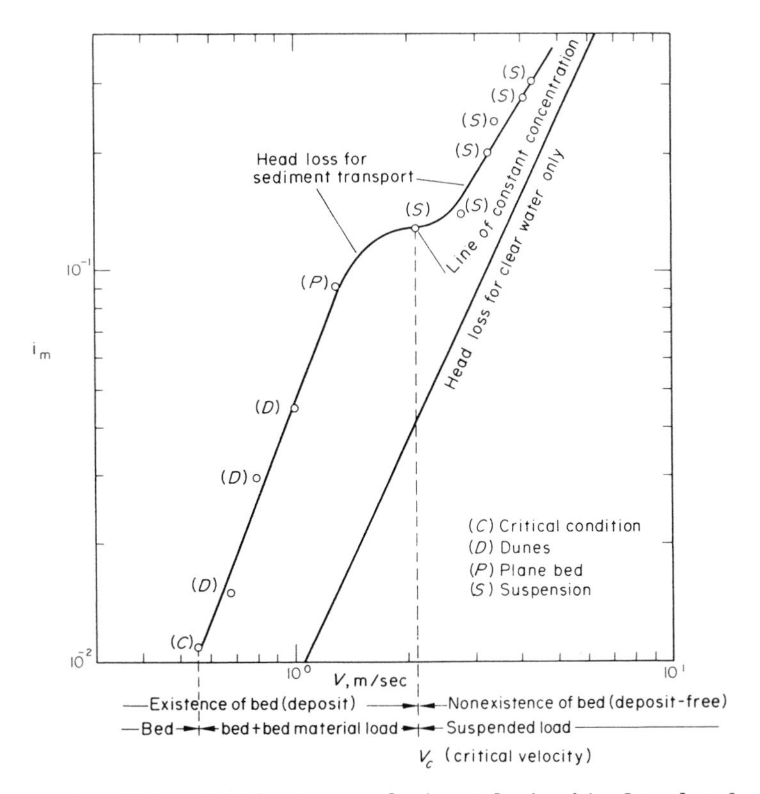

Fig. 12.4. Head loss vs. velocity relationship for closed-conduit flow, for sand with d = 2.00 mm. (After Acaroglu and Graf, 1969.)

12.1 Homogeneous Suspensions in Horizontal Pipe

The homogeneous suspensions can be subdivided into two groups: suspensions of fine-grained sediments at high velocities and the suspension of so-called non-settling slurries which involve particles of less than 30 μm, fibre suspensions, etc.

The basic mechanism of suspension of particles denser than the fluid and coarse enough to settle readily (w > ~ 1 mm/s) is the turbulent exchange of water from lower levels with higher particle concentration with that at higher levels with lower particle concentration. This mixing has to balance the steady downward fall of the particles by gravity. The critical Reynolds number for onset of turbulence is still about 2000 if the apparent viscosity, discussed in Chapter 2, is used. This mode of homogeneous suspension may be looked upon as the asymptote to which the heterogeneous suspensions approach with increasing velocity.

However, the major group of materials in this regime are the very fine-grained slurries and fibrous solids which may also be maintained in suspension in laminar flows. These particles interact

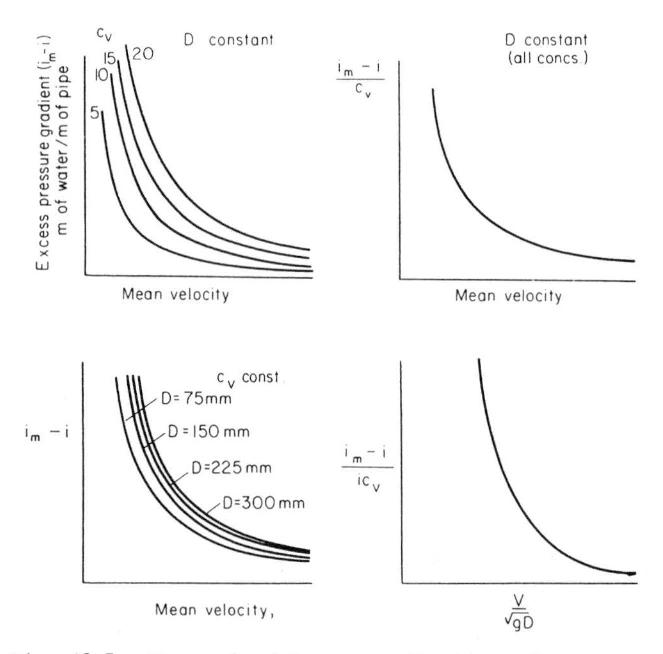

Fig. 12.5. Excess head loss as a function of concentration and pipe size.

with the fluid and both lose their individual characteristics. Such interaction may occur even at very low concentrations. For example, a few per cent by volume of bentonite or of fibrous particles can drastically alter the flow resistance. Because of the change of characteristics the designer must know the properties of the slurry rather than of its components. Even slurries of inert fine materials may not behave as newtonian fluids. The stress-strain features of the mixture can be determined by measuring the shear stress τ versus dy/dy. This will show whether the slurry behaves as a newtonian fluid, is dilatant, pseudoplastic, Bingham fluid, etc. With fine-grained slurries it is also necessary to determine whether the slurry is purely viscous, visco-elastic or has time dependent properties. After that the model to describe the problem can be selected, but this is a topic in its own right and dealt with in books on Rheology. For introductory reading on these slurries see Bain and Bonnington (1970), Wasp et al. (1977), Zandi (1971) and on wood pulp Duffy and Lee (1978).

For inert granular material (sands up to 1.5 mm) experiments have shown that above a certain velocity the head loss, in terms of the column of sediment water mixture in a given pipeline, is approximately the same as that for clear water in terms of the column of water, i.e.

$$f_m = f[1 + (S_s - 1)c_v] \qquad (12.1)$$

where f_m and f are the Darcy-Weisbach friction factors for mixtures and water respectively and c_v is volumetric concentration. In terms of energy gradient i

$$\frac{i_m - i}{c_v i} = S_s - 1 \qquad (12.2)$$

Equations (12.1) and (12.2) are approximate only. Observations are on record where $(i_m - i)/c_v i$ is greater than $(S_s - 1)$ as well as less. Reduction of head loss, compared to clear water, occurs with fine particles, generally less than 0.15 mm but normally much smaller. The best known is the effect of addition on bentonite clay to water. Departures may become significant with increasing concentration. With $c_v > 6\%$ the mixture may no longer have newtonian properties. Yet in some instances good results have been obtained with $c_v \sim 15\text{-}20\%$. The apparent viscosity should be used in calculation, because this reduces the value of Re and may increase f.

The critical velocity separating the homogeneous and heterogeneous flow regions is according to Newitt et al. (1955)

$$V_H = (1800gDw)^{1/3} \qquad (12.3)$$

12.2 Heterogeneous Suspensions in Horizontal Pipe

The heterogeneous suspension regime is the most important mode of transport of granular materials by pipelines, because here the maximum amount of solids is transported per unit energy input. Among the earlier studies the most prominent are those by Durand Condolios (1952). Their formula with various modifications is still the primary one in use. Durand and his coworkers at Sogreah, Grenoble, carried out extensive tests with sands, gravels and coal ranging in particle size from 0.2 to 25 mm in pipelines with D = 40 to 580 mm and c_v from 2 to 23% (50 to 600 g/l). Figure 12.3 is one of their plots of experimental data. They found that the results condensed into a well-defined curve on the $\phi = (i_m - i)/ic_v$ versus $(V^2/gd)\sqrt{C_x}$ plane (Fig. 12.6) on which 310 data points are shown. They concluded that the data are described by

$$\Phi_D = \frac{i_m - i}{ic_v} = K_D \left[\frac{\sqrt{(gD)}}{V}\right]^3 \left[\frac{1}{\sqrt{C_x}}\right]^{3/2} \qquad (12.4)$$

where C_x is the drag coefficient of grains at terminal fall velocity in still water, i.e. $C_x = C_D$

$$C_D = \frac{4}{3}\,\frac{gd(S_s - 1)}{w^2} \qquad (12.5)$$

and $K_D \cong 17.6$. The modification of the $V/\sqrt{gD}$ abscissa of Fig. 12.5 was based on the observation that the pressure drop of the mixture increased with the increasing particle size up to about $d \cong 2$ mm. The only particle parameter which remains constant for

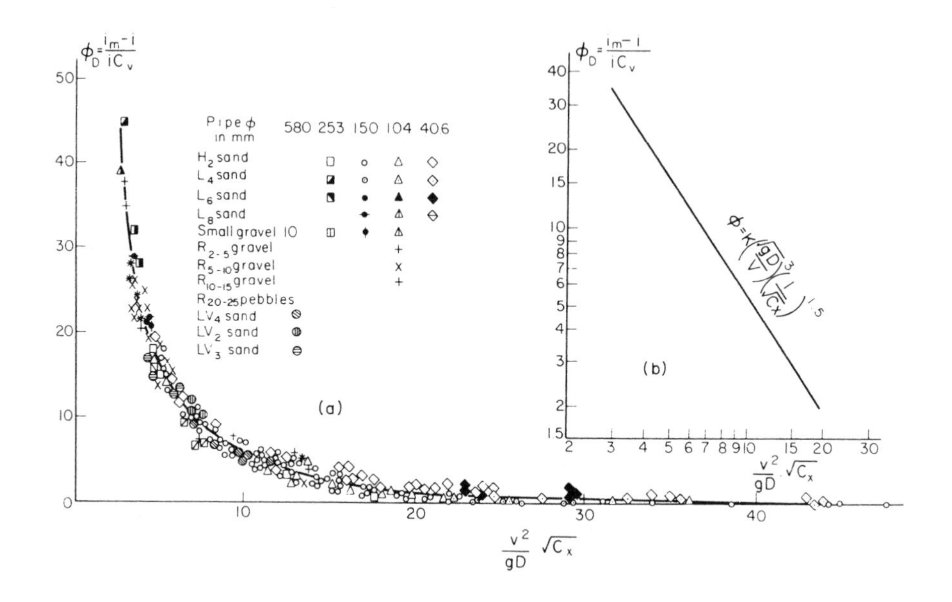

Fig. 12.6. Head losses in pipes with non-deposit flow regimes (after Durand).

larger grains is the drag coefficient. Equation (12.4) is basically

$$\phi_D = Kf_1\left[\frac{V^2}{gD}\right]f_2\left[\frac{w^2}{gd}\right]f_3(S_s - 1) \tag{12.6}$$

where K, and the functions f_1, f_2 and f_3 have to be determined from experiment for the particular set of conditions. By substitution of eqn (12.5) into (12.4), we obtain

$$\phi_D = K\left[\frac{V^2}{gD(S_s - 1)}\sqrt{\left[\frac{gD(S_s - 1)}{w^2}\right]^{\frac{1}{2}}}\right]^{-3/2} \tag{12.7}$$

Newitt et al. (1955) and Bonnington (1959) quoted values of K_D = 150 and K = 121. The latter with S_s = 2.65 becomes 180. Zandi and Govatos (1967) found that Durand's and other data totalling 1452 points were satisfied by

$$\phi_D = 176\left[\frac{\sqrt{gD}}{V}\right]^3\left[\frac{1}{\sqrt{CD}}\right]^{1.5} \tag{12.8}$$

or

$$\phi_D = 81\left[\frac{\sqrt{gD}}{V}\right]^3\left[\frac{S_s - 1}{\sqrt{C_D}}\right]^{1.5} \tag{12.9}$$

They also introduced

$$\psi = \frac{V^2 \sqrt{C_D}}{gD(S_s - 1)} \tag{12.10}$$

and proposed

$$\phi = 280\psi^{-1.93} \; ; \quad \psi < 10 \tag{12.11}$$

$$\phi = 6.3\psi^{-0.354} ; \quad \psi > 10 \tag{12.12}$$

In this notation eqn (12.9) is $\phi = 81\psi^{-1.5}$. Equations (12.9), (12.11) and (12.12) are compared in Fig. 12.7. In terms of the above parameters eqn (12.2) becomes

$$\phi = S_s - 1$$

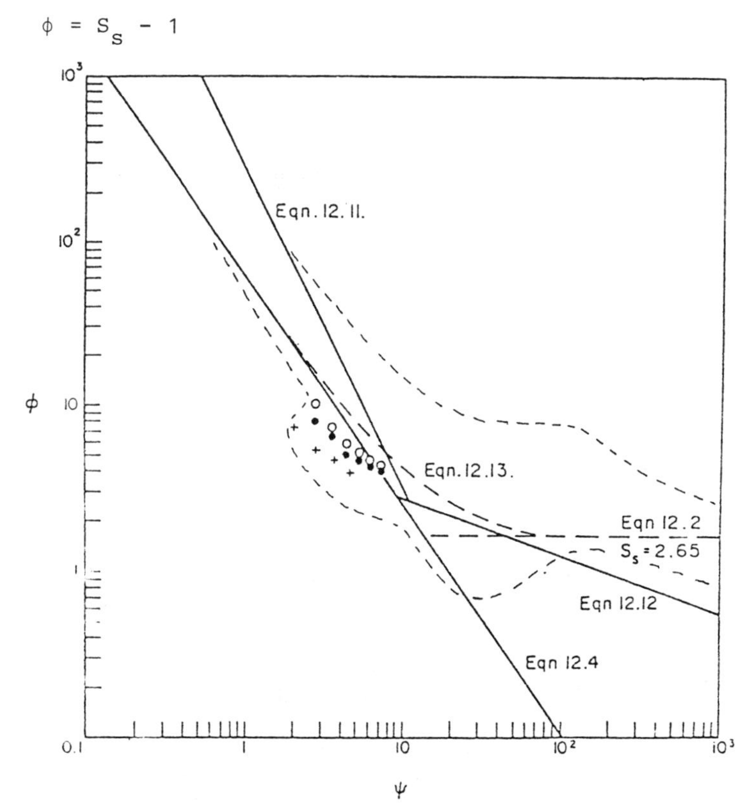

Fig. 12.7. Various transport relationships on the ϕ-ψ plane. Points refer to experimental data: $S_s = 4.7$, $d_{50} = 0.11$ mm, D = 100.2 mm, , $c_V = 13\%$, o, $c_V = 25\%$; D = 311.2 mm; , $c_V = 13\%$, +, $c_V = 25\%$. Results with sands with $S_s = 2.65$ usually cluster close to the functions represented by lines. Thin dotted lines indicate the limits of spread of experimental data.

which is a horizontal line in Fig. 12.7. The discontinuity at the intersection of this line with that by eqn (12.9) is not real. To remove this discontinuity Charles (1970) proposed

$$\phi = K\psi^{-1.5} + (S_s - 1) \qquad (12.13)$$

which provides a smooth transition between ϕ and ψ throughout the heterogeneous and homogeneous regimes. Another result contrary to observation is that as $V \to \infty$ eqn (12.9) shows that $i_m - i \to 0$. When comparing data it has to be kept in mind that $(i_m - i)$ is correlated. This means that when ϕ is small, large differences in ϕ mean relatively small differences in i_m.

It was shown by Babcock (1968) that an improved presentation of data is obtained by plotting $\phi(S_s - 1)$ as ordinate and V^2/gD as abscissa. This will retain the relative positions of data points with respect to their velocities.

As is clear from eqn (12.6) many forms of this equation are possible. For example, Haydon and Stelson (1971) proposed

$$\frac{i_m - i}{c_v i} = 100\left[\frac{gDw(S_s - 1)}{V^2\sqrt{\{gd(S_s - 1)\}}}\right]^{1.3} \qquad (12.14)$$

for D = 25 and 50 mm, S_s = 2.60 - 2.66 and 0.18 < d < 3.7 mm.

Rose and Duckworth (1969) correlate the friction factor, for transport of coarse materials in suspension by liquids and gases, to mass flow ratio m^* (mass flow rate of solids to fluids), d/D, coefficient of restitution ε (unity in liquids), density ratio, influence of inclination of pipe θ and V^2/gD. The solid friction factor is expressed as

$$f_s = f(m^*)f(d/D)f(\varepsilon)\ f(\rho_s/\rho)f(\theta)f(V^2/gD).$$

The fluid friction is the usual

$$f_f = f(Re, k_s/D)$$

relationship, and

$$f_m = f_s + f_f$$

The various functions are given in graphical form. Similar relationships given for the acceleration length of pipe, acceleration pressure, pressure drop for established flow and due to bends. Until a clear analytical theory is developed there will be no answer to the question of which formula is the best. Particularly in the approximate region of $2 < \psi < 50$ and $1 < \phi < 10$ of Fig. 12.7, both the slopes and constant can assume a variety of values. Zandi and Govatos concluded that eqn (12.9) was invalid for solids transported by saltation. Equations (12.11) and (12.12) were fitted to data which excluded points relating to transport in saltation. Yet, scatter of points on

Fig. 12.6 is small and one has no reason to mistrust eqn (12.9) as long as it is applied to conditions within the range upon which the equation is based. Field data seems to indicate that eqn (12.9) may yield good results for concentrations as high as 30%.

Frequently it is necessary to scale up data obtained from tests with small diameter pipes to large diameter pipes. For homogeneous slurries, methods based on rheological data obtained by viscometer tests are given by Cheng (1970), Hanks and Dadia (1971) and Kemblowski and Kolodziejski (1973) and Kazanskij (1978). A design procedure for scaling up pipe-loop test results on slurry at a particular concentration, in both laminar and turbulent regimes, was given by Bowen (1961). The method applies to viscous liquids. The case where slip occurs at the wall is discussed by Kenchington (1974) and Harris and Quader (1971). The simplest forms of the scaling equations are $i_m - i \propto V^x D^y$ and the constants are determined by linear regression from the test data. For example, Thomas, A.D. (private communication) obtained for c = 12% by volume of illmenite ($S_s = 4.5$, $d_{50} \cong$ 165 - 170 μm) $i_m - i = 269\, D^{-0.0261} V^{-1.2} + 527\, D^{-1.2} V^{1.8}$ where i_m and i are in mm H_2O/m, D is in mm and V is in m/s.

Newitt et al. (1955) also subdivided the heterogeneous flow into "suspension" and "flow with a moving bed". Their experiments covered $5.7 < w < 253$ mm/s, $1.18 < S_s < 4.60$ and $0 < c_v < 37\%$ but only one pipe size D = 25.4 mm. For suspension they proposed

$$\frac{i_m - i}{ic_v} = 1100(S_s - 1)\,\frac{gD}{V^2}\,\frac{w}{V} \tag{12.15}$$

and for flow with moving bed

$$\frac{i_m - i}{ic_v} = 66(S_s - 1)\,\frac{gD}{V^2} \tag{12.16}$$

Equation (12.15) can be compared with eqn (12.9). In both $\phi \propto V^{-3}$ but there are minor differences in regard to D,d,w and S_s. Equating eqns (12.15) and (12.16) yields an expression for the transition velocity from suspension to flow with a moving bed

$$V_{tr} = 17w \tag{12.17}$$

Equations (12.15) and (12.16) can also be converted by use of $i = f(1/D)(V^2/2g)$ into

$$f_m = f\left[1 + 1100(S_s - 1)\,\frac{gD}{V^2}\,\frac{w}{V}\,c_v\right] \tag{12.18}$$

and

$$f_m = f\left[1 + 66(S_s - 1)\,\frac{gD}{V^2}\,c_v\right] \tag{12.19}$$

Lazarus and Neilson (1978) proposed for the heterogeneous region a friction factor relationship[2]

$$\frac{f_m}{f} = 1.45\ Y^{-0.4} \tag{12.20}$$

where

$$Y = \frac{\rho}{\rho_s}\frac{V^2}{gD}\left[\frac{\nu}{VDc_w}\right]^{\frac{1}{2}}\left[1000\left[\frac{d}{D}\right]^{0.44}\log\frac{d}{D} + 1.31\right]^{\tanh(1 + c_w)}$$

and c_w is concentration by weight. The heterogeneous regime is said to exist for $0.8 < Y < 2.35$.

Kazanskij (1978) expressed the energy loss in terms of equivalent sand grain roughness k_s

$$\frac{i_m - i}{i\ c_v} = 1.58 \times 10^4 (k_s/D)^{2/3}(gD/V^2)(S_s - 1)[w/(gd)^{\frac{1}{2}}] \tag{12.21}$$

Relationships similar to those by Newitt et al. were also derived by Führböter (1961).

The criterion proposed for separation of saltation from heterogeneous suspension by Zandi and Govatos is the index of flow regime

$$N_I = \frac{V^2\sqrt{C_D}}{c_v Dg(S_s - 1)} \tag{12.22}$$

$N_I < 40$ is said to relate to transport with bed load.

12.3 Limit or Deposit Velocity

The velocity which separates heterogeneous suspension from transport with bed load is not unique. It depends whether the flow is decreasing or increasing in passing from one regime to another, and is lower for decreasing flow. Durand and Condolios introduced V_L as the limiting velocity for deposition. Another criterion in use is the velocity corresponding to minimum pressure drop, denoted here by V_m. Figure 12.8 shows diagrammatically the relationship between V_L and V_m. The limit velocity increases first with concentration, up to $c_v \sim 10\text{-}15\%$, and remains constant thereafter. Above a particle size of about 0.5 mm the particle size has very little effect on the limit velocity. For a given concentration and particle size the start of deposition occurs at constant values of $V_L/\sqrt{gD}$, or

$$\frac{V_L}{\sqrt{\{gD(S_s - 1)\}}} = F_L \tag{12.23}$$

Figure 12.9 shows this relationship as given by Durand and by Condolios and Chapus (1963) and also a relationship between $V_L/\sqrt{(2gD)}$ versus c_v for particles larger than 0.5 mm.

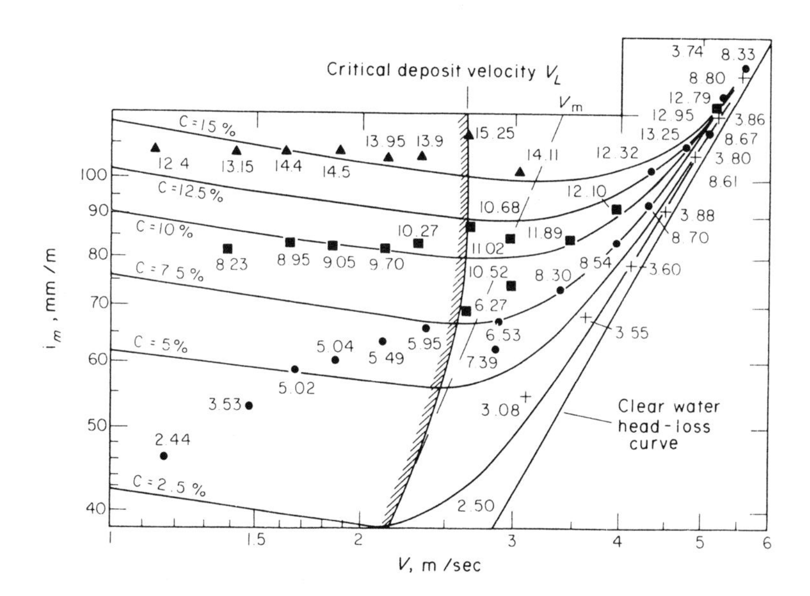

Fig. 12.8. Head loss vs. velocity relationship with equi-concentration lines, for sand graded to 0.4 mm. (After Condolios and Chapus, 1963.)

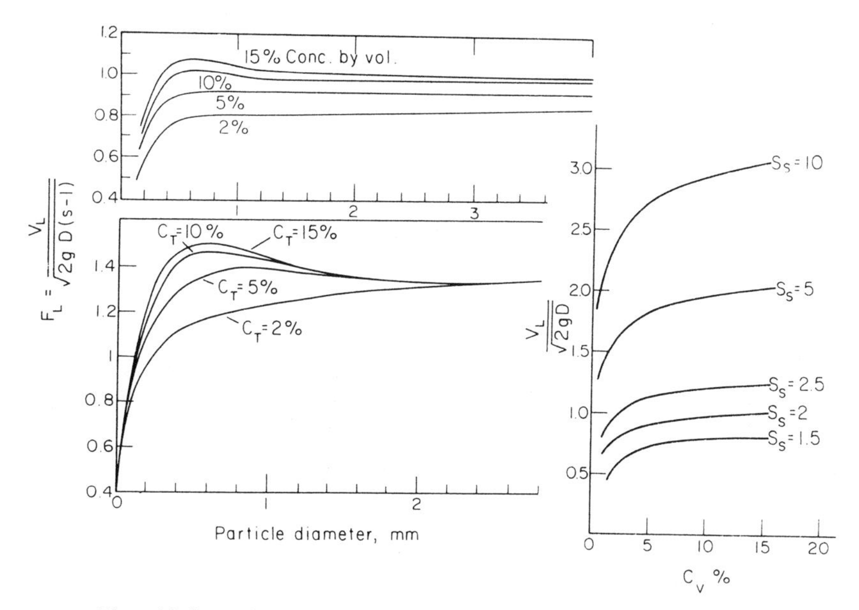

Fig. 12.9. Limit deposit velocity after Condolios and Chapus, after Durand, and as a function of c_V and S_S. (After Bain and Bonnington, 1970).

Yufin (1949) defined V_L as

$$V_L = 14.23d^{0.65}D^{0.54}\exp\{1.36[1 + c_v(S_s - 1)]^{0.5}d^{-0.13}\}$$

$$10 \text{ mm} < D < 455 \text{ mm},\ 0.25 < d < 7.36 \text{ mm, and } 0 < c_v < 30\% \quad (12.24)$$

The limiting deposit velocity also depends on the fluid-particle system, ρ_s/ρ, d/D and c_v as seen from Fig. 12.10 and Fig. 12.11 by Sinclair (1962). The maximum value is correlated by

$$\frac{V^2_{L\ max}}{gd_{85}(S_s - 1)^{0.8}} = f\left[\frac{d_{85}}{D}\right] \quad (12.25)$$

where $f(d_{85}/D) = 650$ in laminar flow and $f(d_{85}/D) = 0.19(d_{85}/D)^{-2}$ in fully turbulent flow.

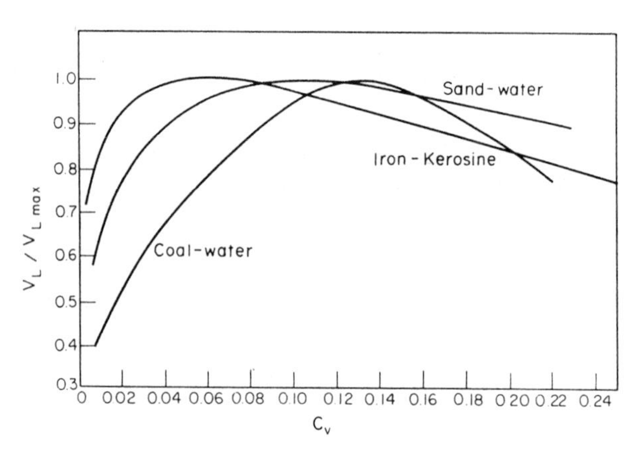

Fig. 12.10. $V_L/V_{L\ max}$ for three solids-liquid suspensions. (After Sinclair, 1962.)

The experiments by Führböter (1961) with sands $0.19 \leq d \leq 0.88$ mm in 300 mm pipe produced data in general agreement with the above limit velocity expressions.

The limit velocity depends:

(i) on particle size $\qquad V_L \propto (d_{50})^{k_1}$

$d_{50} < 0.5$ mm	$0.5 < k_1 < 1$
$0.5 < d < 1.5$ mm	$0 < k_1 < 0.5$
$d > 1.5$ mm	$k_1 = 0$

(ii) on pipe size $\qquad V_L \propto D^{k_2}$

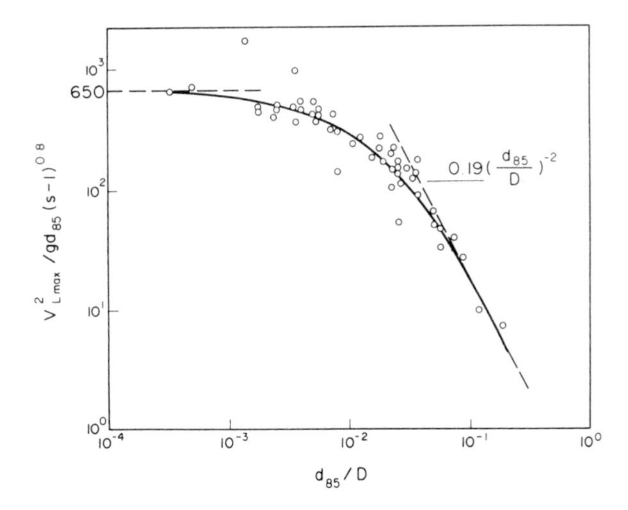

Fig. 12.11. Proposed correlation for maximum limit deposit-velocity. (After Sinclair, 1962.)

where $k_2 \cong 0.5 \pm 0.1$, the positive sign is associated with $d/D > 0.1$ and the negative with $d/D < 0.01$,

(iii) on concentration $\quad V_L \propto c_v^{k_3}$

where k_3 varies from about 0.3 to 0.1 for c_v = 5% to 15% respectively, and on

(iv) specific gravity $\quad V_L \propto (S_s - 1)^{k_4}$

where $k_4 \cong 0.5$.

A more recent discussion of the limit velocity is that by Goedde (1978).

The critical velocity V_m is obtained by substituting $fV^2/2g$ for i in eqn (12.9) and putting $di_m/dV = 0$. This yields, for the minimum head loss,

$$V_m = 3.43c_v^{1/3}\left[\frac{gD(S_s - 1)}{\sqrt{C_D}}\right]^{\frac{1}{2}} \qquad (12.26)$$

If eqn (12.13) is used

$$V_m = 3.43c_v^{1/3}\frac{[gD(S_s - 1)]^{\frac{1}{2}}}{[c_v(S_s - 1) + 1]^{1/3}} \qquad (12.27)$$

For transport with bed load and over a stationary bed in the pipeline reference is made to Graf et al. (1968) which leads to

$$\phi_A = 10.39\psi_A^{-2.52} \qquad (12.28)$$

where

$$\phi_A = \frac{cVR}{\sqrt{[(S_s - 1)gd^3]}} \quad \text{and} \quad \psi_A = \frac{(S_s - 1)d}{SR}$$

R being the hydraulic mean radius corresponding to the cross-section of the pipe where flow takes place and S is the slope of the energy gradient.

12.4 Non-horizontal Pipeline

In vertical pipes the velocity of solids in upward flow is less and in downward flow greater than fluid velocity by the value of the settling velocity, approximately w. Thus, in upward flow the spatial concentration of solids is greater than the discharge concentration. The effect is summarized after Worster and Denny (1955) in Fig. 12.12 in terms of V the bulk velocity, v_w the water velocity, the fall velocity w, concentration delivered c_{vd} and the spatial concentration of solids in the pipeline c_{vp}.

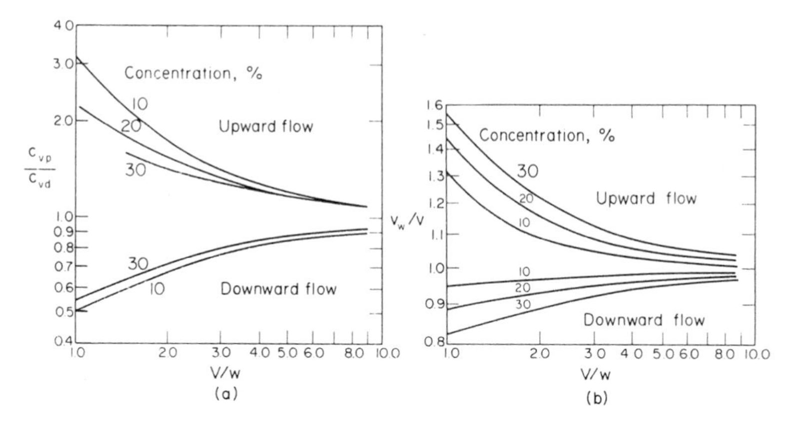

Fig. 12.12. Relationships between concentrations and velocities in vertical pipe.

The pressure of solids in the upward flow causes the local velocity to be greater than the mean velocity, hence, increasing i and causing the static head due to the vertical column of mixture to be greater by $c_{vp}(S_s - 1)$ than that due to fluid alone. Of this only $\pm(c_{vp} - c_{vd})(S_s - 1)$ is wasted energy. The total pressure gradient in excess of the static lift is therefore

$$i_m = i(V_w/V)^2 \pm c_{vp}(S_s - 1) \qquad (12.29)$$

Here $(V_w/V)^2$ is frequently close to unity and $c_{vp} \cong c_v$ leading to

$$i_m \cong i \pm c_v(S_s - 1)$$

where the minus sign refers to downward flow. If the gradient in horizontal pipe is

$$i = i_w + \Delta i_h$$

then for a pipe sloping at angle θ to the horizontal

$$i_m = i + \Delta i_h \cos\theta + c_{vd}(S_s - 1)\sin\theta \qquad (12.30)$$

as seen from Fig. 12.13.

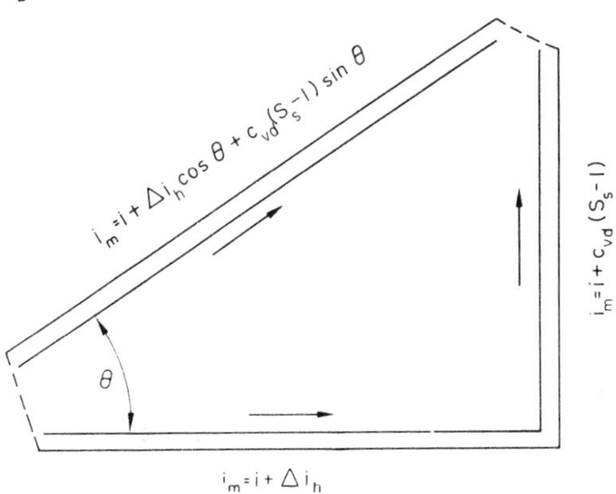

Fig. 12.13. Head loss in inclined pipes.

12.5 Measurement of Transport

Reviews of on-line meters for measurement of sediment laden pipe flows have been presented by Graf (1971), Baker and Hemp (1979) and others. Of the various types the electromagnetic flow meters have been found to give accurate results over a wide range of velocity and concentrations (Faddick et al. 1979). A source of error with these meters is the presence of magnetic material in the slurry, like magnetic or iron sand.

A very simple loop system is described by Einstein and Graf (1966), also known as the counter flow meter. It is an inverted U-loop in which the sediment fall velocity is against the flow direction in one leg and with the velocity in the other. With the symbols in Fig. 12.14:

$$(\Delta h_u - \Delta h_d)/2L = (S_s - 1)(c_v + XZ) \qquad (12.31)$$

$$(\Delta h_u - \Delta h_d)/2L = (S_s - 1)Z + X[1 + (S_s - 1)c_v] \qquad (12.32)$$

where

$$Z = \frac{w}{V} c_v(1 - c_v)$$

$$X = f_m V^2/2gD$$

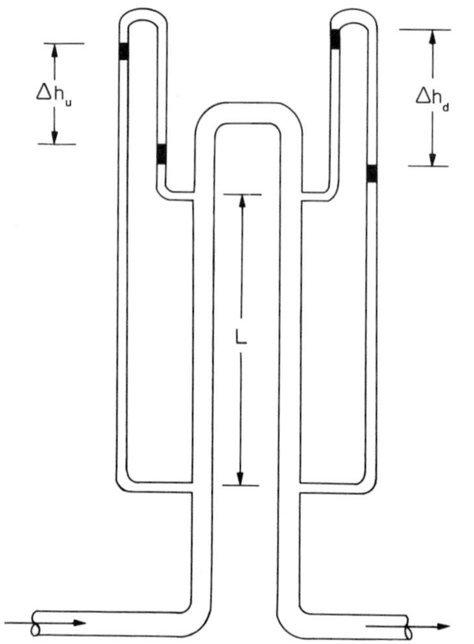

Fig. 12.14. Counterflow meter with U-tube manometers.

and f is obtained from the Moody diagram with the kinematic viscosity of the mixture. The equations can be solved by trial and error for c_V and V.

12.6 Partially Filled Pipes

A common operating condition for most stormwater and sewage pipelines is the partially filled pipe. The flows also carry appreciable amounts of sand, silt and grit. The sediment transport characteristics of a circular pipe at partial flow depths are illustrated in Fig. 12.15, in which both the sediment load and concentration have been normalised by the values of the pipe flowing full. The above study (Fröhlich, 1985) also showed that a pipe with a flat floor at 0.1 diameter elevation is more efficient in deposit free sediment transport than the circular cross section. Writing the sediment transport Q_S proportional to $\eta\rho QS$, i.e. stream power approach and evaluating transport efficiency η yielded:

circular cross section	$\eta = 31.58\ S^{1.17}$
floor at 0.1 D	$\eta = 64.69\ S^{1.07}$
V-groove floor	$\eta = 54.03\ S^{1.5}$

The sediment transport efficiency with 0.1 D floor and full pipe flow is seen to increase about four fold compared to circular cross section whereas the V-groove is very inefficient. The 0.1 D floor reduces the water discharge to about 7.5% or requires a 3% larger diameter for the same flow at the same slope. The

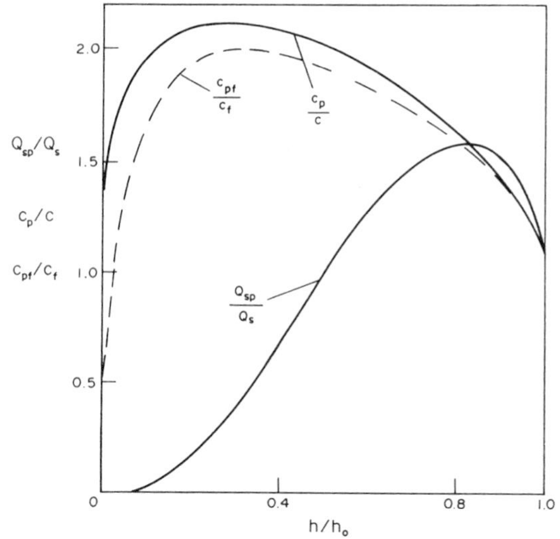

Fig. 12.15. Sediment transport characteristics of a circular pipe: Q_{sp}/Q_s is transport at partial filling of depth h to that in full flowing pipe $h = h_o = D$; c_p/c gives the corresponding concentrations; c_{fp}/c_f refers to concentrations in a pipe with a flat floor at 0.1 diameter normalised with full pipe flow value at $h_o = 0.9$ D.

results show that in pipes where sediment loads constitute a design criterion significant reductions in the required slope can be achieved with the flat-floor cross-sections. The relative concentration distribution for the 0.1 D floor cross-section is dotted in Fig. 12.15.

REFERENCES*

Abt, S.R., Kloberdanz, R.L. & Mendoza, C. (1984) Unified culvert scour determination, *J. Hydraulic Engineering, ASCE*, Vol. 110, 1475-1479.

Abt, S.R., Ruff, J.F. & Doehring, F.K. (1985) Culvert slope effects on outlet scour, *J. Hydraulic Engineering, ASCE*, Vol. 111, 1363-1367.

Ackers, P. and Charlton, F.G. (1970) Meander geometry arising from varying flows, *J. Hydrology*, Vol. 11, 230-252.

Ackers, P. and White, W.R. (1973) Sediment transport: new approach and analysis, *Proc. ASCE*, Vol. 99, HY 11, 2041-2060.

Ackers, P. (1980) Use of sediment transport concepts in stable channel design, *Int. Workshop on Alluvial River Problems*, Roorkee, India.

Adachi, S. (1967) A theory of stability of streams, *Proc. 12th Congress Intern. Assoc. for Hydraulic Research*, Vol. 1, 338-343.

Acaroglu, E.R. and Graf, W.H. (1969) The effect of bed forms on the hydraulic resistance, *Proc. 13th Congress I.A.H.R.*, Kyoto, Vol. 2, 199-202.

Ahmad, M. (1953) Experiments on design and behaviour of spur dikes, *Proc. Congress I.A.H.R.*, Minnesota, 145-159.

Alam, A.M.Z. and Kennedy, J.F. (1969) Friction factors for flow in sand bed channels, *Proc. ASCE*, Vol. 95, HY 6, 1973-1992.

Alfrink, B.J. and van Rijn, L.C. (1983) Two-equation turbulence model for flow in trenches, *J. Hydraulic Engineering,ASCE*, Vol. 109, 941-958.

Allen, J.R.L. (1968) *Current Ripples*, North Holland Publ. Co., 211-220.

Allen, J.R.L. (1970) A quantitative model of grain size and sedimentary structures in lateral deposits, *Journal of Geology*, Vol. 7, 129-146.

Allen, J.R.L. (1970) *Physical Processes of Sedimentation*, Elsevier.

Allen, J. (1985) Field evaluation of beach profile response to wave steepness as predicted by the Dean model, *Coastal Engineering*, Vol. 9, 71-80.

Amoafo, B.D. (1985) Pier scour at a confluence, *University of Canterbury*, New Zealand, Dept. of Civil Engng. Research Report 85-1.

Anderson, A.G. (1953) The characteristics of sediment waves formed by flow in open channels, *Proc. 3rd Midwestern Conf. on Fluid Mech.*, University of Minnesota, Minneapolis, 379-395.

Anderson, A.G. (1967) On development of stream meanders, *Proc. 12th Congress I.A.H.R.*, Fort Collins, Vol. 1, 370-378.

Anderson, A.G. (1974) Scour at bridge waterways - A Review, *U.S. Dept. of Transportation*, Office of Research and Development, Environmental Design and Control Division.

Andrews, E.D. and Parker, G. (1985) The coarse surface layer as a response to gravel mobility, *Proc. Workshop on Gravel-Bed Rivers*, Pingree, Colorado, Chapter 10.

Annambhotla, V.S.S., Sayre, W.W. and Livesey, R.H. (1972) Statistical properties of Missouri River bed forms, *Proc. ASCE*, Vol. 98, WW4, 489-510.

Antsyferov, S.M. and Kos'yan, R.D. (1980) Sediment suspended in

*An addendum of references added in proof follows this list.

stream flow, *Proc. ASCE*, Vol. 106, HY2, 313-330.
Antsyferov, S.M., Basinski, T. and Pykhov, N.V. (1983) Measurements of coastal suspended sediment concentrations, *Coastal Engineering*, Vol. 7, 146-166.
Arulanandan, K., Loganathan, P. and Krone, R.B. (1975) Pore and eroding fluid influences on surface erosion of soil, *Proc. ASCE*, Vol. 101, GTI, 51-66.
Ashida, K. and Tanaka, Y. (1967) A statistical study of sand waves, *Proc. 12th Congress I.A.H.R.*, Fort Collins, Vol. 2, 103-110.
Ashida, K. and Fujita, M. (1986) Stochastic model for particle suspension in open channels, *Journal of Hydroscience and Hydraulic Engineering*, Japan Soc. Civil Engng. Vol. 4, 21-46.
Ashmore, P. and Parker, G. (1983) Confluence scour in coarse braided streams, *Water Resources Research*, Vol. 19, 393-402.
Aubrey, D.G. (1980) The statistical prediction of beach changes in Southern California, *J. Geophysical Research*, Vol. 85, 3264-3276.
Bagnold, R.A. (1936) The movement of desert sand, *Proc. Roy. Soc.* London, Vol. A157, 594-620.
Bagnold, R.A. (1941) *The Physics of Blown Sand and Desert Dunes*, Methuen. Reprint 1973, Chapman and Hall.
Bagnold, R.A. (1946) Motion of waves in shallow water. Interaction of waves and sand bottoms, *Proc. Roy. Soc.* London, Vol. A187, 1-15.
Bagnold, R.A. (1954) Experiments on a gravity-free dispersion of large solid spheres in a newtonian fluid under shear, *Proc. Roy. Soc.* London, Ser. A, Vol. 225, 49-63.
Bagnold, R.A. (1956) The flow of cohesionless grains in fluids, *Phil.Trans. Roy. Soc. A.*, London, Ser. A. Vol. 249 (964), 235-297.
Bagnold, R.A. (1962) Auto-suspension of transported sediment; turbidity currents, *Proc. Roy. Soc. A.*, London, Vol. 265, No. 1322.
Bagnold, R.A. (1963) Mechanics of marine sedimentation: in *The Sea*, edited by M.N. Hill, Interscience Publication, 507-582.
Bagnold, R.A. (1966) An approach to the sediment transport problem from general physics, *U.S. Geol. Survey* Professional Paper, 422-I.
Bagnold, R.A. (1973) The nature of saltation and of "bed-load" transport in water, *Proc. Roy. Soc.* London, Vol. A332, 473-504.
Bagnold, R.A. (1980) An empirical correlation of bedload transport rates in flumes and natural rivers, *Proc. Roy. Soc.* London, Vol. A372, 453-473.
Babcock, H.A. (1968) Heterogeneous flow of heterogeneous solids, *Symposium of Solid-liquid Flow in Pipes*, Univ. of Pennsylvania, 125-148.
Bain, A.G. and Bonnington, S.T. (1970) *The Hydraulic Transport of Solids by Pipeline*, Pergamon Press.
Baker, R.C. and Hemp, J. (1979) A review of concentration meters for granular slurries, *Hydrotransport 6*, Canterbury, England, Vol. 1.
Bakker, W.T. (1974) Sand concentration in an oscillatory flow, *Proc. 14th Coastal Engineering Conf.*, Copenhagen, Vol. 2, 1129-1148.

Bakker, W.T. and Doorn, T. (1978) Near-bottom velocities in wave with a current, *Proc. 16th Coastal Engineering Conf.*, Hamburg, Vol. 2, 1394-1413.
Barnes, Jr. H.H. (1967) Roughness characteristics of natural channels, *U.S. Geological Survey*, Water-Supply Paper 1849.
Barton, J.R. and Lin, P.N. (1955) A study of the sediment transport in alluvial streams, Rep. No. 55 JRB2, Dept. Civil Engng, *Colorado College*, Fort Collins.
Bascom, W.N. (1951) The relationship between sand size and beach-face slope, *Trans. Am. Geophys. Un.*, Vol. 32, No. 6. Also in *Waves and Beaches*, Anchor Books, N.Y. 1980.
Bascom, W.N. (1980) *Waves and Beaches*, Anchor Books, N.Y.
Bathurst, J.C. (1978) Flow resistance of large-scale roughness, *Proc. ASCE*, Vol. 104, HY12, 1587-1603.
Bathurst, J.C., Graf, W.H. and Cao, H.H. (1983) Initiation of sediment transport in steep channels with coarse bed material, In *Mechanics of Sediment Transport*, edited by B.M. Sumer and A. Müller, Belkema, The Netherlands, 207-213.
Bathurst, J. (1985) Flow resistance estimation in mountain rivers, *ASCE J. Hydraulic Engineering*, Vol. 111, No. 4, 625-643.
Battjes, J.A. (1974) Surf similarity, *Proc. 14th Coastal Enggineering Conference*, Copenhagen, Vol. 1, 466-480.
Battjes, J.A. (1975) Modelling of turbulence in the surf zone, *Symposium on Modelling Techniques*, California, A.S.C.E., 1050-1061.
Bell, R.G. (1980) Non-equilibrium bedload transport by steady and non-steady flows, *University of Canterbury*, Dept. Civil Engng., N.Z. Research Report 80/23.
Belly, P.Y. (1964) Sand movement by wind, *U.S. Army Coastal Engineering Research Center* TM No. 1. (see Kadib, A-L).
Beverage, J.P. and Culbertson, J.K. (1964) Hyperconcentrations of suspended sediment, *Proc. ASCE*, Vol. 90, HY6, 117-128.
Binnie, A.M. and Williams, E.E. (1966) Self-induced waves in a moving open channel, *J. Hydraulic Research*, Vol. 4, 17-35.
Bijker, E.W. (1967) Some considerations about scales for coastal models with movable bed, *Delft Hydraulics Laboratory* Publication No. 50, 142.
Bijker, E.W. (1968) Littoral drift as a function of waves and currents, *Proc. 11th Coastal Engineering Conf.* London. *Delft Hydraulics Laboratory* Publication No. 58 (1969).
Bijker, E.W., Kalkwijk, J.P. Th. and Pieters, T. (1974) Mass transport in gravity waves on a sloping bottom, *Proc. 14th Coastal Engineering Conf.* Copenhagen, Vol. 1, 447-465.
Birkemeier, W.A. (1984) Time scales of nearshore profile changes. *Proc. 19th Coastal Engng Conf.*, Houston, Vol. 2, 1507-1521.
Bishop, A.A., Simons, D.B. and Richardson, E.V. (1965) Total bed material transport, *Proc. ASCE*, Vol. 91, HY2, 175-191.
Blandamer, M.J., Burgess, J., Robertson, R.E. and Scott, J.M.W. (1982) Dependence of equilibrium and rate constants on temperature and pressure, *Chemical Review*, Vol. 3, 259-86.
Blasius, H. (1912) Über die Abhängigkeit der Formen der Riffeln und Geschiebebänke vom Gefälle, *Zeitschrift für Bauwesen*.
Blench, T. (1957) *Regime behaviour of canals and rivers*, Butterworth, London.
Blench, T. (1969) *Mobile-bed fluviology*, University of Alberta Press, Canada, (First edition 1966).
Blinco, P.H. and Partheniades, E. (1971) Turbulence character istics in free surface flows over smooth and rough boundaries

J. Hydraulic Research, Vol. 9, 43-71.
Bockris, J. O'M. and Reddy, A.K.N. (1970) *Modern Electrochemistry*, Plenum Press, N.Y.
Boczar-Karakiewicz, B. and Bona, J.L. (1981) Über die Riffbildung an sandingen Küsten durch progressive Wellen, *Tech. Univ. Braunschweig*, Mitteilungen Leichtweiss-Inst. No. 70, 377-420.
Boczar-Karakiewicz, B., Papliaska, B. and Winiecki, J. (1981) Formation of bars by surface waves in shallow water. Laboratory experiments, *Polska Akademia Nauk*, Instytut Budownictwa Wodnego, Gdansk, Rozprawy Hydrotechniczne Zeszut 43.
Boczar-Karakiewicz, B., Long, B.F. and Drapeau, G. (1983) Formation and modification of a system of sand bars by progressive gravity waves in the presence of tides, *Proc. Canadian Coastal Conf.* 1983, Vancouver, 37-51.
Bogardi, J.L. (1965) European concepts of sediment transportation, *Proc. ASCE*, Vol. 91, HY1, 29-54.
Bogardi, J.L. (1974) *Sediment Transport in Alluvial Streams*, Akadémiai Kiado, Budapest.
Bohan, J.P. (1970) Erosion and riprap requirements at culvert and storm-drain outlets, *U.S. Army Corps of Engineers*, Waterways Exp. Stat. Vicksburg, Res. Rep. H-70-2.
Bonasoundas, M. (1973) Strömungsvorgang und Kolkproblem, *Oscar von Miller Institut, Techn. Univ. Munich*, Rep. No. 28.
Bonnefille, R. (1963) Essais de Synthese des lois de debut d' entrainment des sediments sous l'action d'un courant en regime uniforme, *Bull. du CREC*, No. 5, Chatou.
Bonnington, S.T. (1959) Experiments on the hydraulic transport of mixed-sized solids, *British Hydromech. Res. Assoc.*, Rep. RR637.
Borland, W.M. (1971) Reservoir sedimentation, Chapter 29 in *River Mechanics*, edited and published by H.W. Shen, Fort Collins, Colorado.
Boussinesq, J. (1872) Théorie des ondes et des remous quise propagent le long d'un canal rectangulaire horizontal, en communiquant au liquide contenu dans ce canal des vitesses sensiblement pareilles de la surface au fond, *Journal de Mathematiques Pures et Appliquées Deuziéme Série*, Vol. 17, 55-108.
Bowen, R.L. Jr. (1961) Scale-up for non-Newtonian fluid flow, *Chem. Engng.*, Vol. 68, (a series of articles: June 12, 243; June 26, 127; July 10, 147; July 24, 143; Aug 7, 129 and Aug. 21, 119).
Bowen, A.J. (1969) The generation of longshore currents on a plain beach, *J. Marine Res.*, Vol. 27, 206-215.
Brandt, G.H. et al. (1972) An Economic Analysis of Erosion and Sediment Control Methods for Watersheds Undergoing Urbanization (C-1677), Final Report by the *Dow Chemical Co.*, Midland, Michigan for the *US Dept. of Interior.* Contract No. 14-31-0001-3392.
Brenner, H. (1961) The slow motion of a sphere through a viscous fluid towards a plane surface, *Chem. Engng. Sci.*, Vol. 16.
Breusers, H.N.C. (1966) Conformity and time scale in two-dimensional local scour, *Proc. Symp. on model and prototype conformity. Hydr. Res. Lab. Poona, India*, 1-8.
Breusers, H.N.C. (1967) Time scale of two-dimensional local scour, *Proc. 12th Congress I.A.H.R.*, Fort Collins, Vol. 3, 257-282.

Breusers, H.N.C., Nicollet, G. and Shen, H.W. (1977) Local scour around cylindrical piers, *J. Hydraulic Res.* Vol. 15, 211-252.
Brooks, N.H. (1965) Calculation of suspended load discharge from velocity and concentration parameters, *Proc. Federal Inter-Agency Sedimentation Conference*, 1963, Misc. Pub. No. 970. Agricul. Res. Service; Calif. Inst. Techn. Publ. KH-P-16.
Brown, C.B. (1950) Sediment Transport, Chapter XII, *Engineering Hydraulics*, edited by H. Rouse, Wiley & Sons.
Brownlie, W.R. (1981) Prediction of flow depth and sediment discharge in open channels, *Calif. Inst. Techn.*, W.M. Keck Lab. Rep. No. KH-R-43A.
Brownlie, W.R. (1983) Flow depth in sand-bed channels, *J. Hydraulic Engineering, ASCE*, Vol. 109, 959-990. Also as "Flow depth in sand-bed rivers", Calif. Inst. Techn. W.M. Keck Lab. Techn. Memo 81-3, 1981.
Brune, G.M. (1953) Trap efficiency of reservoirs, *Trans. Am. Geophys. Un.*, Vol. 34, No. 3.
Bruschin, J. (1985) Disc. Flow depth in sand-bed channels, *ASCE J. Hydraulic Engineering*, Vol. 111, 736-739.
Bugliarello, G. and Jackson, E.D. (1964) Random walk study of convective diffusion, *Proc. ASCE*, Vol. 90, EM.4, 49-78.
Burt, T.N. and Stevenson, J.R. (1983) Field settling velocity of Thames mud, *Hydr. Research Station*, Wallingford, Report No. IT 251.
Callander, R.A. (1969) Instability and river channels, *J. Fluid Mech.* Vol. 36, Part 3, 464-80.
Callander, R.A. (1978) River meandering, *Ann. Rev. Fluid Mech.*, Vol. 10, 129-158.
Camfield, F.E. (1982) Calculation of trapped reflected waves, *Proc. ASCE J. of Waterway, Port, Coastal and Ocean Division*, Vol. 108, WW1, 109-111.
Camp, T.R. (1946) Sedimentation and the design of settling tanks, *Trans. ASCE*, Vol. 111, 895-958.
Carstens, M.R., Neilson, F.M. and Altinbilek, H.D. (1969) Bed forms generated in laboratory under an oscillatory flow: analytical and experimental study, *U.S. Army Coastal Engineering Research Center*, Tech. Memo No. 28.
Casey, H.J. (1935) Über Geschiebebewegung, *Preuss. Versuchsanstalt für Wasserbau und Schiffbau*, Berlin, Vol. 19.
Cecen, K., Bayazit, M. and Sumer, M. (1969) Distribution of suspended matter and similarity criteria in settling basins, *Proc. 13th Congress I.A.H.R.*, Kyoto, Vol. 4, 251-225.
Celik, I. and Rodi, W. (1984) Simulation of free-surface effects in turbulent channel flows, *Physico-Chemical Hydrodynamics*, Vol. 5, No. 3/4, 217-227.
Celik, I. and Rodi, W. (1984) A deposition-entrainment model for suspended sediment transport, *Sonderforschungsbereich* 210, Report No. SFB 210/T/6, *Univ. of Karlsruhe*, F.R.G.
Celik, I. and Rodi, W. (1985) Mathematical modelling of suspended sediment transport in open channels, *Proc. 21st Congress I.A.H.R.*, Melbourne, Vol. 3, 534-538.
Chabert, J. and Engeldinger, P. (1956) Etude des Affouillements autour des Piles des Ponts, *Laboratoire National d'Hydraulique* Chatou, France.
Chan, K.W., Baird, M.H.I. and Round, G.F. (1972) Behaviour of beds of dense particles in a horizontally oscillating liquid, *Proc. Roy. Soc.*, London, Ser. A, Vol. 330, 537-559.
Chang, H.H. (1979) Geometry of rivers in regime, *Proc. ASCE*, Vol. 105, HY6, 691-706.

Chang, H.H. (1979) Minimum stream power and river channel patterns, *J. Hydrology*, Vol. 41, 303-327.
Chang, H.H. (1980) Stable alluvial canal design, *Proc. ASCE*, Vol. 106, HY5,
Chang, H.H. (1982) Fluvial hydraulics of deltas and alluvial fans, *Proc. ASCE*, Vol. 108, HY11, 1282-1295.
Chang, H.H. (1985) River morphology and thresholds, *J. Hydraulic Engineering, ASCE*, Vol. 111, No. 3, 503-519.
Chang, H.H. (1986) River channel changes: Adjustments of equilibrium, *J. Hydraulic Engineering, ASCE*, Vol. 112, No. 1, 43-55.
Charles, M.E. (1970) Transport of solids by pipelines, Proc. Hydro-transport 1, *British Hydromech. Res. Assoc.*, Paper A33.
Chee, R.K.W. (1982) Live-bed scour at bridge piers, *University of Auckland, N.Z.*, ME. Thesis, Dept. of Civil Engng. Report no. 290.
Chen, Charng-Ning (1974) Evaluation and control of erosion in urbanizing watersheds, *Proc. Nat. Symp. on Urban Rainfall and Runoff and Sediment Control*, Univ. of Kentucky, Lexington.
Chen, Charng-Ning (1975) Design of sediment retension basins, *Proc. National Symposium on Urban Hydrology and Sediment Control*, Published by Office of Research, Univ. of Kentucky, Lexington, College of Engineering.
Chen, Y.H., Lopez, J.L. and Richardson, E.V. (1978) Mathematical modelling of sediment in reservoirs, *Proc. ASCE*, Vol. 104, HY12, 1605-1616.
Cheng, D.C.H. (1970) A design procedure for pipeline flow of non-Newtonian dispersed systems, *First Intern. Conf. on Hydraulic Transport of Solids in Pipes*, Warwick, England, Paper J5.
Chepil, W.S. (1959) Equilibrium of soil grains at the threshold of movement by wind, *Proc. Soil Science Soc. of America*, Vol. 23, 422-438.
Chepil, W.S. and Woodruff, N.P. (1963) The physics of wind erosion and its control, *Advances in Agronomy*, Vol. 15, 211-302.
Chepil, W.S. and Woodruff, N.P. (1965) Wind erosion and transportation, Progress Report Task Committee on Preparation of Sediment Manual, *Proc. ASCE*, Vol. 91, HY2, 267-287.
Chian Min Wu (1973) Scour at downstream end of dams in Taiwan, *Proc. I.A.H.R. Symp. on River Mechanics*, Bangkok, Vol. 1, Paper A13, 137-142.
Chiew, Y.M. (1984) Local scour at bridge piers, Ph.D. Thesis, *Univ. of Auckland*, N.Z. School of Engineering, Rep. No. 355.
Chin, C.O. (1985) Stream bed armouring, *Univ. of Auckland, N.Z.* Ph.D. Thesis, Dept. of Civil Engng. Report No. 403.
Chien, Ning (1956) The present status of research on sediment transport, *Trans ASCE*, Vol. 121, 833-868.
Chitale, S.V. (1966) Design of alluvial channels, *Proc. 6th Congress of ICID, Delhi*, Q20, R.17.
Chitale, S.V. (1976) Shape and size of alluvial channels, *Proc. ASCE*, Vol. 102, HY7, 1003-1011.
Chiu, C.L. and Chen, K.C. (1969) Stochastic hydrodynamics of sediment transport, *Proc. ASCE*, Vol. 95, EM5, 1215-1230.
Clark, J.A. and Markland, E. (1970) Vortex structures in turbulent boundary layers, *Aeronaut J.*, Vol. 74, 243-244.
Clarke, F.R.W. (1962) The action of submerged jets on moveable material, Ph.D. Thesis, *Imperial College*, London.

Clauser, F.H. (1956) The turbulent boundary layer, *Adv. Appl. Mechanics*, Vol. 4, 2-51.
Clifton, H.E. and Dingler, J.R. (1984) Wave-formed structures and paleoenvironmental reconstruction, *Developments in Sedimentology*, Vol. 39, editors B. Greenwood and R.A. Davis Jr. Elsevier Science Publishers, 165-198.
Coffey, F.C. and Nielsen, P. (1984) Aspects of wave current boundary layer flows, *Proc. 19th Coastal Engineering*, Houston, Vol. 2, 2231-2245.
Colby, B.R. and Hembree, C.H. (1955) Computation of total sediment discharge, Niobrara River near Cody, Nebraska, *U.S. Geol. Survey*, Water Supply Paper 1357.
Colby, B.R. (1957) Relationship of unmeasured sediment discharge to mean velocity, *Trans. Am. Geophys. Union*, Vol. 38, No. 5, 708-717.
Colby, B.R. (1964) Practical computations of bed-material discharge, *Proc. ASCE*, Vol. 90, HY2, 217-246.
Colebrook, C.F. and White, C.M. (1937) Experiments with fluid friction in roughened pipes, *Proc. Roy. Soc., London*, Vol. A161, 367-381.
Coleman, N.L. (1967) A theoretical and experimental study of drag and lift forces acting on a sphere resting on a hypothetical stream bed, *Proc. 12th Congress I.A.H.R.*, Fort Collins, Vol. 3, 185-192.
Coleman, N.L. (1969) A new examination of sediment suspension in open channels, *J. Hydraulic Research*, Vol. 7, 67-82.
Coleman, N.L. (1970) Flume studies of the sediment transfer coefficient, *Water Resources Research*, Vol. 6, 801-809.
Coleman, N.L. (1977) Extension of the drag coefficient function for a stationary sphere on a boundary of similar spheres, *La Houille Blanche*, No. 4, 325-328.
Coleman, N.L. (1985) Effects of suspended sediment on open-channel velocity distributions, *Euromech. 192*, Munich/ Neubiberg, F.R. Germany Hochschule der Bundeswehr.
Coles, D. (1956) The law of the wake in the boundary layer, *J. Fluid Mech.*, Vol. 1, 191-226.
Condolios, E. and Chapus, E.E. (1963) Transporting solid material in the pipelines, *Chem. Engng.*, June 24; July 8 and July 24.
Copeland, G.J.M. (1985) Practical radiation stress calculations connected with equations of wave propagation, *Coastal Engineering*, Vol. 9, 195-219.
Corino, E.R. and Brodkey, R.S. (1969) A visual investigation of the wall region in turbulent flow, *J. Fluid Mech.*, Vol. 37, 1-30.
Croad, R.N. (1981) Physics of erosion of cohesive soils, *University of Auckland*, N.Z., Ph.D. Thesis, Dept. Civil Engng. Rep. No. 247.
Dalrumple, R.A. (1978) Rip currents and their causes, *Proc. 16th Coastal Engineering Conf.*, Hamburg, Vol. 2, 1414-1427.
Damle, P.M., Venkatraman, C.P. and Desai, S.C. (1966) Evaluation of scour below ski-jump buckets of spillways. Chapter in: Model and Prototype Conformity, Vol. 1, *Central Water and Power Research Station, Poona, Proc. Golden Jubilee Symposium*, 154-163.
Darwin, G.H. (1883/4) On the formation of ripple-marks, *Proc. Roy. Soc. London.*
Dash, U. (1968) Erosive behaviour of cohesive soils, Ph.D. Thesis, *Purdue University, USA.*

Davis, R.A. Jr. (Editor)(1978) *Coastal Sedimentary Environments*, Springer-Verlag.
Dawson, R. (1975) An activated-state model for clay suspension rheology, *Trans. Soc. Rheology*, Vol. 19, No. 2, 229-244.
deVries, M. (1973) River-bed variations-aggradation and degradation, *Delft Hydraulics Laboratory* Publication No. 107.
Deacon, G.F. (1894) In discussion of paper on "Estuaries" by H.L. Partiot, *Proc. Inst. Civil Engrs.*, *London*, CXVIII, 44-189.
Dean, R.G. (1973) Heuristic models of sand transport in the surf zone. *Proc. Conf. on Eng. Dynamics in the Surf Zone*, Australia, 208-214.
Dean, R.G. (1976) Beach erosion: causes, processes and remedial measures, *CRC Critical Reviews in Environmental Concern*, Vol. 6, 259-296.
Delo, E.A. and Burt, T.N. (1986) The hydraulic engineering characteristics of estuarine muds. *Hydraulic Research, Wallingford*, U.K., Rep. SR77.
Dette, H.H. and Führböter, A. (1974) Field investigations in surf zones, *Proc. 14th Coastal Engineering Conf.*, Copenhagen, Vol. 1, 518-537.
Dietz, J.W. (1969) Kolkbildung im feinem oder leichter Sohlmaterialen bei strömenden Abfluss, *T.U. Karlsruhe*, Mitteilungen Rehbock Flussbaulab., Heft 155, 1-119.
Dietz, J.W. (1972) Modellversuche über die Kolkbildung, *Die Bautechnik*, Vol. 49, 162-168, 240-245.
Dietz, J.W. (1972) Systematische Modellversuche über die Pfeilerkolkbildung. *Mitteilungsblatt der Bundesanstalt für Wasserbau*, Nr. 31, Karlsruhe, FGR, 95-109.
Dietz, J.W. (1973) Kolkbildung an einem kreiszylindrischen Pfeiler, *Die Bautechnik*, Vol. 50, 203-208.
Dingler, J.R. and Inman, D.L. (1976) Wave-formed ripples in nearshore sands, *Proc. 15th Coastal Engineering Conf.*, Honolulu, Vol. 2, 2109-2126.
Dingler, J.R. (1979) The threshold of grain motion under oscillatory flow in a laboratory wave channel, *J. Sediment Petrol*, Vol. 49, 287-294.
Diplas, P. (1986) Bedload transport in gravel-bed streams: some properties, *Proc. 3rd Intern. Symp. on River Sedimentation*, Univ. of Mississippi, School of Engng., Vol. 3, 925-934.
Diplas, P. (1987) Bedload transport in gravel-bed streams, *J. Hydraulic Engineering ASCE*, Vol. 113, 277-292.
Dixit, J.G. (1982) Resuspension potential of deposited kaolinite beds, *Univ. of Florida*, Gainsville, M.Sc. Thesis.
Dobbins, W.E. (1944) Effect of turbulence on sedimentation, *Trans. ASCE*, Vol. 109, 629-678.
Du Boys, M.P. (1879) Le Rhone et les Rivieres a Lit affouillable, *Mem. Doc.*, *Pont et Chaussees*, Ser. 5, Vol. XVIII.
Du Buat, P. (1786) Principes d'hydraulique, 2nd ed. De l'Imprimerie de Monsieur, Paris.
Du Toit, C.G. and Sleath, J.F.A. (1981) Velocity measurements close to rippled beds in oscillatory flow, *J. Fluid Mech.*, Vol. 112, 71-96.
Duffy, G.G. and Lee, P.F.W. (1978) Drag reduction in the turbulent flow of wood pulp suspensions, *Appita*, Vol. 31, 280-286.
Dunn, I.S. (1959) Tractive resistance of cohesive channels, *Proc. ASCE*, Vol. 85, SM3, 1-24.
Durand, R. and Condolis, E. (1952) The hydraulic transport of

coal and solid materials in pipes, *Proc. Colloq. Hydraulic Transportation*, London. Also Durand, R. (1953) Basic relationships of the transportation of solids in pipes - Experimental Research, *Proc. 5th Congress I.A.H.R.*, Minneapolis.
Dury, G.H. (1965) Theoretical implications of underfit streams, *U.S. Geological Survey*, Professional Paper 452-C.
Eagelson, P.S., Dean, R.G. and Peralta, L.A. (1958) The mechanics of the motion of discrete spherical bottom sediment particles due to shoaling waves, *U.S. Army Beach Erosion Board*, TM104; *Proc. ASCE*, Vol. 85, HY10(1959), 53-79; Trans. Vol. 126, Pt1(1961), 1162-1189.
Eagelson, P.S., Glenne, B. and Dracup, J.A. (1963) Equilibrium characteristics of sand beaches, *Proc. ASCE*, Vol. 89, HY1, 35-57.
Einstein, H.A. (1942) Formulae for transportation of bed-load, *Trans. ASCE*, Vol. 107, 561-577.
Einstein, H.A. (1950) The bed-load function for sediment transportation in open channel flows, *U.S. Dept. of Agriculture, Soil Conservation Service*, Washington, D.C., Tech. Bull. No. 1026.
Einstein, H.A. and Barbarossa, N.L. (1952) River channel roughness, *Proc. ASCE*, Vol. 77, Sep. No. 78, *Trans. ASCE*, Vol. 117, 1121-1146.
Einstein, H.A. and Chien, N. (1954) Second approximation to the solution of the suspended load theory, MRD Sediment Series No. 2, *Univ. of California and Missouri River Division, U.S. Corps of Engineers*.
Einstein, H.A. and Chien, N. (1955) Effects of heavy sediment concentration near the bed on velocity and sediment distribution, MRD Sediment Series No. 8, *Univ. of California, Berkeley and U.S. Army Corps of Engineers, Missouri River Div.*
Einstein, H.A. (1964) Sedimentation, Pt. 2, River sedimentation in *Handbook of Applied Hydrology*, ed. Ven Te Chow, McGraw-Hill, Chapter 17, 35-67.
Einstein, H.A. and Graf, W.H. (1966) Loop system for measuring sand-water mixtures, *Proc. ASCE*, Vol. 92, HY1, 1-12.
Engelund, F. (1965) A critique for the occurrence of suspended load, Notules Hydrauliques, *La Houille Blanche*, No. 6, 801-802.
Engelund, F. (1966) On the possibility of formulating a universal spectrum function for dunes, *Techn. Univ. Denmark*, Hydr. Lab. Basic Research Progress Report No. 18.
Engelund, F. (1966) Hydraulic resistance of alluvial streams, *Proc. ASCE*, Vol. 92, HY2, 315-326.
Engelund, F. (1967) Closure of discussion, *Proc. ASCE*, Vol. 93, HY4, 287-296.
Engelund, F. and Hansen, E. (1967) *A monograph on sediment transport in alluvial streams*, Teknisk Forlag, Copenhagen.
Engelund, F. (1970) Instability of erodible beds, *J. Fluid Mech.*, Vol. 42, Pt. 2, 225-244.
Engelund, F. and Skovgaard, O. (1973) On the origin of meandering and braiding in alluvial streams, *J. Fluid Mechanics*, Vol. 59, 289-302.
Engelund, F. (1974) Flow and bed topography in channel bends, *Proc. ASCE*, Vol. 100, HY11, 1631-1648.
Engelund, F. (1974) The development of oblique dunes, *Techn. Univ.*

Denmark, Inst. Hydro-dynamics and Hydr. Engng. Progress Report No. 31, 32 and in Nos. 33, 34 by Fredsøe, J.

Engelund, F. and Fredsøe, J. (1974) Transition from dunes to plane bed in alluvial channels, *Tech. Univ. Denmark*, Inst. of Hydro-dynamics and Hydr. Engng. Series Paper No. 4.

Engelund, F. and Fredsøe, J. (1976) A sediment transport model for straight alluvial channels, *Nordic Hydrology*, Vol. 7, 293-306.

Engelund, F. and Fredsøe, J. (1982) Hydraulic theory of alluvial rivers, *Advances in Hydroscience*, Vol. 13, 187-215.

Engelund, F. and Wan Zhaohui (1984) Instability of hyperconcentrated flow, *J. Hydraulic Engineering*, ASCE, Vol. 110, 219-233.

Ertel, H. (1966) Kinematik und Dynamik formbeständig wandernder Transversaldünen, *Monatsberichte der Deutschen Akademie der Wissenschaften*, zu Berlin, Band 8, Heft 10.

Ettema, R. (1980) Scour at bridge piers, *Univ. of Auckland, N.Z.* School of Engineering, Rep. No. 216.

Exner, F.M. (1920) Zur Physik der Dünen, *Sitz.-Ber. Akad. Wiss.* Wien, Math-naturw. Klasse, Abt. IIa, Band 129.

Exner, F.M. (1925) Über die Wechselwirkung zwischen Wasser und Geschiebe in Flüssen, Sitz-Ber. Akad. Wiss. Wien, Abt. IIa, Band 134.

Exner, F.M. (1931) Zur Dynamik der Bewegungsformen auf der Erdoberfläche, *Ergebnisse der kosmischen Physik*, Band 1.

Falcón, M.A. and Kennedy, J.F. (1983) Flow in alluvial-river curves, *J. Fluid Mechanics*, Vol. 3, 1-16.

Falcón, M. (1984) Secondary flow in curved open channels, *Ann. Rev. Fluid Mechanics*, Vol. 16, 179-193.

Farhoudi, J. and Smith, K.V.H. (1982) Time scale for scour downstream of hydraulic jump, *Proc. ASCE*, Vol. 108, HY10, 1147-1161.

Farhoudi, J. and Smith, K.V.H. (1985) Local scour profiles downstream of hydraulic jump, *J. Hydraulic Res.*, Vol. 23, 343-358.

Fenton, J.D. and Abbott, J.E. (1977) Initial movement of grains on a stream bed: the effect of relative protrusion, *Proc. Roy. Soc., London*, Vol. 352, 523-537.

Fenton, J.D. (1979) A high-order cnoidal wave theory, *J. Fluid Mech.* Vol. 94, Pt. 1, 129-161.

Ferguson, R.I. (1973) Channel pattern and sediment type, *Area*, Vol. 5, 38-41.

Fidleris, V. and Whitmore, R.L. (1961) Experimental determination of the wall effect for spheres falling axially in cylindrical vessels, *Brit. J. Appl. Phys.*, Vol. 12.

Field, W.G. (1971) Flood protection at highway bridge openings, *Univ. of Newcastle*, N.S.W., Engng. Bulletin, CE3.

Fisher, F.H., Williams, R.B. and Dial, O.E. Jr. (1970) Analytic equation of state for water and sea water, Fifth Report of the Joint Panel on Oceanographic Tables and Standards, Kiel, *UNESCO Techn. Papers in Marine Science*, No. 14, Annex II, 10-20.

Fitzgerald, D.M., Penland, S. and Nummedal, D. (1984) Control of barrier island shape by inlet sediment bypassing: East Frisian Islands, West Germany in *Developments in Sedimentology*, edited by Greenwood, B. and Davis, R.A. Jr., Vol. 39, Elsevier Science Publishers, 355-376.
Also Marine Geology, Vol. 60, (1984).

Flaxman, E.M. (1963) Channel stability in undisturbed cohesive soils, *Proc. ASCE*, Vol. 89, HY2, 87-96.
Fleming, G. (1969) Design curves for suspended load estimation, *Proc. Inst. Civ. Engng.*, London, Vol. 43, 1-9.
Fletcher, B.P. and Grace, J.L. Jr. (1974) Practical guidance for design of lined channel expansions at culvert outlets, *U.S. Army Corps. of Engineers*, Waterways Experim. Stat., Vicksburg, Tech. Rep. H-74-9.
Forchheimer, P. (1924) *Hydarulik*, Teubner Verlagsgesellschaft, Leipzig, Berlin.
Fortier, S. and Scobey, F.C. (1926) Permissible canal velocities, *Trans. ASCE*, Vol. 89, 940-984.
Fortnum, B.C.H. (1981) Waves recorded at Dowsing Light Vessel between 1970 and 1979, *Institute of Oceanographic Sciences*, U.K., Report No. 126.
Fredsøe, J. (1975) The friction factor and height-length relations in flow over a dune-covered bed, *Inst. of Hydrodynamic and Hydraulic Engng.* (ISVA), Techn. Univ. Denmark, Progress Report No. 37.
Fredsøe, J. (1978) Meandering and braiding of rivers, *J. Fluid Mechanics*, Vol. 84, 607-624.
Fredsøe, J., Anderson, O.H. and Silberg, S. (1983) Distribution of suspended sediment in large waves, *The Danish Center for Applied Mathematics and Mechanics*, Tech. University of Denmark, Report No. 269.
Fredsøe, J. (1984) Turbulent boundary layer in wave-current motion, *ASCE J. Hydraulic Engineering*, Vol. 110, 1103-1120.
Friedkin, J.F. (1945) Meandering of alluvial rivers, *U.S. Corps. of Engineers*, Waterways Experiment Station.
Frijlink, H.C. (1952) Discussion des Formules de Débit Solide de Kalinske, Einstein et Meyer-Peter et Mueller compte tenue des mesures récentes de transport dans les rivières Néerlandaises, *2me Journal Hydraulique:* Soc. Hydraulique de France, Grenoble, 98-103.
Fröhlich, G.R. (1985) Über den Einfluss der Sohlenform auf den Sand-transport bei geringen Feststoffkonzentrationen in teil- und vollgefüllten Rohrleitungen. Dissertation, Tech. *University Braunschweig, FRG.*
Führböter, A. (1967) Über die Förderung von Sand-Wasser-Gemischen in Rohrleitungen, *Techn. Univ. Hannover*, Mitteilungen Franzius Inst. No. 19.
Führböter, A. (1967) Mechanik der Strömungsriffel, *Tech. Univ. Hannover*, Mitt. Franzius Institut, No. 29.
Führböter, A. (1979) Stormbänke (Grossriffel) und Dünen als Stabilisierungsformen, *Tech. Univ. Braunschweig*, Mitt. Leichtweiss-Institut, No. 67, 155-192.
Führböter, A. (1983) Zur Bildung von makroskopischen Ordnungs-strukturen (Strömungsriffel und Dünen) aus sehr kleinen Zufallsstörungen, *Techn. Univ. Braunschweig*, Mitteilungen des Leichtweiss-Instituts, Heft 79.
Führböter, A. (1983) Über mikrobiologische Einflüsse auf den Erosionsbeginn bei Sandwatten, *Wasser und Boden*, No. 3, 106-116.
Fukuoka, S. (1968) Generation, development and spectrum of sand-waves, *Tokyo Inst. Technology*, Dept. Civil Engng. Tech. Rep. No. 4, 45-55.
Fukuoka, S. and Kikkawa, H. (1971) Characteristics of open-channel

flow with sediment, Techn. Rep. No. 10, *Tokyo Institute of Technology*, Dept. of Civil Engineering.

Galay, V.J. (1983) Causes of river bed degradation, *Water Resources* Research, Vol. 19, 1057-1090.

Galvin, C.J. Jr. (1969) Breaker travel and choice of design wave height, *Proc. ASCE*, Vol. 95, WW2, 175-200.

Garbrecht, G. (1961) Erfahrungswerte über die zulässigen Strömungsgeschwindigkeiten in Flüssen und Kanälen, *Wasser und Boden*, Vol. 5.

Garde, R.J. and Ranga Raju, K.G. (1985) *Mechanics of Sediment Transportation and Alluvial Stream Problems*, Wiley Eastern Ltd.

Gessler, J. (1965) Der Geschiebetriebbeginn bei Mischungen untersucht an natürlichen Abpflästerungserscheinungen in Kanälen, *Mitteilungen der Versuchsanstalt für Wasserbau und Erdbau*, No. 69, E.T.H. Zurich.

Gessler, J. (1970) Self-stabilizing tendencies of alluvial channels, *Proc. ASCE*, Vol. 96, WW2, 235-249.

Gessler, J. (1971) Critical shear stress for sediment mixtures, *Proc. 14th Congress I.A.H.R., Paris*, Vol. 3, No. Cl.

Gessler, J. (1973) Behaviour of sediment mixtures in rivers, *Proc. I.A.H.R. Symposium on River Mechanics*, Bangkok, Vol. 1, 395-406.

Gilbert, G.K. and Murphy, E.C. (1914) The transportation of debris by running water, *U.S. Geological Survey*, Prof. Paper 86.

Gill, M.A. (1968) Rationalization of Lacey's regime flow equations *Proc. ASCE*, Vol. 94, HY4, 983-995.

Goda, Y. (1975) Irregular wave deformation in the surf zone, *Coastal Engng. in Japan*, Vol. 18, 13-26.

Göhren, H. (1975) Zur Dynamik und Morphologie der hohen Sandbänke im Wattenmeer zwischen Jade und Eider, *Die Küste*, Vol. 27, 28-49.

Graf, W.H. and Acaroglu, E.R. (1968) Sediment transport in conveyance systems, Part 1, *Bull. Int. Assoc. Sci. Hydrol.*, XIII, No. 2.

Graf, W.H. (1971) *Hydraulics of Sediment Transport*, McGraw-Hill.

Grant, W.D. and Madsen, O.S. (1982) Movable bed roughness in unsteady oscillatory flow, *J. Geophysical Res.*, Vol. 87, 469-481.

Grass, A.J. (1970) The initial instability of fine sand, *Proc. ASCE*, Vol. 96, HY3, 619-632.

Grass, A.J. (1971) Structural features of turbulent flow over smooth and rough boundaries, *J. Fluid Mech.*, Vol. 50, 233-255.

Grass, A.J. (1981) Sediment transport by waves and currents, *SERC London Centre for Marine Technology*, Rep. No. FL29.

Grim, R.E. (1962) *Applied Clay Mineralogy*, McGraw-Hill.

Grissinger, E.H. (1966) Resistance of selected clay systems to erosion by water, *Water Resources Research*, Vol. 2, 131-138.

Hallermeier, R.J. (1980) Sand motion initiation by water waves: two asymptotes, *Proc. ASCE J. Waterway Port Coastal and Ocean Div.* Vol. 106, WW3, 299-318.

Hallermeier, R.J. (1981) A profile zonation for seasonal sand beaches from wave climate, *Coastal Engineering*, Vol. 4, 253-277.

Hama, F.R. (1954) Boundary layer characteristics for smooth and

rough surfaces, *Trans. Soc. of Naval Architects and Marine Engngs.* Vol. 62, 333-358.
Hanks, R.W. and Dadia, B.H. (1971) Theoretical analysis of the turbulent flow of non-Newtonian slurries in pipes, *J.A.I. Ch.E.*, Vol. 17, 554-557.
Hannah, C.R. (1978) Scour at pile groups, *University of Canterbury, N.Z.*, Civil Engineering Research Rep. No. 78-3, 92.
Happel, J. and Brenner, H. (1965) *Low Reynolds Number Hydrodynamics*, Prentice-Hall.
Harris, J. and Quader, A.K.M.A. (1971) Design procedure for pipelines transporting non-Newtonian fluids and solid-liquid systems, *Pipes and Pipelines*, Vol. 16, No. 415, 307-311.
Hartung, W. (1957) Die Gesetzmässigkeit der Kolkbildung hinter über-strömten Wehren, Dissertation, *Techn. Univ. Braunschweig.*
Hattori, M. (1969) The mechanics of suspended sediment due to wave action, *Coastal Engineering in Japan*, Vol. 12.
Häusler, E. (1983) Spillways and outlets with high energy concentrations, *Intern. Committee on Large Dams, Trans. Intern. Symposium on Layout of Dams in Narrow Gorges*, Rio de Janeiro Vol. 2, 177-194.
Havinga, H. and Urk, A. van (1980) Solving river problems in the Netherlands, *Proc. I.A.H.R. Symposium on River Engng.*, Belgrade.
Hayashi, T., Ozaki, S. and Ichibashi, T. (1956) Study of bed load transport of sediment mixture, *Proc. 24th Japanese Conf. on Hydraulics.*
Hayashi, T. (1970) Formation of dunes and anti-dunes in open channels, *Proc. ASCE*, Vol. 96, HY2, 357-366.
Hayashi, T., Ozaki, S. and Ichibashi, T. (1980) Study on bed load transport of sediment mixtures, *Proc. 24th Japanese Conference on Hydraulics.*
Haydon, J.W. and Stelson, T.E. (1971) Hydraulic conveyance of solids in pipes, *Symposium of Solid-liquid Flow in Pipes*, Univ. of Pennsylvania, 149-163.
Hayter, E.J. (1986) Estuarial sediment bed models, *Estuarine Cohesive Sediment Dynamics*, Lecture Notes on Coastal and Estuarine Studies, Vol. 14, Edited by A.J. Mehta, Springer-Verlag, 326-359.
Hazen, A. (1904) On sedimentation, *Trans. ASCE*, Vol. 53, 45-71.
Hedegaard, I.B. (1985) Wave generated ripples and resulting sediment transport in waves, Institute of Hydrodynamics and Hydraulic Engineering, Techn. *Univ. of Denmark*, Series, Paper No. 36.
Henderson, F.M. (1961) Stability of alluvial channels, *Proc. ASCE*, Vol. 87, HY6, 109.
Hey, R.D. (1979) Flow resistance in gravel-bed rivers, *Proc. ASCE*, Vol. 105, HY4, 365-379.
Hey, R.D., Bathurst, J.C. and Thorne, C.R. (1982) *Gravel-Bed Rivers*, Wiley & Sons.
Heywood, H. (1938) Measurement of the fineness of powdered materials, *Proc. Inst. Mech. Eng.*, Vol. 140.
Hill, H.M., Srinivasan, V.S. and Unny, T.E. (1967) Instability of flat bed in alluvial channels, *ASCE Ann. Meeting* and National Meeting on Water Resources Engineering, New Orleans LA.
Hino, M. (1963) Turbulent flow with suspended particles, *Proc. ASCE*, Vol. 89, HY4, 161-185.

Hino, M. (1968) Equilibrium-range spectra of sand waves formed by flowing water, *J. Fluid Mech.*, Vol. 34, Pt. 3, 565-573.
Hinze, J.O. (1959) *Turbulence, An Introduction to its Mechanism and Theory*, McGraw-Hill, 2nd Ed. (1975).
Hjorth, P. (1972) Lokal Erosion och Erosionsverken vid Arlopps-ledning i Kustnära Områden, *Inst. Vattenbyggnad, Tekn. Hogskolan, Lund*, Bulletin Serie B, No. 21.
Hjorth, P. (1975) Studies on the nature of local scour, *Dept. Water Res. Engng., Lund Inst. of Technology*, Bulletin Series A, No. 46.
Hjulström, (1935) The morphological activity of rivers as illustrated by River Fyris, *Bull. Geol. Inst.* Uppsala, Vol. 25, Chap. III.
Höfer, U. (1984) Beginn der Sedimentbewegung bei Gewässersohlen mit Riffeln oder Dünen, Tech. Berichte Ing. Hydrologie und Hydraulik, Inst. für Wasserbau, *TH Darmstadt*, No. 32.
Homeier, H. and Kramer, J. (1957) Verlagerung der Platen in Riffbogen vor Norderney und ihre Anlandung an den Strand, *Jahresbericht* 1956, Forschungstelle Norderney, Vol. 8, 37-60.
Hom-ma, M. and Horikawa, K. (1962) Suspended sediment due to wave action, *Proc. 8th Coastal Engineering Conf. Mexico*, 168-193. Also in Horikawa (1978), 264-269.
Hopkins, G.R., Vance, R.W. and Kasraie, B. (1983) Scour around bridge piers, *Fedral Highway Administration, Washington D.C.*, Final Report FHWA - RD-78-163. Also FHWA -RD-75-56 (1975) and FHWA -RD-78-103 (1980).
Horikawa, K. and Shen, H.W. (1960) Sand movement by wind - on the characteristics of sand traps, *U.S. Army Corps of Engrs.*, Beach Erosion Board, Tech. Memo No. 119.
Horikawa, K. and Kuo, C-T (1966) A study on wave transformation inside surf zone, *Proc. 10th Coastal Engineering Conf., Tokyo*, Vol. 1, 217-233.
Horikawa, K. and Watanabe, A. (1968) Laboratory study on oscillatory boundary layer flow, *Proc. 11th Coastal Engineering Conf.* London, Vol. 1, 467-486.
Horikawa, K. (1978) *Coastal Engineering*, University of Tokyo Press and A. Halsted Press.
Horikawa, A., Watanabe, A. and Katori, S. (1982) Sediment transport under sheet flow condition, *Proc. 18th Coastal Engineering Conf.*, Cape Town, Vol. 2, 1335-1352.
Horton, R.E. (1945) Erosional development of streams and their drainage basins; hydrological approach to quantitative morphology. *Bull. Geol. Soc. Am.*, Vol. 56, 275-370.
Huang, N.E. (1970) Transport induced by wave motion, *J. Marine Research*, Vol. 28, 35-50.
Hunt, I.A. Jr. (1959) Design of seawalls and breakwaters, *Proc. ASCE*, Vol. 85, WW3.
Hunt, S.N. (1954) The turbulent transport of suspended sediment in open channels, *Proc. Roy. Soc. London*, Vol. A224, 322-335.
Hübbe, H. (1861) Von der Beschaffenheit und dem Verhalten des Sandes im Wasser, *Zeitschrift für Bauwesen.*
Hutchison, D.L. (1972) Physics of erosion of cohesive soils, Ph.D Thesis, *University of Auckland, N.Z.*, Dept. Civil Engng. Report No. 81.
Inglis, Sir. C. (1949) The behaviour and control of rivers and canals, *Central Water-Power-Irrigation and Navigation Research Station, Poona*, Research Publ. No. 13, Part 2.

Ikeda, S. and Asaeda, T. (1981) Sediment suspension with rippled bed, *Saitama Univ. Faculty of Engng.*, Report Vol. 11, 1-19. Also in J. Hydraulic Engng., ASCE, Vol. 109 (1983), 409-423.
Ikeda, S., Parker, G. and Sawai, K. (1981) Bend theory of river meanders, Part 1. Linear development, *J. Fluid Mechanics*, Vol. 112, 363-377.
Inman, D.L. (1957) Wave generated ripples in nearshore sands, *U.S. Army Beach Erosion Board*, Tech, Memo No. 100.
Iribarren, C.R. and Nogales, C. (1949) Protection des Ports, Section II, Comm. 4, *XVIIth Intern. Navig. Congress, Lisbon*, 31-80.
Irmay, S. (1949) On steady flow formulae in pipes and channels, Paper No. III-3, *Third Meeting I.A.H.R.*, *Grenoble*, France.
Itakura, T. and Kishi, T. (1980) Open channel flow with suspended sediments, *Proc. ASCE*, Vol. 106, HY8, 1325-1343.
Izumiya, T. and Isobe, M. (1986) Breaking criterion on non-uniformly sloping beach, *Proc. 20th Coastal Engng, Conf.*, Taiwan, Vol. 1, 318-327.
Jackson, R.G. (1976) Sedimentological and fluid-dynamic implications of the turbulent bursting phenomenon in geophysical flows, *J. Fluid Mech.*, Vol. 77, 531-60.
Jaeggi, M.N.R. (1984) Formation and effects of alternate bars, *ASCE J. Hydraulic Engng.*, Vol. 110, 142-156.
Jaeggi, M.N.R. (1983) Alternierende Kiesbänke, *Mitt. Versuchsanstalt* für Wasserbau Hydrologie und Glaziologie, No. 62, ETH Zürich.
Jaeger, Ch. (1939) Über die Ähnlichkeit bei flussbaulichen Modellversuchen, *Wasserkraft und Wasserwirtschaft*, Vol. 34, No. 23/34.
Jaffe, B.E., Sternberg, R.W. and Sallenger, A.H. (1984) The role of suspended sediment in shore-normal beach profile changes, *Proc. 19th Coastal Engineering Conf.*, Houston, Vol. 2, 1983-1996.
Jain, S.C. and Kennedy, J.F. (1971) The growth of sand waves, *Proc. Intern. Symposium on Stochastic Hydraulics*, Pittsburgh
Jain, S.C. (1981) Riverbed aggradation due to overloading, *Proc. ASCE*, Vol. 107, HY1, 120-124.
James, W.R. (1975) Techniques in evaluating suitability of borrow material for beach nourishment, *U.S. Army Corps of Engineers* Waterways Experiment Station, Vicksburg, Miss. TM-60.
Jansen, R.Ph., Bendegom, L. van, Berg, J. van den, de Vries, M. and Zanen, A. (1979) *Principles of River Engineering. The non-tidal alluvial river*, Pittman.
Jarrett, J.T. (1976) Tidal Prism-Inlet area relationships, GITI Re. 3, *U.S. Army Corps of Engineers*, Waterway Experiment Station, Vicksburg, Miss.
Jensen, P. (1970) Synthèse des principales recherches sur les dunes et les rides fluviales, *Bulletin de la Direction des Etudes et Recherches*, Série A, Nucléaire, Thermique, Hydraulique, No. 2, 5-38.
Johnson, G. (1967) The effect of entrained air on scouring capacity of water jets, *Proc. 12th Congress I.A.H.R.*, Fort Collins, Vol. 3, C26, 218-226.
Johnson, J.W. (1949) Scale effects in hydraulic models involving wave motion, *Trans. Am. Geophysical Union*, Vol. 30, 517-527.
Jonsson, I.G. (1963) Measurements in the turbulent wave boundary layer, *Proc. 10th Congr. Intern. Assoc. for Hydraulic Res.*, London, Vol. 1, 85-92.

Jonsson, I.G. (1967) Wave boundary layers and friction factors, *Proc. 10th Conf. Coastal Engineering, Tokyo*, 1966, Vol. 1, 127-148.

Jonsson, I.G. and Carlsen, N.A. (1976) Experimental and theoretical investigations in an oscillatory rough turbulent boundary layer, *J. Hydr. Res.*, Vol. 14, 45-60.

Jonsson, I.G. (1978) Energy flux and wave action in gravity waves propagating on a current, *J. Hydraulic Research*, Vol. 16, 223-234.

Jonsson, I.G. (1978) A new approach to oscillatory rough turbulent boundary layers, *Inst. of Hydrodynamics and Hydraulic Engineering, Techn. University of Denmark*, Series Paper 17.

Jonsson, I.G., Brink-Kjaer, O. and Thomas, G. (1978) Wave action and set-down for waves on a shear current, *J. Fluid Mech.*, Vol. 87, Part 3, 401-416.

Jonsson, I.G. (1980) A new approach to oscillatory rough turbulent boundary layers, *Ocean Engineering*, Oxford, England, Vol. 7, 109-152.

Kadib, A-L. (1964) Sand transport by wind, studies with sand C (0.145 mm). Addendum II in sand transport by wind by Belly, P-Y. *U.S. Army Corps of Engineers, Coastal Engineering Research Research Center*, Techn. Memo No. 1.

Kajima, R., Shimizu, T., Maruyama, K. and Saito, S. (1982) Experiments on beach profile change with a large wave flume, *Proc. 18th Coastal Engineering Conf.*, Cape Town, Vol. 2, 1385-1404.

Kajiura, K. (1968) A model of the bottom boundary layer in water waves, Bull. Earthquake Res. Inst. *Univ. of Tokyo*, Vol. 76, 75-123.

Kalinske, A.A. (1942) Criteria for determining sand-transport by surface creep and saltation, *Trans. AM. Geophys. Union*, Vol. 23, Pt. 2.

Kalinske, A.A. (1947) Movement of sediment as bed-load in rivers, *Trans. Am. Geophys. Union*, Vol. 28, No. 4, 615-620.

Kalinske, A.A. and Hsia, C.H. (1945) Study of transportation of fine sediments by flowing water, *State University of Iowa*, Studies in Engineering, Bulletin 29.

Kalkanis, G. (1964) Transportation of bed material due to wave action, *U.S. Army Coastal Engineering Research Center*, Techn. Memo, No. 2.

Kamphuis, J.W. (1975) Friction factor under oscillatory waves, *Proc. ASCE J. Waterway Port Coastal and Ocean Div.*, Vol. 101, WW2, 135-144.

Kana, T.W. (1979) Suspended sediment in breaking waves, *University of South Carolina, Dept. of Geology*, Tech. Rep. No. 18-CRD.

Karasev, I.T. (1964) The regimes of eroding channels in cohesive material, *Soviet Hydrology* (Am. Geophys. Union), Vol. 6, 551-579.

Kazanskij, I. (1978) Scale-up effects in hydraulic transport theory and practice, *Hydrotransport 5*, Vol. 2, Hannover, West Germany.

Kazanskij, I. (1981) Über theoretische und praxisbezogene Aspekte des hydraulischen Feststofftransportes, *Mitt. des Franzius-Inst.* Univ. Hannover, Vol. 52.

Kellerhals, R. (1967) Stable channels with gravel-paved beds, *Proc. ASCE*, Vol. 93 (WW), 63-84.
Disc. Nov. 1967, closure Aug. 1968.

Kelly, W.E. and Gularte, R.C. (1981) Erosion resistance of cohesive soils, *Proc. ASCE*, Vol. 107, HY10, 1211-1214.

Kennedy, J.F. (1961) Stationary waves and anti-dunes in alluvial channels, *California Inst. Technology*, W.M. Keck Lab. Rep. No. KH-R-2.

Kennedy, J.F. (1963) The mechanics of dunes and antidunes in erodible-bed channels, *J. Fluid Mech.*, Vol. 16, 512-544.

Kennedy, J.F. (1969) The formation of sediment ripples, dunes and anti-dunes, *Ann. Rev. Fluid Mech.*, Vol. 1, 147-168.

Kennedy, J.F. and Locher, F.A. (1971) Sediment suspension by water waves: waves on beaches and resulting sediment transport, *Proc. of an Advanced Seminar*, Mathematics Research Center Publication, *Univ. of Wisconsin*, Madison Wisc., No. 28, 249-295.

Kennedy, R.G. (1895) The prevention of silting in irrigation canals, *Min. Proc. Inst. Civil Engrs.*, Vol. CXIX.

Kemblowski, Z. and Kolodziejski, J. (1973) Flow resistance of non-Newtonian fluids in transitional turbulent flow, *Int. Chem. Engng.*, Vol. 13, 265-279.

Kenchington, J.M. (1974) The prediction of pressure drop in slurry pipelines, *Multi-phase flow symposium*, Inst. Chem. Engng., at Univ. of Strathclyde, Glasgow.

Kerssens, P.J.M. and van Rijn, L.C. (1977) Model for non-steady suspended sediment transport, *Delft Hydraulic Laboratory* Publication, No. 191.

Kerssens, P.M., Prins, A.D. and van Rijn, L.C. (1979) Model for suspended sediment transport, *Proc. ASCE*, Vol. 105, HY5, 461-476.

Keulegan, G.H. (1938) Laws of turbulent flow in open channels, *Nat. Bureau of Standards, Journal of Research*, Vol. 21.

Kikkawa, H., Ikeda, S. and Kitagawa, A. (1976) Flow and bed topography in curved open channels, *Proc. ASCE*, Vol. 102, HY9, 1327-1342.

King, C.A.M. (1972) *Beaches and Coasts*, 2nd edition, Edward Arnold, London.

Kinsman, B. (1965) *Wind Waves:* Their Generation and Propagation on the Ocean Surface, Prentice-Hall, Inc.

Kirkby, M.J., Morgan, R.P.C. (1980) *Soil Erosion*, Wiley, N.Y.

Kishi, T. (1980) Bed forms and hydraulic relations for alluvial streams, *Application of Stochastic Processes in Sediment Transport*, Chapter 5, Water Res. Publication, edited by H.W. Shen and H. Kikkawa.

Klaassen, G.J., Ognik, H.J.M. and van Rijn, L.C. (1986) DHL-Research on bedforms, resistance to flow and sediment transport, *Proc. 3rd Intern. Symp. on River Sedimentation*, Jackson, Miss.

Klaassen, G.J. and Urk, A. van (1986) Testing expressions for the resistance to flow of bedforms, *Delft Hydraulics Laboratory*, Rijkswaterstaat, TOW Rivers, Report R 657-XVI.

Kline, S.J., Reynolds, W.C., Straub, F.A. and Runstadler, P.W. (1967) The structure of turbulent boundary layers, *J. Fluid Mech.*, Vol. 30, pt. 4, 741-773.

Knight, D.W. and Demetriou, J.D. (1983) Flood plain and main channel flow interaction, *J. Hydraulic Engineering, ASCE*, Vol. 109, 1073-1092.

Kobus, H., Leister, P. and Westrich, B. (1979) Flow field and scouring effects of steady and pulsating jets impinging on a moveable bed, *J. Hydraulic Research*, Vol. 17, 175-192.

Kobus, H. (Editor) (1980) *Hydraulic Modelling*, German Assoc. for Water Resources and Land Improvement, Bull. 7 in cooperation with I.A.H.R., Verlag Paul Paray and Pitman.

Kolkman, P.A. (1982) An artificial ceiling for free surface flow reproduction in scale modelling of local scour, *Proc. Int. Conf. of Hydraulic Modelling of Civil Engineering Structures*, Coventry, 397-410.

Komar, P.D. and Miller, M.C. (1974) Sediment threshold under oscillatory waves, *Proc. 14th Coastal Engineering Conf.*, Copenhagen, Vol. 2, 756-775.

Komar, P.D. (1976) Longshore currents and sand transport on beaches, *Ocean Engineering III, ASCE*, 333-354.

Komura, S. and Simons, D.B. (1967) River bed degradation below dams, *Proc. ASCE*, Vol. 93, HY4, 1-14.
Discussions: Vol. 94, HY1, 336-340; HY2, 589-598; HY3, 757-764; HY5, 1346-1350; Vol. 95, HY3, 1042-1048.

Kondrat'ev, N.E., Lyapin, A.N., Popov, I.V., Pinkovskii, S.I., Fedorov, N.N. and Yakunin, I.I. (1962) *River flow and river channel formation*, from russian (1959) by Israel Program for Scientific Translations, U.S. Dept. of the Interior.

Korteweg, D.J. and de Vries, G. (1895) On the change of form of long waves advancing in a rectangular canal, and on a new type of long stationary waves, *Phil. Mag.*, Series 5, Vol. 39, 422-443.

Kossyan, R.D., Antsyferov, S.M., Dachev, V.Zh. and Pykhov, N.V. (1982) Determination of the absolute concentrations on suspended sediment according to information obtained through bathometers-accumulators, *Interactions of the Atmosphere, Hydrosphere and Litosphere in the Nearshore Zone.* Results of the International Experiments "Kamchiya 78", Bulgarian Academy of Sciences (in Russian).

Kramer, H. (1935) Sand mixtures and sand movement in fluvial models, *Trans. ASCE*, Vol. 100, 798-873.

Kraus, N.C. and Sasaki, T.O. (1979) Influence of wave angle on the longshore current, *Mar. Sci. Commun.*, Vol. 5, 91-126.

Kriebel, D.L. and Dean, R.G. (1985) Numerical simulation of time-dependent beach and dune erosion, *Coastal Engineering*, Vol. 9, 221-245.

Krone, R.B. (1962) Flume studies of the transport of sediment in estuarial shoaling processes, *Univ. of California, Berkeley*, Hydr. Engng. Lab. and Sanit. Engng. Res. Lab. Report.

Krumbein, W.C. (1951) A method for specification of sand for beach fills, *U.S. Army Corps of Engineers*, Beach Erosion Board, TM-102.

Krumbein, W.C. and James, W.R. (1965) A log normal size distribution model for estimating stability of beach fill material, *U.S. Army Corps of Engineers*, Waterways Experiment Station, Vicksburg, Miss., TM-16.

Kubota, S., Horikawa, K. and Hotta, S. (1982) Blown sand on beaches, *Proc. 18th Coastal Engineering Conf.*, Cape Town, Vol. 2, 1181-1198.

Kwan, T.F. (1984) Study of abutment scour, *University of Auckland, N.Z.*, ME Thesis, Dept. Civil Engineering Rep No. 328.

Lacey, G. (1929) Stable channels in alluvium, *Proc. Inst. Civil Engrs.*, Vol. 229, 259-384.

Lacey, G. (1946) A general theory of flow in alluvium, *Inst. of Civil Engrs.*, Vol. 27, 16.

Lacey, G. (1953) Uniform flow in alluvial rivers and canals, *Proc. Inst. Civil Engrs.*, Vol. 237, 421.

Lacey, G. (1958) Flow in alluvial channels with sandy mobile beds, *Proc. Inst. Civil Engrs.*, Vol. 9, 145.

Lambermont, J. and Lebon, G. (1978) Erosion of cohesive soils, *J. Hydraulic Res.*, Vol. 16, No. 1, 28-44.
Lane, E.W. (1935) Stable channels in erodible material, *Proc. ASCE*, Vol. 61, 1307-1326.
Lane, E.W. (1952) Progress Report on Results of Studies on Design of Stable Channels, *Bureau of Reclamation*, Rep. No. Hyd-352.
Lane, E.W. and Kalinske, A.A. (1939) The relation of suspended to bed material in rivers, *Trans. A.G.U.*, Vol. 20, 637.
Lane, E.W. and Kalinske, A.A. (1941) Engineering calculations of suspended sediment, *Trans. Am. Geophys. Union*, Vol. 22, 603-607.
Lane, E.W. (1953) Progress report on studies on the design of stable channels by the Bureau of Reclamation, Proc. ASCE, Sep. No. 280.
Lane, E.W. and Carlson, E.J. (1953) Some factors affecting the stability of canals constructed in coarse granular materials, *Proc. Minnesota Intern. Hydraulics Convention.*
Lane, E.W. (1955) Design of stable channels, *Trans. ASCE*, Vol. 120, Paper 2776.
Lane, E.W. (1958) Sediment engineering as a quantitative science, USBR Report.
Langbein, W.B. et al. (1949) Annual runoff in the United States, U.S. Geological Survey Circ. 52, 14.
Langbein, W.B. and Schumm, S.A. (1958) Yield of sediment in relation to mean annual precipitation, *Trans. A. Geophys. Union*, Vol. 39, 1076-1084.
Lau, Y. Lam (1987) Disc. on the determination of ripple geometry, *J. Hydraulic Engineering, ASCE*, Vol. 113, 128-131.
Laursen, E.M. (1952) Observation on the nature of scour, *Proc. 5th Hydraulic Conference*, Iowa Inst. of Hydr. Res.
Laursen, E.M. and Toch, A. (1956) Scour around bridge piers and abutments, *Iowa Highway Res. Board*, Bulletin No. 4, 60.
Laursen, E.M. (1958) Scour at bridge crossings, *Iowa Highway Res. Board*, Bulletin No. 8.
Laursen, E.M. (1958) Sediment transport mechanics in stable channel design, *Trans. ASCE*, Vol. 123.
Laursen, E.M. (1958) The total sediment load of streams, *Proc. ASCE*, Vol. 84, HY1, 1-36.
Laursen, E.M. (1960) Scour at bridge crossings, *Proc. ASCE*, Vol. 86, HY2, 39-54. Also Trans. ASCE (1962), Vol. 127, Pt1, 166-209.
Laursen, E.M. (1963) Analysis of relief bridge scour, *Proc. ASCE*, Vol. 89, HY3, 93-118.
Lawler, M.T. and Lu, Pau-Chang (1971) The role of lift in the radial migration of particles in a pipe flow, *Advances in Solid-Liquid Flow in Pipes and its Application*, edited by I. Zandi, Pergamon Press.
Lazarus, J.H. and Neilson, I.D. (1978) A generalised correlation for friction head losses of settling mixtures in horizontal smooth pipes, *Hydrotransport 5*, Hannover, West Germany, Vol. 1.
Leopold, L.B. and Wolman, G.M. (1957) River channel patterns: braided, meandering and straight, *U.S. Geol. Survey*, Prof. Paper 282-B.
Leopold, L.B. and Wolman, G.M. (1960) River meanders, *Bulletin Geological Society of America*, Vol. 71, 769-794.
Leopold, L.B. and Langbein, W.B. (1962) The concept of entropy in landscape evolution, *U.S. Geological Survey*, Prof. Paper 500-A.

Leopold, L.B., Wolman, G.M. and Miller, J.P. (1964) *Fluvial Process in Geomorphology*, W.H. Freeman and Co., San Francisco.
Leopold, L.B. and Emmett, W.W. (1976) Bedload measurements, East Fork River, Wyoming, *Proc. National Academy of Sciences*, USA, Vol. 73, 1000-1004.
Leatherman, S.P. (Editor) (1979) *Barrier Islands*, Academic Press, N.Y.
Lee, D.Y. (1979) Resuspension and deposition of Lake Erie sediments, *Case Western Reserve Univ.*, Cleveland, Ohio, M.Sc. Thesis.
Li Huon, (1954) Stability of oscillatory laminar flow along a wall, *U.S. Army Beach Erosion Board*, Techn. Mem. No. 47.
Liang, S.S. and Wang, H. (1973) Sediment transport in random waves, *Univ. of Delaware, College of Marine Studies*, Techn. Rep. 26.
Limerinos, J.T. (1970) Determination of the Manning coefficient for measured bed roughness in natural channels, *U.S. Geol. Survey*, Water Supply Paper 1898-B.
Lindley, E.S. (1919) Regime channels, *Proc. Punjab Engng. Congress*, Vol. 7.
Liou, Y.D. (1970) Hydraulic erodibility of two pure clay systems, Ph.D. Thesis, *Colorado State University*.
List, E.J. (1982) Turbulent jets and plumes, *Annual Review of Fluid Mechanics*, Vol. 14.
Liu, H.K. (1957) Mechanics of sediment-ripple formation, *Proc. ASCE*, Vol. 83, HY2, Paper 1197.
Liu, H.K. and Hwang, S.Y. (1959) Discharge formula for straight alluvial channels, *Proc. ASCE*, Vol. 85, HY11, 65.
Liu, H.K., Chang, F.M. and Skinner, M.M. (1961) Effect of bridge constriction on scour and backwater, *Colorado State Univ.* Engng. Res. Center Rep. CER60HKL22.
Liu, P.L-F. and Dalrymple, R. (1978) Bottom frictional forces and longshore currents due to waves with large angles of incidence, *Journal of Marine Research*, Vol. 36(2), 357-325.
Liu, Vi-Chang (1956) Turbulent dispersion of dynamic particles, *J. Meteorology*, Vol. 13, 399.
Lofquist, F.E.B. (1978) Sand ripple growth in an osciallatory flow water tunnel, *U.S. Army Corps of Engineers, Coastal Engineering Research Center*, Techn. Paper No. 78-5.
Longuet-Higgins, M.S. (1950) A theory on the origin of microseims, *Phil. Trans. Roy. Soc.*, London, A, Vol. 243, 1-35.
Longuet-Higgins, M.S. (1953) Mass transport in water waves, *Phil. Trans. Roy. Soc.*, London, A245, 535-91.
Longuet-Higgins, M.S. and Stewart, R.W. (1964) Radiation stress in water waves: a physical discussion with applications, *Deep-Sea Research*, Pergamon Press, Vol. 11, 529-562.
Longuet-Higgins, M.S. (1970) Longshore currents generated by oblique incident sea waves, Part I and II, *J. Geophysical Research*, Vol. 75, 6778-6801.
Losada, M.A. and Desiré, J.M. (1985) Incipient motion on horizontal granular bed in non-breaking water waves, *Coastal Engineering*, Vol. 9, 357-370.
Lovera, F. and Kennedy, J.F. (1969) Friction-factors for flatbed flows in sand channels, *Proc. ASCE*, Vol. 95, HY4, 1127-1234.
Low, P.F. (1961) Physical chemistry of clay-water interactions, *Adv. Agron.*, Vol. 13, Academic Press.

Low, P.F. (1968) Mineralogy data requirements in soil physical investigations, *Mineralogy in Soil Science and Engineering*, SSSA special Publication No. 3.
Lundgren, H. (1972) Turbulent currents in the presence of waves, *Proc. 13th Coastal Engineering Conf.*, Vancouver, Vol. 1, 623-634.
Lyle, W.M. and Smerdon, E.T. (1965) Relations of compaction and other soil properties to the erosion resistance of soils, *Trans. Am. Soc. Agricultural Engrs.*, Vol. 8, No. 3.
MacDonald, T.C. (1977) Sediment suspension and turbulence in an oscillating flume, *U.S. Army Corps of Engineers, Coastal Engineering Research Center*, Tech. Paper, No. 77-4.
McNown, J.S., Lee H.M., McPherson, M.B. and Engez, S.M. (1948) Influence of boundary proximity on the drag of spheres, *Proc. 7th Int. Congress Appl. Mech.*, London.
McNown, J.S. and Newlin, J. (1951) Drag of spheres in cylindrical boundaries, *Proc. 1st Nat. Congr. Appl. Mech.*
McNown, J.S. (1951) Particles in slow motion, *La Houille Blanche*, Vol. 6, No. 5.
McTigue, D.F. (1981) Mixture theory for suspended sediment transport, *Proc. ASCE*, Vol. 107, HY6, 659-673.
Maddock, T. Jr. (1976) Equations for resistance to flow and sediment transport in alluvial channels, *Water Resources Research*, Vol. 12, No. 1, 11-21.
Madsen, O.S. and Grant, W.D. (1975) The threshold of sediment movement under oscillatory water waves: a discussion, *Journal of Sedimentary Petrology*, Vol. 45, 360-361.
Madsen, O.S. and Grant, W.D. (1976) Quantitative description of sediment transport by waves, *Proc. 15th Coastal Engineering Conf.*, Honolulu, Vol. 2, 1093-1112.
Mahmood, K. (1971) Flow in sand bed channels, *Colorado State Univ.*, Fort Collins, Water Management Tech. Report, No. 11.
Mahmood, K. and Shen, H.W. (1971) The regime concept of sediment-transporting canals and rivers, *River Mechanics*, Chapter 30, Edited and Published by H.W. Shen, Colorado.
Makkaveer, V.M. (1931) On theory of turbulent mode in sediment suspension, Izvestiya Gos. gidrologichekogo Instituta, Moscow, Issue 32, 5-27.
Maner, S.B. (1958) Factors affecting sediment delivery rates in the Red Hills physiographic area, *Trans. Am. Geophys. Union*, Vol. 39, 669-675.
Manohar, M. (1955) Mechanics of bottom sediment movement due to wave action, *U.S. Army Beach Erosion Board*, Tech. Memo No. 75.
Mantz, P.A. (1973) Cohesionless, fine graded, flaked sediment transport by water, *Nature, Physical Science*, Vol. 246, 14-16.
Marchillon, E.K., Clamen, A. and Gauvin, W.H. (1964) Oscillatory motion of freely falling disks, *Physics of Fluids*, Vol. 7, No. 12.
Marshall, C.E. (1964) *The Physical Chemistry and Mineralogy of Soils*, Vol. I: *Soil Materials*, Wiley & Sons.
Martins, R.B.F. (1975) Scouring of rocky river beds by free jet spillways, *Water Power and Dam Construction*, (April), 152-153.
Mash, F.D., Espey, W.H. Jr. and Moore, W.L. (1963) Measurements of the shear resistance of cohesive sediments, *Proc. Federal Inter-Agency Sedimentation Conf.*, Agricult. Res. Service, Misc. Publ. No. 970, Washington D.C., 151-155.

Mason, P.J. (1984) Erosion of plunge pools downstream of dams due to the action of free-trajectory jets, *Proc. Inst. Civil Engrs.*, Vol. 76, Part I, 523-537.

Mason, P.J. and Arumugam, K. (1985) Free jet scour below dams and flip buckets, *J. Hydraulic Engineering, ASCE*, Vol. 111, 220-235.

Matthes, G.H. (1947) Macroturbulence in natural stream flow, *Trans. Am. Geophys. Union*, Vol. 28, No. 2, 255-265.

Mehta, A.J. (1973) Depositional behaviour of cohesive sediments, Ph.D. Thesis, *University of Florida*, Gainesville.

Mehta, A.J. and Partheniades, E. (1975) An investigation of the depositional properties of flocculated fine sediments, *J. Hydraulic Research*, Vol. 13, No. 4, 361-381.

Mehta, A.J., Parchure, T.M., Dixit, J.G. and Ariathurai, R. (1982) Resuspension potential of deposited cohesive sediment beds, in *Estuarine Comparisons*, Edit. V.S. Kennedy, Academic Press, 591-609.

Melville, B.W. (1974) Scour at bridge sites, *Univ. of Auckland, N.Z.* School of Engineering, Report No. 104, or Proc. NZ National Roads Board RRU Seminar on Bridge Design and Research.

Melville, B.W. (1975) Local scour at bridge sites, *Univ. of Auckland, N.Z.* Ph.D. Thesis, Dept. of Civil Engng. Rep. No. 117.

Melville, B.W. and Raudkivi, A.J. (1977) Flow characteristics in local scour at bridge piers, *J. Hydr. Res.*, Vol. 15, 373-380.

Melville, B.W. (1984) Live-bed scour at bridge piers, *J. Hydraulic Engineering, ASCE*, Vol. 110, 1243-1247.

Mendoza, C., Abt, S.R. and Ruff, J.F. (1983) Headwall influence on scour at culvert outlets, *J. Hydraulic Engineering, ASCE*, Vol. 109, 1056-1060.

Mertens, W. (1986) Über die Deltabildung in Stauräumen, *Leichtweiss-Institut, Tech. Univ. Braunschweig*, Dissertation.

Meyer-Peter, E., Favre, H. and Einstein, A. (1934) Neuere Versuchsresultate über den Geschiebetrieb, *Schweiz. Bauzeitung*, 103, No. 13.

Miche, R. (1944) Mouvements ondulatoires de la mer en profondeur constante on décroissante, Ann. des Ponts et Chaussées, 114è Année.

Michell, J.H. (1893) On the highest waves in water, *Philosophical Magazine*, Vol. 36, 5th Series, 430-437.

Migniot, C. (1968) Etude des propriétés physiques de différents sediments trés fins et de leur comportement sous des actions hydro-dynamiques, *La Houille Blanche*, Vol. 27, No. 7, 591-620.

Miller, M.C., McCave, I.N. and Komar, P.D. (1977) Threshold of sediment motion under unidirectional currents, *Sedimentology*, Vol. 24, 507-527.

Miller, M.C. and Komar, P.D. (1980) Oscillation sand ripples generated by laboratory apparatus, *J. Sediment Petrol.*, Vol. 50, 173-182.

Mirtskhoulava, T.E. (1966) Erosional stability of cohesive soils, *J. Hydr. Research*, Vol. 4, 37-49.

Mirtskhoulava, Ts. E. (1981) Estimation of channel stability to scour in cohesive soils, in *Adv. in Sediment Transport*, Editor T. Manthey, Zakad Narodowy imienia Ossolinskich, Wydawnictwo Polskiej Akademii Nauk, Wrocaw. Confer. Nov. 13-18, 1978, in Jabonna, Poland.

Mirsi, R.L., Garde, R.J. and Ranga Raju, K.G. (1984) Bed load transport of coarse nonuniform sediment, *J. Hydraulic Engineering, ASCE*, Vol. 110, 312-328.
Mitchell, J.K. (1976) *Fundamentals of Soil Behaviour*, Wiley & Sons.
Mogridge, G.R. and Kamphuis, J.W. (1972) Experiments on bed form generation by wave action, *Proc. 13th Coastal Engineering Conf.*, Vancouver, 1123-1142.
Montague, C.L. (1986) Influence on biota on erodibility of sediments, in *Estuarine Cohesive Sediment Dynamics*, edited by A.J. Mehta, Lecture Notes on Coastal and Estuarine Studies, Vol. 14, Springer-Verlag, 251-269.
Moore, W.L. and Masch, F.D. (1962) Experiments on the scour resistance of cohesive sediments, *J. Geophys. Res.*, Vol. 67, No. 4, 1437-49.
Moore, W.J. (1972) *Physical Chemistry*, 5th ed., Longman.
Moraes, Carlos de Campos (1970) Experiments of wave reflexion on impermeable slopes, *Proc. 12th Coastal Engineering Conf.*, Washington D.C., Vol. 1, 509-521.
Mosley, M.P. (1976) An experimental study of channel confluences, *J. of Geology*, Vol. 84, 535-562.
Moshagen, H. (1984) Sand waves in free surface flow, Thesis No. P-1-84, *University of Trondheim*, Division of Port and Ocean Engineering.
Mostafa, G. (1957) River bed degradation below large capacity reservoir, *Trans. ASCE*, Vol. 122, 866-895.
Munk, W.H. (1949) The solitary wave theory and its application to surf problems, *Annals of the New York Academy of Sciences* Vol. 52, 376-462.
Musgrave, G.W. (1947) The quantitative evaluation of factors in water erosion, a first approximation, *J. Soil Water Conserv.* Vol. 2, No. 3, 133-138.
Nakato, T., Locher, F.A., Glover, J.R. and Kennedy, J.F. (1977) Wave entrainment of sediment from rippled beds, *Proc. ASCE*, Vol. 103, WW1, 83-100.
Newitt, D.M., Richardson, J.F., Abbott, M. and Turtle, R.B. (1955) Hydraulic conveying of solids in horizontal pipes, *Trans. Inst. Chem. Engng.*, Vol. 33, 93.
Neill, C.R. (1964) River-bed scour, a review for engineers, *Canadian Good Roads' Assoc.*, Tech. Publ. No. 23.
Neill, C.R. (1967) Mean velocity criterion for scour of coarse uniform bed material, *Proc. 12th Congress Intern. Assoc. for Hydraulic Res.*, Fort Collins, Vol. 3, 46-54.
Neill, C.R. and Yalin, M.S. (1969) Quantitative definition of beginning of bed movement, *Proc. ASCE*, Vol. 95, HY1, 585-588.
Neill, C.R. (1975) *Guide to Bridge Hydraulics*, Univ. of Toronto Press.
Nielsen, P. (1981) Dynamics and geometry of wave generated ripples, *J. Geophysical Research*, Vol. 86(C7), 6467-6472.
Nielsen, P. (1983) Analytical determination of nearshore wave height variation due to refraction shoaling and friction, *Coastal Engineering*, Vol. 7, 233-251.
Nielsen, P. (1984) On the motion of suspended sand particles, *J. Geophysical Research*, Vol. 89, C1, 616-626.
Nielsen, P. (1984) Field measurements of time-averaged suspended sediment concentration under waves, *Coastal Engineering*, Vol. 8, 51-72.
Nielsen, P. (1985) Reply to Discussion on paper in J. Geophys.

Res., Vol. 89(1984), 616-626, *J. Geophysical Research*, Vol. 90, C2, 3255-3256.
Nielsen, P. (1986) Suspended sediment concentration under waves, *Coastal Engineering*, Vol. 10, 23-31.
Nordin, C.F. Jr. (1964) Aspects of flow resistance and sediment transport Rio Grande near Bernalillo, New Mexico, *U.S. Geological Survey Water Supply Paper*, 1498-H; also in *Proc. ASCE*, Vol. 97(1971), 101-141.
Nordin, C.F. and Algert, J.H. (1966) Spectral analysis of sand waves, *Proc. ASCE*, Vol. 92, HY5, 95-114.
Nordin, C.F. (1969) A note on scour and fill associated with migrating sand waves, *Tech. Univ. Denmark*, Hydraulic Lab. Basic Research Progress Report, No. 18.
Novak, P. (1956) Study of stilling basins with special regard to their end sill, *Proc. 6th Congress I.A.H.R.*, Hague, Vol. 3, C15.
Novak, P. (1961) Influence of bed load passage on scour and turbulence downstream of a stilling basin, *Proc. 9th Congress of I.A.H.R.*, Dubrovnik, 66-75.
O'Brien, M.P. (1969) Equilibrium flow areas of inlets on sandy coasts, *Proc. ASCE*, Vol. 95, WW1, 43-52.
Odgaard, A.J. (1981) Transverse bed slope in alluvial channel bends, *Proc. ASCE*, Vol. 107, HY12, 1677-1694.
Odgaard, A.J. (1982) Bed characteristics in alluvial channel bends, *Proc. ASCE*, Vol. 108, HY11, 1268-1281.
Odgaard, A.J. Shear-induced secondary currents in channel flows, *Proc. ASCE*, Vol. 110, HY7, 996-1004.
Odgaard, A.J. (1984) Flow and bed topography in alluvial channel bend, *Proc. ASCE*, Vol. 110, HY4, 521-536.
O'Loughlin, E.M. and MacDonald, E.G. (1964) Some roughness-concentration effects on boundary resistance, *La Houille Blanches*, Vol. 19, 773-782.
Oliver, D.R. and Ward, S.G. (1959) Studies of viscosity and sedimentation of suspensions: Part 5. The viscosity of settling suspensions of spherical particles, *Brit. J. Appl. Physics*, Vol. 10, 317-321.
Orgis, H. (1974) Geschiebebetrieb und Bettbildung, *Österreichische Ingenieur-Zeitschrift*, 17. Jahrgang, Heft 9, 285-292.
Owen, P.R. (1964) Saltation of uniform grains in air, *J. Fluid Mech.*, Vol. 20, 225.
Owen, P.R. (1969) Pneumatic transport, *J. Fluid Mech.*, Vol. 39, 707-432.
Owen, M.W. (1970) A detailed study of the settling velocities of an estuary mud, *Hydr. Research Station*, Wallingford, Rep. No. INT 78.
Owen M.W. (1975) Erosion of Avonmouth mud, *Hydraulic Research Station*, Wallingford, Rep. INT 150.
Owen M.W. and Thorn, M.F.C. (1978) Effect of waves on sand transport by currents, *Proc. 16th Coastal Engineering Conf.*, Hamburg, Vol. 2, 1675-1687.
Paintal, A.S. and Garde, R.J. (1964) Discussion of sediment transport mechanics: Suspension of sediment, *Proc. ASCE*, Vol. 90, HY4, 257-265.
Paintal, A.S. (1971) A stochastic model of bed load transport, *J. Hydr. Research*, Vol. 9, No. 4, 527-554.
Parchure, T.M. (1984) Erosional behaviour of deposited cohesive sediments, Ph.D. Thesis, *Univ. of Florida*, Gainesville. Also in Parchure, T.M. and Mehta, A.J. (1985) Erosion of

soft cohesive sediment deposits, *J. Hydraulic Engng. ASCE*, Vol. 111, 1308-1326.
Parker, G. (1976) On the causes and characteristic scales of meandering and braiding in rivers, *J. Fluid Mech.*, Vol. 76, 457-480.
Parker, G. (1978) Self-formed straight rivers with equilibrium banks and mobile bed. Part 1: The sand-silt river, Part 2: The gravel river, *J. Fluid Mech.*, Vol. 89, 109-125 and 127-146.
Parker, G. and Klingeman, P.C. (1982) On why gravel bed streams are paved, *Water Resources Research*, Vol. 18, 1407-1423.
Parker, G., Sawai, K. and Ikeda, S. (1982) Bend theory of river meanders. Part 2, Nonlinear deformation of finite-amplitude bends, *J. Fluid Mech.*, Vol. 115, 303-314.
Parker, G., Diplas, P. and Akiyama, J. (1983) Meander bends of high amplitude, *Proc. ASCE*, Vol. 109, HY10, 1323-1337.
Parker, W.R. (1986) On the observation of cohesive sediment behaviour for engineering purposes, in *Esturial Cohesive Sediment Dynamics*. Lecture Notes on Coastal and Estuarine Studies, edited by A.J. Mehta, Springer-Verlag, Vol. 14, 270-289.
Partheniades, E. (1965) Erosion and deposition of cohesive soils, *Proc. ASCE*, Vol. 91, HY1, 105-139.
Pasenau, H. and Ulrich, J. (1974) Giant and mega ripples in the German Bight and studies of their migration in a testing area, (Lister Tief), *Proc. 14th Coastal Engineering Conf.*, Copenhagen, Vol. 2, 1025-1035.
Pazis, G.C. and Graf, W.H. (1977) Weak sediment transport, *Proc. ASCE*, Vol. 103, HY7, 799-802.
Peirce, T.J., Jarman, R.T. and Turbille, C.M. (1970) An experimental study of silt scouring, *Proc. Inst. Civil. Engrs.*, Vol. 45, 231-243.
Pelnard-Considère, (1954) Essai de Théorie à l'Evolution des Formes de Rivages en Plages de Sables et de Galets, *Quatroéme Journées de l'Hydraulique*, Paris, Question 3, Les Energies de la Mer.
Pemberton, E.L. (1972) Einstein's bedload function applied to channel design and degradation, Ch. 16 in *Sedimentation, Symposium to honour Prof. H.A. Einstein*, Edited and Published by H.W. Shen, Fort Collins, Colorado.
Perry, A.E. and Schofield, W.H. (1969) Rough wall turbulent boundary layers, *J. Fluid Mech.*, Vol. 37, 383-413.
Pettyjohn, E.S. and Christiansen, E.B. (1948) Effect of particle shape of free settling rates of isometric particles, *Chem. Engng. Prog.*, Vol. 44, No. 2.
Phillips, B.C. (1984) Spatial and temporal lag effects in bedload sediment transport, Ph.D. Thesis, *University of Canterbury, N.Z.* Dept. of Civil Engng. Rep. 84-110.
Pien, C.L. (1941) Investigation of turbuelnce and suspended material transportation in open channels, Ph.D. Thesis, *University of Iowa.*
Radoslovich, E.W. and Norrish, K. (1962) The cell dimensions and symmetry of layer lattice silicates I. Some structural considerations, *Am. Mineralogy*, Vol. 47, 599-616.
Rajaratnam, N. (1976) *Turbulent Jets*, Elsevier, Amsterdam.
Rajaratnam, N. and Ahmadi, R.M. (1979) Interaction between main channel and flood plain flows, *Proc. ASCE*, Vol. 105, HY5, 573-588.

Rajaratnam, N. and Ahmadi, R.M. (1981) Hydraulics of channels with flood plains, *J. Hydraulic Research*, Vol. 19, 43-59.
Rajaratnam, N. (1981) Further studies on the erosion of sand beds by plane water jets: Study 1: Erosion of sand beds by obliquely impinging plane turbulent submerged water jets, *University of Alberta*, Dept. of Civil Engng. Report WRE 81.
Rajaratnam, N. (1981) Erosion by plane turbulent jets, *J. Hydraulic Research*, Vol. 19, 339-358.
Rajaratnam, N. (1982) Erosion by submerged circular jets, *Proc. ASCE*, Vol. 108, HY2, 262-267.
Ramette M.(1979) Une approche rationelle de la morphologie fluviale, *La Houille Blanche*, Vol. 34, 491-498.
Ramette, M. (1980) A theoretical approach to fluvial processes, *Int. Symp. on River Sedimentation*, Beijing, China.
Ranga Raju, K.G. (1985) Transport of sediment mixtures, *Proc. 21st Congress I.A.H.R.*, Melbourne, Vol. 6, 35-46.
Rao, M.N. (1971) An erosion study of halloysite clay, *University of Auckland, N.Z.*, Dept. of Civil Engng. Rep. No. 75.
Raudkivi, A.J. (1963) Study of sediment ripple formation, *Proc. ASCE*, Vol. 89, HY6, 15-33.
Raudkivi, A.J. (1965) Turbulence and vorticity in loose boundary hydraulics, *Proc. 2nd Australasian Conf. on Hydraulics and Fluid Mechanics*, Univ. of Auckland, N.Z., A 135-142.
Raudkivi, A.J. (1966) Bed forms in alluvial channels, *J. Fluid Mech.*, Vol. 26, Pt. 3, 507-514.
Raudkivi, A.J. (1967) Analysis of resistance in fluvial channels, *Proc. ASCE*, Vol. 93, HY5, 73-84.
Raudkivi, A.J. and Small, A.F. (1974) Hydroelastic excitation of cylinders, *J. Hydraulic Research*, Vol. 12, 99-131.
Raudkivi, A.J. and Hutchison, D.L. (1974) Erosion of kaolinite clay by flowing water, *Proc. Roy. Soc.*, London, A 337, 537-554.
Raudkivi, A.J. and Callander, R.A. (1975) *Advanced Fluid Mechanics*, An Introduction, Edward Arnold, London.
Raudkivi, A.J. (1976) *Loose Boundary Hydraulics*, 2nd Edition, Pergamon Press.
Raudkivi, A.J. and Ettema, R. (1977) Effect of sediment gradation on clear-water scour, *Proc. ASCE*, Vol. 103, HY10, 1209-1213.
Raudkivi, A.J. and Sutherland, A.J. (1981) Scour at bridge crossings, National Roads Board Research Unit Bulletin No. 54, Wellington, N.Z., p. 100.
Raudkivi, A.J. (1981) *Grundlagen des Sedimenttransports*, Sonderforschungsbereich 79, Univ. of Hannover, Springer-Verlag, 1982.
Raudkivi, A.J. (1983) Thoughts on ripples and dunes, *J. Hydr. Research*. Vol. 21, 315-321.
Raudkivi, A.J. and Tan, S.K. (1984) Erosion of cohesive soils, *Jour. of Hydraulic Research*, Vol. 22, 217-233.
Raudkivi, A.J. (1986) Functional trends of scour at bridge piers, *J. Hydraulic Engineering, ASCE*, Vol. 112, 1-13.
Raudkivi, A.J. (1988) The roughness height under waves, *Jour. of Hydraulic Research*, Vol. 26, 569-584.
Reineck, H.E. and Singh, I.B. (1973) *Depositional Sedimentary Environments*, Springer-Verlag.
Reynolds, A.J. (1976) A decade's investigation of the stability of erodible stream beds, *Nordic Hydrology*, Vol. 7, 161-180.

Ribberink, J.S. and van der Sande, J.T.M. (1985) Aggradation in rivers due to overloading-analytical approaches, *J. Hydraulic Res.*, Vol. 23, 273-283.
Richards, K.J. (1980) The formation of ripples and dunes on an erodible bed, *J. Fluid Mech.*, Vol. 99, 597-618.
Richards, K. (1982) *Rivers; Form and Process in Alluvial Channels*, Methuen, London.
Richardson, E.V., Stevens, M.A. and Simons, D.B. (1975) The design of spurs for river training, *Proc. 16th Congress, I.A.H.R.*, Sao Paulo, Vol. 2, 382-388.
Ripley, H.C. (1927) Relation of depth to curvature of channels, *Trans. ASCE*, Vol. 90, 207-265.
Rodi, W.(1978) Turbulence models and their application in hydraulics: A State of the Art Review, *Sonderforschungsbereich 80*, Report No. SFB 80/T/127, *Univ. of Karlsruhe*, F.R.G.
Roehl, J.W. (1962) Sediment source areas, delivery ratio and influencing morphological factors, *IASH Publication No. 59*, Commission on Land Erosion, 202-213.
Rose, H.E. and Duckworth, R.A. (1969) Transport of solid particles in liquid and gases, *The Engineer*, Vol. 227, 392-396; 430-433; 478-483.
Rouse, H. (1937) Modern conceptions of the mechanics of turbulence, *Trans. ASCE*, Vol. 102, 463-543.
Rouse, H. (1940) Criteria for similarity in transportation of sediment, *Proc. Hydr. Conf. Univ. of Iowa*, Studies in Engineering Bulletin 20, 33-49.
Rouse, H. (1965) Critical analysis of open-channel resistance, *Proc. ASCE*, Vol. 91, HY4, 1-25.
Rottner, J. (1959) A formula for bed load transportation, *La Houille Blanche*, Vol. 4, No. 3, 301-307.
Rozovskii, I.L. (1961) *Flow of Water in Bends of Open Channels*, Academy of Sciences of Ukrainian SSR, Kiev 1957. Translated U.S. Dept. of Commerce No OTS 60-51133.
Rubey, W.W. (1933) Equilibrium conditions in debris-laden streams, *Am. Geophysics Union Trans.*, 14th Ann. Meeting, 497-505.
Ruff, J.F. et al. (1982) Scour at culvert outlets in mixed bed materials, *Federal Highway Admin., USA*, Rep. FHWA/RD-82/011.
Sargunam, A., Riley, P., Arulanandan, K. and Krone, R.B. (1973) Physico-chemical factors in erosion of cohesive soils, *Proc. ASCE*, Vol. 99, 55-558.
Sasaki, T.O. and Horikawa, K. (1978) Observations of near-shore current and edge waves, *Proc. 16th Coastal Engineering Conf.*, Hamburg, Vol. 1, 791-809.
Schiller, L. and Naumann, A. (1933) Über die grundlegenden Berechnungen bei der Schwerkraftaufbereitung, *Zeitschrift d. V.D.I.*, Vol. 77.
Schlichting, H. (1936) Experimentelle Untersuchungen zum Rauhigkeitsproblem, *Ing. - Arch.*, Vol. 7, NACA Tech. Memo 823, 1-34.
Schoklitsch, A. (1932) Kolkbildung unter Überfallstrahlen, *Die Wasserwirtschaft*, page 341.
Schoklitsch, A. (1934) Geschiebetrieb und Geschiebefracht, *Wasserkraft und Wasserwirtschaft*, Jahrgang 39, Heft 4.
Schoklitsch, A. (1935) *Stauraumverlandung und Kolkabwehr*, Julius Springer-Verlag, Wien.
Schoemaker, H.J. (1982) Grundlagen der Dimensionierung nach Erfahrungen beim Betrieb von unverkleideten Bewässerungs-

kanälen, *Techn. Univ. Munich, Inst. Hydraulik und Gewässer kunde*, Mitteilungen Nr. 37, 1-32.
Schumm, S.A. (1960) The shape of alluvial channels in relation to sediment type, *U.S. Geological Survey*, Professional Paper No. 3 52B.
Schumm, S.A. (1965) Quaternary Paleohydrology in "*Quaternary of the United States*", edited by Wright, H.E. Jr. and Frey, D.G., Princeton Univ. Press, 783-794.
Schumm, S.A. (1968) River adjustment to altered hydrologic regimen - Murrumbidgee River and palaeochannels, Australia, *U.S. Geological Survey*, Professional Paper, No. 598.
Schumm, S.A. and Khan, H.R. (1972) Experimental study of channel patterns, *Bull. Geol. Soc. Am.*, Vol. 83, 1755-1770.
Schumm, S.A. (1977) *The Fluvial System*, Wiley-Inter Science Publ.
Scott, T. (1954) Sand movement by waves, *U.S. Army Beach Erosion Board*, Tech. Memo No. 48.
Shalash, M.S.E. (1959) Die Kolkbildung beim Ausfluss unter Schützen, Dissertation, Tech. University, München.
Sheen, S.J. (1964) Turbulence over a sand ripple, Master of Engng. Thesis, *University of Auckland, N.Z.*
Shen, H.W., Ogawa, Y. and Karaki, S.S. (1963) Time variation of bed deformation near bridge piers, *Proc. 10th Congress I.A.H.R.*, Leningrad, Paper No. 3, 14.
Shen, H.W., Schneider, V.R. and Karaki, S. (1969) Local scour around bridge piers, *Proc. ASCE*, Vol. 95, HY6, 1919-1940.
Shen, H.W. and Hung, C.S. (1971) An engineering approach to total bed material load by regression analysis, *Proc. Sedimentation Symposium*, Berkeley.
Sheng, P.Y. (1986) Modelling bottom boundary layer and cohesive sediment dynamics in estuarine and coastal waters, *Estuarine Cohesive Sediment Dynamics*, Lecture Notes on Coastal and Estuarine Studies, Vol. 14, Edited by A.J. Mehta, Springer-Verlag, 360-400.
Shields, A. (1936) Anwendung der Aehnlichkeits-Mechanik und der Turbulenzforschung auf die Geschiebebewegung, *Preussische Versuchsanstalt für Wasserbau und Schiffbau*, Berlin, Heft 26.
Shepard, F.P. and Wanless, H.R. (1971) *Our Changing Coastlines*, McGraw-Hill.
Shinohara, K. and Tsubaki, T. (1959) On the characteristics of sand waves formed upon beds of the open channels and rivers, *Reports of Research Institute for Applied Mechanics, Kyushu University*, Japan, Vol. 7, No. 25.
Shirasuna, T. (1973) Formation of sand waves, *Proc. 15th Congress I.A.H.R.*, Istanbul, Vol. 1, 107-114.
Shore Protection Manual (1984) Coastal Engineering Research Center, Dept. of the Army, *U.S. Army Corps of Engineers*.
Simons, D.B. and Richardson, E.V. (1960) Discussion of resistance properties of sediment laden streams, *Trans. ASCE*, Vol. 125, Part 1, 1170-1172.
Simons, D.B., Richardson, E.V. and Haushild, W.L. (1962) Depth-discharge relations in alluvial channels, *Proc. ASCE*, Vol. 88, HY5, 57-72.
Simons, D.B., Richardson, F.V. and Nordin, C.F., Jr. (1964) Sedimentary structures generated by flow in alluvial channels, *Colorado State Univ.*, Fort Collins, Rep. CER 64, DBS-EVR-CFN 15, Am. Assoc. of Petroleum Geologists Special Publication No. 12.
Simons, D.B. and Stevens, M.A. (1972) Scour control in rock

basins at culvert outlets, Chapter 24 in *River Mechanics*, edited by H.W. Shen, Vol. 2.
Simons, D.B. and Sentürk, F. (1977) *Sediment Transport Technology*, Water Resources Publications, Fort Collins, Colorado.
Skovgaard, O., Jonsson, I.G. and Bertelsen, J.A. (1975) Computation of wave height due to refraction and friction, *Proc. ASCE*, Vol. 101, WW, 15-32.
Sleath, J.F.A. (1974) Stability of laminar flow at seabed, *Proc. ASCE*, Vol. 100, WW2, 105-122.
Sleath, J.F.A. (1975) A contribution to the study of vortex ripples, *Journal of Hydraulic Research*, Vol. 13, 315-328.
Sleath, J.F.A. (1976) On rolling-grain ripples, *Journal of Hydraulic Research*, Vol. 14, 69-81.
Sleath, J.F.A. (1978) Measurements of bed load in oscillatory flow, *Proc. ASCE*, Vol. 104, WW4, 291-307.
Sleath, J.F.A. (1984) Measurements of mass transport over a rough bed, *Proc. 19th Coastal Engineering Conf.*, Houston, Vol. 2, 1149-1160.
Sleath, J.F.A. (1984) *Sea Bed Mechanics*, J. Wiley & Sons.
Smerdon, E.T. and Beasley, R.P. (1961) Critical traction forces in cohesive soils, *Agric. Engng.*, Vol. 42, 26-29.
Smerdon, E.T. (1964) Effect of rainfall on critical tractive force in channels with shallow flow, *Trans. Am. Soc. Agric. Engrs.*, Vol. 7, No. 3, 287-290.
Song, C.C.S. and Yang, C.T. (1982) Minimum energy and energy dissipation rate, *Proc. ASCE*, Vol. 108, HY5, 690-706.
Soni, J.P., Garde, R.J. and Ranga Raju, K.G. (1980) Aggradation in streams due to overloading, *Proc. ASCE*, Vol. 106, HY1, 117-132.
Soulsby, R.L. and Smallman, J.V. (1986) A direct method of calculating bottom orbital velocity under waves, *Hydraulic Research Ltd.*, Wallingford, U.K., Report SR 76.
Soulsby, R.L. (1987) The relative contributions of waves and tidal currents to marine sediment transport, *Hydraulic Research Ltd.*, Wallingford, U.K., Report No. SR 125.
Stanley, D.J. and Swift, D.J.P. (1976) (Editors) *Marine Sediment Transport and Environmental Management*, Wiley & Sons.
Staub, C., Jonsson, I.G. and Svendsen, I.A. (1983) Measurements of instantaneous sediment suspension in oscillatory flow by a new rotating-wheel aparatus, Progress Report No. 58, *ISVA, Tech. Univ. of Denmark*, 41-50. Also *Proc. 19th Coastal Engng. Conf.*, Houston, 1984, Vol. 3, 2310-2321.
Stehr, E. (1975) Grenzschicht-Theoretische Studie über die Gesetze der Strombank- und Riffelbildung, *Hamburger Küstenforschung*, Vol. 34.
Straub, L.G. (1935) Missouri River Report, *U.S. Army Corps of Engineers*, House Document 238, Appendix XV, 73rd U.S. Congress 2nd Session, page 1156. See Vanoni (1975) 58-61.
Straub, L.G. (1950) Chapter XII in *Engineering Hydraulics*, Edited by H. Rouse, Wiley & Sons.
Strickler, A. (1923) Beiträge zur Frage der Geschwindigkeitsformel und der Rauhigkeitszahlen für Ströme, Kanäle und geschlossene Leitungen, *Mitteilungen des eidgenössischen Amtes für Wasserwirtschaft*, Bern, Switzerland.
Sukegawa, N. (1970) Conditions for the occurrence of river meanders, *J. Faculty of Engineering, Univ. of Tokyo*, Vol. 30, 289-306.
Sumer, B.M. and Oguz, B. (1978) Particle motion near the bottom

in turbulent flow in an open channel, *J. Fluid Mech.*, Vol. 86, 109-128.

Sumer, B.M. and Deigaard, R. (1981) Particle motions near the bottom in turbulent flow, Part 2, *J. Fluid Mech.*, Vol. 109, 311-337.

Sunamura, T. and Horikawa, K. (1974) Two-dimensional beach transformation due to waves, *Proc. 14th Coastal Engineering Conf.*, Copenhagen, 920-938.

Sundborg, A. (1956) The River Klaralven. A study of fluvial processes, *Geografiska Annaler*, Ser. A, No. 115, 127-316.

Svendsen, I.A. and Brink-Kjaer, O. (1972) Shoaling cnoidal waves, *Proc. 13th Coastal Engineering Conf.*, Vancouver, Vol. 1, 365-383.

Svendsen, I.A. (1974) Cnoidal waves over a gently sloping bottom, Series Paper No. 6, Inst. Hydrodyn. and Hydraulic Engng. *Techn. University of Denmark.*

Swart, D.H. (1974 a) Offshore sediment transport and equilibrium beach profiles, *Delft Hydraulics Laboratory* Publication No. 131.

Swart, D.H. (1974 b) A schematization of onshore-offshore transport, *Proc. 14th Coastal Engineering Conf.*, Copenhagen, Vol. 2, 884-900.

Swart, D.W. (1976) Predictive equations regarding coastal transports, *Proc. 15th Coastal Engineering Conf.*, Honolulu, Vol. 2, 1113-1132.

Tanaka, S. and Sugimoto, S. (1958) On the distribution of suspended sediment in experimental flume flow, *Memoirs of the Faculty of Engineering, Kobe University*, Japan, No. 5.

Task Committee (1968) Erosion of cohesive sediments, *Proc. ASCE*, Vol. 94, HY4, 1017-1049.

Taylor, G. (1946) Note on R.A. Bagnold's empirical formula for the critical water motion corresponding with the first disturbance of grains on a flat surface, *Proc. Roy. Soc. London*, Vol. A187, 16-18.

Taylor, B.D. and Vanoni, V.A. (1972) Temperature effects in low-transport, flat-bed flows, *Proc. ASCE*, Vol. 98, HY8, 1427-1445.

Taylor, D.B. and Vanoni, V.A. (1972) Temperature effects in high-transport, flat-bed flows, *Proc. ASCE*, Vol. 98, HY12, 2191-2206.

Tchen, C.M. (1947) Mean value and correlation problems connected with the motion of small particles suspended in a turbulent fluid, Ph.D. Thesis, *Delft*, Summarized by Hintze (1959).

Tey, C.B. (1984) Local scour at bridge abutments, *University of Auckland, N.Z.*, M.E. Thesis, Dept. of Civil Engineering Rep. No. 329.

Thompson, E.F. and Harris, D.L. (1972) A wave climatology for U.S. coastal waters, *Proc. Offshore Technology Conf.*, Dallas.

Thompson, S.M. and Campbell, P.L. (1979) Hydraulics of a large channel paved with boulders, *J. Hydr. Res.*, Vol. 17, 341-354.

Thompson, S.M. and Davoren, A. (1983) Local scour at a pier, Ohau River, Otago, New Zealand, *Ministry of Works and Development*, Christchurch, N.Z., Rep. No. WS 756.

Thorn, M.F.C. and Parsons, J.G. (1977) Properties of Grangemouth mud, *Hydraulic Research Station*, Wallingford, U.K., Rep. No. EX 781.

Thorn, M.F.C. (1979) Properties of Belawan mud, *Hydraulic Research Station*, Wallingford, U.K., Rep. No. EX 880.
Thorn, M.F.C. and Parsons, J.G. (1980) Erosion of cohesive sediments in estuaries: An engineering guide, *Proc. 3rd Intern. Symposium on Dredging Techn. BHRA* Bordeaux, France, edited by H.S. Stephens, 349-358.
Thorn, M.F.C. (1981) Physical processes of siltation in tidal channels, *Proc. Hydraulic Modelling Applied to Maritime Engineering Problems*, I.C.E., London, 47-55.
Thornton, E.B. and Guza, R.T. (1982) Energy saturation and phase speeds measured on a natural beach, *J. Geophysical Research*, Vol. 87, 9499-9508.
Thornton, E.B. (1969) Long-shore current and sediment transport, *Univ. of Florida*, Gainesville, Dept. Coastal and Oceanographic Engng., Techn. Rep. No. 5.
Toffaleti, F.B. (1968) A procedure for computation of the total river sand discharge and detailed distribution, bed to surface, *U.S. Army Corps, Vicksburg*, Techn. Rep. No. 5, Also in *Proc. ASCE*, Vol. 95, HY1, 225-248 (1969).
Togashi, H. and Nakamura, T. (1977) An empirical study of tsunami run-up on uniform slopes, *Coastal Engng. in Japan*, Vol. 20, 95-108.
Torobin, L.B. and Gauvin, W.H. (1959-60) Fundamental aspects of solids-gas flow, Parts I-V, *Canadian J. Chem Engng.*, Aug. '59, 129-141; Oct. 167-176; Dec. 224-236; Oct '60, 142-153: Dec. 189-200.
Unsöld, G. (1982) Der Transportbeginn rolligen Schlenmaterials in gleichförmigen turbulenten Strömungen: Eine kritische Überprüfung der Shields-Funktion und ihre experimentelle Erweiterung auf feinkörnige, nicht-bindige Sedimente. Dissertation, Mathematische-Naturwissenschaftlichen Fakultät *Universität Kiel, W. Germany*.
Urk, A. van (1982) Bedforms in relation to hydraulic roughness and unsteady flow in the Rhine branches, *Mechanics of Sediment Transport*, edited by B.M. Sumer and A. Müller, Proc. Euromech. 156, Istanbul, 151-157.
Ursell, F. (1952) Edge waves on a sloping beach, *Proc. Roy. Soc. London*, Vol. A 214, 79-97.
U.S. Bureau of Reclamation (1949) Design and construction of Imperial Dam and Desilting Works, *Boulder Canyon Final Reports*, Part IV, also Vanoni (1975) p.582.
U.S. Bureau of Reclamation (1977) *Design of Small Dams*, Water Resources Technical Publication, U.S. Dept. of Interior.
U.S. Interagency Committee (1957) Some fundamentals of particle size analysis. A study of methods used in measurement and analysis of sediment loads in streams, *Subcommittee on Sedimentation, Interagency Committee on Water Resources*, Report No. 12, St. Anthony Falls Hydr. Laboratory, Minneapolis, Minn.
van Bendegom, L. (1947) Eenige beschouwingen over riviermorphologie en rivierverbetering, *De Ingenieur*, Vol. 59, No. 4, 1-11.
van der Tempel, M. (1972) Interaction forces between condensed bodies in contact, *Advan. Colloid Interface Sci.*, Vol. 3, 137-159.
Van Kestern, W.G.M. and Bakker, W.T. (1984) Near bottom velocities in waves with a current; analytical and numerical computations, *Proc. 19th Coastal Engineering Conf.*, Houston, Vol. 2, 1161-1177.

van Olphen, H. (1966) *Clay Colloid Chemistry*, Interscience.
van Rijn, L.C. (1982) Equivalent roughness of alluvial bed, *Proc. ASCE*, Vol. 108, HY10, 1215-1218. Also as "The prediction of bed forms, alluvial roughness and sediment transport", *Delft Hydraulics Laboratory* Report S 487-III, and van Rijn (1984).
van Rijn, L.C. (1984) Sediment Transport Part I: Bed load transport, *J. Hydraulic Engineering, ASCE*, Vol. 110, 1431-1456.
Part II: Suspended load transport, *J. Hydraulic Engineering ASCE*, Vol. 110, 1613-1641.
Part III: Bed forms and alluvial roughness, *J. Hydraulic Engineering, ASCE*, Vol. 110, 1733-1754.
van Rijn, L.C. (1985) Mathematical models for sediment concentration profiles in steady flow, *Proc. Euromech. 192, Transport of Suspended Solids in Open Channels*, Hochschule der Bundeswehr, 8014 Neubiberg, FRG, B 1, 1-35.
Vanoni, V.A. (1946) Transportation of suspended sediment by water, *Trans. ASCE*, Vol. 111, 67-133.
Vanoni, V.A. and Brooks, N.H. (1957) Laboratory studies of the roughness and suspended load of alluvial streams, Sedimentation Laboratory, *California Institute of Technology*, Rep. E-68.
Vanoni, V.A. and Nomicos, G.N. (1959) Resistance properties of sediment laden streams, *Proc. ASCE*, Vol. 85, HY5, 77-107.
Vanoni, V.A. (Editor)(1975) *Sedimentation Engineering*, ASCE-Manuals and Reports on Engineering Practice - No. 54.
Velikanov, M.A. (1936) Formation of sand ripples on the stream bottom, *Commission de Potamologie*, Sec. 3, Rapport 13, Int. Assoc. Sci. Hydrology.
Velikanov, M.A. (1954) Principle of the gravitational theory of the movement of sediments, *Acad. of Sci. Bull., USSR*, Geophys. Series No. 4, 349-359 (GTS Transl. No. 62-15004).
Velikanov, M.A. (1955) *Dynamics of Alluvial Streams*, Vol. II (Sediment and Flow Bed), State Publishing House for Theoretical and Technical Literature, Moscow (in Russian).
Velikanov, M.A. (1958) *Alluvial Process* (Fundamental Principles), State Publishing House for Physical and Mathematical Literature, Moscow (in Russian).
Vellinga, P. (1983) Predictive computational model for beach and dune erosion during storm surges, *Proc. Coastal Structures* '83, Arlington, Virginia, USA, Published by ASCE.
Veronese, A. (1937) Erosion de fond en avel d'une décharge, *I.A.H.R. Meeting for Hydraulic Works*, Berlin.
Vincent, C.L. and Resio, D.T. (1977) An eigenfunction parameterization of a time sequence of wave spectra, *Coastal Engineering*, Vol. 1, 185-205.
Vincent, C.L. and Corson, W.D. (1980) The geometry of selected U.S. tidal inlets, GITI Rep. No. 20, *U.S. Army Corps of Engrs*, Washington, D.C.
Visser, J. (1970) Measurement of the force of adhesion between submicron carbon-black particles and cellulose film in aqueous solution, *J. Colloid Interface Sci.*, Vol. 34, 26-31.
Visser, P.J. (1984) A mathematical model of uniform longshore currents and the comparison with laboratory data, *Delft Univ. of Techn.* Dept. of Civil Engng., Laboratory of Fluid Mechanics Rep. No. 84-2.
Vlugter, H. (1962) Sediment transportation by running water and

the stable channels in alluvial soils, *Bouw- en Waterbouwkunde, De Ingenieur*, Vol. 74, No. 36, 227-231.
Vollmers, H. and Pernecker, L. (1967) Beginn des Feststofftransportes für feinkörnige Materialien in einer richtungskonstanten Strömung, *Die Wasserwirtschaft*, Heft 6.
Vreugdenhil, C.B. and de Vries, M. (1973) Analytical approaches to non-steady bed load transport, *Delft Hydraulics Laboratory* Res. Rep. S 78-III.
Wan Zhaohui and Xu Yian (1984) The utilization of hyper-concentrated flow and its mechanism, *Proc. 4th Congress Asian and Pacific Division, I.A.H.R.*, Thailand, 1791-1808.
Wan Zhaohui (1985) Bed material movement in hyperconcentrated flow, *J. Hydraulic Engineering, ASCE*, Vol. 111, 987-1002.
Ward, S.G. (1955) Properties of well-defined suspensions of solids in liquids, *J. Oil and Colour Chemists Ass.*, Vol. 38, No. 9.
Wasp, E.J., Kenny, J.P. and Gandhi, R.L. (1977) *Solid-liquid flow slurry pipeline transportation*, Trans. Techn. Publications, Clausthal, W. Germany.
Webby, M.G. (1984) General scour at a constriction, *National Roads Board RRU, N.Z.*, Bridge Design Seminar, Road Res. Bulletin No. 73, 109-118.
Weggel, J.R. (1972) Maximum breaker height, *Proc. ASCE*, Vol. 98, WW4, 529-548.
Westrich, B. and Kobus, H. (1973) Erosion of a uniform sand by continuous and pulsating jets, *Proc. 15th Congress I.A.H.R.*, Istanbul, Vol. 1, 91-98.
Westrich, B. (1974) Erosion eines gleichkörnigen Sandbettes durch stationäre und pulsierende Strahlen. Dissertation, *Tech. Univ. Karlsruhe*, W. Germany.
White, C.M. (1940) The equilibrium of grains on the bed of a stream, *Proc. Roy. Soc. London*, Vol. A 174, No. 958.
White, S.J. (1970) Plane bed threshold of fine grained sediments, Nature, Vol. 228, No. 5267, 152-153.
White, W.R., Paris, E. and Bettess, R. (1979) A new general method for predicting the frictional characteristics of alluvial streams, *Hydraulics Research Station*, Wallingford, U.K., Report No. IT 187.
White, W.R., Milli, H. and Crabbe, A.D. (1973) Sediment transport: An Appraisal of available methods, *Hydraulic Research Station*, Wallingford, U.K., Report No. INT 119, Vol. 1 and 2.
White, W.R. and Day, T.J. (1982) Transport of graded gravel bed materials in *Gravel-Bed Rivers*, Edited by R.D.Hey, Bathurst, J.C. and Thorne, C.R., Wiley Interscience Public.
White, W.R., Bettess, R. and Wang Shiqiang (1987) The frictional characteristics of alluvial streams in the lower and upper regimes, *Proc. Inst. Civil Engrs.* Part 2, Vol. 83, 685-700.
Whitmore, R.L. (1957) The relationship of the viscosity to the settling rate of slurries, *J. Inst. Fuel*, May issue.
Whittaker, J.C. and Schleiss, A. (1984) Scour related to energy dissipators for high head structures, Mitteilungen Nr. 73 of *E.T.H. Versuchsanstalt für Wasserbau*, Zurich.
Wijbenga, J.H.A. and Klaassen, G.J. (1981) Changes in bedform dimensions under steady flow conditions in a straight flume, *Delft Hydraulics Laboratory*, Communications Publ. No. 260.
Williams, G.P. (1970) Flume width and water depth effects in sediment transport experiments, *U.S. Geological Survey*, Professional Paper 562-H.

Willis, J.C., Coleman, N.L. and Ellis, W.M. (1972) Laboratory study of transport of fine sand, *Proc. ASCE*, Vol. 89, HY3, 489-501.

Willmarth, W.W. (1975) Pressure fluctuations beneath turbulent boundary layers, *Annual Review Fluid Mech.*, Vol. 7, 13-38.

Wischmeier, W.H. and Smith, D.D. (1965) Predicting rainfall-erosion losses from cropland east of the Rocky Mountains, *Agriculture Handbook*, No. 282, U.S. Dept. of Agriculture.

Wooding, R.A., Bradley, E.F. and Marshall, J.K. (1973) Drag due to regular arrays of roughness elements of varying geometry, *Boundary-Layer Meteorology*, Vol. 5, 285-308.

Worster, R.C. and Denny, D.F. (1955) The hydraulic transport of solid material in pipes, *Proc. Inst. Mech. Engrs.*, Vol. 169, 32.

Wright, D.L., May, S.K., Short, A.D. and Green, M.O. (1984) Beach and surf zone equilibria and response times, *Proc. 19th Coastal Engineering Conf.*, Houston, Vol. 2, 2150-2164.

Yalin, M.S. and Russell, R.C.H. (1962) Similarity in sediment transport due to waves, *Proc. 8th Coastal Engineering Conf.*, Mexico, 151-167.

Yalin, M.S. (1963) An expression for bed-load transportation, *Proc. ASCE*, Vol. 89, HY3, 221-250.

Yalin, M.S. (1971) *Theory of Hydraulic Models*, MacMillan, N.Y.

Yalin, M.S. (1972) *Mechanics of Sediment Transport*, Pergamon Press.

Yalin, M.S. (1977) On the determination of ripple length, *Proc. ASCE*, Vol. 103, HY4, 439-442.

Yalin, M.S. and Karahan, E. (1979) Steepness of sedimentary dunes, *Proc. ASCE*, Vol. 105, HY4, 381-392.

Yalin, M.S. and Karahan, E. (1978) On the geometry of ripples due to waves, *Proc. 16th Coastal Engineering Conf.*, Hamburg, Vol. 2, 1776-1786.

Yalin, M.S. (1985) On the determination of ripple geometry, *J. Hydraulic Engineering, ASCE*, Vol. 111, 1148-1155.

Yang, C.T. (1971) Potential energy and stream morphology, *Water Resources Research*, Vol. 7, 311-322.

Yang, C.T. (1972) Unit stream power and sediment transport, *Proc. ASCE*, Vol. 98, HY10, 1805-1820.

Yang, C.T. (1973) Incipient motion and sediment transport, *Proc. ASCE*, Vol. 99, HY10, 1679-1704.

Yang, C.T. (1974) Minimum stream power and fluvial hydraulics, *Proc. ASCE*, Vol. 102, HY7, 919.

Yang, C.T. and Stall, J.B. (1976) Applicability of unit stream power equation, *Proc. ASCE*, Vol. 102, HY5, 559-568.

Yang, C.T. (1977) The movement of sediment in rivers, *Geophysical Surveys*, Vol. 3, 39-68.

Yang, C.T. and Song, C.C.S. (1979) Theory of minimum rate of energy dissipation, *Proc. ASCE*, Vol. 105, HY7, 769-784.

Yang, C.T. (1979) Unit stream power equations for total load, *J. Hydrology*, Vol. 40, 123-138.

Yang, C.T. and Molinas, A. (1982) Sediment transport and unit stream power function, *Proc. ASCE*, Vol. 108, HY6, 774-793.

Yang, C.T. (1984) Unit stream power equation for gravel, *J. Hydraulic Engineering, ASCE*, Vol. 110, 1783-1797.

Yong, R.N. and Warkentin, B.P. (1975) *Soil Properties and Behaviour*, Amsterdam, Elsevier Scientific.

Yufin, A.P. (1949) Izvestiya U-ser. Tekh. Nauk., No. 8, 1146. Also quoted in *Proc. ASCE*, Vol. 96, HY7 (1970), 1503-38.

Yücel, Ö. and Graf, W.H. (1973) Bed load deposition in reservoirs,

Proc. 15th Congress I.A.H.R., Istanbul, Vol. 1, 271-278. Also Graf, W.H. (1977) Reservoir sedimentation, *Ecole polytechnique fédérale* de Lausanne, Communication Lab. d'hydraulique No. 35.

Zandi, I. and Govatos, G. (1967) Heterogeneous flow of solids in pipelines, *Proc. ASCE*, Vol. 93, HY3, 145-159.

Zandi, I. (Editor) (1971) *Advances on Solid-Liquid Flow in Pipes and its Application*, Pergamon Press.

Zanke, U. (1976) Über die Naturähnlichkeit von Geschiebeversuchen bei einer Gewässersohle mit Transportkörpen, *Mitt. Franzius-Institut, Techn. Univ. Hannover*, No. 44.

Zanke, U. (1977) Neuer Ansatz zur Berechnung des Transportbeginns von Sedimenten unter Strömungseinfluss, *Mitteilungen des Franzius-Institut, Techn. Univ. Hannover*, Heft 46.

Zanke, U. (1981) Seegang erzeugte Kolke am Bauwerken, Sonderforschungsbereich 79, Teilsprojekt B9, *Techn. Univ. Hannover.*

Zanke, U. (1982) *Grundlagen der Sedimentbewegung*, Springer-Verlag.

Zenkovich, V.P. (1967) *Processes of Coastal Development*, Oliver and Bond, Edinburgh and London.

Zimmerman, C. and Kennedy, J.F. (1978) Transverse bed slopes in curved alluvial streams, *Proc. ASCE*, Vol. 104, HY1, 33-48.

Zimmermann, C. and Naudascher, E. (1979) Sohlausbildung und Sedimentbewegung in Krümmungen alluvialer Gerinne, *Wasserwirtschaft*, Vol. 69, 110-117.

Zingg, A.W. (1953) Wind tunnel studies on the movement of sedimentary materials, *Proc. 5th Hydr. Research*, Iowa Inst. of Hydraulic Research, Bulletin 34, 111-135.

Znameskaya, N.S. (1969) Morphological principle of modelling of river-bed processes, Proc 13th Congress, I.A.H.R., Kyoto, Vol. 5, 1.

ADDENDUM TO REFERENCES

Baker, R.E. (1986) Local scour at bridge piers in uniform sediments, *University of Auckland*, Dept. of Civil Engng. Rep. No. 402.

Battjes, J.A. and Janssen, J.P.F.M. (1978) Energy loss and set-up due to breaking random waves, *Proc. 16th Coastal Engineering Conf.*, Vol. 1, 569-587.

Battjes, J.A. (1986) Energy dissipation in breaking solitary and periodic waves, *Delft University of Technology*, Communications on hydraulic and geotechnical engineering, Report No. 86-5.

Battjes, J.A. (1988) Surf-zone dynamics, *Ann. Rev. Fluid Mech.*, Vol. 20, 257-293.

Birkhoff, G. (1950) *Hydrodynamics*, Princeton Univ. Press.

Bretting, A.E. (1958) Stable channels, *Acta Polytechnica Scandinavica*, Ci 1.

Cebeci, T. and Smith, A.M.O. (1974) *Analysis of turbulent boundary layers*, Academic Press.

Chaleb, K.O. (1929-30) Minutes of *Proc. Inst. Civil Engrs*, London, Vol. 229 Pt. 1, 260, 223, 285.

Chien, Ning (1956) Graphic design of alluvial channels, *Trans. A.S.C.E.*, Vol. 121, 1267-80.

Chien, Ning (1957) A concept of regime theory, *Trans. A.S.C.E.*, Vol. 122, 785-793.

Cunha, Veiga da L. (1967) About the roughness in alluvial channels with comparatively coarse bed material, *Proc. 12th Congress intern. Assoc. for Hydraulic Research*, Vol. 1, 76-84.

Cur (1987) Manual on artificial beach nourishment, *Centre for Civil Engineering Research*, Codes and Specifications, Rep. 130, Gouda, The Netherlands.

Dally, W.R., Dean, R.G. and Dalrymple, R.A. (1985) Wave height variation across beach of arbitrary profile. *J. Geophysical Research*, Vol. 90, No. C6, 11917-27.

Emmerling, R. (1973) The instantaneous structure of the wall pressure under a turbulent boundary layer flow, *Max-Planck-Inst.* für Strömungsforschung, Bericht Nr. 9.

Faddick, R., Pouska, G., Connery, J., Napoli, D. and Punis, G. (1979) Ultrasonic velocity meter, *Hydrotransport 6*, Vol. 1, Canterbury, England.

Fredsøe, J. (1982) Shape and dimensions of stationary dunes in rivers, *Proc. A.S.C.E.*, Vol. 108, HY 8, 932-947.

Führböter, A. (1971) Über die Bedeutung des Lufteinschlages für die Energieumwandlung in Brandungszonen, *Die Küste*, Nr. 21, 34-42.

Gibbs, H.J. (1962) A study of erosion and tractive force characteristics in relation to soil mechanics properties, *U.S. Bureau of Reclamation*, Em-643.

Glover, R.E. and Florey, Q.L. (1951) Stable Channel Profiles, *U.S. Bureau of Reclamation*, Hydraulic Laboratory Report No. Hyd-325.

Goedde, E.T. (1978) To the critical velocity of heterogeneous hydraulic transport, *Hydrotransport 5*, Vol. 2, Hannover, West Germany.

Hwang, L. and Divoky, D. (1970) Breaking wave set-up and decay on gentle slopes, *Proc. 12th Coastal Engng. Conf.*, Vol. 1, 377-389.

Langhaar, H.L. (1962) *Dimensional Analysis and Theory of Models*, Wiley & Sons.

Larson, M. (1988) Quantification of beach profile change, *Lund University, Institute of Science and Technology*, Sweden, Report No. 1008.

Le Méhauté, B. (1962) On non-saturated breakers and wave run-up, *Proc. 12th Coastal Engng. Conf.*, 77-92.

Leliavsky, S. (1955) *An Introduction to Fluvial Hydraulics*, Constable.

Lewin, J. (1976) Initiation of bed forms and meanders in coarse-grained sediment, *Geological Soc. of America Bulletin*, Vol. 87, 281-285.

Lundgren, H. and Jonsson, I.G. (1964) Shear and velocity distribution in shallow channels, *Proc. A.S.C.E.*, Vol. 90, HY1, 1-21.

Novak, P. and Cabelka, J. (1981) *Models in hydraulic engineering: Physical principles and design applications*, Pitman, Boston.

Sedov, L.I. (1959) *Similarity and Dimensional Methods in Mechanics*, Academic Press.

Simons, D.B. and Albertson, M.L. (1963) Uniform water conveyance in alluvial material, *Trans. Am. Soc. Civil. Engrs.*, Vol. 128/I, 65-167.

Sinclair, C.G. (1962) The limit deposit velocity of heterogeneous suspensions, *Proc. Symp. on Interaction between Fluids and Particles*, European Federation of Chem. Engrs., London, 78-86.

Stive, M.J.F. (1984) Energy dissipation in waves breaking on gentle slopes, *Coastal Engineering*, Vol. 8, 99-127.

Sunamura, T. and Takeda, A. (1987) Wave-induced geomorphic response of eroding beaches with special reference to seaward migrating bars, *Proc. of Coastal Sediments '87*, ASCE, 884-900.

Svendsen, I.A., Madsen, P.A. and Hansen, J.B. (1978) Wave characteristics in the surf zone, *Proc. 16th Coastal Engng. Conf.*, 520-539.

Thornton, E.B. and Guza, R.T. (1983) Transformation of wave height distribution, *Journal Geophysical Research*, Vol. 88, 5925-38.

Vellinga, P. (1986) Beach and dune erosion during storm surges, *Delft Hydraulics*, Communication No. 372.

Zeller, J. (1967) Meandering channels in Switzerland, *Intern. Ass. Sci. Hydrology*, Symposium on River Morphology, Bern. (1967) Flussmorphologische Studie zum Mäanderproblem, *Geographica Helvetica*, Vol. 22, No. 2.

Subject Index

Author Index